Map Projections— A Working Manual

By JOHN P. SNYDER

U.S. GEOLOGICAL SURVEY PROFESSIONAL PAPER 1395

Supersedes USGS Bulletin 1532

UNITED STATES GOVERNMENT PRINTING OFFICE, WASHINGTON: 1987

DEPARTMENT OF THE INTERIOR
DONALD PAUL HODEL, *Secretary*

U.S. GEOLOGICAL SURVEY
Dallas L. Peck, *Director*

PREFACE

This publication is a major revision of USGS Bulletin 1532, which is titled *Map Projections Used by the U.S. Geological Survey.* Although several portions are essentially unchanged except for corrections and clarification, there is considerable revision in the early general discussion, and the scope of the book, originally limited to map projections used by the U.S. Geological Survey, now extends to include several other popular or useful projections. These and dozens of other projections are described with less detail in the forthcoming USGS publication *An Album of Map Projections.*

As before, this study of map projections is intended to be useful to both the reader interested in the philosophy or history of the projections and the reader desiring the mathematics. Under each of the projections described, the nonmathematical phases are presented first, without interruption by formulas. They are followed by the formulas and tables, which the first type of reader may skip entirely to pass to the nonmathematical discussion of the next projection. Even with the mathematics, there are almost no derivations and very little calculus. The emphasis is on describing the characteristics of the projection and how it is used.

This professional paper, like Bulletin 1532, is also designed so that the user can turn directly to the desired projection, without reading any other section, in order to study the projection under consideration. However, the list of symbols may be needed in any case, and the random-access feature will be enhanced by a general understanding of the concepts of projections and distortion. As a result of this intent, there is some repetition which will be apparent when the book is read sequentially.

For the more complicated projections, equations are given in the order of usage. Otherwise, major equations are given first, followed by subordinate equations. When an equation has been given previously, it is repeated with the original equation number, to avoid the need to leaf back and forth. Numerical examples, however, are placed in appendix A. It was felt that placing these with the formulas would only add to the difficulty of reading through the mathematical sections.

The equations are frequently taken from other credited or standard sources, but a number of equations have been derived or rearranged for this publication by the author. Further attention has been given to computer efficiency, for example by encouraging the use of nested power series in place of multiple-angle series.

I acknowledged several reviewers of the original manuscript in Bulletin 1532. These were Alden P. Colvocoresses, William J. Jones, Clark H. Cramer, Marlys K. Brownlee, Tau Rho Alpha, Raymond M. Batson, William H. Chapman, Atef A. Elassal, Douglas M. Kinney (ret.), George Y. G. Lee, Jack P. Minta (ret.), and John F. Waananen, all then of the USGS, Joel L. Morrison, then of the University of Wisconsin/Madison, and the late Allen J. Pope of the National Ocean Survey. I remain indebted to them, especially to Dr. Colvocoresses of the USGS, who is the one person most responsible for giving me the opportunity to assemble this work for publication. In addition, Jackie T. Durham and Robert B. McEwen of the USGS have been very helpful with the current volume, and several reviewers, especially Clifford J. Mugnier, a consulting cartographer, have provided valuable critiques which have influenced my revisions. Other users in and out of the USGS have also offered useful comments. For the plotting of all computer-prepared maps, the personnel of the USGS Eastern Mapping Center have been most cooperative.

John P. Snyder

iii

CONTENTS

Page

ILLUSTRATIONS

TABLES

SYMBOLS

If a symbol is not listed here, it is used only briefly and identified near the formulas in which it is given.

Az = azimuth, as an angle measured clockwise from the north.

a = equatorial radius or semimajor axis of the ellipsoid of reference.

b = polar radius or semiminor axis of the ellipsoid of reference.

   = $a(1 - f) = a(1 - e^2)^{1/2}$.

c = great circle distance, as an arc of a circle.

e = eccentricity of the ellipsoid.

   = $(1 - b^2/a^2)^{1/2}$.

f = flattening of the ellipsoid.

h = relative scale factor along a meridian of longitude. (For general perspective projections, h is height above surface of ellipsoid.)

k = relative scale factor along a parallel of latitude.

n = cone constant on conic projections, or the ratio of the angle between meridians to the true angle, called l in some other references.

R = radius of the sphere, either actual or that corresponding to scale of the map.

S = surface area.

x = rectangular coordinate: distance to the right of the vertical line (Y axis) passing through the origin or center of a projection (if negative, it is distance to the left). In practice, a "false" x or "false easting" is frequently added to all values of x to eliminate negative numbers. (Note: Many British texts use X and Y axes interchanged, not rotated, from this convention.)

y = rectangular coordinate: distance above the horizontal line (X axis) passing through the origin or center of a projection (if negative, it is distance below). In practice, a "false" y or "false northing" is frequently added to all values of y to eliminate negative numbers.

z = angular distance from North Pole of latitude ϕ, or $(90° - \phi)$, or colatitude.

z_1 = angular distance from North Pole of latitude ϕ_1, or $(90° - \phi_1)$.

z_2 = angular distance from North Pole of latitude ϕ_2, or $(90° - \phi_2)$.

$\ln$ = natural logarithm, or logarithm to base e, where e = 2.71828.

θ = angle measured counterclockwise from the central meridian, rotating about the center of the latitude circles on a conic or polar azimuthal projection, or beginning due south, rotating about the center of projection of an oblique or equatorial azimuthal projection.

θ' = angle of intersection between meridian and parallel.

λ = longitude east of Greenwich (for longitude west of Greenwich, use a minus sign).

λ_0 = longitude east of Greenwich of the central meridian of the map, or of the origin of the rectangular coordinates (for west longitude, use a minus sign). If ϕ_1 is a pole, λ_0 is the longitude of the meridian extending down on the map from the North Pole or up from the South Pole.

λ' = transformed longitude measured east along transformed equator from the north crossing of the Earth's Equator, when graticule is rotated on the Earth.

ρ = radius of latitude circle on conic or polar azimuthal projection, or radius from center on any azimuthal projection.

ϕ = north geodetic or geographic latitude (if latitude is south, apply a minus sign).

ϕ_0 = middle latitude, or latitude chosen as the origin of rectangular coordinates for a projection.

ϕ' = transformed latitude relative to the new poles and equator when the graticule is rotated on the globe.

ϕ_1, ϕ_2 = standard parallels of latitude for projections with two standard parallels. These are true to scale and free of angular distortion.

ϕ_1 (without ϕ_2) = single standard parallel on cylindrical or conic projections; latitude of central point on azimuthal projections.

ω = maximum angular deformation at a given point on a projection.

---

1. All angles are assumed to be in radians, unless the degree symbol (°) is used.

2. Unless there is a note to the contrary, and if the expression for which the arctan is sought has a numerator over a denominator, the formulas in which arctan is required (usually to obtain a longitude) are so arranged that the Fortran ATAN2 function should be used. For hand calculators and computers with the arctan function but not ATAN2, the following conditions must be added to the limitations listed with the formulas:

   For arctan (A/B), the arctan is normally given as an angle between $-90°$ and $+90°$, or between $-\pi/2$ and $+\pi/2$. If B is negative, add $\pm 180°$ or $\pm \pi$ to the initial arctan, where the $\pm$ takes the sign of A, or if A is zero, the $\pm$ arbitrarily takes a + sign. If B is zero, the arctan is $\pm 90°$ or $\pm \pi/2$, taking the sign of A. Conditions not resolved by the ATAN2 function, but requiring adjustment for almost any program, are as follows:

   (1) If A and B are both zero, the arctan is indeterminate, but may normally be given an arbitrary value of 0 or of λ_0, depending on the projection, and

   (2) If A or B is infinite, the arctan is $\pm 90°$ (or $\pm \pi/2$) or 0, respectively, the sign depending on other conditions. In any case, the final longitude should be adjusted, if necessary, so that it is an angle between $-180°$ (or $-\pi$) and $+180°$ (or $+\pi$). This is done by adding or subtracting multiples of 360° (or 2π) as required.

3. Where division is involved, most equations are given in the form $A = B/C$ rather than $A = \frac{B}{C}$. This facilitates typesetting, and it also is a convenient form for conversion to Fortran programming.

ACRONYMS

American Geographical Society	SOM	Space Oblique Mercator
Geodetic Reference System	SPCS	State Plane Coordinate System
Hotine (form of ellipsoidal) Oblique Mercator	UPS	Universal Polar Stereographic
International Map Committee	USC&GS	United States Coast and Geodetic Survey
International Map of the World	USGS	United States Geological Survey
International Union of Geodesy and Geophysics	UTM	Universal Transverse Mercator
National Aeronautics and Space Administration	WGS	World Geodetic System

acronyms are not listed, since the full name is used through this bulletin.

MAP PROJECTIONS—
A WORKING MANUAL

By John P. Snyder

ABSTRACT

After decades of using only one map projection, the Polyconic, for its mapping program, the U.S. Geological Survey (USGS) now uses several of the more common projections for its published maps. For larger scale maps, including topographic quadrangles and the State Base Map Series, conformal projections such as the Transverse Mercator and the Lambert Conformal Conic are used. Equal-area and equidistant projections appear in the *National Atlas*. Other projections, such as the Miller Cylindrical and the Van der Grinten, are chosen occasionally for convenience, sometimes making use of existing base maps prepared by others. Some projections treat the Earth only as a sphere, others as either ellipsoid or sphere.

The USGS has also conceived and designed several new projections, including the Space Oblique Mercator, the first map projection designed to permit mapping of the Earth continuously from a satellite with low distortion. The mapping of extraterrestrial bodies has resulted in the use of standard projections in completely new settings. Several other projections which have not been used by the USGS are frequently of interest to the cartographic public.

With increased computerization, it is important to realize that rectangular coordinates for all these projections may be mathematically calculated with formulas which would have seemed too complicated in the past, but which now may be programmed routinely, especially if aided by numerical examples. A discussion of appearance, usage, and history is given together with both forward and inverse equations for each projection involved.

INTRODUCTION

The subject of map projections, either generally or specifically, has been discussed in thousands of papers and books dating at least from the time of the Greek astronomer Claudius Ptolemy (about A.D. 150), and projections are known to have been in use some three centuries earlier. Most of the widely used projections date from the 16th to 19th centuries, but scores of variations have been developed during the 20th century. In recent years, there have been several new publications of widely varying depth and quality devoted exclusively to the subject. In 1979, the USGS published *Maps for America*, a book-length description of its maps (Thompson, 1979). The USGS has also published bulletins describing from one to three projections (Birdseye, 1929; Newton, 1985).

In spite of all this literature, there was no definitive single publication on map projections used by the USGS, the agency responsible for administering the National Mapping Program, until the first edition of Bulletin 1532 (Snyder, 1982a). The USGS had relied on map projection treatises published by the former Coast and Geodetic Survey (now the National Ocean Service). These publications did not include sufficient detail for all the major projections now used by the USGS and others. A widely used and outstanding treatise of the Coast and Geodetic Survey (Deetz and Adams, 1934), last revised in 1945, only touches upon the Transverse Mercator, now a commonly used projection for preparing maps. Other projections such as the Bipolar Oblique Conic Conformal, the Miller Cylindrical, and the Van der Grinten, were just being developed, or, if older, were seldom used in 1945. Deetz and Adams predated the extensive use of the computer and

1

pocket calculator, and, instead, offered extensive tables for plotting projections with specific parameters.

Another classic treatise from the Coast and Geodetic Survey was written by Thomas (1952) and is exclusively devoted to the five major conformal projections. It emphasizes derivations with a summary of formulas and of the history of these projections, and is directed toward the skilled technical user. Omitted are tables, graticules, or numerical examples.

In USGS Bulletin 1532 the author undertook to describe each projection which has been used by the USGS sufficiently to permit the skilled, mathematically oriented cartographer to use the projection in detail. The descriptions were also arranged so as to enable a lay person interested in the subject to learn as much as desired about the principles of these projections without being overwhelmed by mathematical detail. Deetz and Adams' (1934) work set an excellent example in this combined approach.

While Bulletin 1532 was deliberately limited to map projections used by the USGS, the interest in the bulletin has led to expansion in the form of this professional paper, which includes several other map projections frequently seen in atlases and geography texts. Many tables of rectangular or polar coordinates have been included for conceptual purposes. For values between points, formulas should be used, rather than interpolation. Other tables list definitive parameters for use in formulas. A glossary as such is omitted, since such definitions tend to be oversimplified by nature. The reader is referred to the index instead to find a more complete description of a given term.

The USGS, soon after its official inception in 1879, apparently chose the Polyconic projection for its mapping program. This projection is simple to construct and had been promoted by the Survey of the Coast, as it was then called, since Ferdinand Rudolph Hassler's leadership of the early 1800's. The first published USGS topographic "quadrangles," or maps bounded by two meridians and two parallels, did not carry a projection name, but identification as "Polyconic projection" was added to later editions. Tables of coordinates published by the USGS appeared in 1904, and the Polyconic was the only projection mentioned by Beaman (1928, p. 167).

Mappers in the Coast and Geodetic Survey, influenced in turn by military and civilian mappers of Europe, established the State Plane Coordinate System in the 1930's. This system involved the Lambert Conformal Conic projection for States of larger east-west extension and the Transverse Mercator for States which were longer from north to south. In the late 1950's, the USGS began changing quadrangles from the Polyconic to the projection used in the State Plane Coordinate System for the principal State on the map. The USGS also adopted the Lambert for its series of State base maps.

As the variety of maps issued by the USGS increased, a broad range of projections became important: The Polar Stereographic for the map of Antarctica, the Lambert Azimuthal Equal-Area for maps of the Pacific Ocean, and the Albers Equal-Area Conic for the *National Atlas* (USGS, 1970) maps of the United States. Several other projections have been used for other maps in the *National Atlas*, for tectonic maps, and for grids in the panhandle of Alaska. The mapping of extraterrestrial bodies, such as the Moon, Mars, and Mercury, involves old projections in a completely new setting. Perhaps the first projection to be originated within the USGS is the Space Oblique Mercator for continuous mapping using imagery from artificial satellites.

It is hoped that this expanded study will assist readers to understand better not only the basis for maps issued by the USGS, but also the principles and formulas for computerization, preparation of new maps, and transference of data between maps prepared on different projections.

MAP PROJECTIONS—GENERAL CONCEPTS

1. CHARACTERISTICS OF MAP PROJECTIONS

The general purpose of map projections and the basic problems encountered have been discussed often and well in various books on cartography and map projections. (Robinson, Sale, Morrison, and Muehrcke, 1984; Steers, 1970; and Greenhood, 1964, are among later editions of earlier standard references.) Every map user and maker should have a basic understanding of projections, no matter how much computers seem to have automated the operations. The concepts will be concisely described here, although there are some interpretations and formulas that appear to be unique.

For almost 500 years, it has been conclusively established that the Earth is essentially a sphere, although a number of intellectuals nearly 2,000 years earlier were convinced of this. Even to the scholars who considered the Earth flat, the skies appeared hemispherical, however. It was established at an early date that attempts to prepare a flat map of a surface curving in all directions leads to distortion of one form or another.

A map projection is a systematic representation of all or part of the surface of a round body, especially the Earth, on a plane. This usually includes lines delineating meridians and parallels, as required by some definitions of a map projection, but it may not, depending on the purpose of the map. A projection is required in any case. Since this cannot be done without distortion, the cartographer must choose the characteristic which is to be shown accurately at the expense of others, or a compromise of several characteristics. If the map covers a continent or the Earth, distortion will be visually apparent. If the region is the size of a small town, distortion may be barely measurable using many projections, but it can still be serious with other projections. There is literally an infinite number of map projections that can be devised, and several hundred have been published, most of which are rarely used novelties. Most projections may be infinitely varied by choosing different points on the Earth as the center or as a starting point.

It cannot be said that there is one "best" projection for mapping. It is even risky to claim that one has found the "best" projection for a given application, unless the parameters chosen are artificially constricting. A carefully constructed globe is not the best map for most applications because its scale is by necessity too small. A globe is awkward to use in general, and a straightedge cannot be satisfactorily used on one for measurement of distance.

The details of projections discussed in this book are based on perfect plotting onto completely stable media. In practice, of course, this cannot be achieved. The cartographer may have made small errors, especially in hand-drawn maps, but a more serious problem results from the fact that maps are commonly plotted and printed on paper, which is dimensionally unstable. Typical map paper can expand over 1 percent with a 60 percent increase in atmospheric humidity, and the expansion coefficient varies considerably in different directions on the same sheet. This is much greater than the variation between common projections on large-scale quadrangles, for example. The use of stable plastic bases for maps is recommended for precision work, but this is not always feasible, and source maps may be available only on paper, frequently folded as well. On large-scale maps, such as topographic quadrangles, measurement on paper maps is facilitated with rectangular grid overprints, which expand with the paper. Grids are discussed later in this book.

The characteristics normally considered in choosing a map projection are as follows:

1. *Area.*—Many map projections are designed to be *equal-area*, so that a coin of any size, for example, on one part of the map covers exactly the same area of the actual Earth as the same coin on any other part of the map. Shapes, angles, and scale must be distorted on most parts of such a map, but there are usually some parts of an equal-area map which are designed to retain these characteristics correctly, or very nearly so. Less common terms used for equal-area projections are *equivalent, homolographic,* or *homalographic* (from the Greek *homalos* or *homos* ("same") and *graphos* ("write")); *authalic* (from the Greek *autos* ("same") and *ailos* ("area")), and *equiareal.*

2. *Shape.*—Many of the most common and most important projections are *conformal* or *orthomorphic* (from the Greek *orthos* or "straight" and *morphē* or "shape"), in that normally the relative local angles about every point on the map are shown correctly. (On a conformal map of the entire Earth there are usually one or more "singular" points at which local angles are still distorted.) Although a large area must still be shown distorted in shape, its small features are shaped essentially correctly. Conformality applies on a point or infinitesimal basis, whereas an equal-area map projection shows areas correctly on a finite, in fact mapwide basis. An important result of conformality is that the local scale in every direction around any one point is constant. Because local angles are correct, meridians intersect parallels at right (90°) angles on a conformal projection, just as they do on the Earth. Areas are generally enlarged or reduced throughout the map, but they are correct along certain lines, depending on the projection. Nearly all large-scale maps of the Geological Survey and other mapping agencies throughout the world are now prepared on a conformal projection. No map can be both equal-area and conformal.

While some have used the term *aphylactic* for all projections which are neither equal-area nor conformal (Lee, 1944), other terms have commonly been used to describe special characteristics:

3. *Scale.*—No map projection shows scale correctly throughout the map, but there are usually one or more lines on the map along which the scale remains true. By choosing the locations of these lines properly, the scale errors elsewhere may be minimized, although some errors may still be large, depending on the size of the area being mapped and the projection. Some projections show true scale between one or two points and every other point on the map, or along every meridian. They are called *equidistant* projections.

4. *Direction.*—While conformal maps give the relative local directions correctly at any given point, there is one frequently used group of map projections, called *azimuthal* (or *zenithal*), on which the directions or azimuths of all points on the map are shown correctly with respect to the center. One of these projections is also equal-area, another is conformal, and another is equidistant. There are also projections on which directions from two points are correct, or on which directions from all points to one or two selected points are correct, but these are rarely used.

5. *Special characteristics.*—Several map projections provide special characteristics that no other projection provides. On the Mercator projection, all rhumb lines, or lines of constant direction, are shown as straight lines. On the Gnomonic projection, all great circle paths—the shortest routes between points on a sphere—are shown as straight lines. On the Stereographic, all small circles, as well as great circles, are shown as circles on the map. Some newer projections are specially designed for satellite mapping. Less useful but mathematically intriguing projections have been designed to fit the sphere conformally into a square, an ellipse, a triangle, or some other geometric figure.

6. *Method of construction.*—In the days before ready access to computers and plotters, ease of construction was of greater importance. With the advent of computers and even pocket calculators, very complicated formulas can be handled almost as routinely as simple projections in the past.

While the above six characteristics should ordinarily be considered in choosing a map projection, they are not so obvious in recognizing a projection. In fact, if the region shown on a map is not much larger than the United States, for example, even a trained eye cannot often distinguish whether the map is equal-area or conformal. It is necessary to make measurements to detect small differences in spacing or location of meridians and parallels, or to make other tests. The type of construction of the map projection is more easily recognized with experience, if the projection falls into one of the common categories.

There are three types of developable[1] surfaces onto which most of the map projections used by the USGS are at least partially geometrically projected. They are the cylinder, the cone, and the plane. Actually all three are variations of the cone. A cylinder is a limiting form of a cone with an increasingly sharp point or apex. As the cone becomes flatter, its limit is a plane.

If a cylinder is wrapped around the globe representing the Earth (see fig. 1), so that its surface touches the Equator throughout its circumference, the meridians of longitude may be projected onto the cylinder as equidistant straight lines perpendicular to the Equator, and the parallels of latitude marked as lines parallel to the Equator, around the circumference of the cylinder and mathematically spaced for certain characteristics. For some cases, the parallels may also be projected geometrically from a common point onto the cylinder, but in the most common cases they are not perspective. When the cylinder is cut along some meridian and unrolled, a cylindrical projection with straight meridians and straight parallels results. The Mercator projection is the best-known example, and its parallels must be mathematically spaced.

If a cone is placed over the globe, with its peak or apex along the polar axis of the Earth and with the surface of the cone touching the globe along some particular parallel of latitude, a conic (or conical) projection can be produced. This time the meridians are projected onto the cone as equidistant straight lines radiating from the apex, and the parallels are marked as lines around the circumference of the cone in planes perpendicular to the Earth's axis, spaced for the desired characteristics. The parallels may not be projected geometrically for any useful conic projections. When the cone is cut along a meridian, unrolled, and laid flat, the meridians remain straight radiating lines, but the parallels are now circular arcs centered on the apex. The angles between meridians are shown smaller than the true angles.

A plane tangent to one of the Earth's poles is the basis for polar azimuthal projections. In this case, the group of projections is named for the function, not the plane, since all common tangent-plane projections of the sphere are azimuthal. The meridians are projected as straight lines radiating from a point, but they are spaced at their true angles instead of the smaller angles of the conic projections. The parallels of latitude are complete circles, centered on the pole. On some important azimuthal projections, such as the Stereographic (for the sphere), the parallels are geometrically projected from a common point of perspective; on others, such as the Azimuthal Equidistant, they are nonperspective.

The concepts outlined above may be modified in two ways, which still provide cylindrical, conic, or azimuthal projections (although the azimuthals retain this property precisely only for the sphere).

1. The cylinder or cone may be secant to or cut the globe at two parallels instead of being tangent to just one. This conceptually provides two standard parallels; but for most conic projections this construction is not geometrically correct. The plane may likewise cut through the globe at any parallel instead of touching a pole, but this is only useful for the Stereographic and some other perspective projections.

---

[1]A developable surface is one that can be transformed to a plane without distortion.

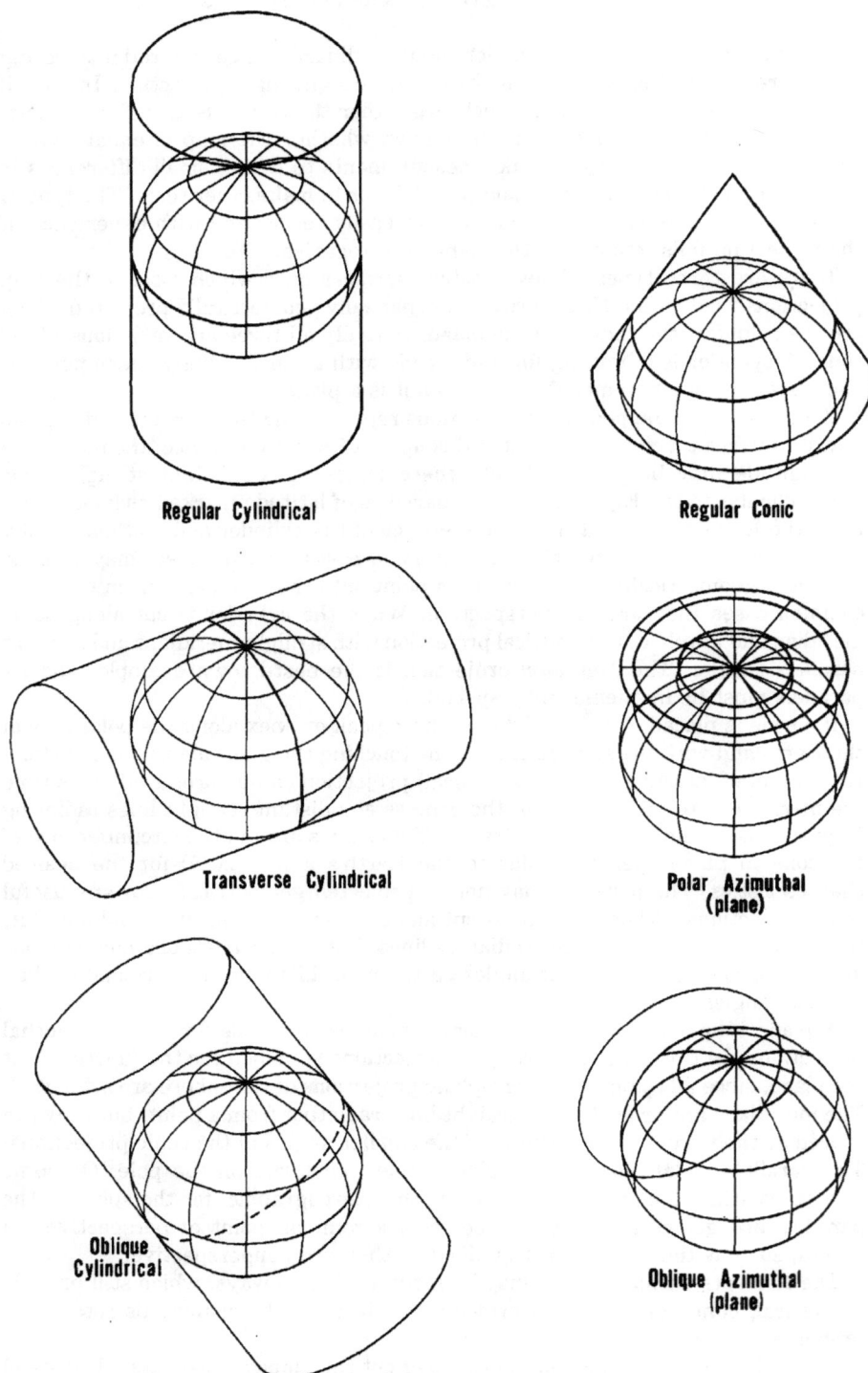

FIGURE 1.—Projection of the Earth onto the three major surfaces. In a few cases, projection is geometric, but in most cases the projection is mathematical to achieve certain features.

2. The axis of the cylinder or cone can have a direction different from that of the Earth's axis, while the plane may be tangent to a point other than a pole (fig. 1). This type of modification leads to important oblique, transverse, and equatorial projections, in which most meridians and parallels are no longer straight lines or arcs of circles. What were standard parallels in the normal orientation now become standard lines not following parallels of latitude.

Other projections resemble one or another of these categories only in some respects. There are numerous interesting pseudocylindrical (or "false cylindrical") projections. They are so called because latitude lines are straight and parallel, and meridians are equally spaced, as on cylindrical projections, but all meridians except the central meridian are curved instead of straight. The Sinusoidal is a frequently used example. Pseudoconic projections have concentric circular arcs for parallels, like conics, but meridians are curved; the Bonne is the only common example. Pseudoazimuthal projections are very rare; the polar aspect has concentric circular arcs for parallels, and curved meridians. The Polyconic projection is projected onto cones tangent to each parallel of latitude, so the meridians are curved, not straight. Still others are more remotely related to cylindrical, conic, or azimuthal projections, if at all.

2. LONGITUDE AND LATITUDE

To identify the location of points on the Earth, a graticule or network of longitude and latitude lines has been superimposed on the surface. They are commonly referred to as meridians and parallels, respectively. The concept of latitudes and longitudes was originated early in recorded history by Greek and Egyptian scientists, especially the Greek astronomer Hipparchus (2nd century, B.C.). Claudius Ptolemy further formalized the concept (Brown, 1949, p. 50, 52, 68).

PARALLELS OF LATITUDE

Given the North and South Poles, which are approximately the ends of the axis about which the Earth rotates, and the Equator, an imaginary line halfway between the two poles, the parallels of latitude are formed by circles surrounding the Earth and in planes parallel with that of the Equator. If circles are drawn equally spaced along the surface of the sphere, with 90 spaces from the Equator to each pole, each space is called a degree of latitude. The circles are numbered from 0° at the Equator to 90° North and South at the respective poles. Each degree is subdivided into 60 minutes and each minute into 60 seconds of arc.

For 2,000 years, measurement of latitude on the Earth involved one of two basic astronomical methods. The instruments and accuracy, but not the principle, were gradually improved. By day, the angular height of the Sun above the horizon was measured. By night, the angular height of stars, and especially the current pole star, was used. With appropriate angular conversions and adjustments for time of day and season, the latitude was obtained. The measuring instruments included devices known as the cross-staff, astrolabe, back-staff, quadrant, sextant, and octant, ultimately equipped with telescopes. They were supplemented with astronomical tables called almanacs, of increasing complication and accuracy. Finally, beginning in the 18th century, the use of triangulation in geodetic surveying meant that latitude on land could be determined with high precision by using the distance from other points of known latitude. Thus measurement of latitude, unlike that of longitude, was an evolutionary development almost throughout recorded history (Brown, 1949, p. 180–207).

MERIDIANS OF LONGITUDE

Meridians of longitude are formed with a series of imaginary lines, all intersecting at both the North and South Poles, and crossing each parallel of latitude at right angles, but striking the Equator at various points. If the Equator is equally divided into 360 parts, and a meridian passes through each mark, 360 degrees of longitude result. These degrees are also divided into minutes and seconds. While the length of a degree of latitude is always the same on a sphere, the lengths of degrees of longitude vary with the latitude (see fig. 2). At the Equator on the sphere, they are the same length as the degree of latitude, but elsewhere they are shorter.

There is only one location for the Equator and poles which serve as references for counting degrees of latitude, but there is no natural origin from which to count degrees of longitude, since all meridians are identical in shape and size. It thus becomes necessary to choose arbitrarily one meridian as the starting point, or prime meridian. There have been many prime meridians in the course of history, swayed by national pride and international influence. For over 150 years, France officially used the meridian through Ferro, an island of the Canaries. Eighteenth-century maps of the American colonies often show longitude from London or Philadelphia. During the 19th century, boundaries of new States were described with longitudes west of a meridian through Washington, D.C., 77°03′ 02.3″ west of the Greenwich (England) Prime Meridian (Van Zandt, 1976, p. 3). The latter was increasingly referenced, especially on seacharts due to the proliferation of

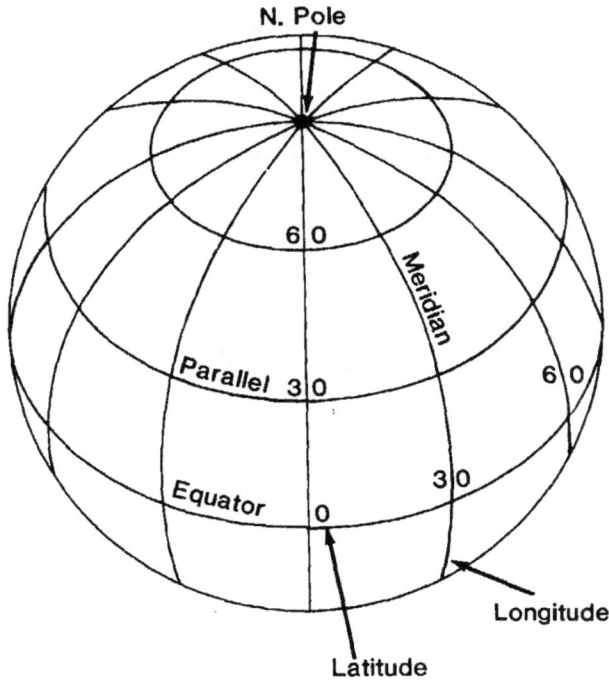

FIGURE 2.—Meridians and parallels on the sphere.

those of British origin. In 1884, the International Meridian Conference, meeting in Washington, agreed to adopt the "meridian passing through the center of the transit instrument at the Observatory of Greenwich as the initial meridian for longitude," resolving that "from this meridian longitude shall be counted in two directions up to 180 degrees, east longitude being plus and west longitude minus" (Brown, 1949, p. 283, 297).

The choice of the prime meridian is arbitrary and may be stated in simple terms. The accurate measurement of the difference in longitude at sea between two points, however, was unattainable for centuries, even with a precision sufficient for the times. When extensive transatlantic exploration from Europe began with the voyages of Christopher Columbus in 1492, the inability to measure east-west distance led to numerous shipwrecks with substantial loss of lives and wealth. Seafaring nations beginning with Spain offered sizable rewards for the invention of satisfactory methods for measuring longitude. It finally became evident that a portable, dependable clock was needed, so that the height of the Sun or stars could be related to the time in order to determine longitude. The study of the pendulum by Galileo, the invention of the pendulum clock by Christian Huygens in 1656, and Robert Hooke's studies of the use of springs in watches in the 1660's provided the basic instrument, but it was not until John Harrison of England responded to his country's substantial reward posted in 1714 that the problem was solved. For five decades, Harrison devised successively more reliable versions of a marine chronometer, which were tested at sea and gradually accepted by the Board of Longitude in painstaking steps from 1765 to 1773. Final compensation required intervention by the King and Parliament (Brown, 1949, p. 208–240; Quill, 1966).

Thus a major obstacle to accurate mapping was overcome. On land, the measurement of longitude lagged behind that of latitude until the development of the clock and the spread of geodetic triangulation in the 18th century made accuracy a reality. Electronic means of measuring distance and angles in the mid- to late-20th century have redefined the meaning of accuracy by orders of magnitude.

CONVENTIONS IN PLOTTING

When constructing meridians on a map projection, the central meridian, usually a straight line, is frequently taken to be a starting point or 0° longitude for calculation purposes. When the map is completed with labels, the meridians are marked with respect to the Greenwich Prime Meridian. The formulas in this book are arranged so that Greenwich longitude may be used directly. All formulas herein use the convention of positive east longitude and north latitude, and negative west longitude and south latitude. Some published tables and formulas elsewhere use positive west longitude, so the reader is urged to use caution in comparing values.

GRIDS

Because calculations relating latitude and longitude to positions of points on a given map can become quite involved, rectangular grids have been developed for the use of surveyors. In this way, each point may be designated merely by its distance from two perpendicular axes on the flat map. The Y axis normally coincides with a chosen central meridian, y increasing north. The X axis is perpendicular to the Y axis at a latitude of origin on the central meridian, with x increasing east. Frequently x and y coordinates are called "eastings" and "northings," respectively, and to avoid negative coordinates may have "false eastings" and "false northings" added.

The grid lines usually do not coincide with any meridians and parallels except for the central meridian and the Equator. Of most interest in the United States are two grid systems: The Universal Transverse Mercator (UTM) Grid is described on p. 57, and the State Plane Coordinate System (SPCS) is described on p. 51. Preceding the UTM was the World Polyconic Grid (WPG), used until the late 1940's and described on p. 127.

Grid systems are normally divided into zones so that distortion and variation of scale within any one zone is held below a preset level. The type of boundaries between grid zones varies. Zones of the WPG and the UTM are bounded by meridians of longitude, but for the SPCS State and county boundaries are used. Some grid boundaries in other countries are defined by lines of constant grid value using a local or an adjacent grid as the basis. This adjacent grid may in turn be based on a different projection and a different reference ellipsoid. A common boundary for non-U.S. offshore grids is an ellipsoidal rhumb line, or line of constant direction on the ellipsoid (see p. 46); the ellipsoidal geodesic, or shortest route (see p. 199) is also used. The plotting of some of these boundaries can become quite complicated (Clifford J. Mugnier, pers. comm., 1985).

3. THE DATUM AND THE EARTH AS AN ELLIPSOID

For many maps, including nearly all maps in commercial atlases, it may be assumed that the Earth is a sphere. Actually, it is more nearly an oblate ellipsoid of revolution, also called an oblate spheroid. This is an ellipse rotated about its shorter axis. The flattening of the ellipse for the Earth is only about one part in three hundred; but it is sufficient to become a necessary part of calculations in plotting accurate maps at a scale of 1:100,000 or larger, and is significant even for 1:5,000,000-scale maps of the United States, affecting plotted shapes by up to 2/3 percent (see p. 27). On small-scale maps, including single-sheet world maps, the oblateness is negligible. Formulas for both the sphere and ellipsoid will be discussed in this book wherever the projection is used or is suitable in both forms.

The Earth is not an exact ellipsoid, and deviations from this shape are continually evaluated. The *geoid* is the name given to the shape that the Earth would assume if it were all measured at mean sea level. This is an undulating surface that varies not more than about a hundred meters above or below a well-fitting ellipsoid, a variation far less than the ellipsoid varies from the sphere. It is important to remember that elevations and contour lines on the Earth are reported relative to the geoid, not the ellipsoid. Latitude, longitude, and all plane coordinate systems, on the other hand, are determined with respect to the ellipsoid.

The choice of the reference ellipsoid used for various regions of the Earth has been influenced by the local geoid, but large-scale map projections are designed to fit the reference ellipsoid, not the geoid. The selection of constants defining the shape of the reference ellipsoid has been a major concern of geodesists since the early 18th century. Two geometric constants are sufficient to define the ellipsoid itself. They are normally expressed either as (1) the semimajor and semiminor axes (or equatorial and polar radii, respectively), (2) the semimajor axis and the flattening, or (3) the semimajor axis and the eccentricity. These pairs are directly interchangeable. In addition, recent satellite-measured reference ellipsoids are defined by the semimajor axis, geocentric gravitational constant, and dynamical form factor, which may be converted to flattening with formulas from physics (Lauf, 1983, p. 6).

In the early 18th century, Isaac Newton and others concluded that the Earth should be slightly flattened at the poles, but the French believed the Earth to be egg-shaped as the result of meridian measurements within France. To settle the matter, the French Academy of Sciences, beginning in 1735, sent expeditions to Peru and Lapland to measure meridians at widely separated latitudes. This established the validity of Newton's conclusions and led to numerous meridian measurements in various locations, especially during the 19th and 20th centuries; between 1799 and 1951 there were 26 determinations of dimensions of the Earth.

The identity of the ellipsoid used by the United States before 1844 is uncertain, although there is reference to a flattening of 1/302. The Bessel ellipsoid of 1841 (see table 1) was used by the Coast Survey from 1844 until 1880, when the bureau adopted the 1866 evaluation by the British geodesist Alexander Ross Clarke using measurements of meridian arcs in western Europe, Russia, India, South Africa, and Peru (Shalowitz, 1964, p. 117–118; Clarke and Helmert, 1911, p. 807–808). This resulted in an adopted equatorial radius of 6,378,206.4 m and a polar radius of 6,356,583.8 m, or an approximate flattening of 1/294.9787.

The Clarke 1866 ellipsoid (the year should be included since Clarke is also known for ellipsoids of 1858 and 1880) has been used for all of North America until a change which is currently underway, as described below.

In 1909 John Fillmore Hayford reported calculations for a reference ellipsoid from U.S. Coast and Geodetic Survey measurements made entirely within the United States. This was adopted by the International Union of Geodesy and Geophysics (IUGG) in 1924, with a flattening of exactly 1/297 and a semimajor axis of exactly 6,378,388 m. This is therefore called the International or the

TABLE 1.—*Some official ellipsoids in use throughout the world*[1]

Name	Date	Equatorial Radius, a meters	Polar Radius b, meters	Flattening f	Use
GRS 80[2]	1980	6,378,137*	6,356,752.3	1/298.257	Newly adopted
WGS 72[3]	1972	6,378,135*	6,356,750.5	1/298.26	NASA; Dept. of Defense; oil companies
Australian	1965	6,378,160*	6,356,774.7	1/298.25*	Australia
Krasovsky	1940	6,378,245*	6,356,863.0	1/298.3*	Soviet Union
Internat'l 1924 Hayford 1909		6,378,388*	6,356,911.9	1/297*	Remainder of the world[†]
Clarke[4]	1880	6,378,249.1	6,356,514.9	1/293.46**	Most of Africa; France
Clarke	1866	6,378,206.4*	6,356,583.8*	1/294.98	North America; Philippines
Airy[4]	1830	6,377,563.4	6,356,256.9	1/299.32**	Great Britain
Bessel	1841	6,377,397.2	6,356,079.0	1/299.15**	Central Europe; Chile; Indonesia
Everest[4]	1830	6,377,276.3	6,356,075.4	1/300.80**	India; Burma; Pakistan; Afghan.; Thailand; etc.

Values are shown to accuracy in excess significant figures, to reduce computational confusion.

[1] Maling, 1973, p. 7; Thomas, 1970, p. 84; Army, 1973, p. 4, endmap; Colvocoresses, 1969, p. 33; World Geodetic, 1974.

[2] Geodetic Reference System. Ellipsoid derived from adopted model of Earth. WGS 84 has same dimensions within accuracy shown.

[3] World Geodetic System. Ellipsoid derived from adopted model of Earth.

[4] Also used in some regions with various modified constants.

* Taken as exact values. The third number (where two are asterisked) is derived using the following relationships: $b = a(1-f)$; $f = 1-b/a$. Where only one is asterisked (for 1972 and 1980), certain physical constants not shown are taken as exact, but f as shown is the adopted value.

** Derived from a and b, which are rounded off as shown after conversions from lengths in feet.

[†] Other than regions listed elsewhere in column, or some smaller areas.

Hayford ellipsoid, and is used in many parts of the world, but it was not adopted for use in North America, in part because of all the work already accomplished using the older datum and ellipsoid (Brown, 1949, p. 293; Hayford, 1909).

There are over a dozen other principal ellipsoids, however, which are still used by one or more countries (table 1). The different dimensions do not only result from varying accuracy in the geodetic measurements (the measurements of locations on the Earth), but the curvature of the Earth's surface (geoid) is not uniform due to irregularities in the gravity field.

Until recently, ellipsoids were only fitted to the Earth's shape over a particular country or continent. The polar axis of the reference ellipsoid for such a region, therefore, normally does not coincide with the axis of the actual Earth, although it is assumed to be parallel. The same applies to the two equatorial planes. The discrepancy between centers is usually a few hundred meters at most. Only satellite-determined coordinate systems, such as the WGS 72 and GRS 80 mentioned below, are considered geocentric. Ellipsoids for the latter systems represent the entire Earth more accurately than ellipsoids determined from ground measurements, but they do not generally give the "best fit" for a particular region.

The reference ellipsoids used prior to those determined by satellite are related to an "initial point" of reference on the surface to produce a datum, the name given to a smooth mathematical surface that closely fits the mean sea-level surface throughout the area of interest. The "initial point" is assigned a latitude, longitude, elevation above the ellipsoid, and azimuth to some point. Once a datum is adopted, it provides the surface to which ground control measurements are referred. The latitude and longitude of all the control points in a given area are then computed relative to the adopted ellipsoid and the adopted "initial point." The projection equations of large-scale maps must use the same ellipsoid parameters as those used to define the local datum; otherwise, the projections will be inconsistent with the ground control.

The first official geodetic datum in the United States was the New England Datum, adopted in 1879. It was based on surveys in the eastern and northeastern states and referenced to the Clarke Spheroid of 1866, with triangulation station Principio, in Maryland, as the origin. The first transcontinental arc of triangulation was completed in 1899, connecting independent surveys along the Pacific Coast. In the intervening years, other surveys were extended to the Gulf of Mexico. The New England Datum was thus extended to the south and west without major readjustment of the surveys in the east. In 1901, this expanded network was officially designated the United States Standard Datum, and triangulation station Meades Ranch, in Kansas, was the origin. In 1913, after the geodetic organizations of Canada and Mexico formally agreed to base their triangulation networks on the United States network, the datum was renamed the North American Datum.

By the mid-1920's, the problems of adjusting new surveys to fit into the existing network were acute. Therefore, during the 5-year period 1927–1932 all available primary data were adjusted into a system now known as the North American 1927 Datum.*** The coordinates of station Meades Ranch were not changed but the revised coordinates of the network comprised the North American 1927 Datum (National Academy of Sciences, 1971, p. 7).

Satellite data have provided geodesists with new measurements to define the best Earth-fitting ellipsoid and for relating existing coordinate systems to the Earth's center of mass. U.S. military efforts produced the World Geodetic System 1966 and 1972 (WGS 66 and WGS 72). The National Geodetic Survey is planning to replace the North American 1927 Datum with a new datum, the North American Datum 1983 (NAD 83), which is Earth-centered based on both satellite and terrestrial data. The IUGG in 1980 adopted a new model of the Earth called the Geodetic Reference System (GRS) 80, from which is derived an ellipsoid which has been adopted for the new North American datum. As a result, the latitude and longitude of almost every point in North America will change slightly, as well as the rectangular coordinates of a given latitude and longitude on a map projection. The difference can reach 300 m. U.S. military agencies are developing a worldwide datum called WGS 84, also based on GRS 80, but with slight differences. For Earth-centered datums, there is no single "origin" like Meades Ranch on the surface. The center of the Earth is in a sense the origin.

For the mapping of other planets and natural satellites, only Mars is treated as an ellipsoid. Other bodies are taken as spheres (table 2), although some irregular satellites have been treated as triaxial ellipsoids and are "mapped" orthographically.

In most map projection formulas, some form of the eccentricity e is used, rather than the flattening f. The relationship is as follows:

$$e^2 = 2f - f^2, \text{ or } f = 1 - (1 - e^2)^{1/2}$$

For the Clarke 1866, e^2 is 0.006768658. For the GRS 80, e^2 is 0.0066943800.

AUXILIARY LATITUDES

By definition, the geographic or geodetic latitude, which is normally the latitude referred to for a point on the Earth, is the angle which a line perpendicular to the surface of the ellipsoid at the given point makes with the plane of the Equator. It is slightly greater in magnitude than the geocentric latitude, except at the Equator and poles, where it is equal. The geocentric latitude is the angle made by a line to the center of the ellipsoid with the equatorial plane.

Formulas for the spherical form of a given map projection may be adapted for use with the ellipsoid by substitution of one of various "auxiliary latitudes" in place of the geodetic latitude. Oscar S. Adams (1921) developed series and other formulas for five substitute latitudes, generally building upon concepts described in the previous century. In using them, the ellipsoidal Earth is, in effect, first transformed to a sphere under certain restraints such as conformality or equal area, and the sphere is then projected onto a plane. If the proper auxiliary

TABLE 2.—*Official figures for extraterrestrial mapping*

[(From Davies, et al., 1983; Davies, Private commun., 1985.) Radius of Moon chosen so that all elevations are positive. Radius of Mars is based on a level of 6.1 millibar atmospheric pressure; Mars has both positive and negative elevations.]

Body	Equatorial radius a* (kilometers)
Earth's Moon	1,738.0
Mercury	2,439.0
Venus	6,051.0
Mars	3,393.4*
Galilean satellites of Jupiter	
Io	1,815
Europa	1,569
Ganymede	2,631
Callisto	2,400
Satellites of Saturn	
Mimas	198
Enceladus	253
Tethys	525
Dione	560
Rhea	765
Titan	2,575
Iapetus	725
Satellites of Uranus	
Ariel	665
Umbriel	555
Titania	800
Oberon	815
Miranda	250
Satellite of Neptune	
Triton	1,600

* Above bodies are taken as spheres except for Mars, an ellipsoid with eccentricity e of 0.101929. Flattening $f = 1 - (1 - e^2)^{1/2}$. Unlisted satellites are taken as triaxial ellipsoids, or mapping is not expected in the near future. Mimas and Enceladus have also been given ellipsoidal parameters, but not for mapping.

latitudes are chosen, the sphere may have either true areas, true distances in certain directions, or conformality, relative to the ellipsoid. Spherical map projection formulas may then be used for the ellipsoid solely with the substitution of the appropriate auxiliary latitudes.

It should be made clear that this substitution will generally not give the projection in its preferred form. For example, using the conformal latitude (defined below) in the spherical Transverse Mercator equations will give a true ellipsoidal, conformal Transverse Mercator, but the central meridian cannot be true to scale. More involved formulas are necessary, since uniform scale on the central meridian is a standard requirement for this projection as commonly used in the ellipsoidal form. For the regular Mercator, on the other hand, simple substitution of the conformal latitude is sufficient to obtain both conformality and an Equator of correct scale for the ellipsoid.

Adams gave formulas for all these auxiliary latitudes in closed or exact form, as well as in series, except for the authalic (equal-area) latitude, which could also have been given in closed form. Both forms are given below. For improved computational efficiency using the series, see equations (3−34) through (3−39). In finding the auxiliary latitude from the geodetic latitude, the closed form may be more useful for computer programs. For the inverse cases, to find geodetic from auxiliary latitudes, most closed forms require iteration, so that the series form is probably preferable. The series form shows more readily the amount of deviation from the geodetic latitude ϕ. The formulas given later for the individual ellipsoidal projections incorporate these formulas as needed, so there is no need to refer back to these for computation, but the various auxiliary latitudes are grouped together here for comparison. Some of Adams' symbols have been changed to avoid confusion with other terms used in this book.

The conformal latitude χ, giving a sphere which is truly conformal in accordance with the ellipsoid (Adams, 1921, p. 18, 84),

$$\chi = 2 \arctan \left\{ \tan (\pi/4 + \phi/2) \left[(1 - e \sin \phi)/(1 + e \sin \phi) \right]^{e/2} \right\} - \pi/2 \tag{3-1}$$

$$= 2 \arctan \left[\left(\frac{1 + \sin \phi}{1 - \sin \phi} \right) \left(\frac{1 - e \sin \phi}{1 + e \sin \phi} \right)^e \right]^{1/2} - \pi/2 \tag{3-1a}$$

$$\begin{aligned} = \phi &- (e^2/2 + 5e^4/24 + 3e^6/32 + 281e^8/5760 + \ldots) \sin 2\phi \\ &+ (5e^4/48 + 7e^6/80 + 697e^8/11520 + \ldots) \sin 4\phi \\ &- (13e^6/480 + 461e^8/13440 + \ldots) \sin 6\phi + (1237e^8/161280 \\ &+ \ldots) \sin 8\phi + \ldots \end{aligned} \tag{3-2}$$

with χ and ϕ in radians. In seconds of arc for the Clarke 1866 ellipsoid,

$$\chi = \phi - 700.0427'' \sin 2\phi + 0.9900'' \sin 4\phi + 0.0017'' \sin 6\phi \tag{3-3}$$

The inverse formula, for ϕ in terms of χ, may be a rapid iteration of an exact rearrangement of (3−1), successively placing the value of ϕ calculated on the left side into the right side of (3−4) for the next calculation, using χ as the first trial ϕ. When ϕ changes by less than a desired convergence value, iteration is stopped.

$$\phi = 2 \arctan \left\{ \tan (\pi/4 + \chi/2) \left[(1 + e \sin \phi)/(1 - e \sin \phi) \right]^{e/2} \right\} - \pi/2 \tag{3-4}$$

The inverse formula may also be written as a series, without iteration (Adams, 1921, p. 85):

$$\begin{aligned} \phi = \chi &+ (e^2/2 + 5e^4/24 + e^6/12 + 13e^8/360 + \ldots) \sin 2\chi \\ &+ (7e^4/48 + 29e^6/240 + 811e^8/11520 + \ldots) \sin 4\chi \\ &+ (7e^6/120 + 81e^8/1120 + \ldots) \sin 6\chi \\ &+ (4279e^8/161280 + \ldots) \sin 8\chi + \ldots \end{aligned} \tag{3-5}$$

or, for the Clarke 1866 ellipsoid, in seconds,

$$\phi = \chi + 700.0420'' \sin 2\chi + 1.3859'' \sin 4\chi + 0.0037'' \sin 6\chi \tag{3-6}$$

Adams referred to χ as the isometric latitude, but this name is now applied to ψ, a separate very nonlinear function of ϕ, which is directly proportional to the spacing of parallels of latitude from the Equator on the ellipsoidal Mercator projection. Another common symbol for isometric latitude is τ. It is also useful for other conformal projections:

$$\psi = \ln \left\{ \tan(\pi/4 + \phi/2) \left[(1 - e \sin \phi)/(1 + e \sin \phi) \right]^{e/2} \right\} \tag{3-7}$$

Because of the rapid variation from ϕ, ψ is not given here in series form. By comparing equations (3−1) and (3−7), it may be seen, however, that

$$\psi = \ln \tan (\pi/4 + \chi/2) \tag{3-8}$$

so that χ may be determined from the series in (3−2) and converted to ψ with (3−8), although there is no particular advantage over using (3−7).

For the inverse of (3−7), to find ϕ in terms of ψ, the choice is between iteration of a closed equation (3−10) and use of series (3−5) with a simple inverse of (3−8):

$$\chi = 2 \arctan e^{\psi} - \pi/2 \tag{3-9}$$

where e is the base of natural logarithms, 2.71828.

For the iteration, apply the principle of successive substitution used in (3−4) to the following, with (2 arctan $e^\psi − \pi/2$) as the first trial ϕ:

$$\phi = 2 \arctan \{e^\psi[(1 + e \sin \phi)/(1 − e \sin \phi)]^{e/2}\} − \pi/2 \qquad (3-10)$$

Note that e and e are not the same.

   The authalic latitude β, on a sphere having the same surface area as the ellipsoid, provides a sphere which is truly equal-area (authalic), relative to the ellipsoid:

$$\beta = \arcsin (q/q_p) \qquad (3-11)$$

where

$$q = (1 − e^2) \{\sin \phi/(1 − e^2 \sin^2 \phi) − (1/(2e)) \ln[(1 − e \sin \phi)/(1 + e \sin \phi)]\} \qquad (3-12)$$

and q_p is q evaluated for a ϕ of 90°. The radius R_q of the sphere having the same surface area as the ellipsoid is calculated as follows:

$$R_q = a(q_p/2)^{1/2} \qquad (3-13)$$

where a is the semimajor axis of the ellipsoid. For the Clarke 1866, R_q is 6,370,997.2 m.

   The equivalent series for β (Adams, 1921, p. 85)

$$\beta = \phi − (e^2/3 + 31e^4/180 + 59e^6/560 + \ldots) \sin 2\phi + (17e^4/360 + 61e^6/1260 + \ldots)$$
$$\sin 4\phi − (383e^6/45360 + \ldots) \sin 6\phi + \ldots \qquad (3-14)$$

where β and ϕ are in radians. For the Clarke 1866 ellipsoid, the formula in seconds of arc is:

$$\beta = \phi − 467.0129'' \sin 2\phi + 0.4494'' \sin 4\phi + 0.0005'' \sin 6\phi \qquad (3-15)$$

   For ϕ in terms of β, an iterative inverse of (3−12) may be used with the inverse of (3−11):

$$\phi = \phi + \frac{(1 − e^2 \sin^2 \phi)^2}{2 \cos \phi} \left[\frac{q}{1 − e^2} − \frac{\sin \phi}{1 − e^2 \sin^2 \phi} + \frac{1}{2e} \ln \left(\frac{1 − e \sin \phi}{1 + e \sin \phi} \right) \right] \qquad (3-16)$$

where

$$q = q_p \sin \beta \qquad (3-17)$$

q_p is found from (3−12) for a ϕ of 90°, and the first trial ϕ is arcsin ($q/2$), used on the right side of (3−16) for the calculation of ϕ on the left side, which is then used on the right side until the change is less than a preset limit. (Equation (3−16) is derived from equation (3−12) using a standard Newton-Raphson iteration.)

   To find ϕ from β with a series:

$$\phi = \beta + (e^2/3 + 31e^4/180 + 517e^6/5040 + \ldots) \sin 2\beta$$
$$+ (23e^4/360 + 251e^6/3780 + \ldots) \sin 4\beta \qquad (3-18)$$
$$+ (761e^6/45360 + \ldots) \sin 6\beta + \ldots$$

or, for the Clarke 1866 ellipsoid, in seconds,

$$\phi = \beta + 467.0127'' \sin 2\beta + 0.6080'' \sin 4\beta + 0.0011'' \sin 6\beta \qquad (3-19)$$

   The rectifying latitude μ (designated ω by Adams), giving a sphere with correct distances along the meridians, requires a series in any case (or a numerical integration which is not shown).

$$\mu = \pi M/2M_p \qquad (3-20)$$

where

$$M = a[(1 - e^2/4 - 3e^4/64 - 5e^6/256 - \dots)\phi - (3e^2/8 + 3e^4/32$$
$$+ 45e^6/1024 + \dots) \sin 2\phi + (15e^4/256 + 45e^6/1024 + \dots) \sin 4\phi$$
$$- (35e^6/3072 + \dots) \sin 6\phi + \dots] \qquad (3-21)$$

and M_p is M evaluated for a ϕ of 90°, for which all sine terms drop out. M is the distance along the meridian from the Equator to latitude ϕ. For the Clarke 1866 ellipsoid, the constants simplify to, in meters,

$$M = 111132.0894\phi° - 16216.94 \sin 2\phi + 17.21 \sin 4\phi - 0.02 \sin 6\phi \qquad (3-22)$$

The first coefficient in (3−21) has been multiplied by $\pi/180$ to use ϕ in degrees. To use μ properly, the radius R_M of the sphere must be $2M_p/\pi$ for correct scale. For the Clarke 1866 ellipsoid, R_M is 6,367,399.7 m. A series combining (3−20) and (3−21) is given by Adams (1921, p. 125):

$$\mu = \phi - (3e_1/2 - 9e_1{}^3/16 + \dots) \sin 2\phi + (15e_1{}^2/16 - 15e_1{}^4/32 + \dots)$$
$$\sin 4\phi - (35e_1{}^3/48 - \dots) \sin 6\phi + (315e_1{}^4/512 - \dots)$$
$$\sin 8\phi + \dots \qquad (3-23)$$

where

$$e_1 = [1 - (1 - e^2)^{1/2}]/[1 + (1 - e^2)^{1/2}] \qquad (3-24)$$

and μ and ϕ are given in radians. For the Clarke 1866 ellipsoid, in seconds,

$$\mu = \phi - 525.3298'' \sin 2\phi + 0.5575'' \sin 4\phi + 0.0007'' \sin 6\phi \qquad (3-25)$$

The inverse of equations (3−23) or (3−25), for ϕ in terms of μ, given M, will be found useful for several map projections to avoid iteration, since a series is required in any case (Adams, 1921, p. 128).

$$\phi = \mu + (3e_1/2 - 27e_1{}^3/32 + \dots) \sin 2\mu + (21e_1{}^2/16 - 55e_1{}^4/32 + \dots)$$
$$\sin 4\mu + (151e_1{}^3/96 - \dots) \sin 6\mu + (1097e_1{}^4/512 - \dots)$$
$$\sin 8\mu + \dots \qquad (3-26)$$

where e_1 is found from equation (3−24) and μ from (3−20), but M is given, not calculated from (3−21). For the Clarke 1866 ellipsoid, in seconds of arc,

$$\phi = \mu + 525.3295'' \sin 2\mu + 0.7805'' \sin 4\mu + 0.0016'' \sin 6\mu \qquad (3-27)$$

The following closed and exact formulas, from which equations (3−20) through (3−25) may be ultimately derived, are given as a matter of interest.

$$M = a (1 - e^2)\int_0^\phi [1/(1 - e^2 \sin^2 \phi)^{3/2}] \, d\phi \qquad (3-27a)$$

Equation (3−27a), the integral of (4−19) in a later chapter, may not be exactly integrated. While Simpson's rule may be used, it is not as satisfactory here as it is in some other cases (equation (27−6a), etc.). However, (3−27a) may be transformed to an elliptic integral of the second kind, for which the arithmetic-geometric-mean (A.G.M.) iteration can provide any desired accuracy within computer programming limitations (Messenger, T.J., pers. commun., 1984; Abramowitz and Stegun, 1964, p. 598−99):

$$M = a [\int_0^\phi (1 - e^2 \sin^2\phi)^{1/2} \, d\phi - e^2 \sin \phi \cos \phi/(1 - e^2 \sin^2 \phi)^{1/2}] \qquad (3-27b)$$

The remaining auxiliary latitudes listed by Adams (1921, p. 84) are more useful for derivation than in substitutions for projections:

The geocentric latitude ϕ_g (designated ψ by Adams) referred to in the first paragraph in this section is simply as follows:

$$\phi_g = \arctan [(1 - e^2) \tan \phi] \qquad (3-28)$$

As a series,

$$\phi_g = \phi - e_2 \sin 2\phi + (e_2{}^2/2) \sin 4\phi - (e_2{}^3/3) \sin 6\phi + \dots \qquad (3-29)$$

TABLE 3.—*Corrections for auxiliary latitudes on the Clarke 1866 ellipsoid*

[Corrections are given, rather than actual values. For example, if the geodetic latitude is 50°N., the conformal latitude is 50° − 11′29.7″ = 49° 48′30.3″ N. For southern latitudes, the corrections are the same, disregarding the sign of the latitude. That is, the conformal latitude for a φ of lat. 50° S. is 49° 48′30.3″ S. From Adams, 1921]

	Geodetic (φ)	Conformal (χ−φ)	Authalic (β−φ)	Rectifying (μ−φ)	Geocentric (φ_g−φ)	Parametric (η−φ)
90°	---------------	0′ 00.0″	0′ 00.0″	0′ 00.0″	0′ 00.0″	0′ 00.0″
85	---------------	− 2 01.9	−1 21.2	−1 31.4	− 2 02.0	−1 00.9
80	---------------	− 4 00.1	−2 40.0	−3 00.0	− 4 00.3	−2 00.0
75	---------------	− 5 50.9	−3 53.9	−4 23.1	− 5 51.3	−2 55.4
70	---------------	− 7 31.0	−5 00.6	−5 38.2	− 7 31.4	−3 45.4
65	---------------	− 8 57.2	−5 58.2	−6 43.0	− 8 57.7	−4 28.6
60	---------------	−10 07.1	−6 44.8	−7 35.4	−10 07.6	−5 03.6
55	---------------	−10 58.5	−7 19.1	−8 14.0	−10 58.9	−5 29.3
50	---------------	−11 29.7	−7 40.1	−8 37.5	−11 30.2	−5 45.0
45	---------------	−11 40.0	−7 47.0	−8 45.3	−11 40.5	−5 50.2
40	---------------	−11 29.1	−7 39.8	−8 37.2	−11 29.4	−5 44.8
35	---------------	−10 57.2	−7 18.6	−8 13.3	−10 57.4	−5 28.9
30	---------------	−10 05.4	−6 44.1	−7 34.5	−10 05.6	−5 03.0
25	---------------	− 8 55.3	−5 57.3	−6 41.9	− 8 55.4	−4 28.0
20	---------------	− 7 29.0	−4 59.7	−5 37.1	− 7 29.1	−3 44.8
15	---------------	− 5 49.2	−3 53.1	−4 22.2	− 5 49.2	−2 54.9
10	---------------	− 3 58.8	−2 39.4	−2 59.3	− 3 58.8	−1 59.6
5	---------------	− 2 01.2	−1 20.9	−1 31.0	− 2 01.2	−1 00.7
0	---------------	0 00.0	0 00.0	0 00.0	0 00.0	0 00.0

where ϕ_g and ϕ are in radians and $e_2 = e^2/(2 - e^2)$. For the Clarke 1866 ellipsoid, in seconds of arc,

$$\phi_g = \phi - 700.44'' \sin 2\phi + 1.19'' \sin 4\phi \qquad (3-30)$$

The reduced or *parametric latitude* η (designated θ by Adams) of a point on the ellipsoid is the latitude on a sphere of radius a for which the parallel has the same radius as the parallel of geodetic latitude ϕ on the ellipsoid through the given point:

$$\eta = \arctan [(1 - e^2)^{1/2} \tan \phi] \qquad (3-31)$$

As a series,

$$\eta = \phi - e_1 \sin 2\phi + (e_1{}^2/2) \sin 4\phi - (e_1{}^3/3) \sin 6\phi + \ldots \qquad (3-32)$$

where e_1 is found from equation (3−24), and η and ϕ are in radians. For the Clarke 1866 ellipsoid, using seconds of arc,

$$\eta = \phi - 350.22'' \sin 2\phi + 0.30'' \sin 4\phi \qquad (3-33)$$

The inverses of equations (3−28) and (3−31) for ϕ in terms of geocentric or reduced latitudes are relatively easily derived and are noniterative. The inverses of series equations (3−29), (3−30), (3−32), and (3−33) are therefore omitted. Table 3 lists the correction for these auxiliary latitudes for each 5° of geodetic latitude.

COMPUTATION OF SERIES

Most of the trigonometric series approximations throughout this book (for example, equations (3−2) and (3−5)) are given in terms of multiple angles. In this arrangement, the coefficients converge to zero more rapidly, but handling by

computer is normally somewhat slower than that occurring with nested trigonometric series. The latter are equivalent to power polynomials and require a minimum number of computations of trigonometric functions from series built into the
software of most computers.

The pertinent series in this book fall into one of three forms (3−34), (3−36) and
(3−38), in which ϕ may be any variable, and $f(\phi)$ is the function:

If $f(\phi) = A \sin 2\phi + B \sin 4\phi + C \sin 6\phi + D \sin 8\phi$ (3−34)

then $f(\phi) = \sin 2\phi \, (A' + \cos 2\phi \, (B' + \cos 2\phi \, (C' + D' \cos 2\phi)))$ (3−35)

where

$$
\begin{aligned}
A' &= A - C \\
B' &= 2B - 4D \\
C' &= 4C \\
D' &= 8D
\end{aligned}
$$

If $f(\phi) = A \sin \phi + B \sin 3\phi + C \sin 5\phi + D \sin 7\phi$ (3−36)

then $f(\phi) = \sin \phi \, (A' + \sin^2\phi \, (B' + \sin^2\phi \, (C' + D' \sin^2\phi)))$ (3−37)

where

$$
\begin{aligned}
A' &= A + 3B + 5C + 7D \\
B' &= -4B - 20C - 56D \\
C' &= 16C + 112D \\
D' &= -64D
\end{aligned}
$$

If $f(\phi) = A + B \cos 2\phi + C \cos 4\phi + D \cos 6\phi + E \cos 8\phi$ (3−38)

then $f(\phi) = A' + \cos 2\phi \, (B' + \cos 2\phi \, (C' + \cos 2\phi \, (D' + E' \cos 2\phi)))$ (3−39)

where

$$
\begin{aligned}
A' &= A - C + E \\
B' &= B - 3D \\
C' &= 2C - 8E \\
D' &= 4D \\
E' &= 8E
\end{aligned}
$$

These are exact equivalents of the series as shown. First the primed coefficients are computed once for the full set of conversions from the original coefficients of (3−34), (3−36), or (3−38), then $\sin 2\phi$ and $\cos 2\phi$ are computed once for
each point in (3−35), or $\sin \phi$ and $\sin^2\phi$ once for each point in (3−37), or $\cos 2\phi$
once for each point in (3−39). Computation of $f(\phi)$ may then proceed from the
innermost nest outward with a speed up to 25−35 percent faster than that with
multiple-angle series.

For more efficient transformation of a great number of points from one set of
coordinates to another, polynomial approximations for the entire projection may
be considered. This is normally only practical for a limited region. For techniques
in determining the polynomial coefficients, the reader is referred to Snyder (1985a,
p. 5−6, 15−24).

4. SCALE VARIATION AND ANGULAR DISTORTION

Since no map projection maintains correct scale throughout, it is important to determine the extent to which it varies on a map. On a world map, qualitative distortion is evident to an eye familiar with maps, after noting the extent to which landmasses are improperly sized or out of shape, and the extent to which meridians and parallels do not intersect at right angles, or are not spaced uniformly along a given meridian or given parallel. On maps of countries or even of continents, distortion may not be evident to the eye, but it becomes apparent upon careful measurement and analysis.

TISSOT'S INDICATRIX

In 1859 and 1881, Nicolas Auguste Tissot published a classic analysis of the distortion which occurs on a map projection (Tissot, 1881; Adams, 1919, p. 153−163; Maling, 1973, p. 64−67). The intersection of any two lines on the Earth is represented on the flat map with an intersection at the same or a different angle. At almost every point on the Earth, there is a right angle intersection of two lines in some direction (not necessarily a meridian and a parallel) which are also shown at right angles on the map. All the other intersections at that point on the Earth will not intersect at the same angle on the map, unless the map is conformal, at least at that point. The greatest deviation from the correct angle is called ω, the maximum angular deformation. For a conformal map, ω is zero. (In some texts, 2ω is used rather than ω.)

Tissot showed this relationship graphically with a special ellipse of distortion called an indicatrix. An infinitely small circle on the Earth projects as an infinitely small, but perfect, ellipse on any map projection. If the projection is conformal, the ellipse is a circle, an ellipse of zero eccentricity. Otherwise, the ellipse has a major axis and minor axis which are directly related to the scale distortion and to the maximum angular deformation.

In figure 3, the left-hand drawing shows a circle representing the infinitely small circular element, crossed by a meridian λ and parallel ϕ on the Earth. The right-hand drawing shows this same element as it may appear on a typical map projection. For general purposes, the map is assumed to be neither conformal nor equal-area. The meridian and parallel may no longer intersect at right angles, but

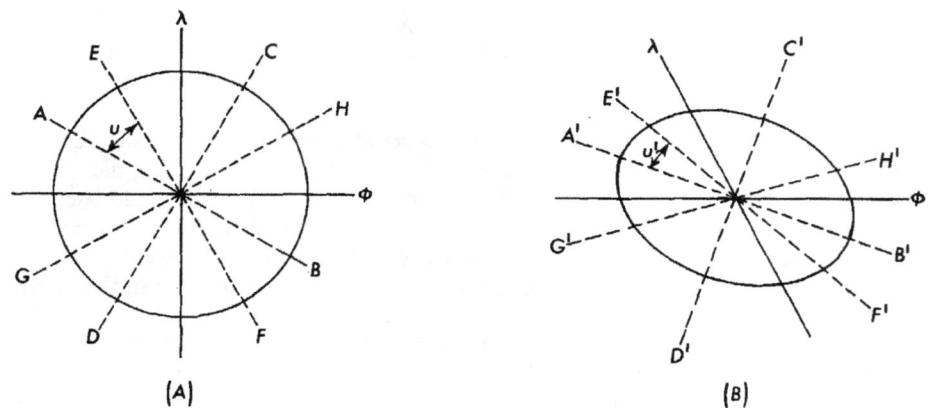

FIGURE 3.—Tissot's Indicatrix. An infinitely small circle on the Earth (A) appears as an ellipse on a typical map (B). On a conformal map, (B) is a circle of the same or of a different size.

there is a pair of axes which intersect at right angles on both Earth (AB and CD) and map ($A'B'$ and $C'D'$). There is also a pair of axes which intersect at right angles on the Earth (EF and GH), but at an angle on the map ($E'F'$ and $G'H'$) farthest from a right angle. The latter case has the maximum angular deformation ω. The orientation of these axes is such that μ + μ' = 90°, or, for small distortions, the lines fall about halfway between $A'B'$ and $C'D'$. The orientation is of much less interest than the size of the deformation. If a and b, the major and minor semiaxes of the indicatrix, are known, then

$$\sin (\omega/2) = |a - b|/(a + b) \tag{4-1}$$

If lines λ and φ coincide with a and b, in either order, as in cylindrical and conic projections, the calculation is relatively simple, using equations (4−2) through (4−6) given below.

Scale distortion is most often calculated as the ratio of the scale along the meridian or along the parallel at a given point to the scale at a standard point or along a standard line, which is made true to scale. These ratios are called "scale factors." That along the meridian is called h and along the parallel, k. The term "scale error" is frequently applied to ($h-1$) and ($k-1$). If the meridians and parallels intersect at right angles, coinciding with a and b in figure 3, the scale factor in any other direction at such a point will fall between h and k. Angle ω may be calculated from equation (4−1), substituting h and k in place of a and b. In general, however, the computation of ω is much more complicated, but is important for knowing the extent of the angular distortion throughout the map.

The formulas are given here to calculate h, k, and ω; but the formulas for h and k are applied specifically to all projections for which they are deemed useful as the projection formulas are given later. Formulas for ω for specific projections have generally been omitted.

Another term occasionally used in practical map projection analysis is "convergence" or "grid declination." This is the angle between true north and grid north (or direction of the Y axis). For regular cylindrical projections this is zero, for regular conic and polar azimuthal projections it is a simple function of longitude, and for other projections it may be determined from the projection formulas by calculus from the slope of the meridian (dy/dx) at a given latitude. It is used primarily by surveyors for fieldwork with topographic maps. Convergence is not discussed further in this work.

DISTORTION FOR PROJECTIONS OF THE SPHERE

The formulas for distortion are simplest when applied to regular cylindrical, conic (or conical), and polar azimuthal projections of the sphere. On each of these types of projections, scale is solely a function of the latitude.

Given the formulas for rectangular coordinates x and y of any cylindrical projection as functions solely of longitude λ and latitude φ, respectively,

$$h = dy/(Rd\phi) \tag{4-2}$$
$$k = dx/(R \cos \phi d\lambda) \tag{4-3}$$

Given the formulas for polar coordinates ρ and θ of any conic projection as functions solely of φ and λ, respectively, where n is the cone constant or ratio of θ to ($\lambda - \lambda_0$),

$$h = -d\rho/(Rd\phi) \tag{4-4}$$
$$k = n\rho/(R \cos \phi) \tag{4-5}$$

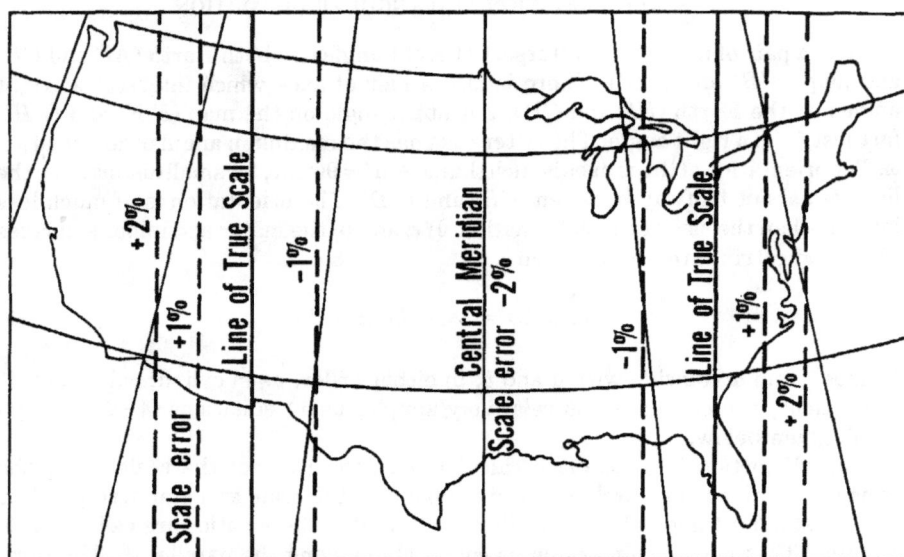

Transverse Mercator Projection

FIGURE 4.—Distortion patterns on common conformal map projections. The Transverse Mercator and
the Stereographic are shown with reduction in scale along the central meridian or at the center of
projection, respectively. If there is no reduction, there is a single line of true scale along the
central meridian on the Transverse Mercator and only a point of true scale at the center of the
Stereographic. The illustrations are conceptual rather than precise, since each base map projection is an identical conic.

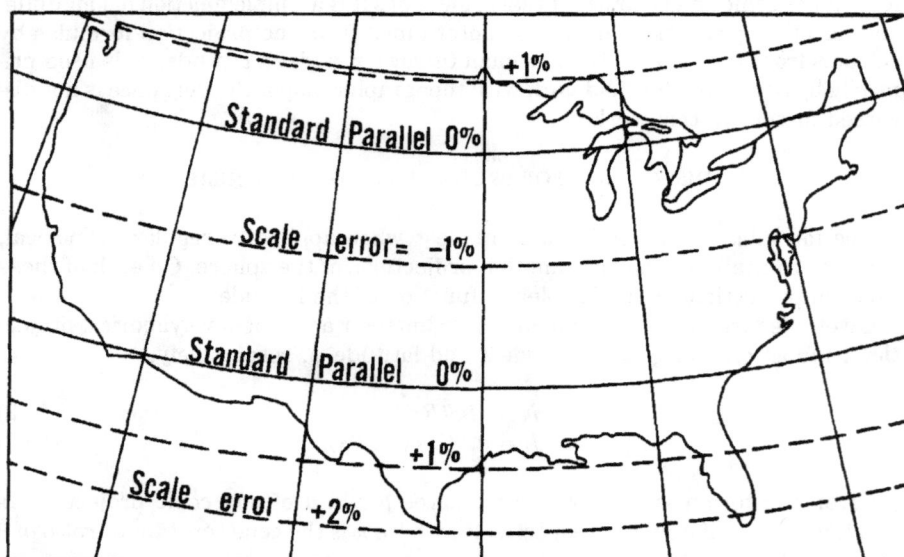

Lambert Conformal Conic Projection

FIGURE 4.—Continued.

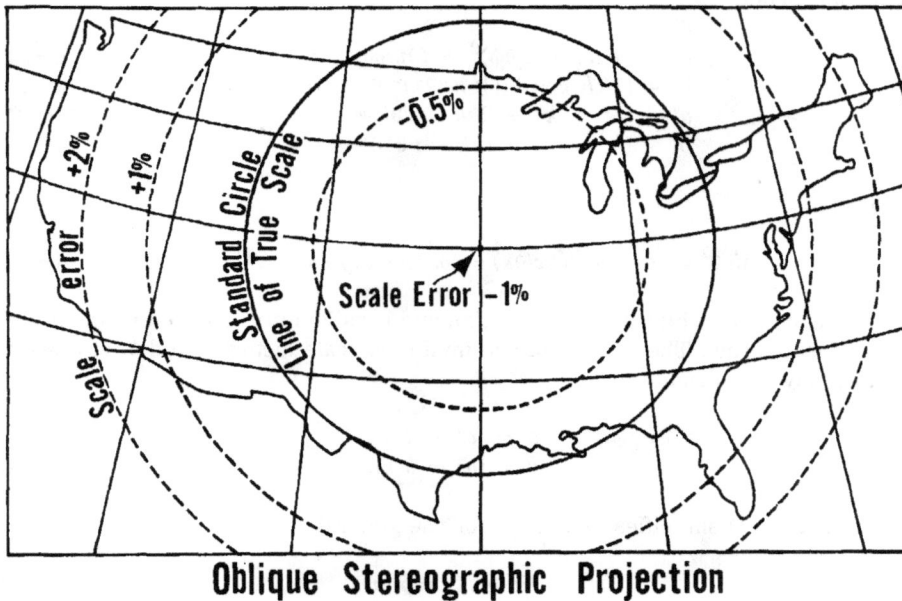

Oblique Stereographic Projection

FIGURE 4.—Continued

Given the formulas for polar coordinates ρ and θ of any polar azimuthal projection as functions solely of ϕ and λ, respectively, equations (4−4) and (4−5) apply, with n equal to 1.0:

$$h = -d\rho/(Rd\phi) \qquad (4-4)$$
$$k = \rho/(R \cos \phi) \qquad (4-6)$$

Equations (4−4) and (4−6) may be adapted to any azimuthal projection centered on a point other than the pole. In this case h' is the scale factor in the direction of a straight line radiating from the center, and k' is the scale factor in a direction perpendicular to the radiating line, all at an angular distance c from the center:

$$h' = d\rho/(Rdc) \qquad (4-7)$$
$$k' = \rho/(R \sin c) \qquad (4-8)$$

An analogous relationship applies to scale factors on oblique cylindrical and conic projections.

For any of the pairs of equations from (4−2) through (4−8), the maximum angular deformation ω at any given point is calculated simply, as stated above,

$$\sin (\omega/2) = |h - k|/(h + k) \qquad (4-9)$$

where $|h-k|$ signifies the absolute value of $(h-k)$, or the positive value without regard to sign. For equations (4−7) and (4−8), h' and k' are used in (4−9) instead of h and k, respectively. In figure 4, distortion patterns are shown for three conformal projections of the United States, choosing arbitrary lines of true scale.

For the general case, including all map projections of the sphere, rectangular coordinates x and y are often both functions of both ϕ and λ, so they must be partially differentiated with respect to both ϕ and λ, holding λ and ϕ, respectively, constant. Then,

$$h = (1/R) [(\partial x/\partial\phi)^2 + (\partial y/\partial\phi)^2]^{1/2} \qquad (4-10)$$
$$k = [1/(R \cos \phi)] [(\partial x/\partial\lambda)^2 + (\partial y/\partial\lambda)^2]^{1/2} \qquad (4-11)$$
$$a' = (h^2 + k^2 + 2hk \sin \theta')^{1/2} \qquad (4-12)$$
$$b' = (h^2 + k^2 - 2hk \sin \theta')^{1/2} \qquad (4-13)$$

where

$$\sin \theta' = [(\partial y/\partial\phi)(\partial x/\partial\lambda) - (\partial x/\partial\phi)(\partial y/\partial\lambda)]/(R^2 hk \cos \phi) \qquad (4-14)$$

θ' is the angle at which a given meridian and parallel intersect, and a' and b' are convenient terms. The maximum and minimum scale factors a and b, at a given point, may be calculated thus:

$$a = (a' + b')/2 \qquad (4-12a)$$
$$b = (a' - b')/2 \qquad (4-13a)$$

Equation (4-1) simplifies as follows for the general case:

$$\sin (\omega/2) = b'/a' \qquad (4-1a)$$

The areal scale factor s:

$$s = hk \sin \theta' \qquad (4-15)$$

For special cases:
(1) $s = hk$ if meridians and parallels intersect at right angles ($\theta' = 90°$);
(2) $h = k$ and $\omega = 0$ if the map is conformal;
(3) $h = 1/k$ on an equal-area map if meridians and parallels intersect at right angles.[2]

DISTORTION FOR PROJECTIONS OF THE ELLIPSOID

The derivation of the above formulas for the sphere utilizes the basic formulas for the length of a given spacing (usually 1° or 1 radian) along a given meridian or a given parallel. The following formulas give the length of a radian of latitude (L_ϕ) and of longitude (L_λ) for the sphere:

$$L_\phi = R \qquad (4-16)$$
$$L_\lambda = R \cos \phi \qquad (4-17)$$

where R is the radius of the sphere. For the length of 1° of latitude or longitude, these values are multiplied by $\pi/180$.

The radius of curvature on a sphere is the same in all directions. On the ellipsoid, the radius of curvature varies at each point and in each direction along a given meridian, except at the poles. The radius of curvature R' in the plane of the meridian is calculated as follows:

$$R' = a(1-e^2)/(1-e^2 \sin^2 \phi)^{3/2} \qquad (4-18)$$

---

[2]Maling (1973, p. 49–81) has helpful derivations of these equations in less condensed forms. There are typographical errors in several of the equations in Maling, but these may be detected by following the derivation closely.

TABLE 4.—*Lengths, in meters, of 1° of latitude and longitude on two ellipsoids of reference*

Latitude (ϕ)	Clarke 1866 ellipsoid		International (Hayford) ellipsoid	
	1° lat.	1° long.	1° lat.	1° long.
90°	111,699.4	0.0	111,700.0	0.0
85	111,690.7	9,735.0	111,691.4	9,735.0
80	111,665.0	19,394.4	111,665.8	19,394.5
75	111,622.9	28,903.3	111,624.0	28,903.5
70	111,565.9	38,188.2	111,567.4	38,188.5
65	111,495.7	47,177.5	111,497.7	47,177.9
60	111,414.5	55,802.2	111,417.1	55,802.8
55	111,324.8	63,996.4	111,327.9	63,997.3
50	111,229.3	71,698.1	111,233.1	71,699.2
45	111,130.9	78,849.2	111,135.4	78,850.5
40	111,032.7	85,396.1	111,037.8	85,397.7
35	110,937.6	91,290.3	110,943.3	91,292.2
30	110,848.5	96,488.2	110,854.8	96,490.4
25	110,768.0	100,951.9	110,774.9	100,954.3
20	110,698.7	104,648.7	110,706.0	104,651.4
15	110,642.5	107,551.9	110,650.2	107,554.8
10	110,601.1	109,640.7	110,609.1	109,643.7
5	110,575.7	110,899.9	110,583.9	110,903.1
0	110,567.2	111,320.7	110,575.5	111,323.9

The length of a radian of latitude is defined as the circumference of a circle of this radius, divided by 2π, or the radius itself. Thus,

$$L_\phi = a(1-e^2)/(1-e^2 \sin^2 \phi)^{3/2} \tag{4-19}$$

For the radius of curvature N of the ellipsoid in a plane perpendicular to the meridian and also perpendicular to a plane tangent to the surface,

$$N = a/(1-e^2 \sin^2\phi)^{1/2} \tag{4-20}$$

Radius N is also the length of the perpendicular to the surface from the surface to the polar axis. The length of a radian of longitude is found, as in equation (4-17), by multiplying N by $\cos \phi$, or

$$L_\lambda = a \cos \phi/(1-e^2 \sin^2\phi)^{1/2} \tag{4-21}$$

The lengths of 1° of latitude and 1° of longitude for the Clarke 1866 and the International ellipsoids are given in table 4. They are found from equations (4-19) and (4-21), multiplied by $\pi/180$ to convert to lengths for 1°.

When these formulas are applied to equations (4-2) through (4-6), the values of h and k for the ellipsoidal forms of the projections are found to be as follows:

For cylindrical projections:

$$\begin{aligned} h &= dy/(R'd\phi) \\ &= (1-e^2 \sin^2\phi)^{3/2} \, dy/[a(1-e^2)d\phi] \end{aligned} \tag{4-22}$$

$$\begin{aligned} k &= dx/(N \cos \phi d\lambda) \\ &= (1-e^2 \sin^2 \phi)^{1/2} \, dx/(a \cos \phi \, d\lambda) \end{aligned} \tag{4-23}$$

For conic projections:

$$\begin{aligned} h &= -d\rho/(R'd\phi) \\ &= -(1-e^2 \sin^2\phi)^{3/2} \, d\rho/[a(1-e^2)d\phi] \end{aligned} \tag{4-24}$$

$$k = n\rho/(N \cos \phi)$$
$$= n\rho(1-e^2 \sin^2 \phi)^{1/2}/(a \cos \phi) \tag{4-25}$$

For polar azimuthal projections:

$$h = -(1-e^2 \sin^2\phi)^{3/2} d\rho/[a(1-e^2)d\phi] \tag{4-24}$$
$$k = \rho(1-e^2 \sin^2\phi)^{1/2}/(a \cos \phi) \tag{4-26}$$

Equations (4−7) and (4−8) do not have ellipsoidal equivalents. Equation (4−9) remains the same for equations (4−22) through (4−26):

$$\sin (\omega/2) = |h-k|/(h+k) \tag{4-9}$$

For the general projection of the ellipsoid, equations (4−10) and (4−11) are similarly modified:

$$h = [(\partial x/\partial \phi)^2 + (\partial y/\partial \phi)^2]^{1/2}(1-e^2 \sin^2\phi)^{3/2}/[a(1-e^2)] \tag{4-27}$$
$$k = [(\partial x/\partial \lambda)^2 + (\partial y/\partial \lambda)^2]^{1/2}(1-e^2 \sin^2\phi)^{1/2}/(a \cos \phi) \tag{4-28}$$

Equations (4−12) through (4−15), (4−12a), (4−13a), and (4−1a), listed for the sphere, apply without change, except that R^2 becomes $a^2(1-e^2)/(1-e^2\sin^2\phi)^2$ in (4−14).

Specific calculations are shown during the discussion of individual projections.

The importance of using the ellipsoid instead of the sphere for designing a projection may be quantitatively evaluated by determining the ratio or product of some of the elementary relationships. The ratio of the differential length of a radian of latitude along a meridian on the sphere to that on the ellipsoid is found by dividing the equation (4−16) by equation (4−19), or

$$C_m = R(1-e^2 \sin^2 \phi)^{3/2}/[a(1-e^2)] \tag{4-29}$$

A related ratio for the length of a radian of longitude along a parallel on the sphere to that on the ellipsoid is found by dividing equation (4−17) by equation (4−21), or

$$C_p = R(1-e^2 \sin^2 \phi)^{1/2}/a \tag{4-30}$$

From these, the local shape factor C_s may be found as the ratio of (4−29) to (4−30):

$$C_s = C_m/C_p = (1-e^2 \sin^2 \phi)/(1-e^2) \tag{4-31}$$

and the area factor C_a is their product:

$$C_a = C_m C_p = R^2(1-e^2 \sin^2 \phi)^2/[a^2(1-e^2)] \tag{4-32}$$

If h and k are calculated for the spherical version of a map projection, the actual scale factors on the spherical version relative to the ellipsoid may be determined by multiplying h by C_m and k by C_p. For normal cylindrical and conic projections and polar azimuthal projections, the conformality or shape factor may be taken as h/k (not the same as ω) multiplied by C_s, and the area scale factor hk may be multiplied by C_a.

Except for C_s, which is independent of R/a, R must be given an arbitrary value such as R_q (see equation (3−13)), R_M (see second sentence following equation (3−22)), or another reasonable balance between the major and minor semiaxes a

TABLE 5.—*Ellipsoidal correction factors to apply to spherical projections based on Clarke 1866 ellipsoid*

Lat. (N&S)	C_m*	C_p	C_t	C_a*
90° _____	0.99548	0.99548	1.00000	0.99099
75 _____	.99617	.99571	1.00046	.99189
60 _____	.99803	.99633	1.00170	.99437
45 _____	1.00058	.99718	1.00341	.99775
30 _____	1.00313	.99802	1.00511	1.00114
15 _____	1.00499	.99864	1.00636	1.00363
0 _____	1.00568	.99887	1.00681	1.00454
Multiply by**	h	k	h/k	hk

* C_m = 1.0 for 48.24° lat. and C_a = 1.0 for 35.32° lat. Values of C_m, C_p, and C_a are based on a radius of 6,370.997 m for the sphere used in the spherical map projection.
** h = scale factor along meridian.

    k = scale factor along parallel of latitude.

    For normal cylindrical and conic projections and polar azimuthal projections:

    h/k = shape factor.

    hk = area scale factor.

    For example, if, on a spherical Albers Equal-Area Conic map projection based on sphere of radius 6,370,997 m, h = 1.00132 and k = 0.99868 at lat. 45° N., this map has an area scale factor of 1.00132 × 0.99868 × 00.99775 = 0.99775, relative to the correct area scale for the Clarke 1966 ellipsoid. If the ellipsoidal Albers were used, this factor would be 1.0.

and b of the ellipsoid. Using R_q and the Clarke 1866 ellipsoid, table 5 shows the magnitude of these corrections. Thus, a conformal projection based on the sphere has the correct shape at the poles for the ellipsoid, but the shape is about $^2/_3$ of 1 percent (0.00681) in error near the Equator (that is, Tissot's Indicatrix is an ellipse with minor axis about $^2/_3$ of 1 percent shorter than the major axis at the Equator when the spherical form is compared to the ellipsoid).

A map extending over a large area will have a scale variation of several percent, which far outweighs the significance of the less-than-1-percent variation between sphere and ellipsoid. A map of a small area, such as a large-scale detailed topographic map, or even a narrow strip map, has a small-enough intrinsic scale variation to make the ellipsoidal correction a significant factor in accurate mapping; e.g., a 7.5-min quadrangle normally has an intrinsic scale variation of 0.0002 percent or less.

CAUCHY-RIEMANN AND RELATED EQUATIONS

Relatively simple equations provide necessary and sufficient conditions for any map projection, spherical or ellipsoidal, to be conformal. These are called the Cauchy-Riemann equations after two 19th-century mathematicians. The concept had been devised, however, during the 18th century. These equations may be written as follows:

$$\partial x/\partial \lambda = \partial y/\partial \psi \qquad (4-33)$$
$$\partial x/\partial \psi = -\partial y/\partial \lambda \qquad (4-34)$$

where ψ is the isometric latitude defined by equation (3−7) for the ellipsoid, or with $e = 0$ in the same equation for the sphere. In the latter case, the above equations simplify to

$$\partial x/(\cos \phi \ \partial \lambda) = \partial y/\partial \phi \qquad (4-35)$$
$$\partial x/\partial \phi = -\partial y/(\cos \phi \ \partial \lambda) \qquad (4-36)$$

For the ellipsoid, equations (4−33) and (4−34) may be written

$$\partial x/(\cos \phi\ \partial \lambda) = (1 - e^2 \sin^2 \phi)\ \partial y/[(1 - e^2)\ \partial \phi] \qquad (4\text{--}37)$$
$$(1 - e^2 \sin^2 \phi) \partial x/[(1 - e^2) \partial \phi] = -\partial y/(\cos \phi\ \partial \lambda) \qquad (4\text{--}38)$$

By substituting x' in place of λ and y' in place of ψ in equations (4–33) and (4–34), conditions are met for conformal transformation of one set of rectangular coordinates (x', y') to another (x, y). That is,

$$\partial x/\partial x' = \partial y/\partial y' \qquad (4\text{--}39)$$
$$\partial x/\partial y' = -\partial y/\partial x' \qquad (4\text{--}40)$$

In this case, if (x', y') represents the transformation of the sphere or ellipsoid onto a flat surface, this transformation must also be conformal. The double transformation is used in a later chapter for the Modified-Stereographic Conformal projections.

Analogous relationships may be obtained for equal-area transformations. The following equation applies to the ellipsoid:

$$(\partial x/\partial \lambda)\ (\partial y/\partial \phi) - (\partial x/\partial \phi)\ (\partial y/\partial \lambda) = a^2\ (1 - e^2)\ \cos \phi/(1 - e^2 \sin^2 \phi)^2 \qquad (4\text{--}41)$$

For the sphere, this simplifies to

$$(\partial x/\partial \lambda)\ (\partial y/\partial \phi) - (\partial x/\partial \phi)\ (\partial y/\partial \lambda) = R^2 \cos \phi \qquad (4\text{--}42)$$

For spherical pseudocylindrical equal-area projections, such as the sinusoidal, the parallels are straight lines parallel to the Equator, so that $(\partial y/\partial \lambda) = 0$. For the many projections in this category, equation (4–42) simplifies further to

$$x = R^2\ \lambda\ \cos \phi/(dy/d\phi) \qquad (4\text{--}43)$$

in which y can be any function of ϕ for a chosen spacing of the parallels.

An equal-area transformation from one set of rectangular coordinates to another must satisfy the following relationship:

$$(\partial x/\partial x')\ (\partial y/\partial y') - (\partial x/\partial y')\ (\partial y/\partial x') = S \qquad (4\text{--}44)$$

where S is the area ratio of the (x,y) map to the (x', y') map.

Most of the above equations (4–33) through (4–44) are difficult to use to derive new projections, although they may be used to determine whether existing projections are conformal or equal-area. Equation (4–43), however, may be fairly readily used to devise new projections which are pseudocylindrical and equal-area. Equation (26–4), discussed later, is a general equation satisfying (4–39) and (4–40), although it is not the only such equation. There is no known general equation satisfying equation (4–44) except in a very elementary way.

5. TRANSFORMATION OF MAP GRATICULES

As discussed later, several map projections have been adapted to showing some part of the Earth for which the lines of true scale have an orientation or location different from that intended by the inventor of the basic projection. This is equivalent to moving or transforming the graticule of meridians and parallels on the Earth so that the "north pole" of the graticule assumes a position different from that of the true North Pole of the Earth. The projection for the sphere may be plotted using the original formulas or graphical construction, but applying them to the new graticule orientation. The actual meridians and parallels may then be plotted by noting their relationship on the sphere to the new graticule, and landforms drawn with respect to the actual geographical coordinates as usual.

In effect, this procedure was used in the past in an often entirely graphical manner. It required considerable care to avoid cumulative errors resulting from the double plotting of graticules. With computers and programmable hand calculators, it now can be a relatively routine matter to calculate directly the rectangular coordinates of the actual graticule in the transformed positions or, with an automatic plotter, to obtain the transformed map directly from the computer.

The transformation most notably has been applied to the azimuthal and cylindrical projections, but in a few cases it has been used with conic, pseudocylindrical, and other projections. While it is fairly straightforward to apply a suitable transformation to the sphere, transformation is much more difficult on the ellipsoid because of the constantly changing curvature. Transformation has been applied to the ellipsoid, however, in important cases under certain limiting conditions.

If either true pole is at the center of an azimuthal map projection, the projection is called the *polar* aspect. If a point on the Equator is made the center, the projection is called the *equatorial* or, less often, *meridian* or *meridional* aspect. If some other point is central, the projection is the *oblique* or, occasionally, *horizon* aspect.

For cylindrical and most other projections, such transformations are called *transverse* or *oblique*, depending on the angle of rotation. In transverse projections, the true poles of the Earth lie on the equator of the basic projection, and the poles of the projection lie on the Equator of the Earth. Therefore, one meridian of the true Earth lies along the equator of the basic projection. The Transverse Mercator projection is the best-known example and is related to the regular Mercator in this manner. For oblique cylindrical projections, the true poles of the Earth lie somewhere between the poles and the equator of the basic projection. Stated another way, the equator of the basic projection is drawn along some great circle route other than the Equator or a meridian of the Earth for the oblique cylindrical aspect. The Oblique Mercator is the most common example. Further subdivisions of these aspects have been made; for example, the transverse aspect may be first transverse, second transverse, or transverse oblique, depending on the positions of the true poles along the equator of the basic projection (Wray, 1974). This has no significance in a transverse cylindrical projection, since the appearance of the map does not change, but for pseudocylindrical projections such as the Sinusoidal, it makes a difference, if the additional nomenclature is desired.

To determine formulas for the transformation of the sphere, two basic laws of spherical trigonometry are used. Referring to the spherical triangle in figure 5, with three points having angles A, B, and C on the sphere, and three great circle arcs a, b, and c connecting them, the Law of Sines declares that

$$\sin A/\sin a = \sin B/\sin b = \sin C/\sin c \qquad (5-1)$$

while by the Law of Cosines,

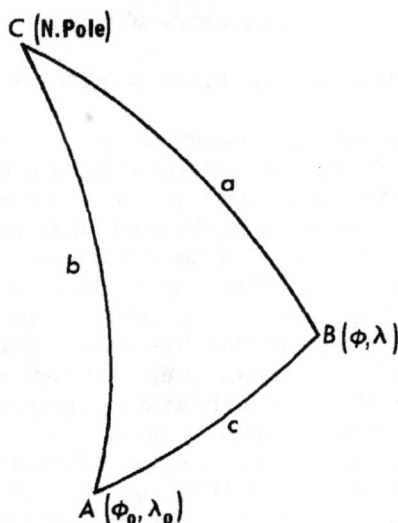

FIGURE 5.—Spherical triangle.

$$\cos c = \cos b \cos a + \sin b \sin a \cos C \qquad (5-2)$$

If C is placed at the North Pole, it becomes the angle between two meridians extending to A and B. If A is taken as the starting point on the sphere, and B the second point, c is the great circle distance between them, and angle A is the azimuth Az east of north which point B bears to point A. When latitude ϕ_1 and longitude λ_0 are used for point A, and ϕ and λ are used for point B, equation $(5-2)$ becomes the following for great circle distance:

$$\cos c = \sin \phi_1 \sin \phi + \cos \phi_1 \cos \phi \cos (\lambda - \lambda_0) \qquad (5-3)$$

While $(5-3)$ is the standard and simplest form of this equation, it is not accurate in practical computation for values of c very close to zero. For such cases, the equation may be rearranged as follows (Sinnott, 1984):

$$\sin (c/2) = \{\sin^2[(\phi - \phi_1)/2] + \cos \phi_1 \cos \phi \sin^2 [(\lambda - \lambda_0)/2]\}^{1/2} \qquad (5-3a)$$

This equation is also exact, and is very accurate in practice for values of c from 0 to nearly 180°.

Equation $(5-1)$ becomes the following for the azimuth:

$$\sin Az = \sin (\lambda - \lambda_0) \cos \phi / \sin c \qquad (5-4)$$

or, with some rearrangement,

$$\cos Az = [\cos \phi_1 \sin \phi - \sin \phi_1 \cos \phi \cos (\lambda - \lambda_0)]/\sin c \qquad (5-4a)$$

or, eliminating c,

$$\tan Az = \cos \phi \sin (\lambda - \lambda_0)/[\cos \phi_1 \sin \phi - \sin \phi_1 \cos \phi \cos (\lambda - \lambda_0)] \qquad (5-4b)$$

Either of the three equations $(5-4)$ through $(5-4b)$ may be used for the azimuth, depending on the form of equation preferred. Equation $(5-4b)$ is usually preferred, since it avoids the inaccuracies of finding an arcsin near 90° or an arccos near 0°. Quadrant adjustment as described under the list of symbols should be employed.

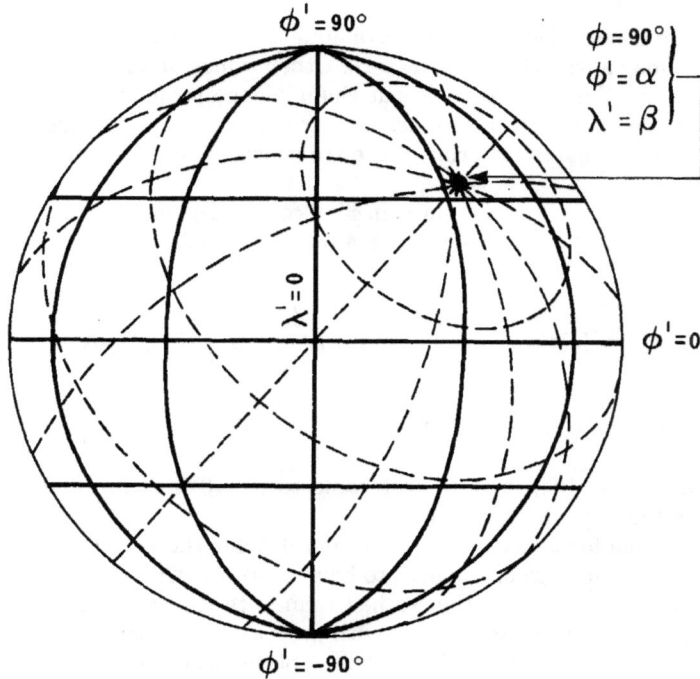

FIGURE 6.—Rotation of a graticule for transformation of projection. Dashed lines show actual longitudes and latitudes (λ and ϕ). Solid lines show the transformed longitudes and latitudes (λ' and ϕ') from which rectangular coordinates (x and y) are determined according to map projection used.

In order to find the latitude ϕ and longitude λ at a given arc distance c and azimuth Az east of north from (ϕ_1, λ_0), the inverse of equations (5–3) and (5–4b) may be used:

$$\phi = \arcsin (\sin \phi_1 \cos c + \cos \phi_1 \sin c \cos Az) \tag{5–5}$$

$$\lambda = \lambda_0 + \arctan [\sin c \sin Az/(\cos \phi_1 \cos c - \sin \phi_1 \sin c \cos Az)] \tag{5–6}$$

Applying these relationships to transformations, without showing some intermediate derivations, formulas (5–7) through (5–8b) are obtained. To place the North Pole of the sphere at a latitude α on a meridian β east of the central meridian ($\lambda' = 0$) of the basic projection (see fig. 6), the transformed latitude ϕ' and transformed longitude λ' on the basic projection which correspond to latitude ϕ and longitude λ of the spherical Earth may be calculated as follows, letting the central meridian λ_0 correspond with $\lambda' = \beta$:

$$\sin \phi' = \sin \alpha \sin \phi - \cos \alpha \cos \phi \cos (\lambda - \lambda_0) \tag{5–7}$$
$$\sin (\lambda' - \beta) = \cos \phi \sin (\lambda - \lambda_0)/\cos \phi' \tag{5–8}$$

or

$$\cos (\lambda' - \beta) = [\sin \alpha \cos \phi \cos (\lambda - \lambda_0) + \cos \alpha \sin \phi]/\cos \phi' \tag{5–8a}$$

or

$$\tan (\lambda' - \beta) = \cos \phi \sin (\lambda - \lambda_0)/[\sin \alpha \cos \phi \cos (\lambda - \lambda_0) + \cos \alpha \sin \phi] \tag{5–8b}$$

Equation (5−8b) is generally preferable to (5−8) or (5−8a) for the reasons stated after equation (5−4b).

These are general formulas for the oblique transformation. (For azimuthal projections, β may always be taken as zero. Other values of β merely have the effect of rotating the X and Y axes without changing the projection.)

The inverse forms of these equations are similar in appearance. To find the geographic coordinates in terms of the transformed coordinates,

$$\sin \phi = \sin \alpha \sin \phi' + \cos \alpha \cos \phi' \cos (\lambda' - \beta) \qquad (5-9)$$
$$\sin (\lambda - \lambda_0) = \cos \phi' \sin (\lambda' - \beta)/\cos \phi \qquad (5-10)$$

or

$$\cos (\lambda - \lambda_0) = [\sin \alpha \cos \phi' \cos (\lambda' - \beta) - \cos \alpha \sin \phi']/\cos \phi \qquad (5-10a)$$

or

$$\tan (\lambda - \lambda_0) = \cos \phi' \sin (\lambda' - \beta)/[\sin \alpha \cos \phi' \cos (\lambda' - \beta) - \cos \alpha \sin \phi'] \quad (5-10b)$$

with equation (5−10b) usually preferable to (5−10) and (5−10a) for the same reasons as those given for (5−4b).

If $\alpha = 0$, the formulas simplify considerably for the transverse or equatorial aspects. It is then more convenient to have central meridian λ_0 coincide with the equator of the basic projection rather than with its meridian β. This may be accomplished by replacing $(\lambda - \lambda_0)$ with $(\lambda - \lambda_0 - 90°)$ and simplifying.

If $\beta = 0$, so that the true North Pole is placed at $(\lambda' = 0, \phi' = 0)$:

$$\sin \phi' = -\cos \phi \sin (\lambda - \lambda_0) \qquad (5-11)$$
$$\cos \lambda' = \sin \phi/[1 - \cos^2 \phi \sin^2(\lambda - \lambda_0)]^{1/2} \qquad (5-12)$$

or

$$\tan \lambda' = - \cos (\lambda - \lambda_0)/\tan \phi \qquad (5-12a)$$

If $\beta = 90°$, placing the true North Pole at $(\lambda' = 90°, \phi' = 0)$:

$$\sin \phi' = - \cos \phi \sin (\lambda - \lambda_0) \qquad (5-13)$$
$$\cos \lambda' = \cos \phi \cos (\lambda - \lambda_0)/[1 - \cos^2 \phi \sin^2 (\lambda - \lambda_0)]^{1/2} \qquad (5-14)$$

or

$$\tan \lambda' = \tan \phi/\cos (\lambda - \lambda_0) \qquad (5-14a)$$

The inverse equations (5−9) through (5−10b) may be similarly altered.

As stated earlier, these formulas may be directly incorporated into the formulas for the rectangular coordinates x and y of the basic map projection for a direct computer or calculator output. If only one or two projections are involved in a package, this may be more efficient. For such transformations of several projections in one software package, it is often easier to calculate the transverse or oblique projection coordinates by first calculating ϕ' and λ' for each point to be plotted (using a general subroutine) and then calculating the rectangular coordinates by inserting ϕ' and λ' into the basic projection formulas. In still other cases, a graphical method has been used.

While these formulas, or their equivalents, will be incorporated into the formulas given later for individual oblique and transverse projections, the concept should help interrelate the various aspects or types of centers of a given projection. The extension of these concepts to the ellipsoid is much more involved technically and in some cases requires approximation. General discussion of this is omitted here.

6. CLASSIFICATION AND SELECTION OF MAP PROJECTIONS

Because of the hundreds of map projections already published and infinite number which are theoretically possible, considerable attention has been given to classification of projections so that the user is not overwhelmed by the numbers and the variety. Generally, the proposed systems classify projections on the basis of property (equal-area, conformal, equidistant, azimuthal, and so forth), type of construction (cylindrical, conical, azimuthal, and so forth), or both. Lee (1944) proposed a combination:

Conical projections
   Cylindric
   Pseudocylindric
   Conic
   Pseudoconic
   Polyconic
   Azimuthal
      Perspective
      Nonperspective

Nonconical projections
   Retroazimuthal (not discussed here)
   Orthoapsidal (not discussed here)
   Miscellaneous

Each of these categories was further subdivided into conformal, authalic (equal-area), and aphylactic (neither conformal nor authalic), but some subdivisions have no examples. This classification is partially used in this book, as the section headings indicate, but the headings are influenced by the number of projections described in each category: Pseudocylindrical projections are included with the "miscellaneous" group, and "space map projections" are given a separate section.

Interest has been shown in some other forms of classification which are more suitable for extensive treatises. In 1962, Waldo R. Tobler provided a simple but all-inclusive proposal (Tobler, 1962). Tobler's classification involves eight categories, four for rectangular and four for polar coordinates. For the rectangular coordinates, category A includes all projections in which both x and y vary with both latitude ϕ and longitude λ, category B includes all in which y varies with both ϕ and λ while x is only a function of λ, C includes those projections in which x varies with both ϕ and λ while y varies only with ϕ, and D is for those in which x is only a function of λ and y only of ϕ. There are very few published projections in category B, but C is usually called pseudocylindrical, D is cylindrical, and A includes nearly all the rest which do not fit the polar coordinate categories.

Tobler's categories A to D for polar coordinates are respectively the same as those for rectangular, except that radius ρ is read for y and angle θ is read for x. The regular conic and azimuthal projections fall into category D, and the pseudoconical (such as Bonne's) into C. Category A may have a few projections like A (rectangular) for which polar coordinates are more convenient than rectangular. There are no well-known projections in B (polar).

Hans Maurer's detailed map projection treatise of 1935 introduced a "Linnaean" classification with five families ("true-circular," "straight-symmetrical," "curved-symmetrical," "less regular," and "combination grids," to quote a translation) subdivided into branches, subbranches, classes, groups, and orders (Maurer, 1935). As Maling says, Maurer's system "is only useful to the advanced student of the subject," but Maurer attempts for map projections what Linnaeus, the Swedish botanist, accomplished for plants and animals in the 18th century (Maling, 1973, p. 98). Other approaches have been taken by Goussinsky (1951) and Starostin (1981).

SUGGESTED PROJECTIONS

Following is a simplified listing of suggested projections. The recommendation can be directly applied in many cases, but other parameters such as the central meridian and parallel or the standard parallels must also be determined. These additional parameters are often chosen by estimation, but they can be chosen by more refined methods to reduce distortion (Snyder, 1985a, p. 93—109). In other cases a more complicated projection may be chosen because of special features in the extent of the region being mapped; the GS50 projection (50-State map) described in this book is an example. Some commonly used projections are not listed in this summary because it is felt that other projections are more suitable for the applications listed, which are not all-inclusive. Some of the projections listed here are not discussed elsewhere in this book.

Region mapped
1. World (Earth should be treated as a sphere)
   A. Conformal (gross area distortion)
       (1) Constant scale along Equator
             Mercator
       (2) Constant scale along meridian
             Transverse Mercator
       (3) Constant scale along oblique great circle
             Oblique Mercator
       (4) Entire Earth shown
             Lagrange
             August
             Eisenlohr
   B. Equal-Area
       (1) Standard without interruption
             Hammer
             Mollweide
             Eckert IV or VI
             McBryde or McBryde-Thomas variations
             Boggs Eumorphic
             Sinusoidal
             misc. pseudocylindricals
       (2) Interrupted for land or ocean
             any of above except Hammer
             Goode Homolosine
       (3) Oblique aspect to group continents
             Briesemeister
             Oblique Mollweide
   C. Equidistant
       (1) Centered on pole
             Polar Azimuthal Equidistant
       (2) Centered on a city
             Oblique Azimuthal Equidistant
   D. Straight rhumb lines
             Mercator
   E. Compromise distortion
             Miller Cylindrical
             Robinson
2. Hemisphere (Earth should be treated as a sphere)
   A. Conformal
             Stereographic (any aspect)

    B. Equal-Area
        Lambert Azimuthal Equal-Area (any aspect)
    C. Equidistant
        Azimuthal Equidistant (any aspect)
    D. Global look
        Orthographic (any aspect)
3. Continent, ocean, or smaller region (Earth should be treated as a sphere for larger continents and oceans and as an ellipsoid for smaller regions, especially at a larger scale)
    A. Predominant east-west extent
      (1) Along Equator
          Conformal: Mercator
          Equal-Area: Cylindrical Equal-Area
      (2) Away from Equator
          Conformal: Lambert Conformal Conic
          Equal-Area: Albers Equal-Area Conic
    B. Predominant north-south extent
        Conformal: Transverse Mercator
        Equal-Area: Transverse Cylindrical Equal-Area
    C. Predominant oblique extent (for example: North America, South America, Atlantic Ocean)
        Conformal: Oblique Mercator
        Equal-Area: Oblique Cylindrical Equal-Area
    D. Equal extent in all directions (for example: Europe, Africa, Asia, Australia, Antarctica, Pacific Ocean, Indian Ocean, Arctic Ocean, Antarctic Ocean)
      (1) Center at pole
          Conformal: Polar Stereographic
          Equal-Area: Polar Lambert Azimuthal Equal-Area
      (2) Center along Equator
          Conformal: Equatorial Stereographic
          Equal-Area: Equatorial Lambert Azimuthal Equal-Area
      (3) Center away from pole or Equator
          Conformal: Oblique Stereographic
          Equal-Area: Oblique Lambert Azimuthal Equal-Area
    E. Straight rhumb lines (principally for oceans)
        Mercator
    F. Straight great-circle routes
        Gnomonic (for less than hemisphere)
    G. Correct scale along meridians
      (1) Center at pole
          Polar Azimuthal Equidistant
      (2) Center along Equator
          Plate Carrée (Equidistant Cylindrical)
      (3) Center away from pole or Equator
          Equidistant Conic

CYLINDRICAL MAP PROJECTIONS

The map projection best known by name is certainly the Mercator—one of the cylindricals. Perhaps easiest to draw, if simple tables are on hand, the regular cylindrical projections consist of meridians which are equidistant parallel straight lines, crossed at right angles by straight parallel lines of latitude, generally not equidistant. Geometrically, cylindrical projections can be partially developed by unrolling a cylinder which has been wrapped around a globe representing the Earth, touching at the Equator, and on which meridians have been projected from the center of the globe (fig. 1). The latitudes can also be perspectively projected onto the cylinder for some projections (such as the Cylindrical Equal-Area and the Gall), but not on the Mercator and several others. When the cylinder is wrapped around the globe in a different direction, so that it is no longer tangent along the Equator, an oblique or transverse projection results, and neither the meridians nor the parallels will generally be straight lines.

7. MERCATOR PROJECTION

SUMMARY

- Cylindrical.
- Conformal.
- Meridians are equally spaced straight lines.
- Parallels are unequally spaced straight lines, closest near the Equator, cutting meridians at right angles.
- Scale is true along the Equator, or along two parallels equidistant from the Equator.
- Loxodromes (rhumb lines) are straight lines.
- Not perspective.
- Poles are at infinity; great distortion of area in polar regions.
- Used for navigation.
- Presented by Mercator in 1569.

HISTORY

The well-known Mercator projection was perhaps the first projection to be regularly identified when atlases of over a century ago gradually began to name projections used, a practice now fairly commonplace. While the projection was apparently used by Erhard Etzlaub (1462–1532) of Nuremberg on a small map on the cover of some sundials constructed in 1511 and 1513, the principle remained obscure until Gerardus Mercator (1512–94) (fig. 7) independently developed it and presented it in 1569 on a large world map of 21 sections totaling about 1.3 by 2 m (Keuning, 1955, p. 17–18).

Mercator, born at Rupelmonde in Flanders, was probably originally named Gerhard Cremer (or Kremer), but he always used the latinized form. To his contemporaries and to later scholars, he is better known for his skills in map and globe making, for being the first to use the term "atlas" to describe a collection of maps in a volume, for his calligraphy, and for first naming North America as such on a map in 1538. To the world at large, his name is identified chiefly with his projection, which he specifically developed to aid navigation. His 1569 map is entitled "Nova et Aucta Orbis Terrae Descriptio ad Usum Navigantium Emendate Accommodata (A new and enlarged description of the Earth with corrections for use in navigation)." He described in Latin the nature of the projection in a large panel covering much of his portrayal of North America:

* * * In this mapping of the world we have [desired] to spread out the surface of the globe into a plane that the places shall everywhere be properly located, not only with respect to their true direction and distance, one from another, but also in accordance with their due longitude and latitude; and further, that the shape of the lands, as they appear on the globe, shall be preserved as far as possible. For this there was needed a new arrangement and placing of meridians, so that they shall become parallels, for the maps hitherto produced by geographers are, on account of the curving and the bending of the meridians, unsuitable for navigation * * *. Taking all this into consideration, we have somewhat increased the degrees of latitude toward each pole, in proportion to the increase of the parallels beyond the ratio they really have to the equator. (Fite and Freeman, 1926, p. 77–78.)

Mercator probably determined the spacing graphically, since tables of secants had not been invented. Edward Wright (ca. 1558–1615) of England later developed the mathematics of the projection and in 1599 published tables of cumulative secants, thereby indicating the spacing from the Equator (Keuning, 1955, p. 18).

FEATURES AND USAGE

The meridians of longitude of the Mercator projection are vertical parallel equally spaced lines, cut at right angles by horizontal straight parallels which are

FIGURE 7.—Gerardus Mercator (1512–94). The inventor of the most famous map projection, which is the prototype for conformal mapping.

increasingly spaced toward each pole so that conformality exists (fig. 8). The spacing of parallels at a given latitude on the sphere is proportional to the secant of the latitude.

The major navigational feature of the projection is found in the fact that a sailing route between two points is shown as a straight line, if the direction or azimuth of the ship remains constant with respect to north. This kind of route is called a loxodrome or rhumb line and is usually longer than the great circle path (which is the shortest possible route on the sphere). It is the same length as a great circle only if it follows the Equator or a meridian. The projection has been standard since 1910 for nautical charts prepared by the former U.S. Coast and Geodetic Survey (now National Ocean Service) (Shalowitz, 1964, p. 302).

The great distortion of area on the Mercator projection of the Earth leads to mistaken concepts when it is the chief basis of world maps seen by students in

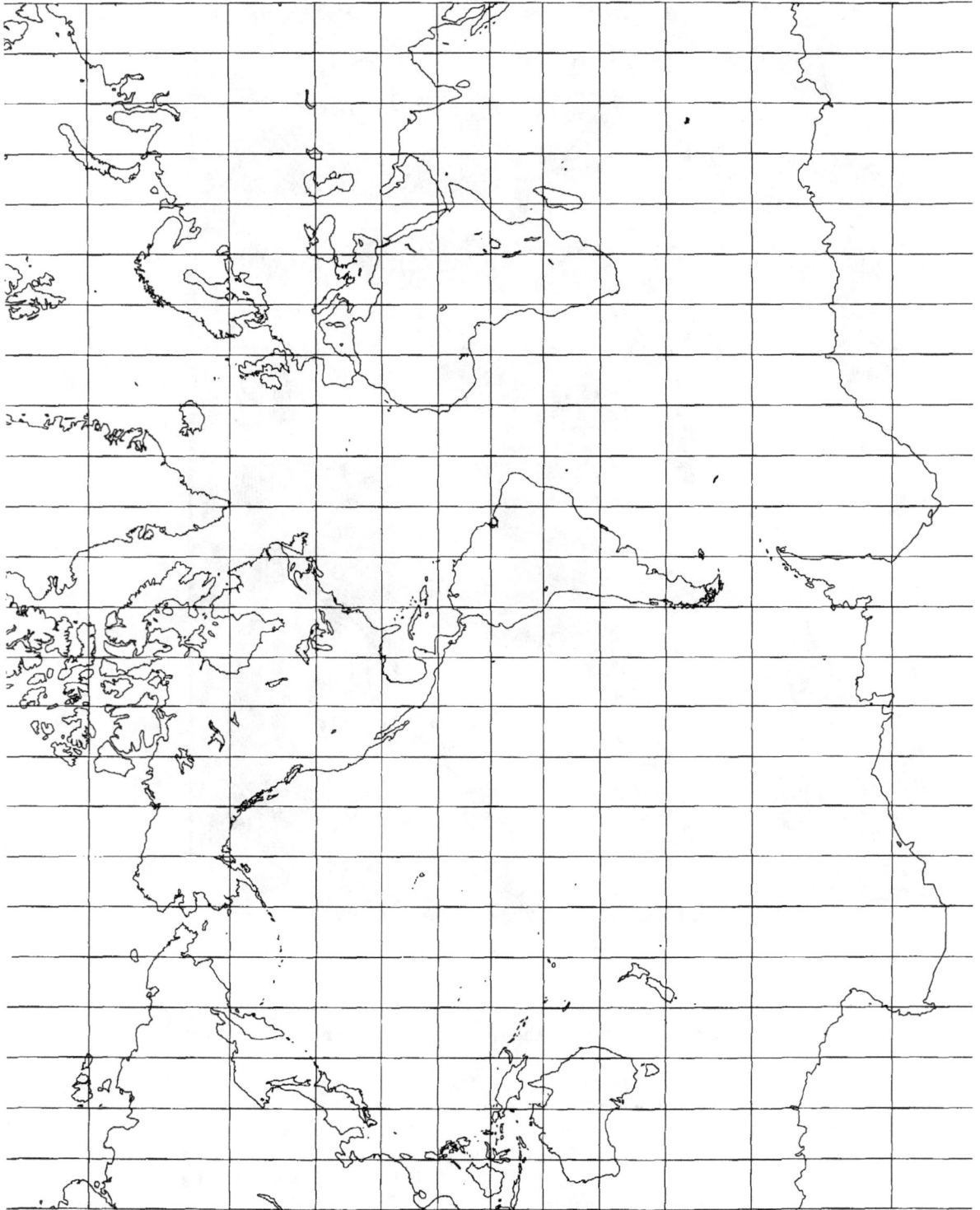

FIGURE 8.—The Mercator projection. The best-known projection. All local angles are shown correctly; therefore, small shapes are essentially true, and it

school. The classic comparison of areas is between Greenland and South America. Greenland appears larger, although it is only one-eighth the size of South America. Furthermore, the North and South Poles cannot be shown, since they are at infinite distance from other parallels on the projection, giving a student an impression they are inaccessible (which of course they seemed to explorers long after the time of Mercator). The last 50 years have seen an increased emphasis on the use of other projections for world maps in published atlases.

Nevertheless, the Mercator projection is fundamental in the development of map projections, especially those which are conformal. It remains a standard navigational tool. It is also especially suitable for conformal maps of equatorial regions. The USGS has recently used it as an inset of the Hawaiian Islands on the 1:500,000-scale base map of Hawaii, for a Bathymetric Map of the Northeast Equatorial Pacific Ocean (although the projection is not stated) and for a Tectonic Map of the Indonesia region, the latter two both in 1978 and at a scale of 1:5,000,000.

The first detailed map of an entire planet other than the Earth was issued in 1972 at a scale of 1:25,000,000 by the USGS Center of Astrogeology, Flagstaff, Arizona, following imaging of Mars by Mariner 9. Maps of Mars at other scales have followed. The mapping of the planet Mercury followed the flybys of Mariner 10 in 1974. Beginning in the late 1960's, geology of the visible side of the Moon was mapped by the USGS in quadrangle fashion at a scale of 1:1,000,000. The four Galilean satellites of Jupiter and several satellites of Saturn were mapped following the Voyager missions of 1979−81. For all these bodies, the Mercator projection has been used to map equatorial portions, but coverage extended in some early cases to lats. 65° N. and S. (table 6).

The cloudy atmosphere of Venus, circled by the Pioneer Venus Orbiter beginning in late 1978, is delaying more precise mapping of that planet, but the Mercator projection alone was used to show altitudes based on radar reflectivity over about 93 percent of the surface.

FORMULAS FOR THE SPHERE

There is no suitable geometrical construction of the Mercator projection. For the sphere, the formulas for rectangular coordinates are as follows:

$$x = R (\lambda - \lambda_0) \tag{7-1}$$
$$y = R \ln \tan (\pi/4 + \phi/2) \tag{7-2}$$

or

$$y = (R/2) [\ln ((1 + \sin \phi)/(1 - \sin \phi))] \tag{7-2a}$$

where R is the radius of the sphere at the scale of the map as drawn, and ϕ and λ are given in radians. There are also several other forms in which equation (7−2) may be written, such as $y = R \operatorname{arcsinh} (\tan \phi) = R \operatorname{arctanh} (\sin \phi) = R \ln (\tan \phi + \sec \phi)$. The X axis lies along the Equator, x increasing easterly. The Y axis lies along the central meridian λ_0, y increasing northerly. If $(\lambda - \lambda_0)$ lies outside the range $\pm 180°$, $360°$ should be added or subtracted so it will fall inside the range. To use ϕ and λ in degrees,

$$x = \pi R (\lambda° - \lambda_0°)/180° \tag{7-1a}$$
$$y = R \ln \tan (45° + \phi°/2) \tag{7-2b}$$

Note that if ϕ is $\pm \pi/2$ or $\pm 90°$, y is infinite. For scale factors, application of equations (4−2), (4−3), and (4−9) to (7−1) and (7−2) or (7−2a) gives results consistent with the conformal feature of the Mercator projection:

TABLE 6.—*Map projections used for extraterrestrial mapping*

[From Batson, private commun., 1985]

Body[1]	Scale	Map format (see below)[2]	Body[1]	Scale	Map f (see b
Moon	1:5,000,000	F	*Galilean satellites of Jupiter*		
	1:2,000,000	K			
	1:1,000,000	K	Io }	1:25,000,000	
Mercury	1:15,000,000	A–1	Europa }	1:15,000,000	
	1:5,000,000	E–1		1:5,000,000	
Venus	1:50,000,000	A–1		1:2,000,000	
	1:25,000,000	B–1	Ganymede }	1:25,000,000	
	1:15,000,000	C	Callisto }	1:15,000,000	
	1:5,000,000	G		1:5,000,000	
Mars	1:25,000,000	A–2		1:2,000,000	
	1:15,000,000	B–2			
	1:5,000,000	D	*Satellites of Saturn*		
	1:2,000,000	H			
	1:500,000	L	Mimas }	1:2,000,000	
			Enceladus }		
			Miranda		
Satellite of Uranus			Tethys }	1:10,000,000	
Ariel	1:10,000,000	A–1	Dione }	1:5,000,000	
	1:5,000,000	B–1	Rhea	1:10,000,000	
				1:5,000,000	
Satellite of Neptune			Iapetus	1:10,000,000	
Triton	(see Ganymede)				

TABLE 6.—*Map projections used for extraterrestrial mapping* - Continued

Map format[2]	Lat. range	Projection[3]	Scale Factor at Lat. N&S[3]		Matching parallel Scale factor at Lat. N&S		Quadrangle size Long. x Lat.		Std. F Lat.,
A–1	57°S–57°N[5]	MER	1.0000	0°	1.7883	56°	360°	114°5	
	55° to pole	PS	1.6354	90	1.7883	56	360	35	
A–2[4]	57°S–57°N[6]	MER	1.0000	0	1.9922	60	360	114[6]	
	55° to pole	PS	1.8589	90	1.9922	60	360	35	
B–1	57°S–57°N	MER	1.0000	0	1.7883	56	180	114	
	55° to pole	PS	1.6354	90	1.7883	56	360	35	
B–2[4]	57°S–57°N	MER	1.0000	0	1.7819	56	180	114	
	55° to pole	PS	1.6298	90	1.7819	56	360	35	
C	57°S–57°N	MER	1.0000	0	1.7883	56	120	57	
	55° to pole	PS	1.6354	90	1.7883	56	360	35	
D[4]	30°S–30°N	MER	1.0000	0	1.1532	30	45	30	
	30°–65°N&S	LCC	1.1259	SP	1.1532	30	60	35	35.8
					1.1611	65			
	65° to pole	PS	1.1067	90	1.1611	65	360	25	
E–1	22°S–22°N[7]	MER	1.0000	0	1.0824	22.5	72	44[7]	
	21°–66°N&S[8]	LCC	1.0494	SP	1.0824	22.5	90	45[8]	28°,
					1.0946	67.5			
	65° to pole	PS	1.0529	90	1.0946	67.5	360	25	
E–2	22°S–22°N	MER	1.0000	13	1.0461	21.34	72	44	
	21°–66°N&S	LCC	1.0000	SP	1.0461	21.34	90	45	30°,
					1.0484	65.19			
	65° to pole	PS	1.0000	90	1.0484	65.19	360	25	
F	50°S–50°N	MER	1.0000	34.06	1.1716	45	180	100	
	45° to pole	PS	1.0000	90	1.1716	45	360	45	
G	25°S–25°N	MER	1.0000	15.90	1.0612	25	40	25	
	25°–75°N&S	LCC	1.0000	SP	1.0612	25	30 (below 50° lat.)	25	34°,
					1.0179	75	60 (above 50° lat.)	25	

TABLE 6.—*Map projections used for extraterrestrial mapping*—Continued

format[2]	Lat. range	Projection[3]	Scale Factor at Lat. N&S [8]		Matching parallel Scale factor at Lat. N&S		Quadrangle size Long. x Lat.		Std. Parallels Lat., Lat.
	75° to pole	PS	1.0000	90	1.0179	75	360	15	
	30°S–30°N	MER	1.0000	27.476	1.0243	30	22.5	15	
	30°–65°N&S	LCC	1.0000	SP	1.0243	30	22.5	17.5 (below 47.5° lat.)	35.83°, 59.17°
					1.0313	65	30	17.5 (above 47.5° lat.)	
	65° to pole	PS	0.9830	90	1.0313	65	45	12.5 (below 77.5° lat.)	
							180	12.5 (above 77.5° lat.)	
	22°S–22°N	MER	1.0000	13	1.0461	21.34	36	22	
	21°–66°N&S	LCC	1.0000	SP	1.0461	21.34	30	22.5 (below 43.5° lat.)	30°, 58°
					1.0484	65.19	45	22.5 (above 43.5° lat.)	
	65° to pole	PS	1.0000	90	1.0484	65.19	90	17.5 (below 82.5° lat.)	
							360	7.5 (above 82.5° lat.)	
	16°S–16°N	MER	1.0000	11.012	1.0211	16	40	32[9]	
	16°–48°N&S	LCC	1.0000	SP	1.0211	16	45	32	21.33°, 42.67°
	48°–80°N&S	LCC	1.0000	SP	none		72	32	53.33°, 74.67°
	80° to pole	PS	1.0000	90	none		360	10	
	82.5°S–82.5°N	TM[10]	0.9960	CM	none		5	5 (below 47.5° lat.)	
							6.67	5 (above 47.5° lat.)	
	82.5° to pole	PS	1.0000	87.5	none		40	5 (below 87.5° lat.)	
							360	2.5 (above 87.5° lat.)	

Notes: [1] Taken as sphere, except for Mars (ellipsoid, eccentricity = 0.101929).
Orthographic projection used for irregular satellites of Mars (Phobos and Deimos), of Jupiter (Amalthea), and Saturn (Hyperion).
Lambert Azimuthal Equal-Area projection used in polar and equatorial aspects for full hemispheres of several planets and satellites.
Oblique Stereographic projection used for basins and other regions of Mars, Moon, etc.
[2] Official format designations use only the letter. Numbers have been added for convenience in this table.
[3] Abbreviations: MER = Mercator, PS = Polar Stereographic, LCC = Lambert Conformal Conic, TM = Transverse Mercator, SP = Standard Parallels.
[4] Scale factors based on Mars ellipsoid.
[5] Venus 1:50,000,000 originally 65°S. to 78°N. Mercator with no polar continuation.
[6] Originally 65°S.–65°N., 130° lat. quad range.
[7] Originally 25°S.–25°N., 50° lat. quad range.
[8] Originally 20°–70°N.&S., 50° lat. quad range.
[9] For Moon 1:1,000,000, quads are 20° long. × 16° lat.
[10] Zones are 20° long. × 75° lat.

$$h = k = \sec \phi = 1/\cos \phi \qquad\qquad (7\text{--}3)$$
$$\omega = 0$$

Normally, for conformal projections, the use of h (the scale factor along a meridian) is omitted, and k (the scale factor along a parallel) is used for the scale factor in any direction. The areal scale factor for conformal projections is k^2 or $\sec^2 \phi$ for the Mercator in spherical form.

The inverse formulas for the sphere, to obtain ϕ and λ from rectangular coordinates, are as follows:

$$\phi = \pi/2 - 2 \arctan (e^{-y/R}) \qquad\qquad (7\text{--}4)$$

or

$$\phi = \arctan[\sinh(y/R)] \qquad\qquad (7\text{--}4\text{a})$$

$$\lambda = x/R + \lambda_0 \qquad\qquad (7\text{--}5)$$

Here $e = 2.7182818 \ldots$, the base of natural logarithms, not eccentricity. These and subsequent formulas are given only in radians, as stated earlier, unless the degree symbol is used. Numerical examples (see p. 266) are given in degrees, showing conversion.

FORMULAS FOR THE ELLIPSOID

For the ellipsoid, the corresponding equations for the Mercator are only a little more involved (see p. 267 for numerical example):

$$x = a(\lambda - \lambda_0) \qquad\qquad (7\text{--}6)$$

$$y = a \ln \left[\tan(\pi/4 + \phi/2) \left(\frac{1 - e \sin \phi}{1 + e \sin \phi} \right)^{e/2} \right] \qquad\qquad (7\text{--}7)$$

or

$$y = (a/2) \ln \left[\left(\frac{1 + \sin \phi}{1 - \sin \phi} \right) \left(\frac{1 - e \sin \phi}{1 + e \sin \phi} \right)^{e} \right] \qquad\qquad (7\text{--}7\text{a})$$

where a is the equatorial radius of the ellipsoid, and e is its eccentricity. Comparing equation (3–7), it is seen that $y = a\psi$. From equations (4–22) and (4–23), it may be found that

$$h = k = (1 - e^2 \sin^2 \phi)^{1/2}/\cos \phi \qquad\qquad (7\text{--}8)$$

and of course $\omega = 0$. The areal scale factor is k^2. The derivation of these equations is shown in Thomas (1952, p. 1, 2, 85–90).

The X and Y axes are oriented as they are for the spherical formulas, and $(\lambda - \lambda_0)$ should be similarly adjusted. Thomas also provides a series equivalent to equation (7–7), slightly modified here for consistency:

$$y/a = \ln \tan (\pi/4 + \phi/2) - (e^2 + e^4/4 + e^6/8 + \ldots) \sin \phi$$
$$+ (e^4/12 + e^6/16 + \ldots) \sin 3\phi - (e^6/80 + \ldots) \sin 5\phi + \ldots \qquad (7\text{--}7\text{b})$$

The inverse formulas for the ellipsoid require rapidly converging iteration, if the closed forms of the equations for finding ϕ are used:

$$\phi = \pi/2 - 2 \arctan \left\{ t[(1 - e \sin \phi)/(1 + e \sin \phi)]^{e/2} \right\} \qquad (7\text{--}9)$$

where

$$t = e^{-y/a} \qquad\qquad (7\text{--}10)$$

TABLE 7.—*Mercator projection: Rectangular coordinates*

Latitude (ϕ)	Sphere ($R=1$)		Clarke 1866 ellipsoid ($a=1$)	
	y	k	y	k
90°	Infinite	Infinite	Infinite	Infinite
85	3.13130	11.47371	3.12454	11.43511
80	2.43625	5.75877	2.42957	5.73984
75	2.02759	3.86370	2.02104	3.85148
70	1.73542	2.92380	1.72904	2.91505
65	1.50645	2.36620	1.50031	2.35961
60	1.31696	2.00000	1.31109	1.99492
55	1.15423	1.74345	1.14868	1.73948
50	1.01068	1.55572	1.00549	1.55263
45	.88137	1.41421	.87658	1.41182
40	.76291	1.30541	.75855	1.30358
35	.65284	1.22077	.64895	1.21941
30	.54931	1.15470	.54592	1.15372
25	.45088	1.10338	.44801	1.10271
20	.35638	1.06418	.35406	1.06376
15	.26484	1.03528	.26309	1.03504
10	.17543	1.01543	.17425	1.01532
5	.08738	1.00382	.08679	1.00379
0	.00000	1.00000	.00000	1.00000
x	0.017453 $(\lambda-\lambda_0)$		0.017453 $(\lambda-\lambda_0)$	

Note: x, y = rectangular coordinates.
ϕ = geodetic latitude.
$(\lambda-\lambda_0)$ = geodetic longitude, measured east from origin in degrees.
k = scale factor, relative to scale at Equator.
R = radius of sphere at scale of map.
a = equatorial radius of ellipsoid at scale of map.
If latitude is negative (south), reverse sign of y.

e is the base of natural logarithms, 2.71828 . . . ,

and the first trial $\phi = \pi/2 - 2 \arctan t$ (7−11)

Inserting the first trial ϕ in the right side of equation (7−9), ϕ on the left side is calculated. This becomes the new trial ϕ, which is used on the right side. The process is repeated until the change in ϕ is less than a chosen convergence factor depending on the accuracy desired. This ϕ is then the final value. For λ,

$$\lambda = x/a + \lambda_0 \quad (7-12)$$

The scale factor is calculated from equation (7−8), using the calculated ϕ.

To avoid the iteration, the series (3−5) may be used with (7−13) in place of (7−9):

$$\phi = \chi + (e^2/2 + 5e^4/24 + e^6/12 + 13e^8/360 + \ldots) \sin 2\chi + (7e^4/48 + 29e^6/240 + 811e^8/11520 + \ldots) \sin 4\chi + (7e^6/120 + 81e^8/1120 + \ldots) \sin 6\chi + (4279e^8/161280 + \ldots) \sin 8\chi + \ldots \quad (3-5)$$

where

$$\chi = \pi/2 - 2 \arctan t \quad (7-13)$$

For improved computational efficiency using the series, see p. 19.

Rectangular coordinates for each 5° of latitude are given in table 7, for both the sphere and the Clarke 1866 ellipsoid, assuming R and a are both 1.0. It should be noted that k for the sphere applies only to the sphere. The spherical projection is not conformal with respect to the ellipsoidal Earth, although the variation is negligible for a map with an equatorial scale of 1:15,000,000 or smaller. It should be noted that any central meridian can be chosen as λ_0 for an existing Mercator map, if forward or inverse formulas are to be used for conversions.

MEASUREMENT OF RHUMB LINES

Since a major feature of the Mercator projection is the straight portrayal of rhumb lines, formulas are given below to determine their true lengths and azimuths. If a straight line on the map connects two points with respective latitudes and longitudes (ϕ_1, λ_1) and (ϕ_2, λ_2), the respective rectangular coordinates (x_1, y_1) and (x_2, y_2) are calculated using equations (7−1) and (7−2) for the sphere or (7−6) and (7−7) for the ellipsoid, inserting the respective subscripts.

For the true (not magnetic) compass bearing or azimuth Az clockwise from north along the rhumb line,

$$Az = \arctan\left[(x_2-x_1)/(y_2-y_1)\right] \tag{7-14}$$

Transposing and using forward and inverse equations for the Mercator, latitude or longitude along the rhumb line may be found for a given longitude or latitude, respectively, knowing the initial point and the azimuth. For example,

$$y_2 = y_1 + (x_2 + x_1)/\tan Az \tag{7-15}$$

in which (x_1, y_1) are calculated for (ϕ_1, λ_1) from (7−6) and (7−7), x_2 is calculated from λ_2 from (7−6), and ϕ_2 is calculated from y_2 using (7−9) and (7−10).

For the true distance s along the rhumb line from ϕ_1 to ϕ_2,

$$s = (M_2-M_1)/\cos Az \tag{7-16}$$

where M_2 and M_1, the distances from the Equator along the meridian, are found for ϕ_2 and ϕ_1, respectively, using equation (3−21) and the same subscripts on M and ϕ:

$$M = a[(1-e^2/4-3e^4/64-5e^6/256- \ldots)\, \phi- (3e^2/8 + 3e^4/32 \\ + 45e^6/1024 + \ldots)\sin 2\phi + (15e^4/256 + 45e^6/1024 + \ldots) \\ \sin 4\phi - (35e^6/3072 + \ldots)\sin 6\phi + \ldots] \tag{3-21}$$

but if $\phi_1 = \phi_2$, equation (7−16) is indeterminate and

$$s = a(\lambda_2-\lambda_1)\cos \phi/(1-e^2\sin^2\phi)^{1/2} \tag{7-17}$$

For the true distance s from initial latitude ϕ_1 to latitude ϕ, equation (7−16) may be used with M instead of M_2. To find (ϕ,λ) corresponding to a given distance s from (ϕ_1, λ_1) along the rhumb line, (7−16) may be inverted to give:

$$M = s \cos Az + M_1 \tag{7-18}$$

M may be converted to ϕ using (3−26),

$$\phi = \mu + (3e_1/2-27e_1^3/32 + \ldots)\sin 2\mu + (21e_1^2/16-55e_1^4/32 + \ldots) \\ \sin 4\mu + (151e_1^3/96- \ldots)\sin 6\mu + (1097e_1^4/512- \ldots)\sin 8\mu + \ldots \tag{3-26}$$

where

$$e_1 = [1-(1-e^2)^{1/2}]/[1+(1-e^2)^{1/2}]\qquad(3-24)$$

and, in a rearrangement of (3−20) and (3−21),

$$\mu = M/[a\,(1-e^2/4-3e^4/64-5e^6/256-\ldots)]\qquad(7-19)$$

Then for longitude λ, rearranging (7−6), (7−7), and (7−14),

$$\lambda = \lambda_1 + \tan Az\,\ln\left[\tan\,(\pi/4+\phi/2)\left(\frac{1-e\,\sin\,\phi}{1+e\,\sin\,\phi}\right)^{e/2}\right]\qquad(7-20)$$

MERCATOR PROJECTION WITH ANOTHER STANDARD PARALLEL

The above formulas are based on making the Equator of the Earth true to scale on the map. Thus, the Equator may be called the standard parallel. It is also possible to have, instead, another parallel (actually two) as standard, with true scale. For the Mercator, the map will look exactly the same; only the scale will be different. If latitude ϕ_1 is made standard (the opposite latitude $-\phi_1$ is also standard), the above forward formulas are adapted by multiplying the right side of equations (7−1) through (7−3) for the sphere, including the alternate forms, by cos ϕ_1. For the ellipsoid, the right sides of equations (7−6), (7−7), (7−8), and (7−7a) are multiplied by cos $\phi_1/(1-e^2\,\sin^2\,\phi_1)^{1/2}$. For inverse equations, divide x and y by the same values before use in equations (7−4) and (7−5) or (7−10) and (7−12). Such a projection is most commonly used for a navigational map of part of an ocean, such as the North Atlantic Ocean, but the USGS has used it for equatorial quadrangles of some extraterrestrial bodies as described in table 6.

8. TRANSVERSE MERCATOR PROJECTION

SUMMARY

- Cylindrical (transverse).
- Conformal.
- Central meridian, each meridian 90° from central meridian, and Equator are straight lines.
- Other meridians and parallels are complex curves.
- Scale is true along central meridian, or along two straight lines equidistant from and parallel to central meridian. (These lines are only approximately straight for the ellipsoid.)
- Scale becomes infinite on sphere 90° from central meridian.
- Used extensively for quadrangle maps at scales from 1:24,000 to 1:250,000.
- Presented by Lambert in 1772.

HISTORY

Since the regular Mercator projection has little error close to the Equator (the scale 10° away is only 1.5 percent larger than the scale at the Equator), it has been found very useful in the transverse form, with the equator of the projection rotated 90° to coincide with the desired central meridian. This is equivalent to wrapping the cylinder around a sphere or ellipsoid representing the Earth so that it touches the central meridian throughout its length, instead of following the Equator of the Earth. The central meridian can then be made true to scale, no matter how far north and south the map extends, and regions near it are mapped with low distortion. Like the regular Mercator, the map is conformal.

The Transverse Mercator projection in its spherical form was invented by the prolific Alsatian mathematician and cartographer Johann Heinrich Lambert (1728−77) (fig. 9). It was the third of seven new projections which he described in 1772 in his classic *Beiträge* (Lambert, 1772). At the same time, he also described what are now called the Cylindrical Equal-Area, the Lambert Conformal Conic, and the Lambert Azimuthal Equal-Area, each of which will be discussed subsequently; others are omitted here. He described the Transverse Mercator as a conformal adaptation of the Sinusoidal projection, then commonly in use (Lambert, 1772, p. 57−58). Lambert's derivation was followed with a table of coordinates and a map of the Americas drawn according to the projection.

Little use has been made of the Transverse Mercator for single maps of continental areas. While Lambert only indirectly discussed its ellipsoidal form, mathematician Carl Friedrich Gauss (1777−1855) analyzed it further in 1822, and L. Krüger published studies in 1912 and 1919 providing formulas suitable for calculation relative to the ellipsoid. It is, therefore, sometimes called the Gauss Conformal or the Gauss-Krüger projection in Europe, but Transverse Mercator, a term first applied by the French map projection compiler Germain, is the name normally used in the United States (Thomas, 1952, p. 91−92; Germain, 1865?, p. 347).

Until recently, the Transverse Mercator projection was not precisely applied to the ellipsoid for the entire Earth. Ellipsoidal formulas were limited to series for relatively narrow bands. In 1945, E. H. Thompson (and in 1962, L. P. Lee) presented exact or closed formulas permitting calculation of coordinates for the full ellipsoid, although elliptic functions, and therefore lengthy series, numerical integrations, and (or) iterations, are involved (Lee, 1976, p. 92−101; Snyder, 1979a, p. 73; Dozier, 1980).

The formulas for the complete ellipsoid are interesting academically, but they are practical only within a band between 4° of longitude and some 10° to 15° of arc distance on either side of the central meridian, because of the much more significant scale errors fundamental to any projection covering a larger area.

FIGURE 9.—Johann Heinrich Lambert (1728–77). Inventor of the Transverse Mercator, the Conformal Conic, the Azimuthal Equal-Area, and other important projections, as well as outstanding
developments in mathematics, astronomy, and physics.

FEATURES

The meridians and parallels of the Transverse Mercator (fig. 10) are no longer
the straight lines they are on the regular Mercator, except for the Earth's Equator,
the central meridian, and each meridian 90° away from the central meridian.
Other meridians and parallels are complex curves.

The spherical form is conformal, as is the parent projection, and scale error is
only a function of the distance from the central meridian, just as it is only a
function of the distance from the Equator on the regular Mercator. The ellipsoidal
form is also exactly conformal, but its scale error is slightly affected by factors
other than the distance alone from the central meridian (Lee, 1976, p. 98).

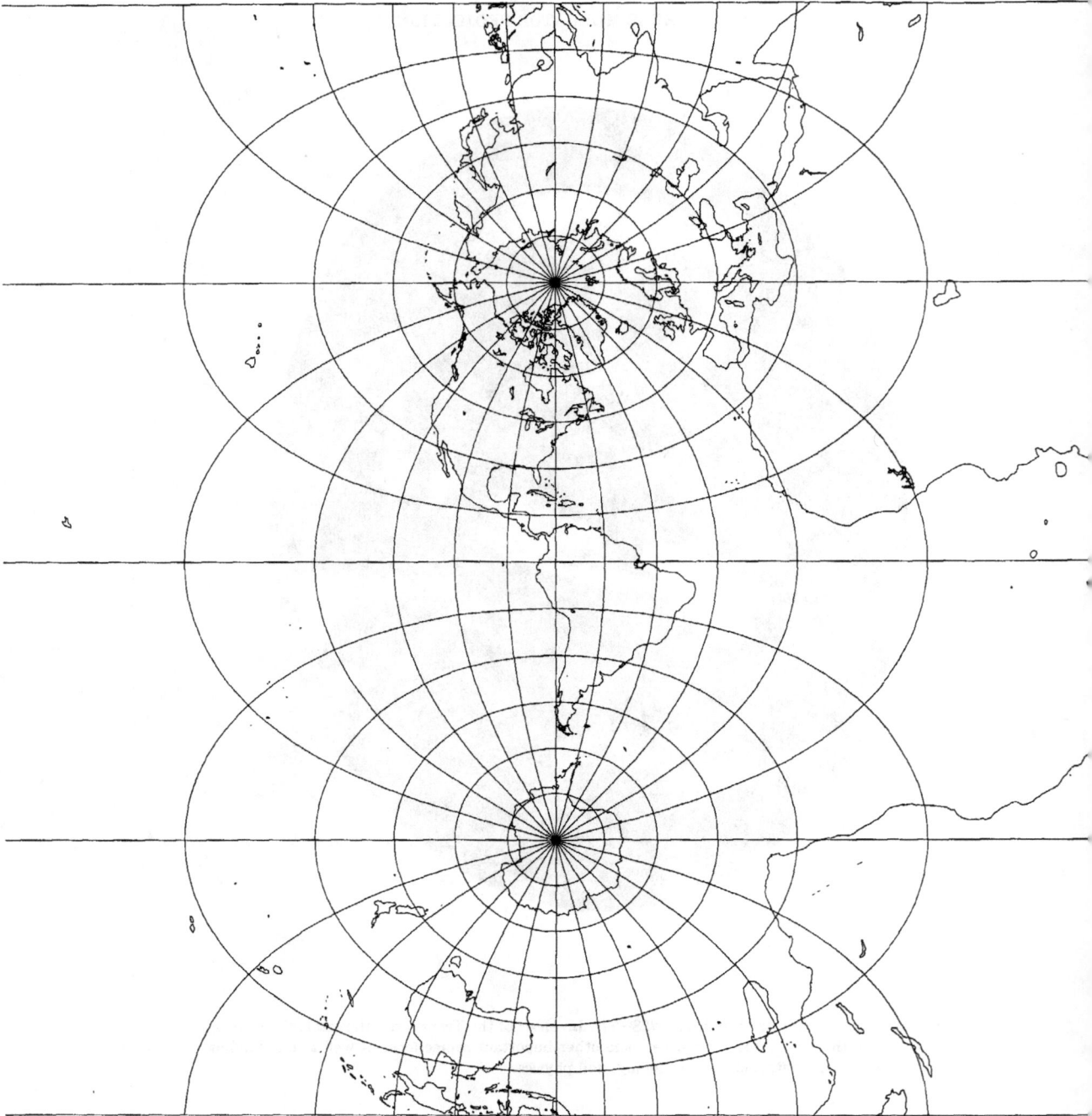

FIGURE 10.—The Transverse Mercator projection. While the regular Mercator has constant scale along the Equator, the Transverse Mercator has constant scale along any chosen central meridian. This projection is conformal and is often used to show regions with greater north-south extent.

The scale along the central meridian may be made true to scale, or deliberately reduced to a slightly smaller constant scale so that the mean scale of the entire map is more nearly correct. There are also forms of the ellipsoidal Transverse Mercator on which the central meridian is not held at a constant scale, but these forms are not used in practice (Lee, 1976, p. 100−101). If the central meridian is mapped at a reduced scale, two straight lines parallel to it and equally spaced from it, one on either side, become true to scale on the sphere. These lines are not perfectly straight on the ellipsoidal form.

With the scale along the central meridian remaining constant, the Transverse Mercator is an excellent projection for lands extending predominantly north and south.

USAGE

The Transverse Mercator projection (spherical or ellipsoidal) was not described by Close and Clarke in their generally detailed article in the 1911 *Encyclopaedia Britannica* because it was "seldom used" (Close and Clarke, 1911, p. 663). Deetz and Adams (1934) favorably referred to it several times, but as a slightly used projection.

The spherical form of the Transverse Mercator has been used by the USGS only recently. In 1979, this projection was chosen for a base map of North America at a scale of 1:5,000,000 to replace the Bipolar Oblique Conic Conformal projection previously used for tectonic and other geologic maps. The scale factor along the central meridian, long. 100° W., is reduced to 0.926. The radius of the Earth is taken at 6,371,204 m, with approximately the same surface area as the International ellipsoid, placing the two straight lines of true design scale 2,343 km on each side of the central meridian.

While its use in the spherical form is limited, the ellipsoidal form of the Transverse Mercator is probably used more than any other one projection for geodetic mapping.

In the United States, it is the projection used in the State Plane Coordinate System (SPCS) for States with predominant north-south extent. (The Lambert Conformal Conic is used for the others, except for the panhandle of Alaska, which is prepared on the Oblique Mercator. Alaska, Florida, and New York use both the Transverse Mercator and the Lambert Conformal Conic for different zones.) Except for narrow States, such as Delaware, New Hampshire, and New Jersey, all States using the Transverse Mercator are divided into two to eight zones, each with its own central meridian, along which the scale is slightly reduced to balance the scale throughout the map. Each zone is designed to maintain scale distortion within 1 part in 10,000. Several States beginning in 1935 also passed legislation establishing the SPCS as a permissible system for recording boundary descriptions or point locations. Several zone changes have occurred for use with the new 1983 datum. They are listed in Appendix C.

In addition to latitude and longitude as the basic frame of reference, the corresponding rectangular grid coordinates in feet are used to designate locations (Mitchell and Simmons, 1945). The parameters for each State are given in table 8. All are based on the Clarke 1866 ellipsoid. It is important to note that, for the metric conversion to feet using this coordinate system, 1 m equals exactly 39.37 in., not the current standard accepted by the National Bureau of Standards in 1959, in which 1 in. equals exactly 2.54 cm. Surveyors continue to follow the former conversion for consistency. The difference is only two parts in a million, but it is enough to cause confusion, if it is not accounted for.

Beginning with the late 1950's, the Transverse Mercator projection was used by the USGS for nearly all new quadrangles (maps normally bounded by meridians and parallels) covering those States using the TM Plane Coordinates, but the

TABLE 8.—*U.S. State plane coordinate systems*

[T indicates Transverse Mercator; L, Lambert Conformal Conic; H, Hotine Oblique Mercator. Modified slightly and updated from Mitchell and Simmons, 1945, p. 45-47]

Area	Projection	Zones	Area	Projection	Zones
Alabama	T	2	Montana	L	3
Alaska	T	8	Nebraska	L	2
	L	1	Nevada	T	3
	H	1	New Hampshire	T	1
Arizona	T	3	New Jersey	T	1
Arkansas	L	2	New Mexico	T	3
California	L	7	New York	T	3
Colorado	L	3		L	1
Connecticut	L	1	North Carolina	L	1
Delaware	T	1	North Dakota	L	2
Florida	T	2	Ohio	L	2
	L	1	Oklahoma	L	2
Georgia	T	2	Oregon	L	2
Hawaii	T	5	Pennsylvania	L	2
Idaho	T	3	Puerto Rico &		
Illinois	T	2	Virgin Islands	L	2
Indiana	T	2	Rhode Island	T	1
Iowa	L	2	Samoa	L	1
Kansas	L	2	South Carolina	L	2
Kentucky	L	2	South Dakota	L	2
Louisiana	L	3	Tennessee	L	1
Maine	T	2	Texas	L	5
Maryland	L	1	Utah	L	3
Massachusetts	L	2	Vermont	T	1
Michigan[1]			Virginia	L	2
obsolete	T	3	Washington	L	2
current	L	3	West Virginia	L	2
Minnesota	L	3	Wisconsin	L	3
Mississippi	T	2	Wyoming	T	4
Missouri	T	3			

Transverse Mercator projection			
Zone	Central meridian	Scale reduction[2]	Origin[3] (latitude)
Alabama			
East	85°50′ W.	1:25,000	30°30′ N.
West	87 30	1:15,000	30 00
Alaska[4]			
2	142 00	1:10,000	54 00
3	146 00	1:10,000	54 00
4	150 00	1:10,000	54 00
5	154 00	1:10,000	54 00
6	158 00	1:10,000	54 00
7	162 00	1:10,000	54 00
8	166 00	1:10,000	54 00
9	170 00	1:10,000	54 00
Arizona			
East	110 10	1:10,000	31 00
Central	111 55	1:10,000	31 00
West	113 45	1:15,000	31 00
Delaware	75 25	1:200,000	38 00
Florida[4]			
East	81 00	1:17,000	24 20
West	82 00	1:17,000	24 20

TABLE 8.—*U.S. State plane coordinate systems*—Continued

Transverse Mercator projection – Continued

Zone	Central meridian	Scale reduction[2]	Origin[3] (latitude)
Georgia			
East _____	82°10′ W.	1:10,000	30°00′ N.
West _____	84 10	1:10,000	30 00
Hawaii			
1 _____	155 30	1:30,000	18 50
2 _____	156 40	1:30,000	20 20
3 _____	158 00	1:100,000	21 10
4 _____	159 30	1:100,000	21 50
5 _____	160 10	0	21 40
Idaho			
East _____	112 10	1:19,000	41 40′
Central _____	114 00	1:19,000	41 40
West _____	115 45	1:15,000	41 40
Illinois			
East _____	88 20	1:40,000	36 40
West _____	90 10	1:17,000	36 40
Indiana			
East _____	85 40	1:30,000	37 30
West _____	87 05	1:30,000	37 30
Maine			
East _____	68 30	1:10,000	43 50
West _____	70 10	1:30,000	42 50
Michigan (old)[4]			
East _____	83 40	1:17,500	41 30
Central _____	85 45	1:11,000	41 30
West _____	88 45	1:11,000	41 30
Mississippi			
East _____	88 50	1:25,000	29 40
West _____	90 20	1:17,000	30 30
Missouri			
East _____	90 30	1:15,000	35 50
Central _____	92 30	1:15,000	35 50
West _____	94 30	1:17,000	36 10
Nevada			
East _____	115 35	1:10,000	34 45
Central _____	116 40	1:10,000	34 45
West _____	118 35	1:10,000	34 45
New Hampshire _	71 40	1:30,000	42 30
New Jersey _____	74 40	1:40,000	38 50
New Mexico			
East _____	104 20	1:11,000	31 00
Central _____	106 15	1:10,000	31 00
West _____	107 50	1:12,000	31 00
New York[4]			
East _____	74 20	1:30,000	40 00
Central _____	76 35	1:16,000	40 00
West _____	78 35	1:16,000	40 00
Rhode Island ____	71 30	1:160,000	41 05
Vermont _____	72 30	1:28,000	42 30

TABLE 8.—*U.S. State plane coordinate systems*—Continued

Transverse Mercator projection – Continued

Zone	Central meridian	Scale reduction[2]	Origin[3] (latitude)
Wyoming			
East _____	105°10′ W.	1:17,000	40°40′ N.
East Central	107 20	1:17,000	40 40
West Central	108 45	1:17,000	40 40
West _____	110 05	1:17,000	40 40

Lambert Conformal Conic projection

Zone	Standard parallels		Origin[5]	
			Long.	Lat.
Alaska[4]				
10 _____	51°50′ N.	53°50′ N.	176°00′ W.[5a]	51°00′ N.
Arkansas				
North _____	34 56	36 14	92 00	34 20
South _____	33 18	34 46	92 00	32 40
California				
I _____	40 00	41 40	122 00	39 20
II _____	38 20	39 50	122 00	37 40
III _____	37 04	38 26	120 30	36 30
IV _____	36 00	37 15	119 00	35 20
V _____	34 02	35 28	118 00	33 30
VI _____	32 47	33 53	116 15	32 10
VII _____	33 52	34 25	118 20	34 08[5b]
Colorado				
North _____	39 43	40 47	105 30	39 20
Central _____	38 27	39 45	105 30	37 50
South _____	37 14	38 26	105 30	36 40
Connecticut _____	41 12	41 52	72 45	40 50[5d]
Florida[4]				
North _____	29 35	30 45	84 30	29 00
Iowa				
North _____	42 04	43 16	93 30	41 30
South _____	40 37	41 47	93 30	40 00
Kansas				
North _____	38 43	39 47	98 00	38 20
South _____	37 16	38 34	98 30	36 40
Kentucky				
North _____	37 58	38 58	84 15	37 30
South _____	36 44	37 56	85 45	36 20
Louisiana				
North _____	31 10	32 40	92 30	30 40
South _____	29 18	30 42	91 20	28 40
Offshore _____	26 10	27 50	91 20	25 40
Maryland _____	38 18	39 27	77 00	37 50[5c]
Massachusetts				
Mainland _____	41 43	42 41	71 30	41 00[5d]
Island _____	41 17	41 29	70 30	41 00[5c]

TABLE 8.—*U.S. State plane coordinate systems*—Continued

Lambert Conformal Conic projection – Continued

Zone	Standard parallels		Origin[5]	
			Long.	Lat.
Michigan (current)[4]				
North	45°29′ N.	47°05′ N.	87°00′ W.	44°47′ N.
Central	44 11	45 42	84 20	43 19
South	42 06	43 40	84 20	41 30
Minnesota				
North	47 02	48 38	93 06	46 30
Central	45 37	47 03	94 15	45 00
South	43 47	45 13	94 00	43 00
Montana				
North	47 51	48 43	109 30	47 00
Central	46 27	47 53	109 30	45 50
South	44 52	46 24	109 30	44 00
Nebraska				
North	41 51	42 49	100 00	41 20
South	40 17	41 43	99 30	39 40
New York[4]				
Long Island	40 40	41 02	74 00	40 30[5f]
North Carolina	34 20	36 10	79 00	33 45
North Dakota				
North	47 26	48 44	100 30	47 00
South	46 11	47 29	100 30	45 40
Ohio				
North	40 26	41 42	82 30	39 40
South	38 44	40 02	82 30	38 00
Oklahoma				
North	35 34	36 46	98 00	35 00
South	33 56	35 14	98 00	33 20
Oregon				
North	44 20	46 00	120 30	43 40
South	42 20	44 00	120 30	41 40
Pennsylvania				
North	40 53	41 57	77 45	40 10
South	39 56	40 58	77 45	39 20
Puerto Rico and Virgin Islands				
1	18 02	18 26	66 26	17 50[5g]
2 (St. Croix)	18 02	18 26	66 26	17 50[5f, g]
Samoa	14°16′ S.	(single)	170 00[5h]	-- --
South Carolina				
North	33°46′ N.	34 58	81 00	33 00
South	32 20	33 40	81 00	31 50
South Dakota				
North	44 25	45 41	100 00	43 50
South	42 50	44 24	100 20	42 20
Tennessee	35 15	36 25	86 00	34 40[5f]

TABLE 8.—*U.S. State plane coordinate systems*—Continued

Lambert Conformal Conic projection – Continued

Zone	Standard parallels		Origin[6]	
			Long.	Lat.
Texas				
North _____	34°39′ N.	36°11′ N.	101°30′ W.	34°00′ N.
North central ___	32 08	33 58	97 30	31 40
Central _____	30 07	31 53	100 20	29 40
South central ___	28 23	30 17	99 00	27 50
South _____	26 10	27 50	98 30	25 40
Utah				
North _____	40 43	41 47	111 30	40 20
Central _____	39 01	40 39	111 30	38 20
South _____	37 13	38 21	111 30	36 40
Virginia				
North _____	38 02	39 12	78 30	37 40
South _____	36 46	37 58	78 30	36 20
Washington				
North _____	47 30	48 44	120 50	47 00
South _____	45 50	47 20	120 30	45 20
West Virginia				
North _____	39 00	40 15	79 30	38 30
South _____	37 29	38 53	81 00	37 00
Wisconsin				
North _____	45 34	46 46	90 00	45 10
Central _____	44 15	45 30	90 00	43 50
South _____	42 44	44 04	90 00	42 00

Hotine Oblique Mercator projection

Zone	Center of projection		Azimuth of central line	Scale[7] reduction
	Long.	Lat.		
Alaska[4]				
1	133°40′ W.[6a]	57°00′ N.	arctan (− ¾)	1:10,000

Great Lakes (U.S. Lake Survey, not State plane coordinates)

1 (Erie, Ont., St. Lawrence R.)	78 00[6b]	44 00	55 40	1:10,000
2 (Huron)	82 30[6c]	43 00	350 37	1:10,000
3 (Michigan)	87 00[6d]	44 00	15 00	1:10,000
4 (Superior, Lake of the Woods)	{88 50 {00.256″[6e]	47 12 21.554″	285 41 42.593″	1:10,000

Note.—All these systems are based on the Clarke 1866 ellipsoid and are based on the 1927 datum. Origin refers to rectangular coordinates. For systems based on 1983 datum, see Appendix C.

[1] The major and minor axes of the ellipsoid are taken at exactly 1.0000382 times those of the Clarke 1866, for Michigan only. This incorporates an average elevation throughout the State of about 800 ft, with limited variation.

[2] Along the central meridian.

[3] At origin, $x = 500,000$ ft, $y = 0$ ft, except for Alaska zone 7, $x = 700,000$ ft; Alaska zone 9, $x = 600,000$ ft; and New Jersey, $x = 2,000,000$ ft.

[4] Additional zones listed in this table under other projection(s).

[5] At origin, $x = 2,000,000$ ft, $y = 0$ ft, except (a) $x = 3,000,000$ ft, (b) $x = 4,186,692.58$, $y = 4,160,926,74$ ft, (c) $x = 800,000$ ft, (d) $x = 600,000$ ft, (e) $x = 200,000$ ft, (f) $y = 100,000$ ft, (g) $x = 500,000$ ft, (h) $x = 500,000$ ft, $y = 0$, but radius to lat. of origin $= -82,000,000$ ft.

[6] At center, (a) $x = 5,000,000$ meters, $y = -5,000,000$ m; (b) $x = -3,950,000$ m, $y = -3,430,000$ m; (c) $x = 1,200,000$ m, $y = -3,500,000$ m; (d) $x = -1,000,000$ m, $y = -4,300,000$ m; (e) $x = 9,000,000$ m, $y = -1,600,000$ m (Berry and Bormanis, 1970).

[7] At central point.

central meridian and scale factor are those of the SPCS zone. Thus, all quadrangles for a given zone may be mosaicked exactly. Beginning in 1977, many USGS maps have been produced on the Universal Transverse Mercator projection (see below). Prior to the late 1950's, the Polyconic projection was used. The change in projection was facilitated by the use of high-precision rectangular-coordinate plotting machines. Some maps produced on the Transverse Mercator projection system during this transition period are identified as being prepared according to the Polyconic projection. Since most quadrangles cover only 7½ minutes (at a scale of 1:24,000) or 15 minutes (at 1:62,500) of latitude and longitude, the difference between the Polyconic and the Transverse Mercator for such a small area is much more significant due to the change of central meridian than due to the change of projection. The difference is still slight and is detailed later under the discussion of the Polyconic projection. The Transverse Mercator is used in many other countries for official topographic mapping as well. The Ordnance Survey of Great Britain began switching from a Transverse Equidistant Cylindrical (the Cassini-Soldner) to the Transverse Mercator about 1920.

The use of the Transverse Mercator for quadrangle maps has been recently extended by the USGS to include the planet Mars. Although other projections are used at smaller scales, quadrangles at scales of 1:1,000,000 and 1:250,000, and covering areas from 200 to 800 km on a side, were drawn to the ellipsoidal Transverse Mercator between lats. 65°N. and S. The scale factor along the central meridian was made 1.0. For the current series, see table 6.

In addition to its own series of larger-scale quadrangle maps, the Army Map Service used the Transverse Mercator for two other major mapping operations: (1) a series of 1:250,000-scale quadrangle maps covering the entire country, and (2) as the geometric basis for the Universal Transverse Mercator (UTM) grid.

The entire area of the United States has been mapped since the 1940's in sections 2° of longitude (between even-numbered meridians, but in 3° sections in Alaska) by 1° of latitude (between each full degree) at a scale of 1:250,000, with the UTM grid superimposed and with some variations in map boundaries at coastlines. These maps were drawn with reference to their own central meridians, not the central meridians of the UTM zones (see below), although the 0.9996 central scale factor was employed. The central meridian of about one-third of the maps coincides with the central meridian of the zone, but it does not for about two-thirds, the "wing" sheets, which therefore do not perfectly match the center sheets. The USGS has assumed publication and revision of this series and is casting new maps using the correct central meridians.

Transverse Mercator quadrangle maps fit continuously in a north-south direction, provided they are prepared at the same scale, with the same central meridian, and for the same ellipsoid. They do not fit exactly from east to west, if they have their own central meridians; although quadrangles and other maps properly constructed at the same scale, using the SPCS or UTM projection, fit in all directions within the same zone.

UNIVERSAL TRANSVERSE MERCATOR PROJECTION

The Universal Transverse Mercator (UTM) projection and grid were adopted by the U.S. Army in 1947 for designating rectangular coordinates on large-scale military maps of the entire world. The UTM is the ellipsoidal Transverse Mercator to which specific parameters, such as central meridians, have been applied. The Earth, between lats. 84° N. and 80° S., is divided into 60 zones each generally 6° wide in longitude. Bounding meridians are evenly divisible by 6°, and zones are numbered from 1 to 60 proceeding east from the 180th meridian from Greenwich with minor exceptions. There are letter designations from south to north (see fig. 11). Thus, Washington, D.C., is in grid zone 18S, a designation covering a quadrangle from long. 72° to 78° W. and from lat. 32° to 40° N. Each of these quadrangles is further subdivided into grid squares 100,000 meters on a side with double-letter designations, including partial squares at the grid boundaries. From lat. 84° N. and 80° S. to the respective poles, the Universal Polar Stereographic (UPS) projection is used instead.

As with the SPCS, each geographic location in the UTM projection is given x and y coordinates, but in meters, not feet, according to the Transverse Mercator projection, using the meridian halfway between the two bounding meridians as the central meridian, and reducing its scale to 0.9996 of true scale (a 1:2,500 reduction). The reduction was chosen to minimize scale variation in a given zone; the variation reaches 1 part in 1,000 from true scale at the Equator. The USGS, for civilian mapping, uses only the zone number and the x and y coordinates, which are sufficient to define a point, if the ellipsoid and the hemisphere (north or south) are known; the 100,000-m square identification is not essential. The lines of true scale are approximately parallel to and approximately 180 km east and west of the central meridian. Between them, the scale is too small; beyond them, it is too great. In the Northern Hemisphere, the Equator at the central meridian is considered the origin, with an x coordinate of 500,000 m and a y of 0. For the Southern Hemisphere, the same point is the origin, but, while x remains 500,000 m, y is 10,000,000 m. In each case, numbers increase toward the east and north. Negative coordinates are thus avoided (Army, 1973, p. 7, endmap). A page of coordinates for the UTM projection is shown in table 9.

The ellipsoidal Earth is used throughout the UTM projection system, but the reference ellipsoid changes with the particular region of the Earth. For all land under United States jurisdiction, the Clarke 1866 ellipsoid is used for the map projection. For the UTM grid superimposed on the map of Hawaii, however, the International ellipsoid is used. The Geological Survey uses the UTM graticule and grid for its 1:250,000- and larger-scale maps of Alaska, and applies the UTM grid lines or tick marks to its quadrangles and State base maps for the other States, although they are generally drawn with different projections or parameters.

FORMULAS FOR THE SPHERE

A partially geometric construction of the Transverse Mercator for the sphere involves constructing a regular Mercator projection and using a transforming map to convert meridians and parallels on one sphere to equivalent meridians and parallels on a sphere rotated to place the equator of one along the chosen central meridian of the other. Such a transforming map may be the equatorial aspect of the Stereographic or other azimuthal projection, drawn twice to the same scale on transparencies. The transparencies may then be superimposed at 90° angles and the points compared.

In an age of computers, it is much more satisfactory to use mathematical formulas. The rectangular coordinates for the Transverse Mercator applied to the sphere (Thomas, 1952, p. 6):

$$x = \tfrac{1}{2}Rk_0 \ln\left[(1 + B)/(1 - B)\right] \qquad (8-1)$$

or

$$x = Rk_0 \operatorname{arctanh} B \qquad (8-2)$$
$$y = Rk_0 \{\arctan\left[\tan\phi/\cos(\lambda - \lambda_0)\right] - \phi_0\} \qquad (8-3)$$
$$k = k_0/(1 - B^2)^{1/2} \qquad (8-4)$$

where

$$B = \cos\phi \sin(\lambda - \lambda_0) \qquad (8-5)$$

(note: If $B = \pm 1$, x is infinite)

TABLE 9.—*Universal Transverse Mercator grid coordinates*

U.T.M. GRID COORDINATES · CLARKE 1866 SPHEROID — METERS

LATITUDE 48°00'00"

Δλ	West of C.M. E	East of C.M. E	N
0°00'00"	500,000.0	500,000.0	5,316,081.3
07 30	490,675.3	509,324.7	5,316,088.9
15 00	481,350.5	518,649.5	5,316,111.6
22 30	472,025.8	527,974.2	5,316,149.4
30 00	462,701.1	537,298.9	5,316,202.3
37 30	453,376.4	546,623.6	5,316,270.3
45 00	444,051.8	555,948.2	5,316,353.5
52 30	434,727.1	565,272.9	5,316,451.7
1 00 00	425,402.5	574,597.5	5,316,565.1
07 30	416,078.0	583,922.0	5,316,693.6
15 00	406,753.5	593,246.5	5,316,837.3
22 30	397,429.0	602,571.0	5,316,996.1
30 00	388,104.5	611,895.5	5,317,169.9
37 30	378,780.2	621,219.8	5,317,359.0
45 00	369,455.9	630,544.1	5,317,563.1
52 30	360,131.6	639,868.4	5,317,782.4
2 00 00	350,807.4	649,192.6	5,318,016.8
07 30	341,483.3	658,516.7	5,318,266.3
15 00	332,159.3	667,840.7	5,318,531.0
22 30	322,835.4	677,164.6	5,318,810.8
30 00	313,511.5	686,488.5	5,319,105.8
37 30	304,187.7	695,812.3	5,319,415.9
45 00	294,864.1	705,135.9	5,319,741.1
52 30	285,540.5	714,459.5	5,320,081.5
3 00 00	276,217.0	723,783.0	5,320,437.0
07 30	266,893.7	733,106.3	5,320,807.7
15 00	257,570.5	742,429.5	5,321,193.6
22 30	248,247.4	751,752.6	5,321,594.6
30 00	238,924.4	761,075.6	5,322,010.8
37 30	229,601.5	770,398.5	5,322,442.1
45 00	220,278.8	779,721.2	5,322,888.6
52 30	210,956.2	789,043.8	5,323,350.3
4 00 00	201,633.8	798,366.2	5,323,827.1

LATITUDE 48°15'00"

Δλ	West of C.M. E	East of C.M. E	N
0°00'00"	500,000.0	500,000.0	5,343,868.4
07 30	490,720.4	509,279.6	5,343,875.9
15 00	481,440.8	518,559.2	5,343,898.6
22 30	472,161.2	527,838.8	5,343,936.3
30 00	462,881.7	537,118.3	5,343,989.2
37 30	453,602.1	546,397.9	5,344,057.2
45 00	444,322.6	555,677.4	5,344,140.2
52 30	435,043.1	564,956.9	5,344,238.4
1 00 00	425,763.7	574,236.3	5,344,351.7
07 30	416,484.3	583,515.7	5,344,480.1
15 00	407,204.9	592,795.1	5,344,623.6
22 30	397,925.6	602,074.4	5,344,782.2
30 00	388,646.3	611,353.7	5,344,955.9
37 30	379,367.1	620,632.9	5,345,144.8
45 00	370,088.0	629,912.0	5,345,348.7
52 30	360,808.9	639,191.1	5,345,567.8
2 00 00	351,529.9	648,470.1	5,345,802.0
07 30	342,251.0	657,749.0	5,346,051.3
15 00	332,972.2	667,027.8	5,346,315.7
22 30	323,693.4	676,306.6	5,346,595.3
30 00	314,414.8	685,585.2	5,346,889.9
37 30	305,136.2	694,863.8	5,347,199.7
45 00	295,857.8	704,142.2	5,347,524.7
52 30	286,579.4	713,420.6	5,347,864.7
3 00 00	277,301.2	722,698.8	5,348,219.9
07 30	268,023.1	731,976.9	5,348,590.3
15 00	258,745.1	741,254.9	5,348,975.8
22 30	249,467.3	750,532.7	5,349,376.4
30 00	240,189.6	759,810.4	5,349,792.2
37 30	230,912.0	769,088.0	5,350,223.1
45 00	221,634.6	778,365.4	5,350,669.2
52 30	212,357.3	787,642.7	5,351,130.4
4 00 00	203,080.2	796,919.8	5,351,606.8

LATITUDE 48°07'30"

Δλ	West of C.M. E	East of C.M. E	N
0°00'00"	500,000.0	500,000.0	5,329,974.7
07 30	490,697.8	509,302.2	5,329,982.3
15 00	481,395.6	518,604.4	5,330,004.9
22 30	472,093.5	527,906.5	5,330,042.7
30 00	462,791.3	537,208.7	5,330,095.6
37 30	453,489.2	546,510.8	5,330,163.6
45 00	444,187.1	555,812.9	5,330,246.7
52 30	434,885.0	565,115.0	5,330,344.9
1 00 00	425,582.9	574,417.1	5,330,458.3
07 30	416,280.9	583,719.1	5,330,586.7
15 00	406,979.0	593,021.0	5,330,730.3
22 30	397,677.0	602,323.0	5,330,889.0
30 00	388,375.2	611,624.8	5,331,062.8
37 30	379,073.4	620,926.6	5,331,251.7
45 00	369,771.6	630,228.4	5,331,455.8
52 30	360,469.9	639,530.1	5,331,675.0
2 00 00	351,168.3	648,831.7	5,331,909.3
07 30	341,866.8	658,133.2	5,332,158.7
15 00	332,565.3	667,434.7	5,332,423.2
22 30	323,264.0	676,736.0	5,332,702.9
30 00	313,962.7	686,037.3	5,332,997.7
37 30	304,661.5	695,338.5	5,333,307.7
45 00	295,360.4	704,639.6	5,333,632.8
52 30	286,059.5	713,940.5	5,333,973.0
3 00 00	276,758.6	723,241.4	5,334,328.4
07 30	267,457.9	732,542.2	5,334,698.9
15 00	258,157.2	741,842.8	5,335,084.6
22 30	248,856.7	751,143.3	5,335,485.4
30 00	239,556.4	760,443.6	5,335,901.4
37 30	230,256.1	769,743.9	5,336,332.5
45 00	220,956.0	779,044.0	5,336,778.8
52 30	211,656.1	788,343.9	5,337,240.3
4 00 00	202,356.3	797,643.7	5,337,716.9

LATITUDE 48°22'30"

Δλ	West of C.M. E	East of C.M. E	N
0°00'00"	500,000.0	500,000.0	5,357,762.3
07 30	490,743.0	509,257.0	5,357,769.9
15 00	481,486.1	518,513.9	5,357,792.5
22 30	472,229.2	527,770.8	5,357,830.3
30 00	462,972.2	537,027.8	5,357,883.1
37 30	453,715.3	546,284.7	5,357,951.0
45 00	444,458.5	555,541.5	5,358,034.1
52 30	435,201.6	564,798.4	5,358,132.2
1 00 00	425,944.8	574,055.2	5,358,245.4
07 30	416,688.0	583,312.0	5,358,373.8
15 00	407,431.3	592,568.7	5,358,517.2
22 30	398,174.6	601,825.4	5,358,675.7
30 00	388,918.0	611,082.0	5,358,849.4
37 30	379,661.5	620,338.5	5,359,038.1
45 00	370,405.0	629,595.0	5,359,242.0
52 30	361,148.6	638,851.4	5,359,460.9
2 00 00	351,892.2	648,107.8	5,359,695.0
07 30	342,636.0	657,364.0	5,359,944.2
15 00	333,379.8	666,620.2	5,360,208.5
22 30	324,123.7	675,876.3	5,360,487.9
30 00	314,867.7	685,132.3	5,360,782.4
37 30	305,611.9	694,388.1	5,361,092.0
45 00	296,356.1	703,643.9	5,361,416.8
52 30	287,100.4	712,899.6	5,361,756.7
3 00 00	277,844.9	722,155.1	5,362,111.7
07 30	268,589.5	731,410.5	5,362,481.9
15 00	259,334.2	740,665.8	5,362,867.2
22 30	250,079.1	749,920.9	5,363,267.6
30 00	240,824.1	759,175.9	5,363,683.1
37 30	231,569.2	768,430.8	5,364,113.9
45 00	222,314.5	777,685.5	5,364,559.7
52 30	213,060.0	786,940.0	5,365,020.7
4 00 00	203,805.6	796,194.4	5,365,496.8

GRID COORDINATES FOR 7.5 MINUTE INTERSECTIONS

and k_0 is the scale factor along the central meridian λ_0. The origin of the coordinates is at (ϕ_0, λ_0). The Y axis lies along the central meridian λ_0, y increasing northerly, and the X axis is perpendicular, through ϕ_0 at λ_0, x increasing easterly.

The inverse formulas for (ϕ, λ) in terms of (x, y):

$$\phi = \arcsin\ [\sin D/\cosh\ (x/Rk_0)] \qquad (8-6)$$
$$\lambda = \lambda_0 + \arctan\ [\sinh\ (x/Rk_0)/\cos D] \qquad (8-7)$$

where

$$D = y/(Rk_0) + \phi_0, \text{ using radians} \qquad (8-8)$$

Rectangular coordinates for the sphere are shown in table 10. Only one octant (quadrant of a hemisphere) needs to be listed, since all other octants are identical except for sign change. See p. 268 for numerical examples.

FORMULAS FOR THE ELLIPSOID

For the ellipsoidal form, the most practical form of the equations is a set of series approximations which converge rapidly to the correct centimeter or less at full scale in a zone extending 3° to 4° of longitude from the central meridian. Beyond this, the forward series as given here is accurate to about a centimeter at 7° longitude, but the inverse series does not have sufficient terms for this accuracy. The forward series may be used with meter accuracy to 10° of longitude. (Many additional terms for use to 24° of longitude may be found in Army (1962).) Coordinate axes are the same as they are for the spherical formulas above. The for-

TABLE 10.—*Transverse Mercator projection: Rectangular coordinates for the sphere*

[Radius of the Earth is 1.0 unit. Longitude measured from central meridian. *y* coordinate is in parentheses under *x* coordinate. Origin of rectangular coordinates at Equator and central meridian. *x* increases east; *y* increases north. One octant of globe is given; other octants are symmetrical]

Long. Lat.	0°	10°	20°	30°	40°
90°	0.0000	0.0000	0.0000	0.0000	0.0000
	(1.57080)	(1.57080)	(1.57080)	(1.57080)	(1.57080)
80	.00000	.03016	.05946	.08704	.11209
	(1.39626)	(1.39886)	(1.40659)	(1.41926)	(1.43653)
70	.00000	.05946	.11752	.17271	.22349
	(1.22173)	(1.22662)	(1.24125)	(1.26545)	(1.29888)
60	.00000	.08704	.17271	.25541	.33320
	(1.04720)	(1.05380)	(1.07370)	(1.10715)	(1.15438)
50	.00000	.11209	.22349	.33320	.43943
	(.87266)	(.88019)	(.90311)	(.94239)	(.99951)
40	.00000	.13382	.26826	.40360	.53923
	(.69813)	(.70568)	(.72891)	(.76961)	(.83088)
30	.00000	.15153	.30535	.46360	.62800
	(.52360)	(.53025)	(.55094)	(.58800)	(.64585)
20	.00000	.16465	.33320	.50987	.69946
	(.34907)	(.35401)	(.36954)	(.39786)	(.44355)
10	.00000	.17271	.35051	.53923	.74644
	(.17453)	(.17717)	(.18549)	(.20086)	(.22624)
0	.00000	.17543	.35638	.54931	.76291
	(.00000)	(.00000)	(.00000)	(.00000)	(.00000)

mulas below are only slightly modified from those presented in standard references to provide mm accuracy at full scale (Army, 1973, p. 5−7; Thomas, 1952, p. 2−3). (See p. 269 for numerical examples.)

$$x = k_0 N[A + (1 - T + C)A^3/6 + (5 - 18T + T^2 + 72C - 58e'^2)A^5/120] \qquad (8-9)$$

$$y = k_0 \{M - M_0 + N \tan \phi \, [A^2/2 + (5 - T + 9C + 4C^2) \\ A^4/24 + (61 - 58T + T^2 + 600C - 330e'^2)A^6/720]\} \qquad (8-10)$$

$$k = k_0[1 + (1 + C)A^2/2 + (5 - 4T + 42C + 13C^2 - 28e'^2)A^4/24 \\ + (61 - 148T + 16T^2)A^6/720] \qquad (8-11)$$

where k_0 = scale on central meridian (e.g., 0.9996 for the UTM projection)

$$e'^2 = e^2/(1 - e^2) \qquad (8-12)$$

$$N = a/(1 - e^2 \sin^2 \phi)^{1/2} \qquad (4-20)$$

$$T = \tan^2\phi \qquad (8-13)$$

$$C = e'^2 \cos^2 \phi \qquad (8-14)$$

$$A = (\lambda - \lambda_0) \cos \phi, \text{ with } \lambda \text{ and } \lambda_0 \text{ in radians} \qquad (8-15)$$

$$M = a[(1 - e^2/4 - 3e^4/64 - 5e^6/256 - \ldots) \phi - (3e^2/8 + 3e^4/32 \\ + 45e^6/1024 + \ldots) \sin 2\phi + (15e^4/256 + 45e^6/1024 \\ + \ldots) \sin 4\phi - (35e^6/3072 + \ldots) \sin 6\phi + \ldots] \qquad (3-21)$$

with ϕ in radians. M is the true distance along the central meridian from the Equator to ϕ. See equation (3−22) for a simplification for the Clarke 1866 ellipsoid.

$M_0 = M$ calculated for ϕ_0, the latitude crossing the central meridian λ_0 at the origin of the x, y coordinates.

Note: If $\phi = \pm \pi/2$, all equations should be omitted except (3−21), from which M and M_0 are calculated. Then $x = 0$, $y = k_0(M - M_0)$, $k = k_0$.

TABLE 10.—*Transverse Mercator projection: Rectangular coordinates for the sphere*—Continued

Long. Lat.	50°	60°	70°	80°	90°
90°	0.0000	0.0000	0.0000	0.0000	0.0000
	(1.57080)	(1.57080)	(1.57080)	(1.57080)	(1.57080)
80	.13382	.15153	.16465	.17271	.17543
	(1.45794)	(1.48286)	(1.51056)	(1.54019)	(1.57080)
70	.26826	.30535	.33320	.35051	.35638
	(1.34097)	(1.39078)	(1.44695)	(1.50768)	(1.57080)
60	.40360	.46360	.50987	.53923	.54931
	(1.21544)	(1.28976)	(1.37584)	(1.47087)	(1.57080)
50	.53923	.62800	.69946	.74644	.76291
	(1.07616)	(1.17355)	(1.29132)	(1.42611)	(1.57080)
40	.67281	.79889	.90733	.98310	1.01068
	(.91711)	(1.03341)	(1.18375)	(1.36673)	(1.57080)
30	.79889	.97296	1.13817	1.26658	1.31696
	(.73182)	(.85707)	(1.03599)	(1.27864)	(1.57080)
20	.90733	1.13817	1.38932	1.62549	1.73542
	(.51522)	(.62923)	(.81648)	(1.12564)	(1.57080)
10	.98310	1.26658	1.62549	2.08970	2.43625
	(.26773)	(.33904)	(.47601)	(.79305)	(1.57080)
0	1.01068	1.31696	1.73542	2.43625	
	(.00000)	(.00000)	(.00000)	(.00000)	Inf.

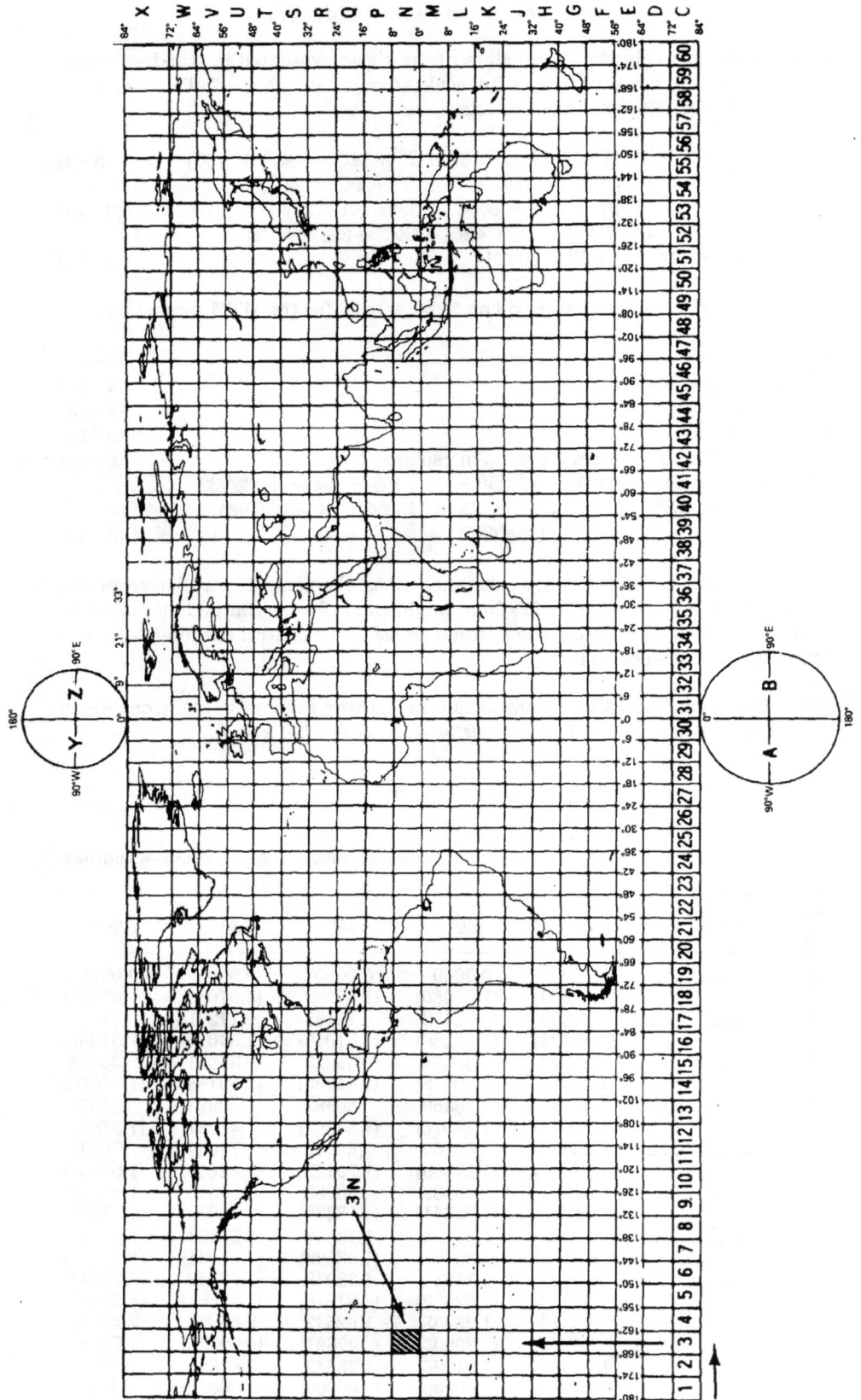

FIGURE 11.—Universal Transverse Mercator (UTM) grid zone designations for the world shown on a horizontally expanded Equidistant Cylindrical projection index map.

TABLE 11.—*Universal Transverse Mercator projection: Location of points with given scale factor*

[x coordinates in meters at various latitudes. Based on inversion of equation (8-16), using Clarke 1866 ellipsoid. Values are on or to right of central meridian (x=500,000 m). For coordinates left of central meridian, subtract values of x from 1,000,000 m. Latitude is north or south]

Lat.	Scale factor					
	0.9996	0.9998	1.0000	1.0002	1.0004	1.0006
80°	500,000	627,946	680,943	721,609	755,892	786,096
70	500,000	627,871	680,836	721,478	755,741	785,927
60	500,000	627,755	680,673	721,278	755,510	785,668
50	500,000	627,613	680,472	721,032	755,226	785,352
40	500,000	627,463	680,260	720,772	754,925	785,015
30	500,000	627,322	680,060	720,528	754,643	784,700
20	500,000	627,207	679,898	720,329	754,414	784,443
10	500,000	627,132	679,792	720,199	754,264	784,276
0	500,000	627,106	679,755	720,154	754,212	784,218

Equation (8−11) for k may also be written as a function of x and ϕ:

$$k = k_0[1 + (1 + e'^2 \cos^2 \phi)x^2/(2k^2_0 N^2)] \qquad (8-16)$$

These formulas are somewhat more precise than those used to compute the State Plane Coordinate tables, which were adapted to use desk calculators of 30−40 years ago. Table 11 shows the variation of k with x.

To obtain UTM or SPCS coordinates, the appropriate "false easting" is added to x and "false northing" added to y after calculation using (8−9) and (8−10).

For the *inverse formulas* (Army, 1973, p. 6, 7, 46; Thomas, 1952, p. 2−3):

$$\phi = \phi_1 - (N_1 \tan \phi_1/R_1)[D^2/2 - (5 + 3T_1 + 10C_1 - 4C_1^2 - 9e'^2)D^4/24$$
$$+ (61 + 90T_1 + 298C_1 + 45T_1^2 - 252e'^2 - 3C_1^2)D^6/720] \qquad (8-17)$$
$$\lambda = \lambda_0 + [D - (1 + 2T_1 + C_1)D^3/6 + (5 - 2C_1 + 28T_1$$
$$- 3C_1^2 + 8e'^2 + 24T_1^2)D^5/120]/\cos \phi_1 \qquad (8-18)$$

where ϕ_1 is the "footpoint latitude" or the latitude at the central meridian which has the same y coordinate as that of the point (ϕ, λ).

It may be found from equation (3−26):

$$\phi_1 = \mu + (3e_1/2 - 27e_1^3/32 + \ldots) \sin 2\mu + (21e_1^2/16$$
$$- 55e_1^4/32 + \ldots) \sin 4\mu + (151e_1^3/96 + \ldots) \sin 6\mu + (1097e_1^4/512 - \ldots)$$
$$\sin 8\mu + \ldots \qquad (3-26)$$

where

$$e_1 = [1-(1-e^2)^{1/2}]/[1 + (1-e^2)^{1/2}] \qquad (3-24)$$
$$\mu = M/[a(1-e^2/4 - 3e^4/64 - 5e^6/256 - \ldots)] \qquad (7-19)$$
$$M = M_0 + y/k_0 \qquad (8-20)$$

with M_0 calculated from equation (3−21) or (3−22) for the given ϕ_0.

For improved computational efficiency using series (3−21) and (3−26), see p. 19. From ϕ_1, other terms below are calculated for use in equations (8−17) and (8−18). (If $\phi_1 = \pm\pi/2$, (8−12), (8−21) through (8−25), (8−17) and (8−18) are omitted, but $\phi = \pm90°$, taking the sign of y, while λ is indeterminate, and may be called λ_0. Also, $k = k_0$.)

$$e'^2 = e^2/(1-e^2) \tag{8-12}$$
$$C_1 = e'^2 \cos^2 \phi_1 \tag{8-21}$$
$$T_1 = \tan^2 \phi_1 \tag{8-22}$$
$$N_1 = a/(1-e^2 \sin^2 \phi_1)^{1/2} \tag{8-23}$$
$$R_1 = a(1-e^2)/(1-e^2 \sin^2 \phi_1)^{3/2} \tag{8-24}$$
$$D = x/(N_1 k_0) \tag{8-25}$$

To convert from tabular rectangular coordinates to ϕ and λ, it is necessary to subtract any "false easting" from x and "false northing" from y before inserting x and y into the inverse formulas. To convert coordinates measured on an existing map, the correct central meridian must be used for the Y axis on the Transverse Mercator, but the X axis may cross it perpendicularly at any latitude chosen by the user.

"MODIFIED TRANSVERSE MERCATOR" PROJECTION

In 1972, the USGS devised a projection specifically for the revision of a 1954 map of Alaska which, like its predecessors, was based on the Polyconic projection. The projection was drawn to a scale of 1:2,000,000 and published at 1:2,500,000 (map "E") and 1:1,584,000 (map "B"). Graphically prepared by adapting coordinates for the Universal Transverse Mercator projection, it is identified as the "Modified Transverse Mercator" projection. It resembles the Transverse Mercator in a very limited manner and cannot be considered a cylindrical projection. It approximates an Equidistant Conic projection for the ellipsoid in actual construction. Because of the projection name, it is listed here. The projection was also used in 1974 for a base map of the Aleutian-Bering Sea Region published at the 1:2,500,000 scale.

The basis for the name is clear from an unpublished 1972 description of the projection, in which it is also stressed that the "latitudinal lines are parallel" and the "longitudinal lines are straight." The computations

were taken from the AMS Technical Manual #21 (Universal Transverse Mercator) based on the Clarke 1866 Spheroid.*** The projection was started from a N−S central construction line of the 153° longitude which is also the centerline of Zone 5 from the UTM tables. Along this line each even degree latitude was plotted from book values. At the plotted point for the 64° latitude, a perpendicular to the construction line (153°) was plotted. From the center construction line for each degree east and west for 4° (the limits of book value of Zone #5) the curvature of latitude was plotted. From this 64° latitude, each 2° latitude north to 70° and south to 54° was constructed parallel to the 64° latitude line. Each degree of longitude was plotted on the 58° and 68° latitude line. Through corresponding degrees of longitude along these two lines of latitude a straight line (line of longitude) was constructed and projected to the limits of the map. This gave a small projection 8° in width and approximately 18° in length. This projection was repeated east and west until a projection of some 72° in width was attained.

For transferring data to and from the Alaska maps, it was necessary to determine projection formulas for computer programming. Since it appeared to be unnecessarily complicated to derive formulas based on the above construction, it was decided to test empirical formulas with actual coordinates. After careful measurements of coordinates for graticule intersections were made in 1979 on the stable-base map, it was determined that the parallels very closely approximate concentric circular arcs, spaced in proportion to their true distances on the ellipsoid, while the meridians are nearly equidistant straight lines radiating from the center of the circular arcs. Two parallels have a scale equal to that along the meridians. The Equidistant Conic projection for the ellipsoid with two standard parallels was then applied to these coordinates as the closest approximation among projections with available formulas. After various trial values for scale and standard parallels were tested, the empirical formulas below (equations (8−26) through (8−32)) were obtained. These agree with measured values within 0.005 inch at mapping scale for 44 out of 58 measurements made on the map and within 0.01 inch for 54 of them.

FORMULAS FOR THE "MODIFIED TRANSVERSE MERCATOR" PROJECTION

The "Modified Transverse Mercator" projection was found to be most closely equivalent to an Equidistant Conic projection for the Clarke 1866 ellipsoid, with the scale along the meridians reduced to 0.9992 of true scale and the standard parallels at lat. 66.09° and 53.50° N. (also at 0.9992 scale factor). For the Alaska Map "E" at 1:2,500,000, using long. 150° W. as the central meridian and lat. 58° N. as the latitude of the origin on the central meridian, the general formulas (Snyder, 1978a, p. 378) reduce with the above parameters to the following, giving x and y in meters at the map scale. The Y axis lies along the central meridian, y increasing northerly, and the X axis is perpendicular at the origin, x increasing easterly.

For the forward formulas:

$$x = \rho \sin \theta \qquad (8-26)$$
$$y = 1.5616640 - \rho \cos \theta \qquad (8-27)$$

where

$$\theta° = 0.8625111(\lambda° + 150°) \qquad (8-28)$$
$$\rho = 4.1320402 - 0.04441727\phi° + 0.0064816 \sin 2\phi \qquad (8-29)$$

For the inverse formulas:

$$\lambda° = (1/0.8625111) \arctan [x/(1.5616640 - y)] - 150° \qquad (8-30)$$
$$\phi° = (4.1320402 + 0.0064816 \sin 2\phi - \rho)/0.04441727 \qquad (8-31)$$

where

$$\rho = [x^2 + (1.5616640 - y)^2]^{1/2} \qquad (8-32)$$

For Alaska Map "B" at a scale of 1:1,584,000, the same formulas may be used, except that x and y are (2,500/1,584) times the values obtained from (8-26) and (8-27). For the inverse formulas, the given x and y must be divided by (2,500/1,584) before insertion into (8-30) and (8-32).

The equation for ϕ, (8-31), involves iteration by successive substitution. If an initial ϕ of 60° is inserted into the right side, ϕ on the left may be calculated and substituted into the right in place of the previous trial ϕ. Recalculations continue until the change in ϕ is less than a preset convergence. If λ as calculated is less than −180°, it should be added to 360° and labeled East Longitude.

Formulas to adjust x and y for the map inset of the Aleutian Islands are omitted here, but the coordinates above are rotated counterclockwise 29.79° and transposed +0.798982 m for x and +0.347600 m for y.

9. OBLIQUE MERCATOR PROJECTION

SUMMARY

- Cylindrical (oblique).
- Conformal.
- Two meridians 180° apart are straight lines.
- Other meridians and parallels are complex curves.
- Scale on the spherical form is true along chosen central line, a great circle at an oblique angle, or along two straight lines parallel to central line. The scale on the ellipsoidal form is similar, but varies slightly from this pattern.
- Scale becomes infinite 90° from the central line.
- Used for grids on maps of the Alaska panhandle, for mapping in Switzerland, Madagascar, and Borneo and for atlas maps of areas with greater extent in an oblique direction.
- Developed 1900–50 by Rosenmund, Laborde, Hotine, and others.

HISTORY

There are several geographical regions such as the Alaska panhandle centered along lines which are neither meridians nor parallels, but which may be taken as great circle routes passing through the region. If conformality is desired in such cases, the Oblique Mercator is a projection which should be considered.

The historical origin of the Oblique Mercator projection does not appear to be sharply defined, although it is a logical generalization of the regular and Transverse Mercator projections. Apparently, Rosenmund (1903) made the earliest published reference, when he devised an ellipsoidal form which is used for topographic mapping of Switzerland. The projection was not mentioned in the detailed article on "Map Projections" in the 1911 *Encyclopaedia Britannica* (Close and Clarke, 1911) or in Hinks' brief text (1912). Laborde applied the Oblique Mercator to the ellipsoid for the topographic mapping of Madagascar in 1928 (Young, 1930; Laborde, 1928). H. J. Andrews (1935, 1938) proposed the spherical forms for maps of the United States and Eurasia. Hinks presented seven world maps on the Oblique Mercator, with poles located in several different positions, and a consequent variety in the regions shown more satisfactorily (Hinks, 1940, 1941).

A study of conformal projections of the ellipsoid by British geodesist Martin Hotine (1898–1968), published in 1946–47, is the basis of the U.S. use of the ellipsoidal Oblique Mercator, which Hotine called the "rectified skew orthomorphic" (Hotine, 1947, p. 66–67). The Hotine approach has limitations, as discussed below, but it provides closed formulas which have been adapted for U.S. mapping of suitable zones. One of its limitations is overcome by a recent series form of the ellipsoidal Oblique Mercator (Snyder, 1979a, p. 74), but other limitations result instead. This later form resulted from development of formulas for the continuous mapping of satellite images, using the Space Oblique Mercator projection (to be discussed later).

While Hotine projected the ellipsoid conformally onto an "aposphere" of constant total curvature and thence to a plane, J. H. Cole (1943, p. 16–30) projected the ellipsoid onto a "conformal sphere," using conformal latitudes (described earlier) to make the sphere conformal with respect to the ellipsoid, then plotted the spherical Oblique Mercator from this intermediate sphere. Rosenmund's system for Switzerland is a more complex double projection through a conformal sphere (Rosenmund, 1903; Bolliger, 1967). Laborde combined the conformal sphere with a complex-algebra transformation of the Oblique Mercator (Reignier, 1957, p. 130).

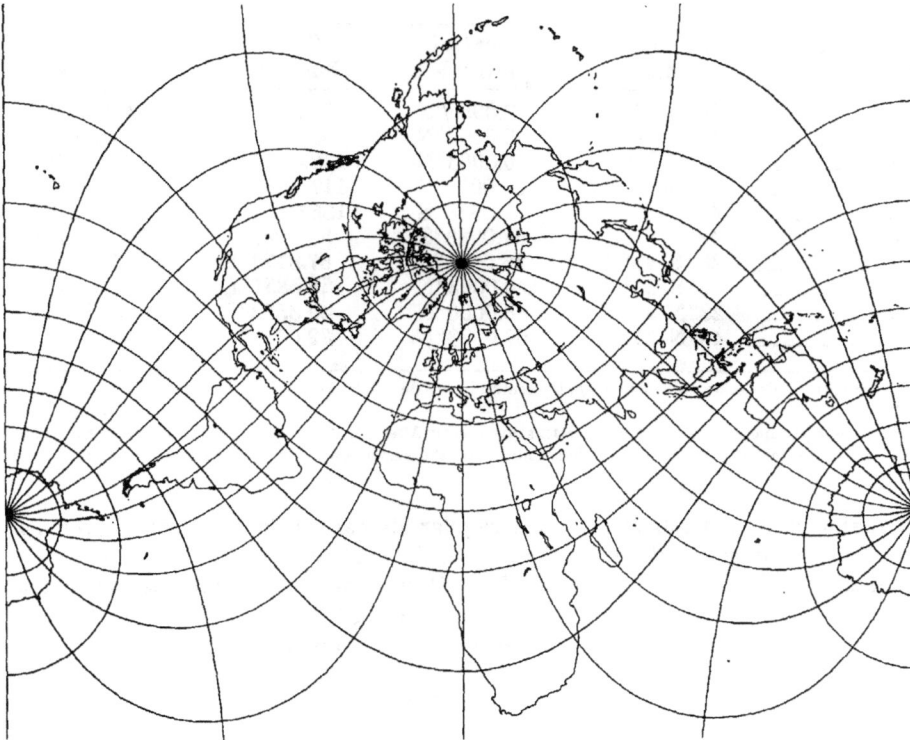

FIGURE 12.—Oblique Mercator projection with the center of projection at lat. 45° N. on the central meridian. A straight line through the point and, in this example, perpendicular to the central meridian is true to scale. The projection is conformal and has been used for regions lying along a line oblique to meridians.

FEATURES

The Oblique Mercator for the sphere is equivalent to a regular Mercator projection which has been altered by wrapping a cylinder around the sphere so that it touches the surface along the great circle path chosen for the central line, instead of along the Earth's Equator. A set of transformed meridians and parallels relative to the great circle may be plotted bearing the same relationship to the rectangular coordinates for the Oblique Mercator projection, as the geographic meridians and parallels bear to the regular Mercator. It is, therefore, possible to convert the geographic meridians and parallels to the transformed values and then to use the regular Mercator equations, substituting the transformed values in place of the geographic values. This is the procedure for the sphere, although combined formulas are given below, but it becomes much more complicated for the ellipsoid. The advent of present-day computers and programmable pocket calculators make these calculations feasible for sphere or ellipsoid.

The resulting Oblique Mercator map of the world (fig. 12) thus resembles the regular Mercator with the landmasses rotated so that the poles and Equator are no longer in their usual positions. Instead, two points 90° away from the chosen great circle path through the center of the map are at infinite distance off the map. Normally, the Oblique Mercator is used only to show the region near the central line and for a relatively short portion of the central line. Under these conditions, it looks similar to maps of the same area using other projections, except that careful scale measurements will show differences.

TABLE 12.—*Hotine Oblique Mercator projection parameters used for Landsat 1, 2, and 3 imagery*

HOM zone	Limiting latitudes	Central latitude	Central longitude[1]	Azimuth of axis
1	48°N–81°N	67.0983°N	81.9700°W	24.7708181°
2	23°N–48°N	36.0000°N	99.2750°W	14.3394883°
3	23°S–23°N	0.0003°N	108.5069°W	13.001443°
4	23°S–48°S	36.0000°S	117.7388°W	14.33948832°
5	48°S–81°S	67.0983°S	135.0438°W	24.7708181°
6	48°S–81°S	67.0983°S	85.1220°E	−24.7708181°
7	23°S–48°S	36.0000°S	67.8170°E	−14.33948832°
8	23°S–23°N	0.0003°N	58.5851°E	−13.001443°
9	23°N–48°N	36.0000°N	49.3532°E	−14.33948832°
10	48°N–81°N	67.0983°N	32.0482°E	−24.7708181°

[1] For path 31. For other path numbers p, the central longitude is decreased (west is negative) by $(360°/251) \times (p - 31)$.

Note: These parameters are used with equations given under Alternate B of ellipsoidal Oblique Mercator formulas, with ϕ_0 the central latitude, λ_0 the central longitude, and α_0 the azimuth of axis east of north. Scale factor k_0 at center is 1.0.

It should be remembered that the regular Mercator is in fact a limiting form of the Oblique Mercator with the Equator as the central line, while the Transverse Mercator is another limiting form of the Oblique with a meridian as the central line. As with these limiting forms, the scale along the central line of the Oblique Mercator may be reduced to balance the scale throughout the map.

USAGE

The Oblique Mercator projection is used in the spherical form for a few atlas maps. For example, the National Geographic Society uses it for atlas and sheet maps of Hawaii, the West Indies, and New Zealand. The spherical form is being used by the USGS for maps of North and South America and Australasia in a new set of 1:10,000,000-scale maps of Hydrocarbon Provinces. For North America, the central scale factor is 0.968, and the transformed pole is at lat. 10°N, long. 10°E. For South America, these numbers are 0.974, 10°N., and 30°E., respectively; for Australasia, 0.978, 55°N, and 160°W. These parameters were chosen after a least-squares analysis of over 100 points on each continent to determine optimum parameters for a common conformal projection.

In the ellipsoidal form it was used, as mentioned above, by Rosenmund for Switzerland and Laborde for Madagascar. Hotine used it for Malaya and Borneo and Cole for Italy. It is used in the Hotine form by the USGS for grid marks on zone 1 (the panhandle) of Alaska, using the State Plane Coordinate System as adapted to this projection by Erwin Schmid of the former Coast and Geodetic Survey. The Hotine form was also adopted by the U.S. Lake Survey for mapping of the five Great Lakes, the St. Lawrence River, and the U.S.-Canada Border Lakes west to the Lake of the Woods (Berry and Bormanis, 1970). Four zones are involved; see table 8 for parameters of these and the Alaska zones.

More recently, the Hotine form was adapted by John B. Rowland (USGS) for mapping Landsat 1, 2, and 3 satellite imagery in two sets of five discontinuous zones from north to south (table 12). The central line of the latter is only a close approximation to the satellite groundtrack, which does not follow a great circle route on the Earth; instead, it follows a path of constantly changing curvature. Until the mathematical implementation of the Space Oblique Mercator (SOM) projection, the Hotine Oblique Mercator (HOM) was probably the most suitable projection available for mapping Landsat type data. In addition to Landsat, the HOM projection has been used to cast Heat Capacity Mapping Mission (HCMM) imagery since 1978. NOAA (National Oceanic and Atmospheric Administration) has also cast some weather satellite imagery on the HOM to make it compatible with Landsat in the polar regions which are beyond Landsat coverage (above lat. 82°).

The parameters for a given map according to the Oblique Mercator projection may be selected in various ways. If the projection is to be used for the map of a smaller region, two points located near the limits of the region may be selected to lie upon the central line, and various constants may be calculated from the latitude and longitude of each of the two points. A second approach is to choose a central point for the map and an azimuth for the central line, which is made to pass through the central point. A third approach, more applicable to the map of a large portion of the Earth's surface, treated as spherical, is to choose a location on the original sphere of the pole for a transformed sphere with the central line as the equator. Formulas are given for each of these approaches, for sphere and ellipsoid.

FORMULAS FOR THE SPHERE

Starting with the forward equations, for rectangular coordinates in terms of latitude and longitude (see p. 272 for numerical examples):

1. Given two points to lie upon the central line, with latitudes and longitudes (ϕ_1, λ_1) and (ϕ_2, λ_2) and longitude increasing easterly and relative to Greenwich. The pole of the oblique transformation at (ϕ_p, λ_p) may be calculated as follows:

$$\lambda_p = \arctan\,[(\cos\phi_1\sin\phi_2\cos\lambda_1 - \sin\phi_1\cos\phi_2\cos\lambda_2)/$$
$$(\sin\phi_1\cos\phi_2\sin\lambda_2 - \cos\phi_1\sin\phi_2\sin\lambda_1)] \qquad (9-1)$$
$$\phi_p = \arctan\,[-\cos(\lambda_p - \lambda_1)/\tan\phi_1] \qquad (9-2)$$

The Fortran ATAN2 function or its equivalent should be used with equation (9-1), but not with (9-2). The other pole is located at $(-\phi_p, \lambda_p \pm \pi)$. Using the positive (northern) value of ϕ_p, the following formulas give the rectangular coordinates for point (ϕ, λ), with k_0 the scale factor along the central line:

$$x = Rk_0 \arctan\,\{[\tan\phi\cos\phi_p + \sin\phi_p\sin(\lambda - \lambda_0)]/\cos(\lambda - \lambda_0)\} \qquad (9-3)$$
$$y = (R/2)k_0\ln[(1+A)/(1-A)] \qquad (9-4)$$

or

$$y = Rk_0\,\text{arctanh}\,A \qquad (9-4a)$$
$$k = k_0/(1-A^2)^{1/2} \qquad (9-5)$$

where

$$A = \sin\phi_p\sin\phi - \cos\phi_p\cos\phi\sin(\lambda - \lambda_0) \qquad (9-6)$$

With these formulas, the origin of rectangular coordinates lies at

$$\phi_0 = 0$$
$$\lambda_0 = \lambda_p + \pi/2 \qquad (9-6a)$$

and the X axis lies along the central line, x increasing easterly. The transformed poles are y equals infinity.

2. Given a central point (ϕ_c, λ_c) with longitude increasing easterly and relative to Greenwich, and azimuth β east of north of the central line through (ϕ_c, λ_c), the pole of the oblique transformation at (ϕ_p, λ_p) may be calculated as follows:

$$\phi_p = \arcsin (\cos \phi_c \sin \beta) \tag{9-7}$$
$$\lambda_p = \arctan [- \cos \beta/(- \sin \phi_c \sin \beta)] + \lambda_c \tag{9-8}$$

These values of ϕ_p and λ_p may then be used in equations (9–3) through (9–6) as before.

3. For an extensive map, ϕ_p and λ_p may be arbitrarily chosen by eye to give the pole for a central line passing through a desired portion of the globe. These values may then be directly used in equations (9–3) through (9–6) without intermediate calculation.

For the inverse formulas, equations (9–1) and (9–2) or (9–7) and (9–8) must first be used to establish the pole of the oblique transformation, if it is not known already. Then,

$$\phi = \arcsin [\sin \phi_p \tanh (y/Rk_0) + \cos \phi_p \sin (x/Rk_0)/\cosh (y/Rk_0)] \tag{9-9}$$
$$\lambda = \lambda_0 + \arctan \{[\sin \phi_p \sin (x/Rk_0) - \cos \phi_p \sinh (y/Rk_0)]/\cos (x/Rk_0)\} \tag{9-10}$$

FORMULAS FOR THE ELLIPSOID

These are the formulas provided by Hotine, slightly altered to use a positive eastern longitude (he used positive western longitude), to simplify calculations of hyperbolic functions, and to use symbols consistent with those of this bulletin. The central line is a geodesic, or the shortest route on an ellipsoid, corresponding to a great circle route on the sphere.

It is customary to provide rectangular coordinates for the Hotine in terms either of (u, v) or (x, y). The (u, v) coordinates are similar in concept to the (x, y) calculated for the foregoing spherical formulas, with u corresponding to x for the spherical formulas, increasing easterly from the origin along the central line, but v corresponds to $-y$ for the spherical formulas, so that v increases southerly in a direction perpendicular to the central line. For the Hotine, x and y are calculated from (u, v) as "rectified" coordinates with the Y axis following the meridian passing through the center point, and increasing northerly as usual, while the X axis lies east and west through the same point. The X and Y axes thus lie in directions like those of the Transverse Mercator, but the scale-factor relationships remain those of the Oblique Mercator.

The normal origin for (u, v) coordinates in the Hotine Oblique Mercator is approximately at the intersection of the central line with the Earth's Equator. Actually it occurs at the crossing of the central line with the equator of the "aposphere," and is, thus, a rather academic location. The "aposphere" is a surface with a constant "total" curvature based on the curvature along the meridian and perpendicular thereto on the ellipsoid at the chosen central point for the projection. The ellipsoid is conformally projected onto this aposphere, then to a plane. As a result, the Hotine is perfectly conformal, but the scale along the central line is true only at the chosen central point along that line or along a relatively flat elliptically shaped line approximately centered on that point, if the scale of the central point is arbitrarily reduced to balance scale over the map. The variation in scale along the central line is extremely small for a map extending less than 45° in arc, which includes most existing usage of the Hotine. A longer central line suggests the use of a different set of formulas, available as a limiting form of the Space Oblique Mercator projection. On Rosenmund's (1903), Laborde's

(1928), and Cole's (1943) versions of the ellipsoidal Oblique Mercator, the central line is a great circle arc on the intermediate conformal sphere, not a geodesic. As on Hotine's version, this central line is not quite true to scale except at one or two chosen points.

The projection constants may be established for the Hotine in one of two ways, as they were for the spherical form. Two desired points, widely separated on the map, may be made to fall on the central line of the projection, or the central line may be given a desired azimuth through a selected central point. Taking these approaches in order:

Alternate A, with the central line passing through two given points.
Given:

    a and e for the reference ellipsoid.

    k_0 = scale factor at the selected center of the map, lying on the central line.

    ϕ_0 = latitude of selected center of the map.

    (ϕ_1, λ_1) = latitude and longitude (east of Greenwich is positive) of the first point which is to lie on the central line.

    (ϕ_2, λ_2) = latitude and longitude of the second point which is to lie on the central line.

    (ϕ, λ) = latitude and longitude of the point for which the coordinates are desired.

There are limitations to the use of variables in these formulas: To avoid indeterminates and division by zero, ϕ_0 or ϕ_1 cannot be $\pm \pi/2$, ϕ_1 cannot be zero or equal to ϕ_2 (although ϕ_2 may be zero), and ϕ_2 cannot be $-\pi/2$. Neither ϕ_0, ϕ_1, nor ϕ_2 should be $\pm \pi/2$ in any case, since this would cause the central line to pass through the pole, for which the Transverse Mercator or polar Stereographic would probably be a more suitable choice. A change of 10^{-7} radian in variables from these special values will permit calculation of an otherwise unsatisfactory condition.

It is also necessary to place both (ϕ_1, λ_1) and (ϕ_2, λ_2) on the ascending portion, or both on the descending portion, of the central line, relative to the Earth's Equator. That is, the central line should not pass through a maximum or minimum between these two points.

If e is zero, the Hotine formulas give coordinates for the spherical Oblique Mercator.

Because of the involved nature of the Hotine formulas, they are given here in an order suitable for calculation, and in a form eliminating the use of hyperbolic functions as given by Hotine in favor of single calculations of exponential functions to save computer time. The corresponding Hotine equations are given later for comparison (see p. 274 for numerical examples).

$$B = [1 + e^2 \cos^4 \phi_0/(1-e^2)]^{1/2} \tag{9-11}$$

$$A = aBk_0(1-e^2)^{1/2}/(1-e^2 \sin^2 \phi_0) \tag{9-12}$$

$$t_0 = \tan (\pi/4 - \phi_0/2)/[(1 - e \sin \phi_0)/(1 + e \sin \phi_0)]^{e/2} \tag{9-13}$$

or $$= \left[\left(\frac{1 - \sin \phi_0}{1 + \sin \phi_0} \right) \left(\frac{1 + e \sin \phi_0}{1 - e \sin \phi_0} \right)^e \right]^{1/2} \tag{9-13a}$$

t_1 = same as (9–13), using ϕ_1 in place of ϕ_0.

t_2 = same as (9–13), using ϕ_2 in place of ϕ_0.

$$D = B(1-e^2)^{1/2}/[\cos \phi_0(1 - e^2 \sin^2 \phi_0)^{1/2}] \tag{9-14}$$

If $\phi_0 = 0$, D may calculate to slightly less than 1.0 and create a problem in the next step. If $D^2 < 1$, it should be made 1.

$$E = [D \pm (D^2 - 1)^{1/2}]t_0{}^B, \text{ taking the sign of } \phi_0 \tag{9-15}$$

$$H = t_1{}^B \tag{9-16}$$

$$L = t_2{}^B \tag{9-17}$$
$$F = E/H \tag{9-18}$$
$$G = (F - 1/F)/2 \tag{9-19}$$
$$J = (E^2 - LH)/(E^2 + LH) \tag{9-20}$$
$$P = (L - H)/(L + H) \tag{9-21}$$
$$\lambda_0 = (\lambda_1 + \lambda_2)/2 - \arctan\{J \tan [B(\lambda_1 - \lambda_2)/2]/P\}/B \tag{9-22}$$
$$\gamma_0 = \arctan\{\sin [B(\lambda_1 - \lambda_0)]/G\} \tag{9-23}$$
$$\alpha_c = \arcsin [D \sin \gamma_0] \tag{9-24}$$

To prevent problems when straddling the 180th meridian with λ_1 and λ_2, before calculating (9−22), if $(\lambda_1 - \lambda_2) < -180°$, subtract 360° from λ_2. If $(\lambda_1 - \lambda_2) > 180°$, add 360° to λ_2. Also adjust λ_0 and $(\lambda_1 - \lambda_0)$ to fall between ±180° by adding or subtracting 360°. The Fortran ATAN2 function is not to be used for equations (9−22) and (9−23). The above equations (9−11) through (9−24) provide constants for a given map and do not involve a specific point (ϕ,λ). Angle α_c is the azimuth of the central line as it crosses latitude ϕ_0, measured east of north. For point (ϕ, λ), calculate the following:

$$t = \text{same as equation (9-13), but using } \phi \text{ in place of } \phi_0.$$

If $\phi = \pm\pi/2$, do not calculate t, but go instead to (9−30).

$$Q = E/t^B \tag{9-25}$$
$$S = (Q - 1/Q)/2 \tag{9-26}$$
$$T = (Q + 1/Q)/2 \tag{9-27}$$
$$V = \sin [B(\lambda - \lambda_0)] \tag{9-28}$$
$$U = (-V \cos \gamma_0 + S \sin \gamma_0)/T \tag{9-29}$$
$$v = A \ln [(1-U)/(1+U)]/2B \tag{9-30}$$

Note: If $U = \pm1$, v is infinite; if $\phi = \pm\pi/2$, $v = (A/B) \ln \tan (\pi/4 \mp \gamma_0/2)$

$$u = A \arctan \{(S \cos \gamma_0 + V \sin \gamma_0)/\cos [B(\lambda-\lambda_0)]\}/B \tag{9-31}$$

Note: If $\cos [B(\lambda-\lambda_0)] = 0$, $u = AB(\lambda-\lambda_0)$. If $\phi = \pm\pi/2$, $u = A\phi/B$.

Care should be taken that $(\lambda-\lambda_0)$ has 360° added or subtracted, if the 180th meridian falls between, since multiplication by B eliminates automatic correction with the sin or cos function.

The scale factor:

$$k = A \cos (Bu/A)(1-e^2\sin^2\phi)^{1/2}/\{a \cos \phi \cos [B(\lambda-\lambda_0)]\} \tag{9-32}$$

If "rectified" coordinates (x, y) are desired, with the origin at a distance (x_0, y_0) from the origin of the (u,v) coordinates, relative to the (X,Y) axes (see fig. 13):

$$x = v \cos \alpha_c + u \sin \alpha_c + x_0 \tag{9-33}$$
$$y = u \cos \alpha_c - v \sin \alpha_c + y_0 \tag{9-34}$$

The formulas given by Hotine and essentially repeated in Thomas (1952, p. 7−9), modified for positive east longitude, u and v increasing in the directions shown in figure 13, and symbols consistent with the above, relate to the foregoing formulas as follows:[3]

---

[3]Hotine uses positive west longitude, x corresponding to u here, and y corresponding to $-v$ here. Thomas uses positive west longitude, y corresponding to u here, and x corresponding to $-v$ here. In calculations of Alaska zone 1, west longitude is positive, but u and v agree with u and v, respectively, here.

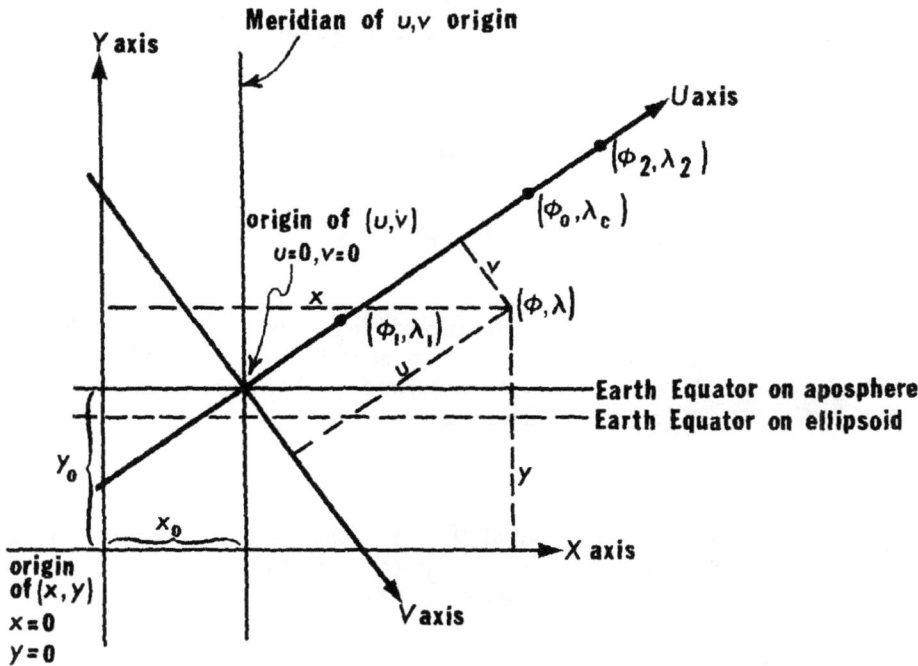

FIGURE 13.—Coordinate system for the Hotine Oblique Mercator projection.

Equivalent to (9−11):

$$e'^2 = e^2/(1-e^2)$$
$$B = (1 + e'^2 \cos^4 \phi_0)^{1/2}$$

Equivalent to (9−12):

$$R'_0 = a(1-e^2)/(1-e^2 \sin^2 \phi_0)^{3/2}$$
$$N_0 = a/(1-e^2 \sin^2 \phi_0)^{1/2}$$
$$A = Bk_0(R'_0 N_0)^{1/2}$$

Other formulas:

$$r_0 = N_0 \cos \phi_0$$
$$\psi_n = \ln \{\tan (\pi/4 + \phi_n/2)[(1-e \sin \phi_n)/(1+e \sin \phi_n)]^{e/2}\}$$

Note: ψ_n is equivalent to $(-\ln t_n)$ using equation (9−13).

$$C = \pm \operatorname{arccosh} (A/r_0) - B\psi_0$$

Note: C is equivalent to $\ln E$, where E is found from equation (9−15); D, from (9−14), is (A/r_0).

$$\tan [\tfrac{1}{2}B(\lambda_1 + \lambda_2) - B\lambda_0] = \frac{\tan [\tfrac{1}{2}B(\lambda_1 - \lambda_2)] \tanh [\tfrac{1}{2}B(\psi_1 + \psi_2) + C]}{\tanh [\tfrac{1}{2}B(\psi_1 - \psi_2)]}$$

The tanh in the numerator is J from equation (9−20), while the tanh in the denominator is P from (9−21). The entire equation is equivalent to (9−22).

$$\tan \gamma_o = \sin [B(\lambda_1 - \lambda_0)]/\sinh (B\psi_1 + C)$$

This equation is equivalent to (9–23), the sinh being equivalent to G from (9–19).

$$\tanh (Bv/Ak_0) = \{\cos \gamma_0 \sin [B(\lambda - \lambda_0)] - \sin \gamma_0 \sinh (B\psi + C)\}/\cosh (B\psi + C)$$

This equation is equivalent to (9–30), with S the sinh function and T the cosh function.

$$\tan (Bu/Ak_0) = \{\cos \gamma_0 \sinh (B_\psi + C) + \sin \gamma_0 \sin [B(\lambda - \lambda_0)]\}/\cos [B(\lambda - \lambda_0)]$$

This equation is equivalent to (9–31).

Alternate B. The following equations provide constants for the Hotine Oblique Mercator projection to fit a given central point and azimuth of the central line through the central point. Given: a, e, k_0, ϕ_0, and (ϕ, λ) as for alternate A, but instead of (ϕ_1, λ_1) and (ϕ_2, λ_2), λ_c and α_c are given,

where

(ϕ_0, λ_c) = latitude and longitude (east of Greenwich is positive), respectively, of the selected center of the map, falling on the central line.

α_c = angle of azimuth east of north, for the central line as it passes through the center of the map (ϕ_0, λ_c).

Limitations: ϕ_0 cannot be zero or $\pm \pi/2$, and the central line cannot be at a maximum or minimum latitude at ϕ_0. If $e = 0$, these formulas also give coordinates for the spherical Oblique Mercator. As with alternate A, these formulas are given in the order of calculation and are modified to minimize exponential computations. Several of these equations are the same as some of the equations for alternate A:

$$B = [1 + e^2 \cos^4 \phi_0/(1 - e^2)]^{1/2} \qquad (9–11)$$
$$A = aBk_0 (1 - e^2)^{1/2}/(1 - e^2\sin^2 \phi_0) \qquad (9–12)$$
$$t_0 = \tan(\pi/4 - \phi_0/2)/[(1 - e\sin \phi_0)/(1 + e\sin \phi_0)]^{e/2} \qquad (9–13)$$
$$D = B(1 - e^2)^{1/2}/[\cos \phi_0 (1 - e^2 \sin^2 \phi_0)^{1/2}] \qquad (9–14)$$

If $\phi_0 = 0$, D may calculate to slightly less than 1.0 and create a problem in the next step. If $D^2 < 1$, it should be made 1.

$$F = D \pm (D^2 - 1)^{1/2}, \text{ taking the sign of } \phi_0 \qquad (9–35)$$
$$E = Ft_0{}^B \qquad (9–36)$$
$$G = (F - 1/F)/2 \qquad (9–19)$$
$$\gamma_0 = \arcsin (\sin \alpha_c/D) \qquad (9–37)$$
$$\lambda_0 = \lambda_c - [\arcsin (G \tan \gamma_0)]/B \qquad (9–38)$$

The values of u and v for center point (ϕ_0, λ_c) may be calculated directly at this point:

$$u_{(\phi_0, \lambda_c)} = \pm (A/B) \arctan [(D^2 - 1)^{1/2}/\cos \alpha_c], \text{ taking the sign of } \phi_0. \qquad (9–39)$$
$$v_{(\phi_0, \lambda_c)} = 0$$

These are the constants for a given map. Equations (9–25) through (9–32) for alternate A may now be used in order, following calculation of the above constants.

The inverse equations for the Hotine Oblique Mercator projection on the ellipsoid may be shown with few additional formulas. To determine ϕ and λ from x and y, or from u and v, the same parameters of the map must be given, except for ϕ and λ, and the constants of the map are found from the above equations (9-11) through (9-24) for alternate A or (9-11) through (9-38) for alternate B. Then, if x and y are given in accordance with the definitions for the forward equations, they must first be converted to (u, v):

$$v = (x - x_0) \cos \alpha_c - (y - y_0) \sin \alpha_c \qquad (9-40)$$
$$u = (y - y_0) \cos \alpha_c + (x - x_0) \sin \alpha_c \qquad (9-41)$$

If (u, v) are given, or calculated as just above, the following steps are performed in order:

$$Q' = e^{-(Bv/A)} \qquad (9-42)$$

where $e = 2.71828\ldots$, the base of natural logarithms

$$S' = (Q' - 1/Q')/2 \qquad (9-43)$$
$$T' = (Q' + 1/Q')/2 \qquad (9-44)$$
$$V' = \sin(Bu/A) \qquad (9-45)$$
$$U' = (V'\cos\gamma_0 + S'\sin\gamma_0)/T' \qquad (9-46)$$
$$t = \{E/[(1 + U')/(1-U')]^{1/2}\}^{1/B} \qquad (9-47)$$

But if $U' = \pm 1$, $\phi = \pm 90°$, taking the sign of U', λ may be called λ_0, and equations (7-9) and (9-48) below are omitted.

$$\phi = \pi/2 - 2 \arctan\{t[(1 - e \sin\phi)/(1 + e \sin\phi)]^{e/2}\} \qquad (7-9)$$

Equation (7-9) is solved by iteration, using $\phi = (\pi/2 - 2 \arctan t)$ as the first trial ϕ on the right side, and using the successive calculations of ϕ on the left side as successive values of ϕ on the right side, until the change in ϕ is less than a chosen convergence value.

$$\lambda = \lambda_0 - \arctan[(S' \cos\gamma_0 - V' \sin\gamma_0)/\cos(Bu/A)]/B \qquad (9-48)$$

Since the arctan (found as the ATAN2 function) is divided by B, it is necessary to add or subtract 360° properly, before the division.

To avoid the iteration, the series (3-5) may be used with (7-13) in place of (7-9):

$$\phi = \chi + (e^2/2 + 5e^4/24 + e^6/12 + 13e^8/360 + \ldots) \sin 2\chi + \\ (7e^4/48 + 29e^6/240 + 811e^8/11520 + \ldots)\sin 4\chi + (7e^6/120 + 81e^8/1120 + \ldots) \\ \sin 6\chi + (4279e^8/161280 + \ldots) \sin 8\chi + \ldots \qquad (3-5)$$

where

$$\chi = \pi/2 - 2 \arctan t \qquad (7-13)$$

For improved computational efficiency using this series, see p. 19.

The equivalent inverse equations as given by Hotine are as follows, following the calculation of constants using the same formulas as those given in his forward equations:

$$\tan[B(\lambda - \lambda_0)] = [\sin\gamma_0 \sin(Bu/A) + \cos\gamma_0 \sinh(Bv/A)]/\cos(Bu/A)$$
$$\tanh(B\psi + C) = [\cos\gamma_0 \sin(Bu/A) - \sin\gamma_0 \sinh(Bv/A)]/\cosh(Bv/A)$$

10. CYLINDRICAL EQUAL-AREA PROJECTION

SUMMARY

- Cylindrical.
- Equal-area.
- Meridians on normal aspect are equally spaced straight lines.
- Parallels on normal aspect are unequally spaced straight lines, closest near the poles, cutting meridians at right angles.
- On transverse aspect, central meridian, each meridian 90° from central meridian, and Equator are straight lines. Other meridians and parallels are complex curves.
- On oblique aspect, two meridians 180° apart are straight lines. Other meridians and parallels are complex curves.
- On normal aspect, scale is true along Equator, or along two parallels equidistant from the Equator.
- On transverse aspect, scale is true along central meridian, or along two straight lines equidistant from and parallel to central meridian. (These lines are only approximately straight for the ellipsoid.)
- On oblique aspect, scale is true along chosen central line, an oblique great circle, or along two straight lines parallel to central line. Scale on ellipsoidal form is similar, but varies slightly from this pattern.
- An orthographic projection of sphere onto cylinder.
- Substantial shape and scale distortion near points 90° from central line.
- Normal and transverse aspects presented by Lambert in 1772.

HISTORY AND USAGE

The fourth of the seven projections proposed by Johann Heinrich Lambert (1772, p. 71−72) and occasionally given his name, is the Cylindrical Equal-Area (fig. 14). In the same work (p. 72−73), he described its transverse aspect (fig. 16), which has hardly been used. Even the normal aspect has seldom been used except as a textbook example of the most easily constructed equal-area projection, but several modifications of the normal aspect have been published.

These modifications consist of compressing the projection from east to west and expanding it in the same ratio from north to south, thereby moving the parallel of no distortion from the Equator to other latitudes. The earliest such modification is from Scotland: James Gall's Orthographic Cylindrical, not the same as his preferred Stereographic Cylindrical, both of which were originated in 1855, has standard parallels of 45° N. and S. (Gall, 1885). Walther Behrmann (1910) of Germany chose 30°, based on certain overall distortion criteria (fig. 15). Very similar later projections were offered by Trystan Edwards of England in 1953 and Arno Peters of Germany in 1967; they were presented as revolutionary and original concepts, rather than as modifications of these prior projections with standard parallels at about 37° and 45°−47°, respectively (Maling, 1966, 1974).

The oblique Cylindrical Equal-Area projection has been proposed with particular parameters for maps of Eurasia and Africa (Thornthwaite, 1927) and of air routes of the British Commonwealth (Poole, 1934). Different parameters are used for fig. 17. The ellipsoidal form of the oblique and transverse aspects has apparently been developed only recently (Snyder, 1985b).

FEATURES

Like other regular cylindricals, the graticule of the normal Cylindrical Equal-Area projection consists of straight equally spaced vertical meridians perpendicular to straight unequally spaced horizontal parallels. To achieve equality of area, the parallels are spaced from the Equator in proportion to the sine of the latitude. This is the simplest equal-area projection.

The normal Cylindrical Equal-Area for the sphere is a true perspective projection onto a cylinder tangent at the Equator: The meridians are projected from the center of the sphere, and the parallels are projected with lines parallel to the equatorial plane, or orthographically from infinity. Modifications such as Behrmann's, described above, are perspective projections onto a secant cylinder. For oblique and transverse aspects, the projection may be perspectively cast on a cylinder tangent or secant at an oblique angle, or centered on a meridian.

There is no distortion of area anywhere on the projections, and no distortion of scale and shape at the standard parallels of the normal aspect, or at the standard lines of the oblique or transverse aspects. There is extreme shape and scale distortion 90° from the central line, or at the poles on the normal aspect. These are the points which have infinite area and linear scale on the various aspects of the Mercator projection. This distortion, even on the modifications described above, is so great that there has been little use of any of the forms for world maps by professional cartographers, and many of them have strongly criticized the intensive promotion in the noncartographic community which has accompanied the presentation of one of the recent modifications.

The meridians and parallels of the transverse and oblique aspects which are straight or curved on the Mercator projection are straight or curved, respectively, on the Cylindrical Equal-Area, except that the curves are differently shaped.

In spite of the shape distortion in some portions of a world map, the projection is well suited for equal-area mapping of regions which are predominantly north-south in extent, or which have an oblique central line, or which lie near the Equator. This is true in the same sense that for mid-latitude regions which extend predominantly east-west, the Albers Equal-Area Conic projection is recommended for equal-area mapping. Actually, the normal Cylindrical Equal-Area is the limiting form of the Albers when the Equator or two parallels symmetrical about the Equator are made standard. If such regions to be mapped are smaller than the United States, the ellipsoidal form should be considered.

FORMULAS FOR THE SPHERE

The geometric construction of the Cylindrical Equal-Area projection has been described above. The forward formulas for the normal aspect are as follows, given R, ϕ_s, λ_0, ϕ, and λ, to find x and y (see p. 278 for numerical examples):

$$x = R (\lambda - \lambda_0) \cos \phi_s \qquad (10-1)$$
$$y = R \sin \phi / \cos \phi_s \qquad (10-2)$$
$$h = \cos \phi / \cos \phi_s \qquad (10-2a)$$
$$k = 1/h \qquad (10-2b)$$

where ϕ_s is the standard parallel (N. or S.), or the Equator in Lambert's original form. The X axis lies along the Equator, x increasing easterly. The Y axis lies along the central meridian λ_0, y increasing northerly, and the origin is ($\phi = 0^0$, λ_0). If ($\lambda - \lambda_0$) lies outside the range $\pm 180°$, $360°$ should be added or subtracted so that it will fall inside the range.

For the transverse aspect, given h_0 instead of ϕ_s,

$$x = (R/h_0) \cos \phi \sin (\lambda - \lambda_0) \qquad (10-3)$$
$$y = R h_0 \{\arctan [\tan \phi / \cos (\lambda - \lambda_0)] - \phi_0\} \qquad (8-3)$$

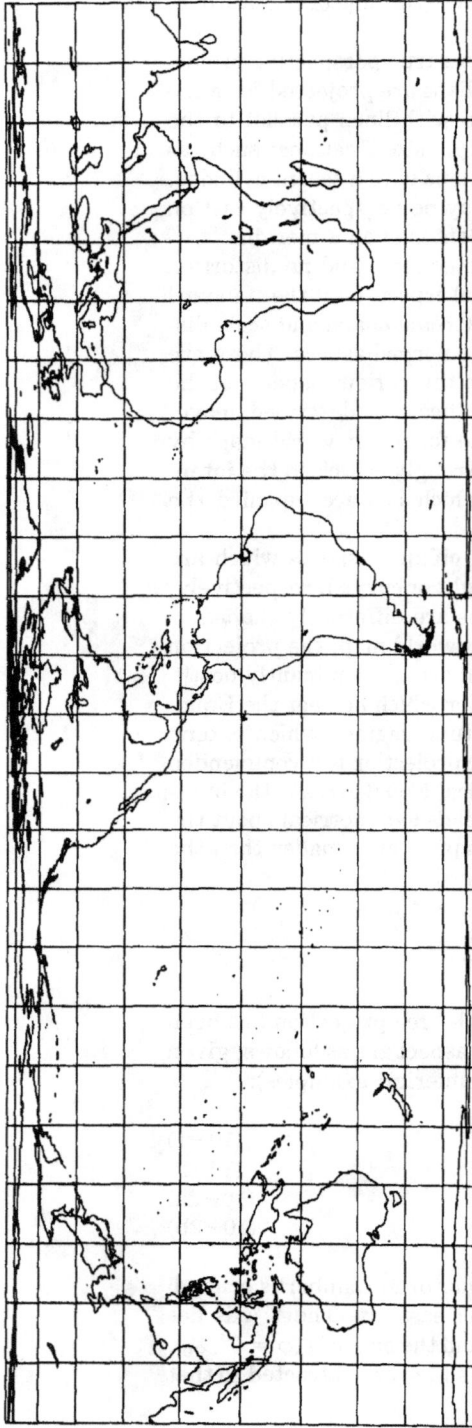

FIGURE 14.—Lambert Cylindrical Equal-Area projection. Standard parallel is the Equator. Seldom used in this form, but is suitable for equal-area strips near the Equator.

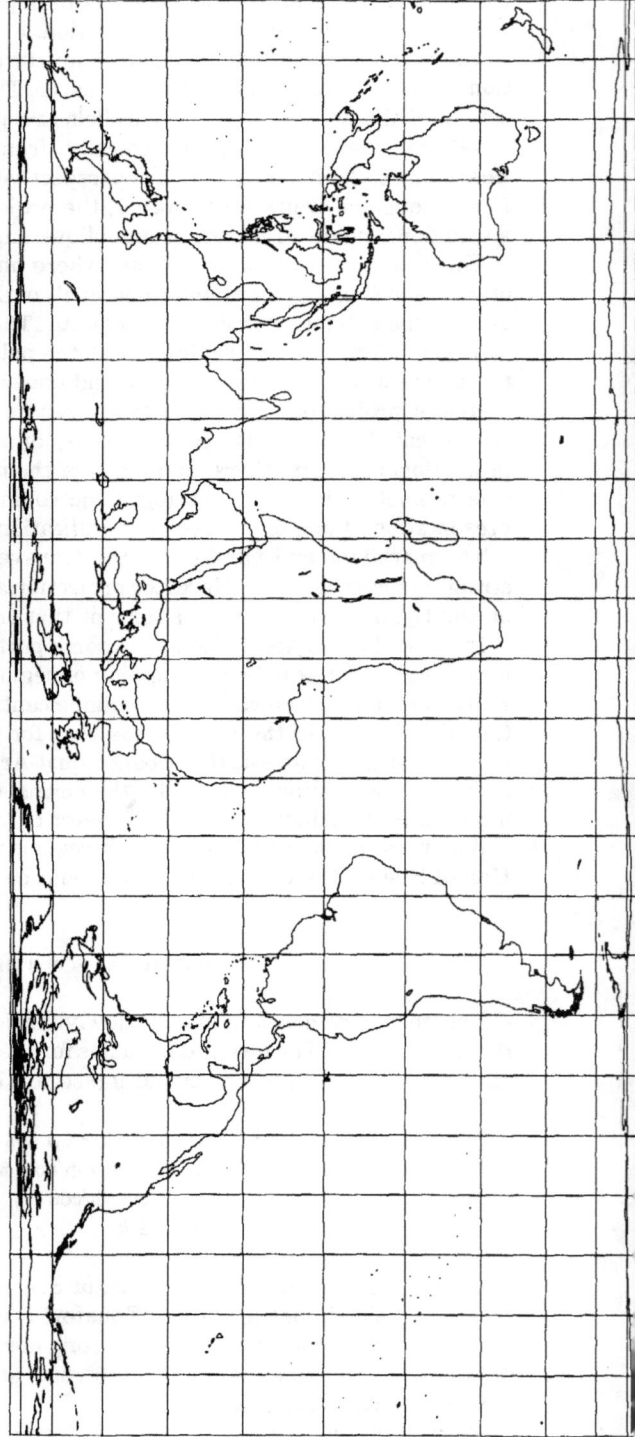

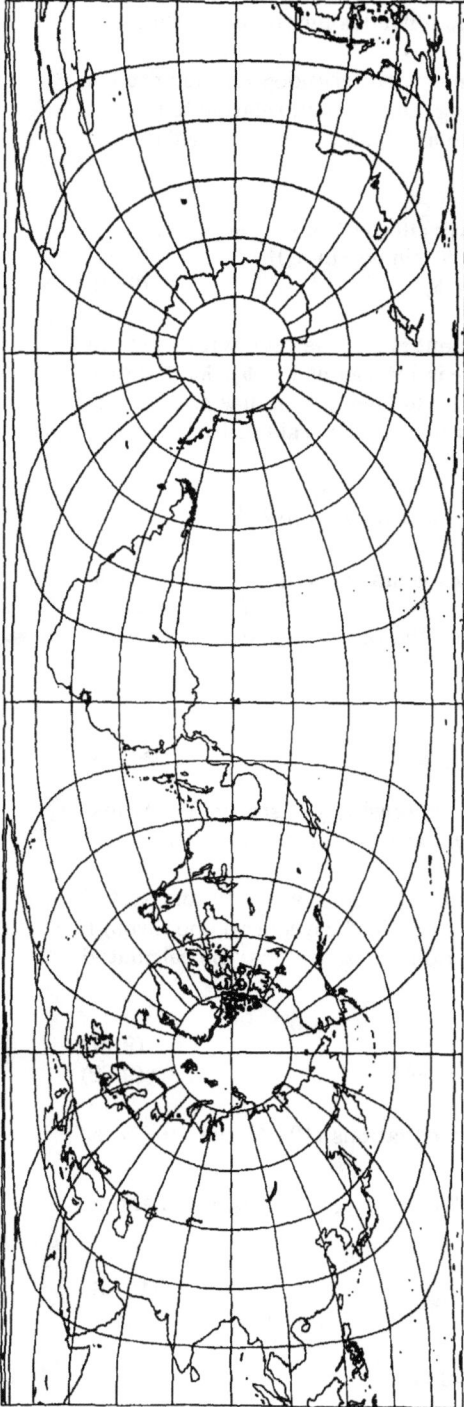

FIGURE 16.—Transverse Cylindrical Equal-Area projection. The central meridian, long. 90° W., as well as long. 90° E., coincides with the equator of the base projection.

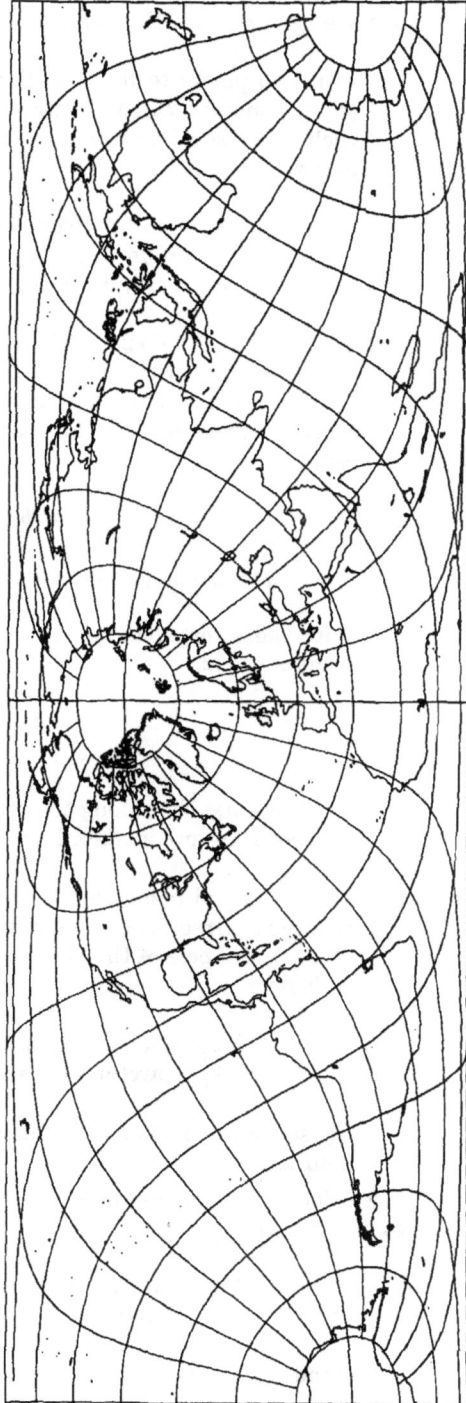

FIGURE 17.—Oblique Cylindrical Equal-Area projection, with central oblique great circle inclined 60° to the Earth's Equator. No distortion along this central line.

where h_0 is the scale factor (normally 1.0) along the central meridian λ_0. The origin of the coordinates is at (ϕ_0, λ_0). The Y axis lies along the central meridian λ_0, y increasing northerly, and the X axis is perpendicular, through ϕ_0 at λ_0, x increasing easterly.

For the oblique aspect, the alternatives used for the Oblique Mercator projection are used here, with modification only in the formulas for the y coordinates:

1. Given two points to lie upon the central line, with latitudes and longitudes (ϕ_1, λ_1) and (ϕ_2, λ_2), and longitude increasing easterly and relative to Greenwich, the pole of the oblique transformation at (ϕ_p, λ_p) may be calculated as follows:

$$\lambda_p = \arctan\ [(\cos \phi_1 \sin \phi_2 \cos \lambda_1 - \sin \phi_1 \cos \phi_2 \cos \lambda_2)/$$
$$(\sin \phi_1 \cos \phi_2 \sin \lambda_2 - \cos \phi_1 \sin \phi_2 \sin \lambda_1)] \qquad (9-1)$$
$$\phi_p = \arctan\ [- \cos\ (\lambda_p - \lambda_1)/\tan \phi_1] \qquad (9-2)$$

The Fortran ATAN2 function or its equivalent should be used with equation (9–1), but not with (9–2). The other pole is located at ($- \phi_p$, $\lambda_p \pm 180°$). Using the positive (northern) value of ϕ_p, the following formulas provide the rectangular coordinates for point (ϕ, λ), with h_0 as the scale factor along the central line:

$$x = Rh_0 \arctan\ [[\tan \phi \cos \phi_p + \sin \phi_p \sin\ (\lambda - \lambda_0)]\ /$$
$$\cos\ (\lambda - \lambda_0)] \qquad (10-4)$$
$$y = (R/h_0)\ [\sin \phi_p \sin \phi - \cos \phi_p \cos \phi \sin\ (\lambda - \lambda_0)] \qquad (10-5)$$

With these formulas for the oblique aspect, the origin of rectangular coordinates lies at

$$\phi_0 = 0$$
$$\lambda_0 = \lambda_p + \pi/2 \qquad (9-6a)$$

and the X axis lies along the central line, x increasingly easterly. The transformed poles are straight lines at $y = R$ and are as long as the central line.

2. Given a central point (ϕ_z, λ_z) with longitude increasing easterly and stated relative to Greenwich, and azimuth γ east of north of the central line through (ϕ_z, λ_z), the pole of the oblique transformation at (ϕ_p, λ_p) may be calculated as follows:

$$\phi_p = \arcsin\ (\cos \phi_z \sin \gamma) \qquad (9-7)$$
$$\lambda_p = \arctan\ [-\cos \gamma/(-\sin \phi_z \sin \gamma)] + \lambda_z \qquad (9-8)$$

These values of ϕ_p and λ_p may be used in equations (10–4) and (10–5) as before.

For the inverse formulas, first for the normal aspect, given R, ϕ_s, λ_0, x, and y, to find ϕ and λ:

$$\phi = \arcsin\ [(y/R) \cos \phi_s] \qquad (10-6)$$
$$\lambda = x/(R \cos \phi_s) + \lambda_0 \qquad (10-7)$$

For the transverse aspect, given h_0 instead of ϕ_s,

$$\phi = \arcsin\ \{[1 - (h_0\ x/R)^2]^{1/2} \sin D\} \qquad (10-8)$$
$$\lambda = \lambda_0 + \arctan\ \{(h_0\ x/R)/[[1 - (h_0\ x/R)^2]^{1/2} \cos D]\} \qquad (10-9)$$

where

$$D = y / (Rh_0) + \phi_0, \text{ using radians} \tag{10-10}$$

For the oblique aspect, equations (9−1) and (9−2) or (9−7) and (9−8) must first be used to establish the pole of the oblique transformation, if it is not known already. Then

$$\phi = \arcsin \{(yh_0/R) \sin \phi_p + [1-(yh_0/R)^2]^{1/2} \cos \phi_p \sin [x/(Rh_0)]\} \tag{10-11}$$

$$\lambda = \lambda_0 + \arctan \{[[1-(yh_0/R)^2]^{1/2} \sin \phi_p \sin [x/(Rh_0)] - (yh_0/R) \cos \phi_p]/[[1-(yh_0/R)^2]^{1/2} \cos [x/(Rh_0)]]\} \tag{10-12}$$

Note that the above equations for the oblique aspect may be used for the transverse aspect, letting $\phi_p = 0^0$, except that the axes are rotated 90°.

FORMULAS FOR THE ELLIPSOID

In the following formulas, the ellipsoid is transformed onto the authalic sphere, but the scale along the desired central line is made constant by variably compressing the scale along this central line to match that along the same path on the ellipsoid. To retain correct area, the distances perpendicular to the central line are increased by the same ratio. For the oblique aspect, the central line is not a geodesic, but is instead an oblique great circle on the authalic sphere.

For the forward formulas using the normal aspect, given a, e, ϕ_s, λ_0, ϕ, and λ, to find x and y (see p. 281 for numerical examples), the equations are given in the order of computation:

$$k_0 = \cos \phi_s/(1-e^2 \sin^2 \phi_s)^{1/2} \tag{10-13}$$

$$q = (1-e^2) \{\sin \phi/(1-e^2 \sin^2 \phi) - [1/(2e)] \ln [(1-e \sin \phi)/(1+e \sin \phi)]\} \tag{3-12}$$

$$x = a k_0 (\lambda-\lambda_0) \tag{10-14}$$

$$y = a q/(2k_0) \tag{10-15}$$

For the transverse aspect, the subsequent formulas for the oblique aspect may be used, but the following are simpler for the transverse alone. Given a, e, h_0, λ_0, ϕ_0, ϕ, and λ, to find x and y, first q is calculated from ϕ using equation (3−12) above. Then

$$\beta = \arcsin (q/q_p) \tag{3-11}$$

where β is the authalic latitude corresponding to ϕ, and q_p is found as q from equation (3−12) for a ϕ of 90°.

$$\beta_c = \arctan [\tan \beta/\cos (\lambda-\lambda_0)] \tag{10-16}$$

$$q_c = q_p \sin \beta_c \tag{10-17}$$

$$\phi_c = \phi_c + \frac{(1 - e^2 \sin^2 \phi_c)^2}{2 \cos \phi_c} \left[\frac{q_c}{1 - e^2} - \frac{\sin \phi_c}{1 - e^2 \sin^2 \phi_c} + \frac{1}{2e} \ln \left(\frac{1 - e \sin \phi_c}{1 + e \sin \phi_c} \right) \right] \tag{3-16}$$

Equation (3−16) requires iteration by successive substitution, using $\arcsin (q_c/2)$ as the first trial ϕ_c on the right side, calculating ϕ_c on the left side, substituting this new ϕ_c on the right side, etc., until the change in ϕ_c is negligible. This does not converge if $\beta_c = \pm 90°$, but then $\phi_c = \beta_c$.

$x = a \cos \beta \cos \phi_c \sin (\lambda - \lambda_0)/[h_0 \cos \beta_c (1 - e^2 \sin^2 \phi_c)^{1/2}]$ (10–18)

$M_c = a [(1 - e^2/4 - 3e^4/64 - 5e^6/256 - \ldots)\phi_c$
$\qquad - (3e^2/8 + 3e^4/32 + 45e^6/1024 + \ldots) \sin 2\phi_c$
$\qquad + (15e^4/256 + 45e^6/1024 + \ldots) \sin 4\phi_c$
$\qquad - (35e^6/3072 + \ldots) \sin 6\phi_c + \ldots]$ (3–21)

$y = h_0 (M_c - M_0)$ (10–19)

where h_0 is the scale factor along the central meridian λ_0, and β_c and ϕ_c are authalic and geodetic "footpoint" latitudes, respectively, with the same y value at the central meridian as the point (ϕ, λ). Constant M_0 is the value of M_c calculated from (3–21) with latitude of origin ϕ_0 in place of ϕ_c. To avoid iteration, equations (10–17) and (3–16) may be replaced with the following series:

$\phi_c = \beta_c + (e^2/3 + 31e^4/180 + 517e^6/5040 + \ldots) \sin 2\beta_c$
$\qquad + (23e^4/360 + 251e^6/3780 + \ldots) \sin 4\beta_c$
$\qquad + (761e^6/45360 + \ldots) \sin 6\beta_c + \ldots$ (3–18)

For the oblique aspect, the location of the pole (ϕ_p, λ_p) may be given, or it may be computed as described under the section on formulas for the sphere above. Points ϕ_1, ϕ_2, ϕ_p and ϕ_z, however, are replaced in equations (9–1), (9–2), (9–7) and (9–8) with β_1, β_2, β_p and β_z, respectively, and β_p is finally converted to ϕ_p, using equations (10–17) and (3–16), or just (3–18), and subscripts p instead of c.

If the ellipsoid is either the Clarke 1866 or the International, Fourier constants may be taken from table 13. If it is a different ellipsoid, coefficients should be calculated as described after these formulas. They may be converted to the specific coefficients for the pole in use as follows:

$B = b + a_2 \cos 2\phi_p + a_4 \cos 4\phi_p + a_6 \cos 6\phi_p + \ldots$ (10–20)

$A_n = b_n + a'_{n2} \cos 2\phi_p + a'_{n4} \cos 4\phi_p + a'_{n6} \cos 6\phi_p + \ldots$ (10–21)

where

$n = 2$ and 4.

From ϕ, β is determined using equations (3–12) and (3–11) above, and, if β_p was not obtained earlier, it is calculated by substituting ϕ_p for ϕ in (3–12) and obtaining β_p from (3–11). Then,

$\lambda' = \arctan \{[\cos \beta_p \sin \beta - \sin \beta_p \cos \beta \cos (\lambda - \lambda_p)]/$
$\qquad\qquad [\cos \beta \sin (\lambda - \lambda_p)]\}$ (10–22)

$x = ah_0 [B\lambda' + A_2 \sin 2\lambda' + A_4 \sin 4\lambda' + A_6 \sin 6\lambda' + \ldots]$ (10–23)

$F = B + 2A_2 \cos 2\lambda' + 4A_4 \cos 4\lambda' + 6A_6 \cos 6\lambda' + \ldots$ (10–24)

$y = (aq_p/2)[\sin \beta_p \sin \beta + \cos \beta_p \cos \beta \cos (\lambda - \lambda_p)]/(h_0 F)$ (10–25)

The axes are as stated for the corresponding aspect of the spherical form. For more efficient computation of series (10–23) and (10–24) see p. 19.

For the inverse formulas for the ellipsoid, the normal aspect will be discussed first. Given a, e, ϕ_s, λ_0, x, and y, to find ϕ and λ (see p. 284 for numerical examples), k_0 is determined from (10–13), and

$\beta = \arcsin [2yk_0/(aq_p)]$ (10–26)

where q_p is found from (3–12), using 90° for ϕ, then ϕ is found from β using (10–17) and (3–16), or just (3–18), without subscripts, these equations being listed under the forward equations above.

TABLE 13.—*Fourier coefficients for oblique and transverse Cylindrical Equal-Area projection for the ellipsoid*

General coefficients:

Coefficient	Clarke 1866 Ellipsoid	International Ellipsoid
b	0.9991507126	0.9991565046
a_2	−0.0008471537	−0.0008413907
a_4	0.0000021283	0.0000020994
a_6	−0.0000000054	−0.0000000053
b_2	−0.0001412090	−0.0001402483
a'_{22}	−0.0001411258	−0.0001401661
a'_{24}	0.0000000839	0.0000000827
a'_{26}	0.0000000006	0.0000000006
b_4	−0.0000000435	−0.0000000429
a'_{42}	−0.0000000579	−0.0000000571
a'_{44}	−0.0000000144	−0.0000000142
a'_{46}	0.0000000000	0.0000000000

Coefficients for specific pole latitudes (Clarke 1866 ellipsoid):

ϕ_p	B	A_2	A_4
0°	0.9983056818	−0.0002822502	−0.0000001158
15	0.9984181201	−0.0002633856	−0.0000001008
30	0.9987260769	−0.0002118145	−0.0000000652
45	0.9991485842	−0.0001412929	−0.0000000290
60	0.9995732199	−0.0000706875	−0.0000000073
75	0.9998854334	−0.0000189486	−0.0000000005
90	1.0	0.0	0.0

Coefficients for specific pole latitudes (International ellipsoid):

ϕ_p	B	A_2	A_4
0°	0.9983172080	−0.0002803311	−0.0000001142
15	0.9984288886	−0.0002615944	−0.0000000995
30	0.9987347648	−0.0002103733	−0.0000000644
45	0.9991544051	−0.0001403310	−0.0000000287
60	0.9995761449	−0.0000702060	−0.0000000072
75	0.9998862200	−0.0000188195	−0.0000000005
90	1.0	0.0	0.0

ϕ_p = latitude of pole of oblique aspect (0° for transverse, 90° for normal).
B, A_n, b, etc. = Fourier coefficients (see text for use).
Note: B is used with λ' in radians. A_6 = −0.0000000001 for ϕ_p = 0° to 20°, but is zero to ten places at higher values of ϕ_p.
Clarke 1866 ellipsoid: semimajor axis a = 6378206.4 m; eccentricity squared e^2 = 0.006768658.
International ellipsoid: a = 6378388 m; e^2 = 0.006722670.

$$\lambda = \lambda_0 + x/(a\ k_0) \qquad (10-27)$$

For the transverse aspect, given a, e, h_0, λ_0, x, and y, to find ϕ and λ:

$$M_c = M_0 + y/h_0 \qquad (10-28)$$

where M_0 is found from ϕ_0 using (3−21) and changing subscripts c to o.

$$\mu_c = M_c/[a(1-e^2/4-3e^4/64-5e^6/256-\ldots)] \qquad (7-19)$$
$$e_1 = [1-(1-e^2)^{1/2}]/[1+(1-e^2)^{1/2}] \qquad (3-24)$$
$$\phi_c = \mu_c + (3e_1/2-27e_1^3/32+\ldots)\sin 2\mu_c + (21e_1^2/16-55e_1^4/32+\ldots)$$
$$\sin 4\mu_c + (151e_1^3/96-\ldots)\sin 6\mu_c + (1097e_1^4/512-\ldots)$$
$$\sin 8\mu_c + \ldots \qquad (3-26)$$

Authalic latitude β_c is determined for ϕ_c using equations (3−12) and (3−11), adding subscripts c to both β and ϕ.

$$\beta' = -\arcsin [h_0\, x \cos \beta_c\, (1-e^2 \sin^2 \phi_c)^{1/2}/(a \cos \phi_c)] \qquad (10-29)$$
$$\beta = \arcsin (\cos \beta' \sin \beta_c) \qquad (10-30)$$
$$\lambda = \lambda_0 - \arctan (\tan \beta'/\cos \beta_c) \qquad (10-31)$$

Latitude ϕ is found from β using (10−17) and (3−16), or just (3−18), all without subscripts c.

For the oblique aspect, given a, e, h_0, ϕ_p, λ_p, x, and y, to find ϕ and λ, Fourier coefficients are determined as described above for the forward oblique ellipsoidal formulas, while the pole location (ϕ_p, λ_p) may be determined if not provided, as described for the forward oblique spherical formulas, and q_p is found from (3−12) using 90° for ϕ. From x, λ' is determined from an iterative inverse of (10−23):

$$\lambda' = [x/(ah_0)-A_2 \sin 2\lambda'-A_4 \sin 4\lambda'-A_6 \sin 6\lambda'- \ldots\ ]/B \qquad (10-32)$$

Using a first trial $\lambda' = x/(ah_0B)$, λ' may be found by successive substitution of trial values into the right side of this equation and solving for a new λ' until the change in λ' is negligible.

Equation (10−24) above is used to find F from λ'. Then,

$$\beta' = \arcsin [2Fh_0y/(aq_p)] \qquad (10-33)$$
$$\beta = \arcsin (\sin \beta_p \sin \beta' + \cos \beta_p \cos \beta' \sin \lambda') \qquad (10-34)$$
$$\lambda = \lambda_p + \arctan [\cos \beta' \cos \lambda'/(\cos \beta_p \sin \beta'-\sin \beta_p \cos \beta' \sin \lambda')] \qquad (10-35)$$

As before, ϕ is found from β using (10−17) and (3−16), or just (3−18), all without subscripts c.

For the determination of Fourier coefficients, if they are not already provided, equation (10−23) above is equivalent to the following equation which requires numerical integration:

$$x/(a\, h_0) = \int_0^{\lambda'} F\, d\lambda' \qquad (10-36)$$

where

$$F = \{\sin^2 \beta_p \cos^2 \phi_c/[(1-e^2 \sin^2 \phi_c) \cos^4 \beta_c]$$
$$+ (1-e^2 \sin^2 \phi_c)\, q_p{}^2 \cos^2 \beta_p \cos^2 \lambda'/(4 \cos^2 \phi_c)\}^{1/2} \qquad (10-37)$$

In order to compute coefficients B and A_n in (10−23),

$$B = (2/\pi)\int_0^{\pi/2} F\, d\lambda' \qquad (10-38)$$
$$A_n = [4/(\pi n)] \int_0^{\pi/2} F \cos n\lambda'\, d\lambda' \qquad (10-39)$$

where n is 2, 4, and 6, successively. To compute coefficients which apply regardless of the value of ϕ_p, equations (10−38) and (10−39) may be rewritten as equations (10−20) and (10−21), where

$$b = (2/\pi) \int_0^{\pi/2} B\, d\phi_p \qquad (10-40)$$
$$a_n = (4/\pi) \int_0^{\pi/2} B \cos n\phi_p\, d\phi_p \qquad (10-41)$$
$$b_n = (2/\pi) \int_0^{\pi/2} A_n\, d\phi_p \qquad (10-42)$$
$$a'_{nm} = (4/\pi) \int_0^{\pi/2} A_n \cos m\, \phi_p\, d\phi_p \qquad (10-43)$$

and n has the values 2 and 4, while $m = 2$, 4, and 6. To determine the coefficients from (10−40) through (10−43), double numerical integration is involved, but this involves a relatively modest computer program: Choosing an interval of 9° (sufficient for 10-place accuracy) in both ϕ_p and λ', and starting with both ϕ_p and λ' at 0°, F is calculated from (10−37) as described below for each 9° of λ' from 0° to 90°, and the various values of F summed in accordance with Simpson's rule as applied to equations (10−38) and (10−39). Thus B, A_2, A_4, and A_6 are computed for $\phi_p = 0°$. Similarly, the constants B and A_n are computed for each 9° of ϕ_p to and including 90°, and the various values are summed by applying Simpson's rule to (10−40) through (10−43), to obtain b, a_2, etc.

To compute F from equation (10−37) for a given λ', first β_p is found from ϕ_p using (3−12) and (3−11), subscripting ϕ and β with p. Then,

$$\beta_c = \arcsin (\cos \beta_p \sin \lambda') \qquad (10-44)$$

Now ϕ_c is found from β_c using (10−17) and (3−16) or just (3−18). All variables for (10−37) are now known, except that it is indeterminate if $\phi_p = 0°$ at the same time that $\lambda' = 90°$. In that case, $F = (q_p/2)^{1/2}$.

11. MILLER CYLINDRICAL PROJECTION

SUMMARY

- Neither equal-area nor conformal.
- Used only in spherical form.
- Cylindrical.
- Meridians and parallels are straight lines, intersecting at right angles.
- Meridians are equidistant; parallels spaced farther apart away from Equator.
- Poles shown as lines.
- Compromise between Mercator and other cylindrical projections.
- Used for world maps.
- Presented by Miller in 1942.

HISTORY AND FEATURES

The need for a world map which avoids some of the scale exaggeration of the Mercator projection has led to some commonly used cylindrical modifications, as well as to other modifications which are not cylindrical. The earliest common cylindrical example was developed by James Gall of Edinburgh about 1855 (Gall, 1885, p. 119–123). His meridians are equally spaced, but the parallels are spaced at increasing intervals away from the Equator. The parallels of latitude are actually projected onto a cylinder wrapped about the sphere, but cutting it at lats. 45° N. and S., the point of perspective being a point on the Equator opposite the meridian being projected. It is used in several British atlases, but seldom in the United States. The Gall projection is neither conformal nor equal-area, but has a blend of various features. Unlike the Mercator, the Gall shows the poles as lines running across the top and bottom of the map.

What might be called the American version of the Gall projection is the Miller Cylindrical projection (fig. 18), presented in 1942 by Osborn Maitland Miller (1897–1979) of the American Geographical Society, New York (Miller, 1942). Born in Perth, Scotland, and educated in Scotland and England, Miller came to the Society in 1922. There he developed several improved surveying and mapping techniques. An expert in aerial photography, he developed techniques for converting high-altitude photographs into maps. He led or joined several expeditions of explorers and advised leaders of others. He retired in 1968, having been best known to cartographers for several map projections, including the Bipolar Oblique Conic Conformal, the Oblated Stereographic, and especially his Cylindrical projection.

Miller had been asked by S. Whittemore Boggs, Geographer of the U.S. Department of State, to study further alternatives to the Mercator, the Gall, and other cylindrical world maps. In his presentation, Miller listed four proposals, but the one he preferred, and the one used, is a fairly simple mathematical modification of the Mercator projection. Like the Gall, it shows visible straight lines for the poles, increasingly spaced parallels away from the Equator, equidistant meridians, and is not equal-area, equidistant along meridians, nor conformal. While the standard parallels, or lines true to scale and free of distortion, on the Gall are at lats. 45° N. and S., on the Miller only the Equator is standard. Unlike the Gall, the Miller is not a perspective projection.

The Miller Cylindrical projection is used for world maps and in several atlases, including the *National Atlas of the United States* (USGS, 1970, p. 330–331).

As Miller (1942) stated,

the practical problem considered here is to find a system of spacing the parallels of latitude such that an acceptable balance is reached between shape and area distortion. By an "acceptable" balance is meant one which to the uncritical eye does not obviously depart from the familiar shapes of the land

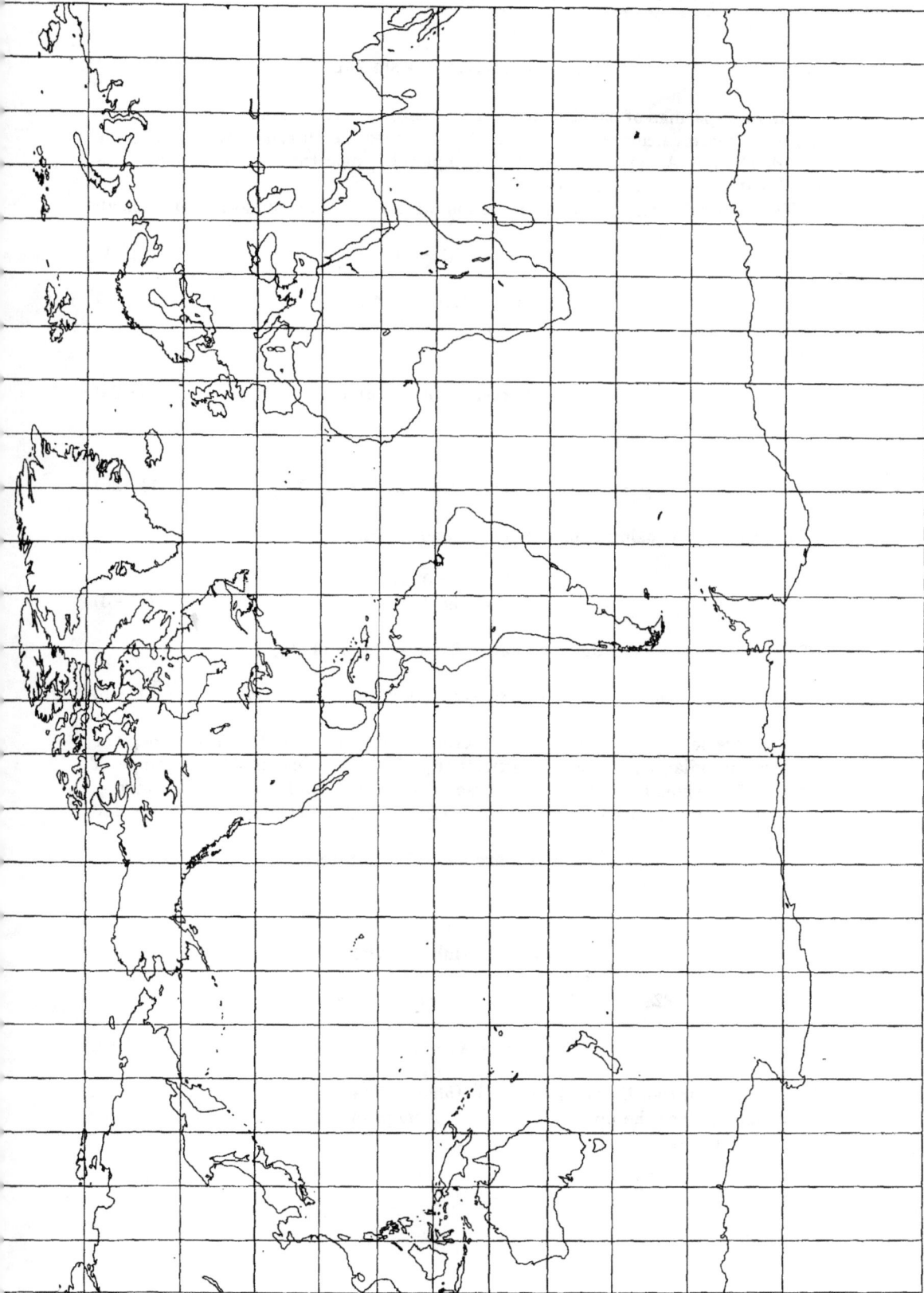

FIGURE 18.—The Miller Cylindrical projection. A projection resembling the Mercator, but having less relative area distortion in polar regions. Neither conformal nor equal-area.

areas as depicted by the Mercator projection but which reduces areal distortion as far as possible under these conditions * * *. After some experimenting, the [Modified Mercator (b)] was judged to be the most suitable for Mr. Boggs's purpose * * *.

FORMULAS FOR THE SPHERE

Miller's spacings of parallels from the Equator are the same as if the Mercator spacings were calculated for 0.8 times the respective latitudes, with the result divided by 0.8. As a result, the spacing of parallels near the Equator is very close to the Mercator arrangement.

The forward formulas, then, are as follows (see p. 287 for numerical examples):

$$x = R(\lambda - \lambda_0) \tag{11-1}$$

$$y = R[\ln \tan (\pi/4 + 0.4\phi)]/0.8 \tag{11-2}$$

or

$$y = R[\text{arcsinh} (\tan 0.8\phi)]/0.8 \tag{11-2a}$$

or

$$y = (R/1.6) \ln ((1 + \sin 0.8\phi)/(1 - \sin 0.8\phi)) \tag{11-2b}$$

The scale factor, using equations (4-2) and (4-3),

$$h = \sec 0.8\phi \tag{11-3}$$
$$k = \sec \phi \tag{11-4}$$

The maximum angular deformation ω, from equation (4-9),

$$\sin \tfrac{1}{2}\omega = (\cos 0.8\phi - \cos \phi)/(\cos 0.8\phi + \cos \phi) \tag{11-5}$$

The X axis lies along the Equator, x increasing easterly. The Y axis lies along the central meridian λ_0, y increasing northerly. If $(\lambda - \lambda_0)$ lies outside the range of $\pm 180°$, $360°$ should be added or subtracted so that it will fall inside the range.

The *inverse* equations are easily derived from equations (11-1) through (11-2a):

$$\phi = 2.5 \arctan e^{(0.8y/R)} - 5\pi/8 \tag{11-6}$$

or

$$\phi = \arctan [\sinh (0.8y/R)]/0.8 \tag{11-6a}$$

where e is 2.71828 . . . , the base of natural logarithms.

$$\lambda = \lambda_0 + x/R \tag{11-7}$$

Rectangular coordinates are given in table 14. There is no basis for an ellipsoidal equivalent, since the projection is used for maps of the entire Earth and not for local areas at large scale.

TABLE 14.—*Miller Cylindrical projection: Rectangular coordinates*

[Radius of sphere=1.0]

ϕ	y	h	k	ω
90° _____	2.30341	3.23607	Infinite	180.00°
85 _____	2.04742	2.66947	11.47371	77.00
80 _____	1.83239	2.28117	5.75877	51.26
75 _____	1.64620	2.00000	3.86370	37.06
70 _____	1.48131	1.78829	2.92380	27.89
65 _____	1.33270	1.62427	2.36620	21.43
60 _____	1.19683	1.49448	2.00000	16.64
55 _____	1.07113	1.39016	1.74345	12.95
50 _____	.95364	1.30541	1.55572	10.04
45 _____	.84284	1.23607	1.41421	7.71
40 _____	.73754	1.17918	1.30541	5.82
35 _____	.63674	1.13257	1.22077	4.30
30 _____	.53962	1.09464	1.15470	3.06
25 _____	.44547	1.06418	1.10338	2.07
20 _____	.35369	1.04030	1.06418	1.30
15 _____	.26373	1.02234	1.03528	.72
10 _____	.17510	1.00983	1.01543	.32
5 _____	.08734	1.00244	1.00382	.08
0 _____	.00000	1.00000	1.00000	.00

x _____ $0.017453 (\lambda° - \lambda_0°)$

Note: $x, y =$ rectangular coordinates.

$\phi =$ geodetic latitude.

$(\lambda° - \lambda_0°) =$ geodetic longitude, measured east from origin in degrees.

$h =$ scale factor along meridian.

$k =$ scale factor along parallel.

$\omega =$ maximum angular deformation, degrees.

Origin of coordinates at intersection of Equator with λ_0, X axis increases east, Y axis increases north. For southern (negative) ϕ, reverse sign of y.

12. EQUIDISTANT CYLINDRICAL PROJECTION

SUMMARY

- Cylindrical.
- Neither equal-area nor conformal.
- Meridians and parallels are equidistant straight lines, intersecting at right angles.
- Poles shown as lines.
- Used for world or regional maps.
- Very simple construction.
- Used only in spherical form.
- Presented by Eratosthenes (B.C.) or Marinus (A.D. 100).

HISTORY AND FEATURES

While the Equidistant Cylindrical projection has received limited use by the USGS and generally has limited value, it is probably the simplest of all map projections to construct and one of the oldest. The meridians and parallels are all equidistant straight parallel lines, the two sets crossing at right angles.

The projection originated probably with Eratosthenes (275?–195? B.C.), the scientist and geographer noted for his fairly accurate measure of the size of the Earth. Claudius Ptolemy credited Marinus of Tyre with the invention about A.D. 100 stating that, while Marinus had previously evaluated existing projections, the latter had chosen "a manner of representing the distances which gives the worst results of all." Only the parallel of Rhodes (lat. 36°N.) was made true to scale on the world map, which meant that the meridians were spaced at about four-fifths of the spacing of the parallels for the same degree interval (Keuning, 1955, p. 13).

Ptolemy approved the use of the projection for maps of smaller areas, however, with spacing of meridians to provide correct scale along the central parallel. All the Greek manuscript maps for the *Geographia*, dating from the 13th century, use the Ptolemy modification. It was used for some maps until the 18th century, but is now used primarily for a few maps on which distortion is considered less important than the ease of displaying special information. The projection is given a variety of names such as Equidistant Cylindrical, Rectangular, La Carte Parallélogrammatique, Die Rechteckige Plattkarte, and Equirectangular (Steers, 1970, p. 135–136). It was called the projection of Marinus by Nordenskiöld (1889).

If the Equator is made the standard parallel, true to scale and free of distortion, the meridians are spaced at the same distances as the parallels, and the graticule appears square. This form is often called the Plate Carrée or the Simple Cylindrical projection.

The USGS uses the Equidistant Cylindrical projection for index maps of the conterminous United States, with insets of Alaska, Hawaii, and various islands on the same projection. One is entitled "Topographic Mapping Status and Progress of Operations (7½- and 15-minute series)," at an approximate scale of 1:5,000,000. Another shows the status of intermediate-scale quadrangle mapping. Neither the scale nor the projection is marked, to avoid implying that the maps are suitable for normal geographic information. Meridian spacing is about four-fifths of the spacing of parallels, by coincidence the same as that chosen by Marinus. The Alaska inset is shown at about half the scale and with a change in spacing ratios. Individual States are shown by the USGS on other index maps using the same projection and spacing ratio to indicate the status of aerial photography.

The projection was chosen largely for ease in computerized plotting. While the boundaries on the base map may be as difficult to plot on this projection as on the others, the base map needs to be prepared only once. Overlays of digital information, which may then be printed in straight lines, may be easily updated without the use of cartographic and photographic skills. The 4:5 spacing ratio is a convenience based on computer line and character spacing and is not an attempt to achieve a particular standard parallel, which happens to fall near lat. 37° N.

FORMULAS FOR THE SPHERE

The formulas for rectangular coordinates are almost as simple to use as the geometric construction. Given R, λ_0, ϕ_1, λ, and ϕ for the forward solution, x and y are found thus:

$$x = R\,(\lambda - \lambda_0)\,\cos\phi_1 \qquad (12-1)$$
$$y = R\phi \qquad (12-2)$$
$$h = 1 \qquad (12-3)$$
$$k = \cos\phi_1/\cos\phi \qquad (12-4)$$

The X axis coincides with the Equator, with x increasing easterly, while the Y axis follows the central meridian λ_0, y increasing northerly. It is necessary to adjust $(\lambda - \lambda_0)$, if it is beyond the range $\pm 180°$, by adding or subtracting 360°. The standard parallel is ϕ_1 (also $-\phi_1$). For the inverse formulas, given R, λ_0, ϕ_1, x, and y, to find ϕ and λ:

$$\phi = y/R \qquad (12-5)$$
$$\lambda = \lambda_0 + x/(R\cos\phi_1) \qquad (12-6)$$

Numerical examples are omitted in the appendix, due to simplicity. It must be remembered, as usual, that angles above are given in radians.

13. CASSINI PROJECTION

SUMMARY

- Cylindrical.
- Neither equal-area nor conformal.
- Central meridian, each meridian 90° from central meridian, and Equator are straight lines.
- Other meridians and parallels are complex curves.
- Scale is true along central meridian, and along lines perpendicular to central meridian. Scale is constant but not true along lines parallel to central meridian on spherical form, nearly so for ellipsoid.
- Used for topographic mapping formerly in England and currently in a few other countries.
- Devised by C. F. Cassini de Thury in 1745 for the survey of France.

HISTORY

Although the Cassini projection has been largely replaced by the Transverse Mercator, it is still in limited use outside the United States and was one of the major topographic mapping projections until the early 20th century. It was first developed by César François Cassini de Thury (1714–1784), grandson of Jean Dominique Cassini. The latter was an outstanding Italian-born astronomer who changed his given names from Giovanni Domenico after being hired in 1669 for astronomical research in Paris, and soon thereafter to begin the survey of France. Cassini de Thury was the third of four generations involved in this project, the first detailed survey of a nation. In 1745 he devised the projection which, with some modifications, still bears the family name and was used for official topographic maps of France until its replacement by the Bonne projection in 1803.

Instead of showing meridians and parallels (except for the central meridian), Cassini employed a system of squares with rectangular grid coordinates, the meridian through Paris serving as one axis. The scale along this central meridian was made correct according to the surveyed distance, thus approximately correcting for the ellipsoid (Craig, 1882, p. 80; Reignier, 1957, p. 98–99). Mathematical analysis by J. G. von Soldner in the early 19th century led to more accurate ellipsoidal formulas, and the name Cassini-Soldner is often used for the form used in topographic mapping.

FEATURES

The spherical form of the Cassini projection (fig. 19) bears the same relation to the Equidistant Cylindrical or Plate Carrée projection that the spherical Transverse Mercator bears to the regular Mercator. Instead of having the straight meridians and parallels of the Equidistant Cylindrical, the Cassini has complex curves for each, except for the Equator, the central meridian, and each meridian 90° away from the central meridian, all of which are straight.

There is no distortion along the central meridian, if it is maintained at true scale, which is the usual case. If it is given a reduced scale factor, the lines of true scale are two straight lines on the map parallel to and equidistant from the central meridian. There is no distortion along them instead. This alternative is rare enough that it is ignored in the discussion and formulas below.

By making a given point (such as Washington, D.C.) the pole on an oblique Equidistant Cylindrical projection, the bearing and distance from that point to any other on the Earth can be read directly as two rectangular coordinates (Botley, 1951). This provides the same information as the oblique Azimuthal

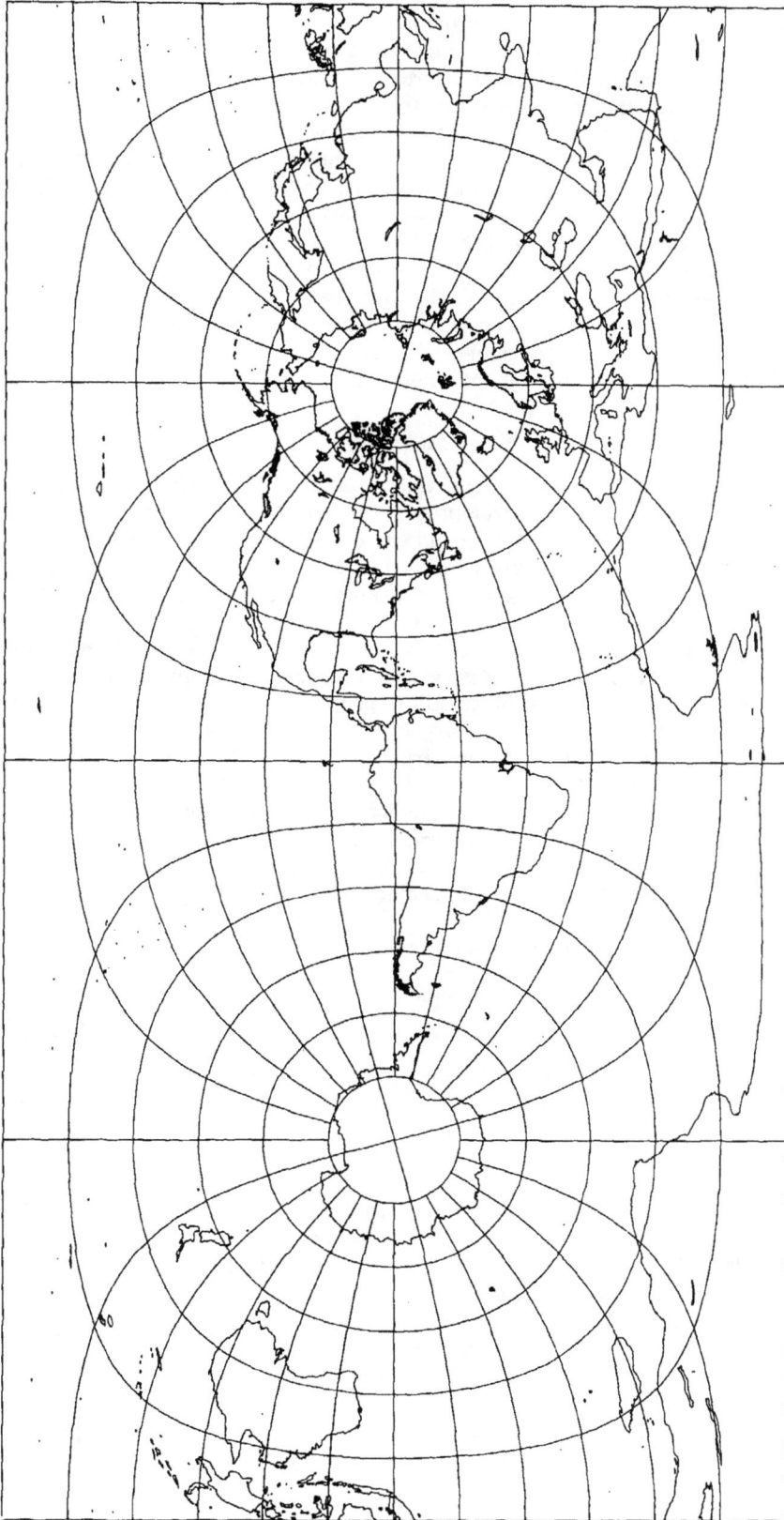

FIGURE 19.—Cassini projection. A transverse Equidistant Cylindrical projection used for large-scale mapping in France, England, and several other countries. Largely replaced by conformal mapping.

Equidistant projection centered on the same point. The oblique cylindrical has the advantage of offering rectangular instead of polar coordinates, but the map is much more distorted near the chosen point.

The scale is correct along the central meridian and also along any straight line perpendicular to the central meridian. It gradually increases in a direction parallel to the central meridian, as the distance from that meridian increases, but the scale is constant along any straight line on the map which is parallel to the central meridian. Therefore, the Cassini is more suitable for regions predominantly north–south in extent, such as Great Britain, than for regions extending in other directions. In this respect, it is also like the Transverse Mercator. The projection is neither equal-area nor conformal, but it has a compromise of both features.

The ellipsoidal form is computed from series which are essentially modifications of those for the ellipsoidal form of the Transverse Mercator and are suitable within only a few degrees to either side of the central meridian. The scale characteristics described above for the spherical form apply to the ellipsoidal form, except that the lines of constant scale paralleling the central meridian are not quite straight.

USAGE

There has been little usage of the spherical version of the Cassini, but the ellipsoidal Cassini-Soldner version was adopted by the Ordnance Survey for the official survey of Great Britain during the second half of the 19th century (Steers, 1970, p. 229). Many of these maps were prepared at a scale of 1:2,500. The Cassini-Soldner was also used for the detailed mapping of many German states during the same period.

Beginning about 1920, the Ordnance Survey began to change to the Transverse Mercator because of the difficulty of measuring scale and direction on the Cassini. Nevertheless, there are several maps still in print which are based on the older projection in Great Britain, and the projection is used in a few other countries such as Cyprus, Czechoslovakia, Denmark, the Federal Republic of Germany, and Malaysia (Clifford J. Mugnier, personal comm., 1985).

A system equivalent to an oblique Equidistant Cylindrical or oblique Cassini projection was used in early coordinate transformations for ERTS (now Landsat) satellite imagery, but it was changed in 1978 to the Hotine Oblique Mercator, and in 1982 to the Space Oblique Mercator projection.

FORMULAS FOR THE SPHERE

For the forward formulas, given R, ϕ_0, λ_0, ϕ, and λ, to find x and y:

$$x = R \arcsin B \tag{13-1}$$
$$y = R \{\arctan [\tan \phi/\cos (\lambda-\lambda_0)] - \phi_0\} \tag{13-2}$$
$$h' = 1/(1-B^2)^{1/2} \tag{13-3}$$

where

$$B = \cos \phi \sin (\lambda-\lambda_0) \tag{8-5}$$

and λ_0 is the central meridian. The origin of the coordinates is at (ϕ_0, λ_0). The Y axis lies along the central meridian λ_0, y increasing northerly, and the X axis is perpendicular, through ϕ_0 at λ_0, x increasing easterly. Equation (13-2) is similar to corresponding equation (8-3) for the spherical Transverse Mercator projection. The scale factor is h' in a direction parallel to the central meridian, while it is 1 in a direction perpendicular to this meridian.

The inverse formulas for (ϕ, λ) in terms of (x, y):

$$\phi = \arcsin [\sin D \cos (x/R)] \tag{13-4}$$
$$\lambda = \lambda_0 + \arctan [\tan (x/R)/\cos D] \tag{13-5}$$

where

$$D = y/R + \phi_0 \qquad (13-6)$$

with ϕ_0 and D in radians. See p. 288 for numerical examples.

FORMULAS FOR THE ELLIPSOID

For the ellipsoidal form, a set of series approximations is given for use in a zone extending 3° to 4° of longitude from the central meridian. Coordinate axes are the same as they are for the spherical formulas above. The formulas below are adapted from those provided by Clifford J. Mugnier (pers. commun., 1979; see also Clark and Clendinning, 1944).

$$x = N [A - TA^3/6 - (8 - T + 8C)TA^5/120] \qquad (13-7)$$
$$y = M - M_0 + N \tan \phi [A^2/2 + (5 - T + 6C)A^4/24] \qquad (13-8)$$
$$s = 1 + x^2 \cos^2 Az (1 - e^2 \sin^2 \phi)^2/[2a^2(1 - e^2)] \qquad (13-9)$$

where

$$N = a/(1 - e^2 \sin^2 \phi)^{1/2} \qquad (4-20)$$
$$T = \tan^2 \phi \qquad (8-13)$$
$$A = (\lambda - \lambda_0) \cos \phi, \text{ with } \lambda \text{ and } \lambda_0 \text{ in radians} \qquad (8-15)$$
$$C = e^2 \cos^2 \phi/(1 - e^2) \qquad (8-14)$$
$$M = a [(1 - e^2/4 - 3e^4/64 - 5e^6/256 - \ldots) \phi - (3e^2/8$$
$$+ 3e^4/32 + 45e^6/1024 + \ldots) \sin 2\phi + (15e^4/256$$
$$+ 45e^6/1024 + \ldots) \sin 4\phi - (35e^6/3072 + \ldots) \sin 6\phi + \ldots] \qquad (3-21)$$

with ϕ in radians. M is the true distance along the central meridian from the Equator to ϕ.

$M_0 = M$ calculated for ϕ_0, the latitude crossing the central meridian λ_0 at the origin of the x, y coordinates. The choice of ϕ_0 does not affect the shape of the projection.

s = the scale factor at an azimuth Az east of north for a given ϕ and x.

For the inverse formulas:

$$\phi = \phi_1 - (N_1 \tan \phi_1/R_1)[D^2/2 - (1 + 3T_1) D^4/24] \qquad (13-10)$$
$$\lambda = \lambda_0 + [D - T_1 D^3/3 + (1 + 3T_1) T_1 D^5/15]/\cos \phi_1 \qquad (13-11)$$

where ϕ_1 is the "footpoint latitude" or the latitude at the central meridian which has the same y coordinate as that of the point (ϕ, λ).
It may be found as follows:

$$\phi_1 = \mu_1 + (3e_1/2 - 27e_1^3/32 + \ldots) \sin 2\mu_1 + (21e_1^2/16$$
$$- 55e_1^4/32 + \ldots) \sin 4\mu_1 + (151e_1^3/96 + \ldots) \sin 6\mu_1$$
$$+ (1097e_1^4/512 - \ldots) \sin 8\mu_1 + \ldots \qquad (3-26)$$

where

$$e_1 = [1 - (1 - e^2)^{1/2}]/[1 + (1 - e^2)^{1/2}] \qquad (3-24)$$
$$\mu_1 = M_1/[a(1 - e^2/4 - 3e^4/64 - 5e^6/256 - \ldots)] \qquad (7-19)$$
$$M_1 = M_0 + y \qquad (13-12)$$

with M_0 calculated from equation (3-21) for the given ϕ_0. For improved computational efficiency using series (3-26), see p. 19.

From ϕ_1, other terms below are calculated for use in equations (13-10) and (13-11). (If $\phi_1 = \pm \pi/2$, $\phi = \pm 90°$, taking the sign of y, while λ is indeterminate, and may be called λ_0.)

$$T_1 = \tan^2 \phi_1 \qquad (8-22)$$
$$N_1 = a/(1 - e^2 \sin^2 \phi_1)^{1/2} \qquad (8-23)$$
$$R_1 = a (1 - e^2)/(1 - e^2 \sin^2 \phi_1)^{3/2} \qquad (8-24)$$
$$D = x/N_1 \qquad (13-13)$$

CONIC MAP PROJECTIONS

Cylindrical projections are used primarily for complete world maps, or for maps along narrow strips of a great circle arc, such as the Equator, a meridian, or an oblique great circle. To show a region for which the greatest extent is from east to west in the temperate zones, conic projections are usually preferable to cylindrical projections.

Normal conic projections are distinguished by the use of arcs of concentric circles for parallels of latitude and equally spaced straight radii of these circles for meridians. The angles between the meridians on the map are smaller than the actual differences in longitude. The circular arcs may or may not be equally spaced, depending on the projection. The Polyconic projection and oblique conic projections have characteristics different from these.

The name "conic" originates from the fact that the more elementary conic projections may be derived by placing a cone on the top of a globe representing the Earth, the apex or tip in line with the axis of the globe, and the sides of the cone touching or tangent to the globe along a specified "standard" latitude which is true to scale and without distortion (see fig. 1). Meridians are drawn on the cone from the apex to the points at which the corresponding meridians on the globe cross the standard parallel. Other parallels are then drawn as arcs centered on the apex in a manner depending on the projection. If the cone is cut along one meridian and unrolled, a conic projection results. A secant cone results if the cone cuts the globe at two specified parallels. Meridians and parallels can be marked on the secant cone somewhat as above, but this will not result in any of the common conic projections with two standard parallels. They are derived from various desired scale relationships instead, and the spacing of the meridians as well as the parallels is not the same as the projection onto a secant cone.

There are three important classes of conic projections: the equidistant (or simple), the conformal, and the equal-area. The Equidistant Conic, with parallels equidistantly spaced, originated in a rudimentary form with Claudius Ptolemy. It eventually developed into commonly used present-day forms which have one or two standard parallels selected for the area being shown. It is neither conformal nor equal-area, but north-south scale along all meridians is correct, and the projection can be a satisfactory compromise for errors in shape, scale, and area, especially when the map covers a small area. It is primarily used in the spherical form, although the ellipsoidal form is available and useful. The USGS uses the Equidistant Conic in an approximate form for a map of Alaska, identified as a "Modified Transverse Mercator" projection, and also in the limiting equatorial form: the Equidistant Cylindrical. Both are described earlier.

The Lambert Conformal Conic projection with two standard parallels is used frequently for large- and small-scale maps. The parallels are more closely spaced near the center of the map. The Lambert has also been used slightly in the oblique form. The Albers Equal-Area Conic with two standard parallels is used for sectional maps of the U.S. and for maps of the conterminous United States. The Albers parallels are spaced more closely near the north and south edges of the map. There are some conic projections, such as perspective conics, which do not fall into any of these three categories, but they are rarely used.

The useful conic projections may be geometrically constructed only in a limited sense, using polar coordinates which must be calculated. After a location is chosen, usually off the final map, for the center of the circular arcs which will represent parallels of latitude, meridians are constructed as straight lines radiating from this center and spaced from each other at an angle equal to the product of the cone constant times the difference in longitude. For example, if a 10° graticule is planned, and the cone constant is 0.65, the meridian lines are spaced at 10° times 0.65 or 6.5°. Each parallel of latitude may then be drawn as a circular arc with a radius previously calculated from formulas for the particular conic projection.

14. ALBERS EQUAL-AREA CONIC PROJECTION

SUMMARY

- Conic.
- Equal-Area.
- Parallels are unequally spaced arcs of concentric circles, more closely spaced at the north and south edges of the map.
- Meridians are equally spaced radii of the same circles, cutting parallels at right angles.
- There is no distortion in scale or shape along two standard parallels, normally, or along just one.
- Poles are arcs of circles.
- Used for equal-area maps of regions with predominant east-west expanse, especially the conterminous United States.
- Presented by Albers in 1805.

HISTORY

One of the most commonly used projections for maps of the conterminous United States is the equal-area form of the conic projection, using two standard parallels. This projection was first presented by Heinrich Christian Albers (1773–1833), a native of Lüneburg, Germany, in a German periodical of 1805 (Albers, 1805; Bonacker and Anliker, 1930). The Albers projection was used for a German map of Europe in 1817, but it was promoted for maps of the United States in the early part of the 20th century by Oscar S. Adams of the Coast and Geodetic Survey as "an equal-area representation that is as good as any other and in many respects superior to all others" (Adams, 1927, p. 1).

FEATURES AND USAGE

The Albers is the projection exclusively used by the USGS for sectional maps of all 50 States of the United States in the *National Atlas* of 1970, and for other U.S. maps at scales of 1:2,500,000 and smaller. The latter maps include the base maps of the United States issued in 1961, 1967, and 1972, the Tectonic Map of the United States (1962), and the Geologic Map of the United States (1974), all at 1:2,500,000. The USGS has also prepared a U.S. base map at 1:3,168,000 (1 inch = 50 miles).

Like other normal conics, the Albers Equal-Area Conic projection (fig. 20) has concentric arcs of circles for parallels and equally spaced radii as meridians. The parallels are not equally spaced, but they are farthest apart in the latitudes between the standard parallels and closer together to the north and south. The pole is not the center of the circles, but is normally an arc itself.

If the pole is taken as one of the two standard parallels, the Albers formulas reduce to a limiting form of the projection called Lambert's Equal-Area Conic (not discussed here, and not to be confused with his Conformal Conic, to be discussed later). If the pole is the only standard parallel, the Albers formulas simplify to provide the polar aspect of the Lambert Azimuthal Equal-Area (discussed later). In both of these limiting cases, the pole is a point. If the Equator is the one standard parallel, the projection becomes Lambert's Cylindrical Equal-Area (discussed earlier), but the formulas must be modified. None of these extreme cases applies to the normal use of the Albers, with standard parallels in the temperate zones, such as usage for the United States.

Scale along the parallels is too small between the standard parallels and too large beyond them. The scale along the meridians is just the opposite, and in fact

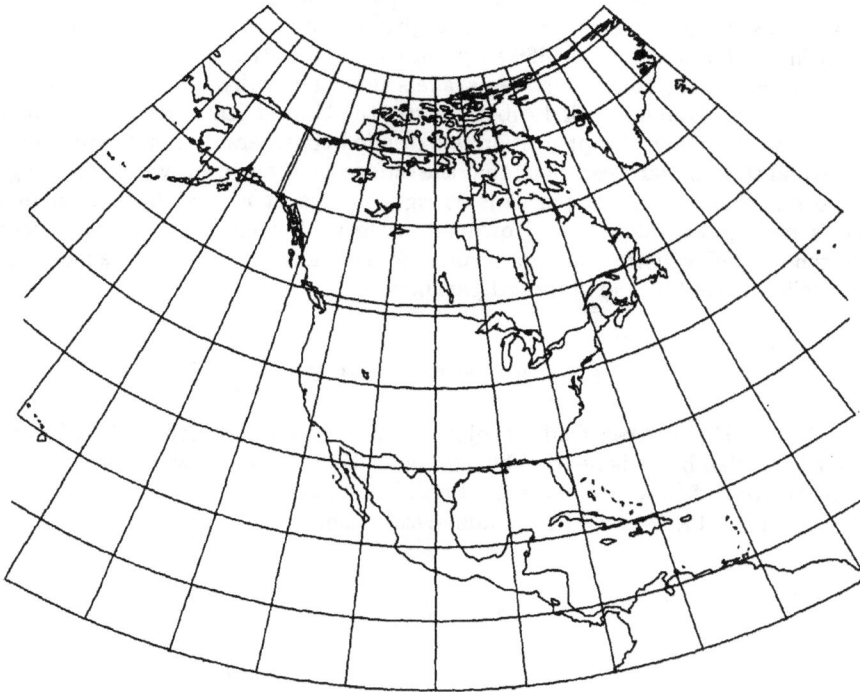

FIGURE 20.—Albers Equal-Area Conic projection, with standard parallels 20° and 60° N. This illustration includes all of North America to show the change in spacing of the parallels. When used for maps of the 48 conterminous States standard parallels are 29.5° and 45.5° N.

the scale factor along meridians is the reciprocal of the scale factor along parallels, to maintain equal area. An important characteristic of all normal conic projections is that scale is constant along any given parallel.

To map a given region, standard parallels should be selected to minimize variations in scale. Not only are standard parallels correct in scale along the parallel; they are correct in every direction. Thus, there is no angular distortion, and conformality exists along these standard parallels, even on an equal-area projection. They may be on opposite sides of, but not equidistant from, the Equator. Deetz and Adams (1934, p. 79, 91) recommended in general that standard parallels be placed one-sixth of the displayed length of the central meridian from the northern and southern limits of the map. Hinks (1912, p. 87) suggested one-seventh instead of one-sixth. Others have suggested selecting standard parallels of conics so that the maximum scale error (1 minus the scale factor) in the region between them is equal and opposite in sign to the error at the upper and lower parallels, or so that the scale factor at the middle parallel is the reciprocal of that at the limiting parallels. Tsinger in 1916 and Kavrayskiy in 1934 chose standard parallels so that least-square errors in linear scale were minimal for the actual land or country being displayed on the map. This involved weighting each latitude in accordance with the land it contains (Maling, 1960, p. 263–266).

The standard parallels chosen by Adams for Albers maps of the conterminous United States are lats. 29.5° and 45.5°N. These parallels provide "for a scale error slightly less than 1 per cent in the center of the map, with a maximum of 1¼ per cent along the northern and southern borders" (Deetz and Adams, 1934, p. 91). For maps of Alaska, the chosen standard parallels are lats. 55° and 65°N., and for Hawaii, lats. 8° and 18°N. In the latter case, both parallels are south of the islands, but they were chosen to include maps of the more southerly Canal Zone and especially the Philippine Islands. These parallels apply to all maps prepared by the USGS on the Albers projection, originally using Adams's published tables of coordinates for the Clarke 1866 ellipsoid (Adams, 1927).

Without measuring the spacing of parallels along a meridian, it is almost impossible to distinguish an unlabeled Albers map of the United States from other conic forms. It is only when the projection is extended considerably north and south, well beyond the standard parallels, that the difference is apparent without scaling.

Since meridians intersect parallels at right angles, it may at first seem that there is no angular distortion. However, scale variations along the meridians cause some angular distortion for any angle other than that between the meridian and parallel, except at the standard parallels.

FORMULAS FOR THE SPHERE

The Albers Equal-Area Conic projection may be constructed with only one standard parallel, but it is nearly always used with two. The forward formulas for the sphere are as follows, to obtain rectangular or polar coordinates, given R, ϕ_1, ϕ_2, ϕ_0, λ_0, ϕ, and λ (see p. 291 for numerical examples):

$$x = \rho \sin \theta \tag{14-1}$$
$$y = \rho_0 - \rho \cos \theta \tag{14-2}$$

where

$$\rho = R(C - 2n \sin \phi)^{1/2}/n \tag{14-3}$$
$$\theta = n(\lambda - \lambda_0) \tag{14-4}$$
$$\rho_0 = R(C - 2n \sin \phi_0)^{1/2}/n \tag{14-3a}$$
$$C = \cos^2 \phi_1 + 2n \sin \phi_1 \tag{14-5}$$
$$n = (\sin \phi_1 + \sin \phi_2)/2 \tag{14-6}$$

ϕ_0, λ_0 = the latitude and longitude, respectively, for the origin of the rectangular coordinates.

ϕ_1, ϕ_2 = standard parallels.

The Y axis lies along the central meridian λ_0, y increasing northerly. The X axis intersects perpendicularly at ϕ_0, x increasing easterly. If $(\lambda - \lambda_0)$ exceeds the range $\pm 180°$, $360°$ should be added or subtracted to place it within the range. Constants n, C, and ρ_0 apply to the entire map, and thus need to be calculated only once. If only one standard parallel ϕ_1 is desired (or if $\phi_1 = \phi_2$), $n = \sin \phi_1$. By contrast, a geometrically secant cone requires a cone constant n of $\sin [(\phi_1 + \phi_2)/2]$, slightly but distinctly different from equation (14-6). If the projection is designed primarily for the Northern Hemisphere, n and ρ are positive. For the Southern Hemisphere, they are negative. The scale along the meridians, using equation (4-4),

$$h = \cos \phi/(C - 2n \sin \phi)^{1/2} \tag{14-7}$$

If equation (4-5) is used, k will be found to be the reciprocal of h, satisfying the requirement for an equal-area projection when meridians and parallels intersect at right angles. The maximum angular deformation may be calculated from equation (4-9). It may be seen from equation (14-7), and indeed from equations (4−4) and (4−5), that distortion is strictly a function of latitude, and not of longitude. This is true of any regular conic projection.

For the inverse formulas for the sphere, given R, ϕ_1, ϕ_2, ϕ_0, λ_0, x, and y: n, C and ρ_0 are calculated from equations (14−6), (14−5), and (14−3a), respectively. Then,

$$\phi = \arcsin\left\{[C-(\rho n/R)^2]/(2n)\right\} \tag{14−8}$$
$$\lambda = \lambda_0 + \theta/n \tag{14−9}$$

where

$$\rho = [x^2+(\rho_0-y)^2]^{1/2} \tag{14−10}$$
$$\theta = \arctan[x/(\rho_0-y)] \tag{14−11}$$

Note: to use the ATAN2 Fortran function, if n is negative, the signs of x, y, and ρ_0 (given a negative sign by equation (14−3a)) must be reversed before inserting them in equation (14−11).

FORMULAS FOR THE ELLIPSOID

The formulas displayed by Adams and most other writers describing the ellipsoidal form include series, but the equations may be expressed in closed forms which are suitable for programming, and involve no numerical integration or iteration in the forward form. Nearly all published maps of the United States based on the Albers use the ellipsoidal form because of the use of tables for the original base maps. (Adams, 1927, p. 1−7; Deetz and Adams, 1934, p. 93−99; Snyder, 1979a, p. 71). Given a, e, ϕ_1, ϕ_2, ϕ_0, λ_0, ϕ, and λ (see p. 292 for numerical examples):

$$x = \rho \sin\theta \tag{14−1}$$
$$y = \rho_0 - \rho\cos\theta \tag{14−2}$$

where

$$\rho = a(C-nq)^{1/2}/n \tag{14−12}$$
$$\theta = n(\lambda-\lambda_0) \tag{14−4}$$
$$\rho_0 = a(C-nq_0)^{1/2}/n \tag{14−12a}$$
$$C = m_1{}^2 + nq_1 \tag{14−13}$$
$$n = (m_1{}^2-m_2{}^2)/(q_2-q_1) \tag{14−14}$$
$$m = \cos\phi/(1-e^2\sin^2\phi)^{1/2} \tag{14−15}$$
$$q = (1-e^2)\{\sin\phi/(1-e^2\sin^2\phi) - [1/(2e)]$$
$$\ln[(1-e\sin\phi)/(1+e\sin\phi)]\} \tag{3−12}$$

with the same subscripts 1, 2, or none applied to m and ϕ in equation (14−15), and 0, 1, 2, or none applied to q and ϕ in equation (3−12), as required by equations (14−12), (14−12a), (14−13), (14−14), and (14−17). As with the spherical case, ρ and n are negative, if the projection is centered in the Southern Hemisphere. For the scale factor, modifying (4−25):

$$k = \rho n/am \tag{14−16}$$
$$= (C-nq)^{1/2}/m \tag{14−17}$$
$$h = 1/k \tag{14−18}$$

While many ellipsoidal equations apply to the sphere if e is made zero, equation (3−12) becomes indeterminate. Actually, if $e=0$, $q=2\sin\phi$. If $\phi_1 = \phi_2$, equation

(14−14) is indeterminate regardless of e, but $n = \sin \phi_1$. The axes and limitations on $(\lambda - \lambda_0)$ are the same as those stated for the spherical formulas. Here, too, constants n, C, and ρ_0 need to be determined just once for the entire map.

For the inverse formulas for the ellipsoid, given a, e, ϕ_1, ϕ_2, ϕ_0, λ_0, x, and y: n, C, and ρ_0 are calculated from equations (14−14), (14−13), and (14−12a); respectively. Then,

$$\phi = \phi + \frac{(1-e^2 \sin^2 \phi)^2}{2 \cos \phi} \left[\frac{q}{1-e^2} - \frac{\sin \phi}{1-e^2 \sin^2 \phi} + \frac{1}{2e} \ln \left(\frac{1-e \sin \phi}{1+e \sin \phi} \right) \right] \qquad (3-16)$$

$$\lambda = \lambda_0 + \theta/n \qquad (14-9)$$

where

$$q = (C - \rho^2 n^2/a^2)/n \qquad (14-19)$$
$$\rho = [x^2 + (\rho_0 - y)^2]^{1/2} \qquad (14-10)$$
$$\theta = \arctan [x/(\rho_0 - y)] \qquad (14-11)$$

To use the Fortran ATAN2 function, if n is negative, the signs of x, y, and ρ_0 must be reversed before insertion into equation (14−11). Equation (3−16) involves iteration by first trying $\phi = \arcsin (q/2)$ on the right side, calculating ϕ on the left side, substituting this new ϕ on the right side, etc., until the change in ϕ is negligible. If

$$q = \pm \{1 - [(1-e^2)/2e] \ln [(1-e)/(1+e)]\} \qquad (14-20)$$

iteration does not converge, but $\phi = \pm 90°$, taking the sign of q.

Instead of the iteration, a series may be used for the inverse ellipsoidal formulas:

$$\phi = \beta + (e^2/3 + 31e^4/180 + 517e^6/5040 + \ldots) \sin 2\beta + (23e^4/360$$
$$+ 251e^6/3780 + \ldots) \sin 4\beta + (761e^6/45360 + \ldots) \sin 6\beta + \ldots \qquad (3-18)$$

where β, the authalic latitude, adapting equations (3−11) and (3−12), is found thus:

$$\beta = \arcsin (q/\{1 - [(1-e^2)/2e] \ln [(1-e)/(1+e)]\}) \qquad (14-21)$$

but q is still found from equation (14−19). Equations (14−9), (14−10), and (14−11) also apply unchanged. For improved computational efficiency using the series, see p. 19.

Polar coordinates for the Albers Equal-Area Conic are given for both the spherical and ellipsoidal forms, using standard parallels of lat. 29.5° and 45.5° N. (table 15). A graticule extended to the North Pole is shown in figure 20.

To convert coordinates measured on an existing map, the user may choose any meridian for λ_0 and therefore for the Y axis, and any latitude for ϕ_0. The X axis then is placed perpendicular to the Y axis at ϕ_0.

TABLE 15.—*Albers Equal-Area Conic projection: Polar coordinates*

[Standard parallels: 29.5° and 45.5° N]

Lat.	Projection for sphere ($R = 6,370,997$ m) ($n = 0.6028370$)			Projection for Clarke 1866 ellipsoid ($a = 6,378,206.4$ m) ($n = 0.6029035$)		
	ρ	h	k	ρ	h	k
52°	6,693,511	0.97207	1.02874	6,713,781	0.97217	1.02863
51	6,801,923	.97779	1.02271	6,822,266	.97788	1.02263
50	6,910,941	.98296	1.01733	6,931,335	.98303	1.01727
49	7,020,505	.98760	1.01255	7,040,929	.98765	1.01251
48	7,130,555	.99173	1.00834	7,150,989	.99177	1.00830
47	7,241,038	.99538	1.00464	7,261,460	.99540	1.00462
46	7,351,901	.99857	1.00143	7,372,290	.99858	1.00143
45.5	7,407,459	1.00000	1.00000	7,427,824	1.00000	1.00000
45	7,463,094	1.00132	.99868	7,483,429	1.00132	.99869
44	7,574,570	1.00365	.99636	7,594,829	1.00364	.99637
43	7,686,282	1.00558	.99445	7,706,445	1.00556	.99447
42	7,798,186	1.00713	.99292	7,818,233	1.00710	.99295
41	7,910,244	1.00832	.99175	7,930,153	1.00828	.99178
40	8,022,413	1.00915	.99093	8,042,164	1.00911	.99097
39	8,134,656	1.00965	.99044	8,154,230	1.00961	.99048
38	8,246,937	1.00983	.99027	8,266,313	1.00978	.99031
37	8,359,220	1.00970	.99040	8,378,379	1.00965	.99044
36	8,471,472	1.00927	.99082	8,490,394	1.00923	.99086
35	8,583,660	1.00855	.99152	8,602,328	1.00852	.99155
34	8,695,753	1.00757	.99249	8,714,149	1.00753	.99252
33	8,807,723	1.00632	.99372	8,825,828	1.00629	.99375
32	8,919,539	1.00481	.99521	8,937,337	1.00479	.99523
31	9,031,175	1.00306	.99694	9,048,649	1.00305	.99696
30	9,142,602	1.00108	.99892	9,159,737	1.00107	.99893
29.5	9,198,229	1.00000	1.00000	9,215,189	1.00000	1.00000
29	9,253,796	.99887	1.00114	9,270,575	.99887	1.00113
28	9,364,731	.99643	1.00358	9,381,141	.99645	1.00357
27	9,475,383	.99378	1.00626	9,491,411	.99381	1.00623
26	9,585,731	.99093	1.00915	9,601,361	.99097	1.00911
25	9,695,749	.98787	1.01227	9,710,969	.98793	1.01222
24	9,805,417	.98463	1.01561	9,820,216	.98470	1.01554
23	9,914,713	.98119	1.01917	9,929,080	.98128	1.01908
22	10,023,616	.97757	1.02294	10,037,541	.97768	1.02283

Note: ρ = radius of latitude circle, meters.

h = scale factor along meridians.

k = scale factor along parallels.

R = assumed radius of sphere.

a = assumed semimajor axis of ellipsoid.

n = cone constant, or ratio of angle between meridians on map to true angle.

15. LAMBERT CONFORMAL CONIC PROJECTION

SUMMARY

- Conic.
- Conformal.
- Parallels are unequally spaced arcs of concentric circles, more closely spaced near the center of the map.
- Meridians are equally spaced radii of the same circles, thereby cutting parallels at right angles.
- Scale is true along two standard parallels, normally, or along just one.
- Pole in same hemisphere as standard parallels is a point; other pole is at infinity.
- Used for maps of countries and regions with predominant east-west expanse.
- Presented by Lambert in 1772.

HISTORY

The Lambert Conformal Conic projection (fig. 21) was almost completely over-looked between its introduction and its revival by the U.S. Coast and Geodetic Survey (Deetz, 1918b), although France had introduced an approximate version, calling it "Lambert," for battle maps of the First World War (Mugnier, 1983). It was the first new projection which Johann Heinrich Lambert presented in his *Beiträge* (Lambert, 1772), the publication which contained his Transverse Mercator described previously. In some atlases, particularly British, the Lambert Conformal Conic is called the "Conical Orthomorphic" projection.

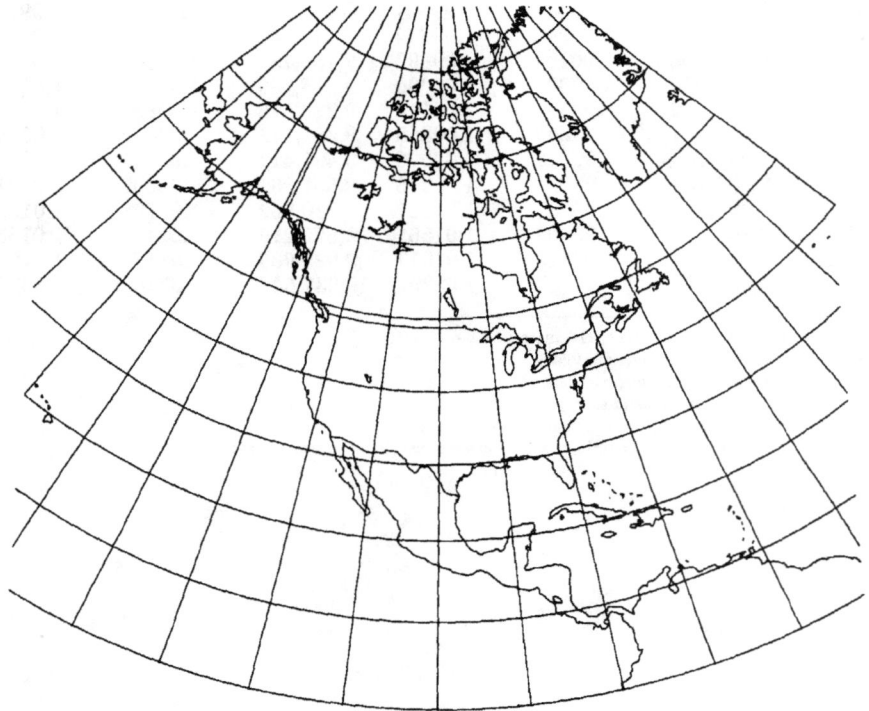

FIGURE 21.—Lambert Conformal Conic projection, with standard parallels 20° and 60° N. North America is illustrated here to show the change in spacing of the parallels. When used for maps of the conterminous United States or individual States, standard parallels are 33° and 45° N.

Lambert developed the regular Conformal Conic as the oblique aspect of a family containing the previously known polar Stereographic and regular Mercator projections. As he stated,

Stereographic representations of the spherical surface, as well as Mercator's nautical charts, have the peculiarity that all angles maintain the sizes that they have on the surface of the globe. This yields the greatest similarity that any plane figure can have with one drawn on the surface of a sphere. The question has not been asked whether this property occurs only in the two methods of representation mentioned or whether these two representations, so different in appearances, can be made to approach each other through intermediate stages. * * * if there are stages intermediate to these two representations, they must be sought by allowing the angle of intersection of the meridians to be arbitrarily larger or smaller than its value on the surface of the sphere. This is the way in which I shall now proceed (Lambert, 1772, p. 28, translation by Tobler).

Lambert then developed the mathematics for both the spherical and ellipsoidal forms for two standard parallels and included a small map of Europe as an example (Lambert, 1772, p. 28–38, 87–89).

FEATURES

Many of the comments concerning the appearance of the Albers and the selection of its standard parallels apply to the Lambert Conformal Conic when an area the size of the conterminous United States or smaller is considered. As stated before, the spacing of the parallels must be measured to distinguish among the various conic projections for such an area. If the projection is extended toward either pole and the Equator, as on a map of North America, the differences become more obvious. Although meridians are equally spaced radii of the concentric circular arcs representing parallels of latitude, the parallels become further apart as the distance from the central parallels increases. Conformality fails at each pole, as in the case of the regular Mercator. The pole in the same hemisphere as the standard parallels is shown on the Lambert Conformal Conic as a point. The other pole is at infinity. Straight lines between points approximate great circle arcs for maps of moderate coverage, but only the Gnomonic projection rigorously has this feature and then only for the sphere.

Two parallels may be made standard or true to scale, as well as conformal. It is also possible to have just one standard parallel. Since there is no angular distortion at any parallel (except at the poles), it is possible to change the standard parallels to just one, or to another pair, just by changing the scale applied to the existing map and calculating a pair of standard parallels fitting the new scale. This is not true of the Albers, on which only the original standard parallels are free from angular distortion.

If the standard parallels are symmetrical about the Equator, the regular Mercator results (although formulas must be revised). If the only standard parallel is a pole, the polar Stereographic results.

The scale is too small between the standard parallels and too large beyond them. This applies to the scale along meridians, as well as along parallels, or in any other direction, since they are equal at any given point. Thus, in the State Plane Coordinate Systems (SPCS) for States using the Lambert, the choice of standard parallels has the effect of reducing the scale of the central parallel by an amount which cannot be expressed simply in exact form, while the scale for the central meridian of a map using the Transverse Mercator is normally reduced by a simple fraction. The scale is constant along any given parallel. While it equals the nominal scale at the standard parallels, it actually changes most slowly in a north-south direction at a parallel nearly halfway between the two standard parallels.

USAGE

It was only a couple of decades after the Coast and Geodetic Survey began publishing tables for the Lambert Conformal Conic projection (Deetz, 1918a,

1918b) that the projection was adopted officially for the SPCS for States of predominantly east-west expanse. The prototype was the North Carolina Coordinate System, established in 1933. Within a year or so, similar systems were devised for many other States, while a Transverse Mercator system was prepared for the remaining States. One or more zones is involved in the system for each State (see table 8) (Mitchell and Simmons, 1945, p. vi). In addition, the Lambert is used for the Aleutian Islands of Alaska, Long Island in New York, and northwestern Florida, although the Transverse Mercator (and Oblique Mercator in one case) is used for the rest of each of these States.

The Lambert Conformal Conic is used for the 1:1,000,000-scale regional world aeronautical charts, the 1:500,000-scale sectional aeronautical charts, and 1:500,000-scale State base maps (all 48 contiguous States[4] have the same standard parallels of lat. 33° and 45° N., and thus match). Also cast on the Lambert are most of the 1:24,000-scale 7½-minute quadrangles prepared after 1957 which lie in zones for which the Lambert is the base for the SPCS. In the latter case, the standard parallels for the zone are used, rather than parameters designed for the individual quadrangles. Thus, all quadrangles for a given zone may be mosaicked exactly. (The projection used previously was the Polyconic, and some recent quadrangles are being produced to the Universal Transverse Mercator projection.)

The Lambert Conformal Conic has also been adopted as the official topographic projection for some other countries. It appears in *The National Atlas* (USGS, 1970, p. 116) for a map of hurricane patterns in the North Atlantic, and the Lambert is used by the USGS for a map of the United States showing all 50 States in their true relative positions. The latter map is at scales of both 1:6,000,000 and 1:10,000,000, with standard parallels 37° and 65° N.

In 1962, the projection for the International Map of the World at a scale of 1:1,000,000 was changed from a modified Polyconic to the Lambert Conformal Conic between lats. 84° N. and 80° S. The polar Stereographic projection is used in the remaining areas. The sheets are generally 6° of longitude wide by 4° of latitude high. The standard parallels are placed at one-sixth and five-sixths of the latitude spacing for each zone of 4° latitude, and the reference ellipsoid is the International (United Nations, 1963, p. 9–27). This specification has been subsequently used by the USGS in constructing several maps for the IMW series.

Perhaps the most recent new topographic use for the Lambert Conformal Conic projection by the USGS is for middle latitudes of the 1:1,000,000-scale geologic series of the Moon and for some of the maps of Mercury, Mars, and Jupiter's satellites (see table 6).

FORMULAS FOR THE SPHERE

For the projection as normally used, with two standard parallels, the equations for the sphere may be written as follows: Given R, ϕ_1, ϕ_2, ϕ_0, λ_0, ϕ, and λ (see p. 295 for numerical examples):

$$x = \rho \sin \theta \tag{14-1}$$
$$y = \rho_0 - \rho \cos \theta \tag{14-2}$$

where

$$\rho = RF/\tan^n (\pi/4 + \phi/2) \tag{15-1}$$
$$\theta = n(\lambda - \lambda_0) \tag{14-4}$$
$$\rho_0 = RF/\tan^n (\pi/4 + \phi_0/2) \tag{15-1a}$$

---

[4]For Hawaii, the standard parallels are lats. 20° 40' and 23° 20' N.; the corresponding base map was not prepared for Alaska.

$$F = \cos \phi_1 \tan^n (\pi/4 + \phi_1/2)/n \qquad (15-2)$$

$$n = \ln (\cos \phi_1/\cos \phi_2)/\ln[\tan (\pi/4 + \phi_2/2)/\tan (\pi/4 + \phi_1/2)] \quad (15-3)$$

ϕ_0, λ_0 = the latitude and longitude for the origin of the rectangular coordinates.
ϕ_1, ϕ_2 = standard parallels.

The Y axis lies along the central meridian λ_0, y increasing northerly; the X axis intersects perpendicularly at ϕ_0, x increasing easterly. If $(\lambda-\lambda_0)$ exceeds the range $\pm 180°$, $360°$ should be added or subtracted. Constants n, F, and ρ_0 need to be determined only once for the entire map.

If only one standard parallel ϕ_1 is desired, equation (15–3) is indeterminate, but $n = \sin \phi_1$. The scale along meridians or parallels, using equations (4–4) or (4–5),

$$k = h = \cos \phi_1 \tan^n(\pi/4 + \phi_1/2)/[\cos \phi \tan^n(\pi/4 + \phi/2)] \qquad (15-4)$$

The maximum angular deformation $\omega = 0$, since the projection is conformal. As with the other regular conics, k is strictly a function of latitude. For a projection centered in the Southern Hemisphere, n and ρ are negative.

For the inverse formulas for the sphere, given R, ϕ_1, ϕ_2, ϕ_0, λ_0, x, and y: n, F, and ρ_0 are calculated from equations (15–3), (15–2), and (15–1a), respectively. Then,

$$\phi = 2 \arctan (RF/\rho)^{1/n} - \pi/2 \qquad (15-5)$$
$$\lambda = \theta/n + \lambda_0 \qquad (14-9)$$

where

$$\rho = \pm[x^2 + (\rho_0-y)^2]^{1/2}, \text{ taking the sign of } n \qquad (14-10)$$
$$\theta = \arctan [x/(\rho_0-y)] \qquad (14-11)$$

The Fortran ATAN2 function does not apply to equation (15–5), but when it is used for equation (14–11), and n is negative, the signs of x, y, and ρ_0 (negative from equation (15–1a)) must be reversed before insertion into the equation. If $\rho = 0$, equation (15–5) involves division by zero, but ϕ is $\pm 90°$, taking the sign of n.

The standard parallels normally used for maps of the conterminous United States are lats. 33° and 45° N., which give approximately the least overall error within those boundaries. The ellipsoidal form is used for such maps, based on the Clarke 1866 ellipsoid (Adams, 1918).

The standard parallels of 33° and 45° were selected by the USGS because of the existing tables by Adams (1918), but Adams chose them to provide a maximum scale error between latitudes 30.5° and 47.5° of one-half of 1 percent. A maximum scale error of 2.5 percent occurs in southernmost Florida (Deetz and Adams, 1934, p. 80). Other standard parallels would reduce the maximum scale error for the United States, but at the expense of accuracy in the center of the map.

FORMULAS FOR THE ELLIPSOID

The ellipsoidal formulas are essential when applying the Lambert Conformal Conic to mapping at a scale of 1:100,000 or larger and important at scales of 1:5,000,000. Given a, e, ϕ_1, ϕ_2, ϕ_0, λ_0, ϕ, and λ (see p. 296 for numerical examples):

$$x = \rho \sin \theta \qquad (14-1)$$
$$y = \rho_0 - \rho \cos \theta \qquad (14-2)$$

$$k = \rho n/(am) \tag{14-16}$$
$$= m_1 t^n/(m t_1{}^n) \tag{15-6}$$

where

$$\rho = aFt^n \tag{15-7}$$
$$\theta = n(\lambda - \lambda_0) \tag{14-4}$$
$$\rho_0 = aF t_0{}^n \tag{15-7a}$$
$$n = (\ln m_1 - \ln m_2)/(\ln t_1 - \ln t_2) \tag{15-8}$$
$$m = \cos \phi/(1 - e^2 \sin^2 \phi)^{1/2} \tag{14-15}$$
$$t = \tan (\pi/4 - \phi/2)/[(1 - e \sin \phi)/(1 + e \sin \phi)]^{e/2} \tag{15-9}$$

or

$$= \left[\left(\frac{1 - \sin \phi}{1 + \sin \phi} \right) \left(\frac{1 + e \sin \phi}{1 - e \sin \phi} \right)^e \right]^{1/2} \tag{15-9a}$$

$$F = m_1/(n t_1{}^n) \tag{15-10}$$

with the same subscripts 1, 2, or none applied to m and ϕ in equation (14-15), and 0, 1, 2, or none applied to t and ϕ in equation (15-9), as required by equations (15-6), (15-7), and (15-8). As with other conics, a negative n and ρ result for projections centered in the Southern Hemisphere. If $\phi = \pm 90°$, ρ is zero for the same sign as n and infinite for the opposite sign. If $\phi_1 = \phi_2$, for the Lambert with a single standard parallel, equation (15-8) is indeterminate, but $n = \sin \phi_1$. Origin and orientation of axes for x and y are the same as those for the spherical form. Constants n, F, and ρ_0 may be determined just once for the entire map.

When the above equations for the ellipsoidal form are used, they give values of n and ρ slightly different from those in the accepted tables of coordinates for a map of the United States, according to the Lambert Conformal Conic projection. The discrepancy is 35-50 m in the radius and 0.0000035 in n. The rectangular coordinates are correspondingly affected. The discrepancy is less significant when it is realized that the radius is measured to the pole, and that the distance from the 50th parallel to the 25th parallel on the map at full scale is only 12 m out of 2,800,000 or 0.0004 percent. For calculating convenience 60 years ago, the tables were, in effect, calculated using instead of equation (15-9),

$$t = \tan (\pi/4 - \phi_g/2) \tag{15-9b}$$

where ϕ_g is the geocentric latitude, or, as shown earlier,

$$\phi_g = \arctan [(1 - e^2)\tan \phi] \tag{3-28}$$

In conventional terminology, the t of equation (15-9) is usually written as $\tan \frac{1}{2}Z$, where Z is the colatitude of the conformal latitude χ (see equation (3-1)).

For the existing tables, then, ϕ_g, the geocentric latitude, was used for convenience in place of χ, the conformal latitude (Adams, 1918, p. 6-9, 34). A comparison of series equations (3-3) and (3-30), or of the corresponding columns in table 3, shows that the two auxiliary latitudes χ and ϕ_g are numerically very nearly the same.

There may be much smaller discrepancies found between coordinates as calculated on modern computers and those listed in tables for the SPCS. This is due to the slightly reduced (but sufficient) accuracy of the desk calculators of 30-40 years ago and the adaptation of formulas to be more easily utilized by them. To obtain SPCS coordinates, the appropriate "false easting" is added to x after calculation using (14-1).

The inverse formulas for ellipsoidal coordinates, given a, e, ϕ_1, ϕ_2, ϕ_0, λ_0, ϕ, and λ: n, F, and ρ_0 are calculated from equations (15-8), (15-10), (15-7a), respectively. Then,

$$\phi = \pi/2 - 2 \arctan \{t[(1-e\sin\phi)/(1+e\sin\phi)]^{e/2}\} \qquad (7-9)$$

where

$$
\begin{aligned}
t &= (\rho/aF)^{1/n} & (15-11) \\
\rho &= \pm[x^2+(\rho_0-y)^2]^{1/2}, \text{ taking the sign of } n. & (14-10) \\
\lambda &= \theta/n+\lambda_0 & (14-9) \\
\theta &= \arctan[x/(\rho_0-y)] & (14-11)
\end{aligned}
$$

As with the spherical formulas, the Fortran ATAN2 function does not apply to equation (7-9), but for equation (14-11), if n is negative, the signs of x, y, and ρ_0 must be reversed.

Equation (7-9) involves rapidly converging iteration: Calculate t from (15-11). Then, assuming an initial trial ϕ equal to $(\pi/2 - 2 \arctan t)$ in the right side of equation (7-9), calculate ϕ on the left side. Substitute the calculated ϕ into the right side, calculate a new ϕ, etc., until ϕ does not change significantly from the preceding trial value of ϕ.

To avoid iteration, series (3-5) may be used with (7-13) in place of (7-9):

$$
\begin{aligned}
\phi = \chi &+ (e^2/2 + 5e^4/24 + e^6/12 + 13e^8/360 + \ldots)\sin 2\chi \\
&+ (7e^4/48 + 29e^6/240 + 811e^8/11520 + \ldots) \\
&\quad \sin 4\chi + (7e^6/120 + 81e^8/1120 + \ldots)\sin 6\chi \\
&\quad\quad + (4279e^8/161280 + \ldots)\sin 8\chi + \ldots \qquad (3-5)
\end{aligned}
$$

where

$$\chi = \pi/2 - 2\arctan t \qquad (7-13)$$

For improved computational efficiency using the series, see p. 19.

If rectangular coordinates for maps based on the tables using geocentric latitude are to be converted to latitude and longitude, the inverse formulas are the same as those above, except that equation (15-9b) is used instead of (15-9) for calculations leading to n, F, and ρ_0, and equation (7-9), or (3-5) and (7-13), is replaced with the following which does not involve iteration:

$$\phi = \arctan[\tan\phi_g/(1-e^2)] \qquad (15-13)$$

where

$$\phi_g = \pi/2 - 2\arctan t \qquad (15-14)$$

and t is calculated from equation (15-11).

Polar coordinates for the Lambert Conformal Conic are given for both the spherical and ellipsoidal forms, using standard parallels of 33° and 45° N. (table 16). The data based on the geocentric latitude are given for comparison. A graticule extended to the North Pole is shown in figure 21.

To convert from tabular rectangular coordinates to ϕ and λ, it is necessary to subtract any "false easting" from x and "false northing" from y before inserting x and y into the inverse formulas. To convert coordinates measured on an existing Lambert Conformal Conic map (or other regular conic projection), the user may choose any meridian for λ_0 and therefore for the Y axis, and any latitude for ϕ_0. The X axis then is placed perpendicular to the Y axis at ϕ_0.

TABLE 16.—*Lambert Conformal Conic projection: Polar coordinates*

[Standard parallels: 33° and 45° N]

Lat.	Projection for sphere (R = 6,370,997 m) (n = 0.6304777)			Projection for Clarke 1866 ellipsoid (a = 6,378,206.4 m) Conformal lat.[1] (n = 0.6304965)			Geocentric lat.[2] (n = 0.6305000)
	ρ	k	k²	ρ	k	k²	ρ
52°	6,359,534	1.02222	1.04494	6,379,530	1.02215	1.04480	—
51	6,472,954	1.01787	1.03606	6,493,008	1.01781	1.03595	6,492,973
50	6,585,914	1.01394	1.02807	6,606,007	1.01389	1.02798	6,605,970
49	6,698,458	1.01040	1.02091	6,718,571	1.01037	1.02084	6,718,537
48	6,810,631	1.00725	1.01456	6,830,746	1.00723	1.01451	6,830,708
47	6,922,475	1.00448	1.00898	6,942,573	1.00446	1.00894	6,942,534
46	7,034,030	1.00206	1.00413	7,054,092	1.00206	1.00412	7,054,052
45	7,145,336	1.00000	1.00000	7,165,344	1.00000	1.00000	7,165,303
44	7,256,432	.99828	.99656	7,276,367	.99828	.99657	7,276,330
43	7,367,355	.99689	.99379	7,387,198	.99690	.99381	7,387,158
42	7,478,142	.99582	.99167	7,497,873	.99584	.99170	7,497,833
41	7,588,828	.99508	.99018	7,608,429	.99510	.99022	7,608,384
40	7,699,449	.99464	.98932	7,718,900	.99467	.98936	7,718,857
39	7,810,038	.99452	.98907	7,829,321	.99454	.98911	7,829,278
38	7,920,631	.99470	.98942	7,939,726	.99472	.98946	7,939,680
37	8,031,259	.99517	.99036	8,050,148	.99519	.99040	8,050,107
36	8,141,957	.99594	.99190	8,160,619	.99596	.99193	8,160,581
35	8,252,757	.99700	.99402	8,271,174	.99702	.99404	8,271,129
34	8,363,692	.99836	.99672	8,381,843	.99836	.99673	8,381,798
33	8,474,793	1.00000	1.00000	8,492,660	1.00000	1.00000	8,492,614
32	8,586,092	1.00193	1.00386	8,603,656	1.00192	1.00385	8,603,610
31	8,697,622	1.00415	1.00831	8,714,863	1.00413	1.00827	8,714,820
30	8,809,415	1.00665	1.01335	8,826,313	1.00662	1.01328	8,826,267
29	8,921,502	1.00944	1.01897	8,938,038	1.00940	1.01888	8,937,986
28	9,033,915	1.01252	1.02520	9,050,070	1.01246	1.02507	9,050,021
27	9,146,686	1.01589	1.03203	9,162,440	1.01581	1.03186	9,162,396
26	9,259,848	1.01954	1.03947	9,275,181	1.01944	1.03927	9,275,132
25	9,373,433	1.02349	1.04754	9,388,326	1.02337	1.04729	9,388,277
24	9,487,474	1.02774	1.05625	9,501,906	1.02759	1.05595	9,501,859
23	9,602,003	1.03228	1.06560	9,615,955	1.03211	1.06525	9,615,911
22	9,717,054	1.03712	1.07563	9,730,506	1.03692	1.07521	9,730,456

[1] Based on rigorous equations using conformal latitude.

[2] Based on geocentric latitude as given in Adams (1918, p. 34) and Deetz and Adams (1934, p. 84).

Notes: ρ = radius of latitude circles, meters.

k^2 = scale factor (areal).

a = assumed semimajor axis of ellipsoid.

R = assumed radius of sphere.

16. EQUIDISTANT CONIC PROJECTION

SUMMARY

- Conic.
- Equidistant.
- Parallels, including poles, are arcs of concentric circles, equally spaced for the sphere, at true spacing for the ellipsoid.
- Meridians are equally spaced radii of the same circles, thereby cutting parallels at right angles.
- Scale is true along all meridians and along one or two standard parallels.
- Used for maps of small countries and regions and of larger areas with predominant east-west expanse.
- Rudimentary form developed by Claudius Ptolemy about A.D. 150.

HISTORY

The simplest kind of conic projection is the Equidistant Conic, often called Simple Conic, or just Conic projection. It is the projection most likely to be found in atlases for maps of small countries, with its equally spaced straight meridians and equally spaced circular parallels. A rudimentary version was described by the astronomer and geographer Claudius Ptolemy about A.D. 150. Probably born in Greece about A.D. 90, he spent most of his life in or near Alexandria, Egypt, and died about A.D. 168. His greatest works were the *Almagest*, describing his scientific theories, and the *Geographia*, which dwelt on mapmaking. These were revived in the 15th century as the most authoritative existing standards.

In developing this projection, Ptolemy did not discuss cones, and a cone would not properly fit his specifications, but he said (*Geographia*, Book 1, ch. 20):

When we cast a glance upon the middle of the northern quarter of the globe in which the greatest part of the oikumene [or ecumene, or inhabited world] lies, then the meridians give the impression of being straight lines if we turn the globe thus that the meridians successively come out of their sideward situation right before the spectator, so that the eye comes in their plane. The parallels give clearly the impression of arcs of circles which turn their convex side to the south (Keuning, 1955, p. 9).

Ptolemy's conic projection extends from latitudes approximating 63°N. to 16°S. Although meridians north of the Equator fan out as straight radii from the center of the circular parallels, they break at the Equator to connect with straight lines to points along the southernmost parallel which are the same distance apart as corresponding points at 16°N.

Johannes Ruysch (?–1533) modified this approach to continue meridians as straight radii below the Equator in a world map of 1508, and Gerardus Mercator made other modifications in the mid-16th century. The Equidistant Conic with two standard parallels is credited to Joseph Nicolas De l'Isle (1688–1768), of an illustrious French mapmaking family. He used it for a map of Russia in 1745. There were differences in his approach, however, which resulted in meridians which are not radii of the circular arcs representing the circles.

Several Scot (Murdoch), Swiss (Euler), English (Everett), and Russian (Vitkovskiy, Kavrayskiy, and others) mathematicians published papers between 1758 and 1934 describing means of selecting the two standard parallels so that distortion is minimized using various criteria. Each of them used the same basic conic projection with concentric circular parallels and straight meridians for radii (Snyder, 1978a). The name of one of them, V. V. Kavrayskiy (or Kavraisky), has been mistakenly applied in some U.S. literature to the basic projection, but his contribution did not occur until 1934.

FEATURES

The Equidistant Conic projection (fig. 22) is neither conformal (like the Lambert Conformal Conic) nor equal-area (like the Albers), but it serves as a compromise between them. The Lambert parallels are more widely spaced away from the central parallel, and the Albers parallels become closer together. The parallels on the Equidistant Conic remain equally spaced on the spherical version (as they are on the sphere) and nearly so on the ellipsoidal version (with the same spacing as the distances along the meridians on the ellipsoid).

As on other normal conics, the meridians are equally spaced radii of the concentric circular arcs which form the parallels. The meridians are spaced at equal angles which are less than the true angles between the meridians; the ratio is called the cone constant, as it is on other conic projections. The poles are normally also plotted as circular arcs.

Either one or two parallels may be made standard or true to scale. There is no shape, area, or scale distortion along the standard parallels. While meridians are at correct scale everywhere, the scale along the parallels between the standard parallels (if there are two) is too small, and the scale along parallels beyond the standard parallel(s) is too great.

If the one standard parallel is the Equator, the Equidistant Conic projection becomes the Plate Carrée form of the Equidistant Cylindrical, but the formulas must be changed. If the two standard parallels are symmetrical about the Equator, the Equirectangular results. If the standard parallel is the pole, the Azimuthal Equidistant projection is obtained.

FIGURE 22.—Equidistant Conic projection, with standard parallels 20° and 60° N. All of North America is included to show that parallels remain equidistant. Compare figures 20 and 21.

USAGE

The Equidistant Conic projection is commonly used in the spherical form in atlases for maps of small countries. Its only use by the USGS has been in an approximate ellipsoidal form for Alaska Maps "B" and "E," but the projection name applied is "Modified Transverse Mercator" (see p. 63), due to the original manner of construction. The formulas for the ellipsoidal version were apparently first published in Snyder (1978a), although there may be several de facto usages of the ellipsoidal form such as the above. For example, the New Mexico Planning Survey in effect devised such a projection in 1936 for the mapping of that State, calling it a "Modified Conic Projection" (Thomas E. Henderson, pers. comm., 1985).

FORMULAS FOR THE SPHERE

For the Equidistant Conic projection with two standard parallels, given R, ϕ_1, ϕ_2, ϕ_0, λ_0, ϕ, and λ, to find x and y (see p. 298 for numerical examples):

$$x = \rho \sin \theta \qquad (14-1)$$
$$y = \rho_0 - \rho \cos \theta \qquad (14-2)$$

where

$$\rho = R (G - \phi) \qquad (16-1)$$
$$\theta = n (\lambda - \lambda_0) \qquad (14-4)$$
$$\rho_0 = R (G - \phi_0) \qquad (16-2)$$
$$G = (\cos \phi_1)/n + \phi_1 \qquad (16-3)$$
$$n = (\cos \phi_1 - \cos \phi_2)/(\phi_2 - \phi_1) \qquad (16-4)$$

ϕ_0, λ_0 = the latitude and longitude for the origin of the rectangular coordinates.
ϕ_1, ϕ_2 = standard parallels.

The Y axis lies along the central meridian λ_0, y increasing northerly; the X axis intersects perpendicularly at ϕ_0, x increasing easterly. If $(\lambda - \lambda_0)$ exceeds the range $\pm 180°$, $360°$ should be added or subtracted. Constants n, G, and ρ_0 need to be determined only once for the entire map.

If only one standard parallel ϕ_1 is desired, equation (16-4) is indeterminate, but $n = \sin \phi_1$. The scale h along meridians is 1.0. Along parallels, using equation (4-5), the scale is

$$k = (G - \phi)n/\cos \phi \qquad (16-5)$$

The maximum angular deformation may be calculated from equation (4-9). As on other regular conics, distortion is only a function of latitude.

For the inverse formulas for the sphere, given R, ϕ_1, ϕ_2, ϕ_0, λ_0, x, and y, to find ϕ and λ: n, G, and ρ_0 are calculated from equations (16-4), (16-3), and (16-2), respectively. Then,

$$\phi = G - \rho/R \qquad (16-6)$$
$$\lambda = \lambda_0 + \theta/n \qquad (14-9)$$

where

$$\rho = \pm [x^2 + (\rho_0 - y)^2]^{1/2}, \text{taking the sign of } n \qquad (14-10)$$
$$\theta = \arctan [x/(\rho_0 - y)] \qquad (14-11)$$

To use the ATAN2 function, if n is negative, the signs of x, y, and ρ_0 (given a negative sign by equation (16−2)) must be reversed before inserting them in equation (14−11).

FORMULAS FOR THE ELLIPSOID

For mapping of regions smaller than the United States at scales greater than 1:5,000,000, using the Equidistant Conic projection, the ellipsoidal formulas should be considered. Given a, e, ϕ_1, ϕ_2, ϕ_0, λ_0, ϕ, and λ, to find x and y (see p. 299 for numerical examples):

$$x = \rho \sin \theta \qquad (14-1)$$
$$y = \rho_0 - \rho \cos \theta \qquad (14-2)$$
$$k = \rho n/(am) \qquad (14-16)$$
$$= (G-M/a)n/m \qquad (16-7)$$

where

$$\rho = a\,G-M \qquad (16-8)$$
$$\theta = n\,(\lambda-\lambda_0) \qquad (14-4)$$
$$\rho_0 = a\,G-M_0 \qquad (16-9)$$
$$n = a(m_1-m_2)/(M_2-M_1) \qquad (16-10)$$
$$m = \cos \phi/(1-e^2 \sin^2 \phi)^{1/2} \qquad (14-15)$$
$$G = m_1/n + M_1/a \qquad (16-11)$$
$$M = a\,[(1-e^2/4-3e^4/64-5e^6/256-\ldots)\phi$$
$$- (3e^2/8+3e^4/32+45e^6/1024+\ldots)\sin 2\phi$$
$$+ (15e^4/256+45e^6/1024+\ldots)\sin 4\phi$$
$$- (35e^6/3072+\ldots)\sin 6\phi + \ldots] \qquad (3-21)$$

with the same subscripts 1, 2, or none applied to m and ϕ in equation (14−15), and 0, 1, 2, or none applied to M and ϕ in equation (3−21). For improved computational efficiency using the series, see p. 19. As with other conics, a negative n and ρ result for projections centered in the Southern Hemisphere. If $\phi_1 = \phi_2$, for the Equidistant Conic with a single standard parallel, equation (16−10) is indeterminate, but $n = \sin \phi_1$. Origin and orientation of axes for x and y are the same as those for the spherical form. Constants n, G, and ρ_0 may be determined just once for the entire map.

For the inverse formulas for the ellipsoid, given a, e, ϕ_1, ϕ_2, ϕ_0, λ_0, x, and y, to find ϕ and λ: n, G, and ρ_0 are calculated from equations (16−10), (16−11), and (16−9), respectively. Then

$$\phi = \mu + (3e_1/2-27e_1^3/32+ \ldots) \sin 2\mu + (21e_1^2/16-55e_1^4/32+ \ldots)$$
$$\sin 4\mu + (151e_1^3/96- \ldots)\sin 6\mu + (1097e_1^4/512- \ldots)\sin 8\mu+ \ldots \qquad (3-26)$$

where

$$e_1 = [1 - (1-e^2)^{1/2}]/[1+(1-e^2)^{1/2}] \qquad (3-24)$$
$$\mu = M/[a(1-e^2/4-3e^4/64-5e^6/256- \ldots)] \qquad (7-19)$$
$$M = a\,G - \rho \qquad (16-12)$$
$$\rho = \pm\,[x^2+(\rho_0-y)^2]^{1/2}, \text{ taking the sign of } n \qquad (14-10)$$
$$\lambda = \lambda_0 + \theta/n \qquad (14-9)$$
$$\theta = \arctan\,[x/(\rho_0-y)] \qquad (14-11)$$

To use the ATAN2 function, if n is negative, the signs of x, y, and ρ_0, must be reversed before inserting them in equation (14−11). For improved computational efficiency using the series (3−26), see p. 19.

Polar coordinates for the Equidistant Conic projection for a map of the United States, assuming standard parallels of lat. 29.5° and 45.5°N., are listed in table 17 for both the spherical and ellipsoidal forms. A graticule extended to the North Pole is shown in figure 22.

To convert coordinates measured on an existing Equidistant Conic map, the user may choose any meridian for λ_0 and therefore for the Y axis, and any latitude for ϕ_0. The X axis then is placed perpendicular to the Y axis at ϕ_0.

TABLE 17.—*Equidistant Conic projection: Polar coordinates*

[Standard parallels: 29.5°, 45.5°N]

Lat.	Projection for sphere (R = 6,370,997 m) (n = 0.6067854)		Projection for Clarke 1866 ellipsoid (a = 6,378,206.4 m) (n = 0.6068355)	
	ρ	k	ρ	k
52°	6,636,493	1.02665	6,656,864	1.02656
51	6,747,688	1.02120	6,768,123	1.02113
50	6,858,883	1.01628	6,879,362	1.01622
49	6,970,078	1.01186	6,990,581	1.01182
48	7,081,272	1.00792	7,101,781	1.00790
47	7,192,467	1.00444	7,212,961	1.00442
46	7,303,662	1.00138	7,324,122	1.00137
45.5	7,359,260	1.00000	7,379,695	1.00000
45	7,414,857	0.99872	7,435,263	0.99873
44	7,526,052	.99646	7,546,384	.99648
43	7,637,247	.99457	7,657,485	.99460
42	7,748,442	.99304	7,768,566	.99307
41	7,859,637	.99186	7,879,628	.99189
40	7,970,831	.99101	7,990,671	.99105
39	8,082,026	.99048	8,101,694	.99052
38	8,193,221	.99026	8,212,697	.99030
37	8,304,416	.99035	8,323,682	.99039
36	8,415,611	.99073	8,434,648	.99077
35	8,526,806	.99140	8,545,594	.99144
34	8,638,001	.99235	8,656,523	.99239
33	8,749,196	.99358	8,767,433	.99361
32	8,860,390	.99508	8,878,325	.99511
31	8,971,585	.99685	8,989,199	.99687
30	9,082,780	.99889	9,100,056	.99889
29.5	9,138,378	1.00000	9,155,478	1.00000
29	9,193,975	1.00118	9,210,896	1.00117
28	9,305,170	1.00373	9,321,720	1.00371
27	9,416,365	1.00654	9,432,527	1.00651
26	9,527,560	1.00960	9,543,318	1.00955
25	9,638,755	1.01291	9,654,093	1.01285
24	9,749,949	1.01648	9,764,854	1.01640
23	9,861,144	1.02030	9,875,600	1.02020
22	9,972,339	1.02437	9,986,332	1.02425

Note: ρ = radius of latitude circles, meters.
h = scale factor along meridians = 1.0.
k = scale factor along parallels.
R = assumed radius of sphere.
a = assumed semimajor axis of ellipsoid.
n = cone constant, or ratio of angle between meridians on map to true angle.

17. BIPOLAR OBLIQUE CONIC CONFORMAL PROJECTION

SUMMARY

- Two oblique conic projections, side-by-side, but with poles 104° apart.
- Conformal.
- Meridians and parallels are complex curves, intersecting at right angles.
- Scale is true along two standard transformed parallels on each conic projection, neither of these lines following any geographical meridian or parallel.
- Very small deviation from conformality, where the two conic projections join.
- Specially developed for a map of the Americas.
- Used only in spherical form.
- Presented by Miller and Briesemeister in 1941.

HISTORY

A "tailor-made" projection is one designed for a certain geographical area. O. M. Miller used the term for some projections which he developed for the American Geographical Society (AGS) or for their clients. The Bipolar Oblique Conic Conformal projection, developed with William A. Briesemeister, was presented in 1941 and designed specifically for a map of North and South America constructed in several sheets by the AGS at a scale of 1:5,000,000 (Miller, 1941).

It is an adaptation of the Lambert Conformal Conic projection to minimize scale error over the two continents by accommodating the fact that North America tends to curve toward the east as one proceeds from north to south, while South America tends to curve in the opposite direction. Because of the relatively small scale of the map, the Earth was treated as a sphere. To construct the map, a great circle arc 104° long was selected to cut through Central America from southwest to northeast, beginning at lat. 20° S. and long. 110° W. and terminating at lat. 45° N. and the resulting longitude of about 19°59′36″ W.

The former point is used as the pole and as the center of transformed parallels of latitude for an Oblique Conformal Conic projection with two standard parallels (at polar distances of 31° and 73°) for all the land in the Americas southeast of the 104° great circle arc. The latter point serves as the pole and center of parallels for an identical projection for all land northwest of the same arc. The inner and outer standard parallels of the northwest portion of the map, thus, are tangent to the outer and inner standard parallels, respectively, of the southeast portion, touching at the dividing line (104°−31°=73°).

The scale of the map was then increased by about 3.5 percent, so that the linear scale error along the central parallels (at a polar distance of 52°, halfway between 31° and 73°) is equal and opposite in sign (−3.5 percent) to the scale error along the two standard parallels (now +3.5 percent) which are at the normal map limits. Under these conditions, transformed parallels at polar distances of about 36.34° and 66.58° are true to scale and are actually the standard transformed parallels.

The use of the Oblique Conformal Conic projection was not original with Miller and Briesemeister. The concept involves the shifting of the graticule of meridians and parallels for the regular Lambert Conformal Conic so that the pole of the projection is no longer at the pole of the Earth. This is the same principle as the transformation for the Oblique Mercator projection. The bipolar concept is unique, however, and it has apparently not been used for any other maps.

FEATURES AND USAGE

The Geological Survey has used the North American portion of the map for the Geologic Map (1965), the Basement Map (1967), the Geothermal Map, and the Metallogenic Map, all retaining the original scale of 1:5,000,000. The Tectonic

Map of North America (1969) is generally based on the Bipolar Oblique Conic Conformal, but there are modifications near the edges. An oblique conic projection about a single transformed pole would suffice for either one of the continents alone, but the AGS map served as an available base map at an appropriate scale. In 1979, the USGS decided to replace this projection with the Transverse Mercator for a map of North America.

The projection is conformal, and each of the two conic projections has all the characteristics of the Lambert Conformal Conic projection, except for the important difference in location of the pole, and a very narrow band near the center. While meridians and parallels on the oblique projection intersect at right angles because the map is conformal, the parallels are not arcs of circles, and the meridians are not straight, except for the peripheral meridian from each transformed pole to the nearest normal pole.

The scale is constant along each circular arc centered on the transformed pole for the conic projection of the particular portion of the map. Thus, the two lines at a scale factor of 1.035, that is, both pairs of the official standard transformed parallels, are mapped as circular arcs forming the letter "S." The 104° great circle arc separating the two oblique conic projections is a straight line on the map, and all other straight lines radiating from the poles for the respective conic projections are transformed meridians and are therefore great circle routes. These straight lines are not normally shown on the finished map.

At the juncture of the two conic projections, along the 104° axis, there is actually a slight mathematical discontinuity at every point except for the two points at which the transformed parallels of polar distance 31° and 73° meet. If the conic projections are strictly followed, there is a maximum discrepancy of 1.6 mm at the 1:5,000,000 scale at the midpoint of this axis, halfway between the poles or between the intersections of the axis with the 31° and 73° transformed parallels. In other words, a meridian approaching the axis from the south is shifted up to 1.6 mm along the axis as it crosses. Along the axis, but beyond the portion between the lines of true scale, the discrepancy increases markedly, until it is over 240 mm at the transformed poles. These latter areas are beyond the needed range of the map and are not shown, just as the polar areas of the regular Lambert Conformal Conic are normally omitted. This would not happen if the Oblique Equidistant Conic projection were used.

The discontinuity was resolved by connecting the two arcs with a straight line tangent to both, a convenience which leaves the small intermediate area slightly nonconformal. This adjustment is included in the formulas below.

FORMULAS FOR THE SPHERE

The original map was prepared by the American Geographical Society, in an era when automatic plotters and easy computation of coordinates were not yet present. Map coordinates were determined by converting the geographical coordinates of a given graticule intersection to the transformed latitude and longitude based on the poles of the projection, then to polar coordinates according to the conformal projection, and finally to rectangular coordinates relative to the selected origin.

The following formulas combine these steps in a form which may be programmed for the computer. First, various constants are calculated from the above parameters, applying to the entire map. Since only one map is involved, the numerical values are inserted in formulas, except where the numbers are transcendental and are referred to by symbols.

If the southwest pole is at point A, the northeast pole is at point B, and the center point on the axis is C,

$$\lambda_B = -110° + \arccos\{[\cos 104° - \sin(-20°)\sin 45°]/ \atop [\cos(-20°)\cos 45°]\} \tag{17-1}$$

= −19°59'36" long., the longitude of B (negative is west long.)

$$n = (\ln \sin 31° - \ln \sin 73°)/[\ln \tan(31°/2) - \ln \tan(73°/2)] \tag{17-2}$$

= 0.63056, the cone constant for both conic projections

$$F_0 = R \sin 31°/[n \tan^n(31°/2)] \tag{17-3}$$

= 1.83376 R, where R is the radius of the globe at the scale of the map.
For the 1:5,000,000 map, R was taken as 6,371,221 m, the radius of a sphere having a volume equal to that of the International ellipsoid.

$$k_0 = 2/[1 + nF_0 \tan^n 26°/(R \sin 52°)] \tag{17-4}$$

= 1.03462, the scale factor by which the coordinates are multiplied to balance the errors

$$F = k_0 F_0 \tag{17-5}$$

= 1.89725 R, a convenient constant

$$Az_{AB} = \arccos\{[\cos(-20°)\sin 45° - \sin(-20°)\cos 45° \cos \atop (\lambda_B + 110°)]/\sin 104°\} \tag{17-6}$$

= 46.78203°, the azimuth east of north of B from A

$$Az_{BA} = \arccos\{[\cos 45° \sin(-20°) - \sin 45° \cos(-20°) \cos \atop (\lambda_B + 110°)]/\sin 104°\} \tag{17-7}$$

= 104.42834°, the azimuth west of north of A from B

$$T = \tan^n(31°/2) + \tan^n(73°/2) \tag{17-8}$$

= 1.27247, a convenient constant

$$\rho_c = \tfrac{1}{2}FT \tag{17-9}$$

= 1.20709 R, the radius of the center point of the axis from either pole

$$z_c = 2 \arctan (T/2)^{1/n} \tag{17-10}$$

= 52.03888°, the polar distance of the center point from either pole

Note that z_c would be exactly 52°, if there were no discontinuity at the axis. The values of ϕ_c, λ_c, and Az_c are calculated as if no adjustment were made at the axis due to the discontinuity. Their use is completely arbitrary and only affects positions of the arbitrary X and Y axes, not the map itself. The adjustment is included in formulas for a given point.

$$\phi_c = \arcsin[\sin(-20°)\cos z_c + \cos(-20°)\sin z_c \cos Az_{AB}] \tag{17-11}$$

= 17°16'28" N. lat., the latitude of the center point, on the southern-cone side of the axis

$$\lambda_c = \arcsin(\sin z_c \sin Az_{AB}/\cos \phi_c) - 110° \tag{17-12}$$

= −73°00'27" long., the longitude of the center point, on the southern-cone side of the axis

$$Az_c = \arcsin[\cos(-20°)\sin Az_{AB}/\cos \phi_c] \tag{17-13}$$

= 45.81997°, the azimuth east of north of the axis at the center point, relative to meridian λ_c on the southern-cone side of the axis

The remaining equations are given in the order used, for calculating rectangular coordinates for various values of latitude ϕ and longitude λ (measured east from Greenwich, or with a minus sign for the western values used here). There are some conditional transfers and adjustments which would apply only if a map extending well beyond the regions of interest were to be plotted; these are omitted to avoid unnecessary complication. It must be established first whether point (ϕ, λ) is north or south of the axis, to determine which conic projection is involved. With these formulas, it is done by comparing the azimuth of point (ϕ, λ) with the azimuth of the axis, all as viewed from B (see p. 301 for numerical examples):

$$z_B = \arccos[\sin 45° \sin \phi + \cos 45° \cos \phi \cos(\lambda_B - \lambda)] \tag{17-14}$$

= polar distance of (ϕ, λ) from pole B

$Az_B = \arctan \{\sin (\lambda_B - \lambda)/[\cos 45° \tan \phi - \sin 45° \cos (\lambda_B - \lambda)]\}$ (17−15)
   = azimuth of (ϕ, λ) west of north, viewed from B

If Az_B is greater than Az_{BA} (from equation (17−7)), go to equation (17−23). Otherwise proceed to equation (17−16) for the projection from pole B.

$\rho_B = F \tan^n \tfrac{1}{2} z_B$ (17−16)
$k = \rho_B n/(R \sin z_B)$ (17−17)
   = scale factor at point (ϕ, λ), disregarding
      small adjustment near axis
$\alpha = \arccos \{[\tan^n \tfrac{1}{2} z_B + \tan^n \tfrac{1}{2}(104° - z_B)]/T\}$ (17−18)

If $|n (Az_{BA} - Az_B)|$ is less than α,

$$\rho_B' = \rho_B/\cos [\alpha - n (Az_{BA} - Az_B)]$$ (17−19)

If the above expression is equal to or greater than α,

$$\rho_B' = \rho_B.$$ (17−20)

Then

$$x' = \rho_B' \sin [n (Az_{BA} - Az_B)]$$ (17−21)
$$y' = \rho_B' \cos [n (Az_{BA} - Az_B)] - \rho_c$$ (17−22)

using constants from equations (17−2), (17−3), (17−7), and (17−9) for rectangular coordinates relative to the axis. To change to nonskewed rectangular coordinates, go to equations (17−32) and (17−33). The following formulas give coordinates for the projection from pole A.

$z_A = \arccos [\sin (-20°) \sin \phi + \cos (-20°) \cos \phi \cos (\lambda + 110°)]$ (17−23)
   = polar distance of (ϕ, λ) from pole A
$Az_A = \arctan \{\sin (\lambda + 110°)/[\cos (-20°) \tan \phi - \sin (-20°)$
            $\cos (\lambda + 110°)]\}$ (17−24)
   = azimuth of (ϕ, λ) east of north, viewed from A
$\rho_A = F \tan^n \tfrac{1}{2} z_A$ (17−25)
$k = \rho_A n/R \sin z_A$ = scale factor at point (ϕ, λ) (17−26)
$\alpha = \arccos \{[\tan^n \tfrac{1}{2} z_A + \tan^n \tfrac{1}{2}(104° - z_A)]/T\}$ (17−27)

If $|n (Az_{AB} - Az_A)|$ is less than α,

$$\rho_A' = \rho_A/\cos [\alpha + n (Az_{AB} - Az_A)]$$ (17−28)

If the above expression is equal to or greater than α,

$$\rho_A' = \rho_A$$ (17−29)

Then

$$x' = \rho_A' \sin [n (Az_{AB} - Az_A)]$$ (17−30)
$$y' = -\rho_A' \cos [n (Az_{AB} - Az_A)] + \rho_c$$ (17−31)
$$x = -x' \cos Az_c - y' \sin Az_c$$ (17−32)
$$y = -y' \cos Az_c + x' \sin Az_c$$ (17−33)

where the center point at (ϕ_c, λ_c) is approximately the origin of (x, y) coordinates, the Y axis increasing due north and the X axis due east from the origin. (The meridian and parallel actually crossing the origin are shifted by about 3' of arc, due to the adjustment at the axis, but their actual values do not affect the calculations here.)

For the inverse formulas for the Bipolar Oblique Conic Conformal, the constants for the map must first be calculated from equations (17–1)–(17–13). Given x and y coordinates based on the above axes, they are then converted to the skew coordinates:

$$x' = -x \cos Az_c + y \sin Az_c \qquad (17-34)$$
$$y' = -x \sin Az_c - y \cos Az_c \qquad (17-35)$$

If x' is equal to or greater than zero, go to equation (17–36). If x' is negative, go to equation (17–45).

$$\rho_B' = [x'^2 + (\rho_c + y')^2]^{1/2} \qquad (17-36)$$
$$Az_B' = \arctan [x'/(\rho_c + y')] \qquad (17-37)$$

Let

$$\rho_B = \rho_B' \qquad (17-38)$$
$$z_B = 2 \arctan (\rho_B/F)^{1/n} \qquad (17-39)$$
$$\alpha = \arccos \{[\tan^n \tfrac{1}{2}z_B + \tan^n \tfrac{1}{2}(104° - z_B)]/T\} \qquad (17-40)$$

If $|Az_B'|$ is equal to or greater than α, go to equation (17–42). If $|Az_B'|$ is less than α, calculate

$$\rho_B = \rho_B' \cos (\alpha - Az_B') \qquad (17-41)$$

and use this value to recalculate equations (17–39), (17–40), and (17–41), repeating until ρ_B found in (17–41) changes by less than a predetermined convergence. Then,

$$Az_B = Az_{BA} - Az_B'/n \qquad (17-42)$$

Using Az_B and the final value of z_B,

$$\phi = \arcsin (\sin 45° \cos z_B + \cos 45° \sin z_B \cos Az_B) \qquad (17-43)$$
$$\lambda = \lambda_B - \arctan \{\sin Az_B/[\cos 45°/\tan z_B - \sin 45° \cos Az_B]\} \qquad (17-44)$$

The remaining equations are for the southern cone only (negative x'):

$$\rho_A' = [x'^2 + (\rho_c - y')^2]^{1/2} \qquad (17-45)$$
$$Az_A' = \arctan [x'/(\rho_c - y')] \qquad (17-46)$$

Let

$$\rho_A = \rho_A' \qquad (17-47)$$
$$z_A = 2 \arctan (\rho_A/F)^{1/n} \qquad (17-48)$$
$$\alpha = \arccos \{[\tan^n \tfrac{1}{2}z_A + \tan^n \tfrac{1}{2}(104° - z_A)]/T\} \qquad (17-49)$$

If $|Az_A'|$ is equal to or greater than α, go to equation (17–51). If $|Az_A'|$ is less than α, calculate

$$\rho_A = \rho_A' \cos (\alpha + Az_A') \qquad (17-50)$$

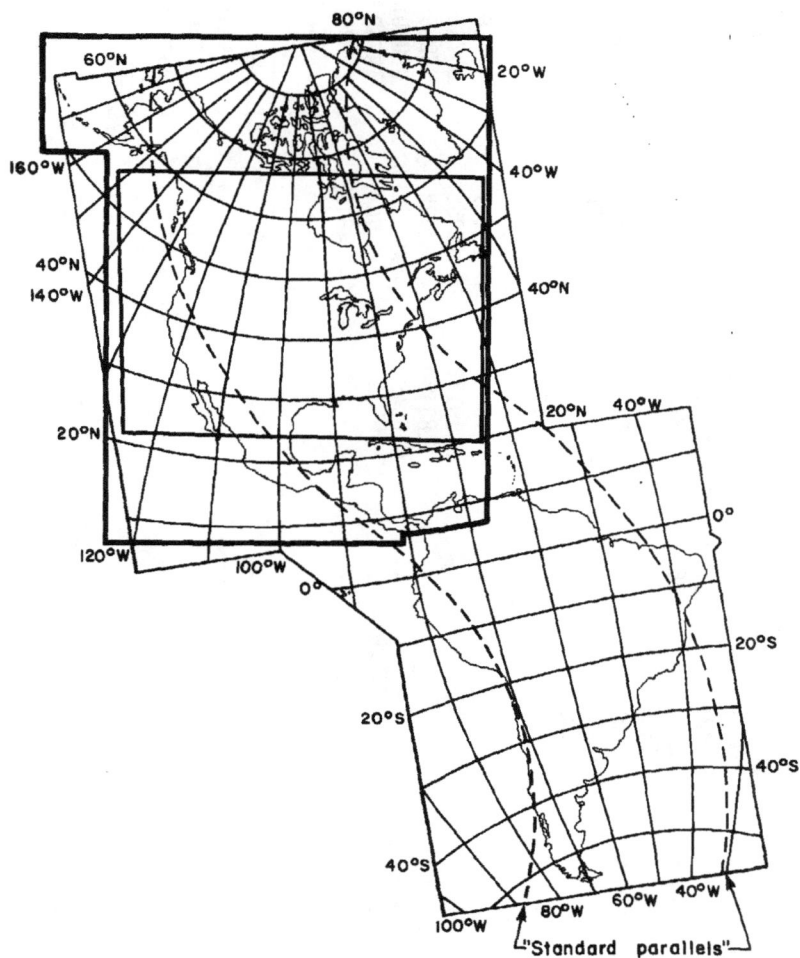

FIGURE 23.—Bipolar Oblique Conic Conformal projection used for various geologic maps. The American Geographical Society, under O. M. Miller, prepared the base map used by the USGS. (Prepared by Tau Rho Alpha.)

and use this value to recalculate equations (17−48), (17−49), and (17−50), repeating until ρ_A found in equation (17−50) changes by less than a predetermined convergence. Then,

$$Az_A = Az_{AB} - Az_A{}'/n \qquad (17-51)$$

Using Az_A and the final value of z_A,

$$\phi = \arcsin\left[\sin(-20°)\cos z_A + \cos 20° \sin z_A \cos Az_A\right] \qquad (17-52)$$

$$\lambda = \arctan\left\{\sin Az_A/[\cos(-20°)/\tan z_A - \sin(-20°)\cos Az_A]\right\} - 110° \qquad (17-53)$$

Equations (17−17) or (17−26) may be used for calculating k after ϕ and λ are determined.

A table of rectangular coordinates is given in table 18, based on a radius R of 1.0, while a graticule is shown in figure 23.

TABLE 18.—*Bipolar Oblique Conic Conformal projection: Rectangular coordinates*

[R=1.0. y coordinates in parentheses below x coordinates. Solid line separates the portions formed from the two transformed poles. Origin at approximately lat. 17°15' N., long. 73°02' W., with Y axis due north at that point only]

Lat.	100°	110°	120°	130°	140°	150°	160°	170°
90°	---	---	---	---	---	---	---	-0.14576 (1.24309)
80	-0.18175 (1.07737)	-0.20973 (1.08634)	-0.23569 (1.09992)	-0.25892 (1.11769)	-0.27876 (1.13914)	-0.29464 (1.16367)	-0.30608 (1.19057)	-.31273 (1.21904)
70	.22126 (.91246)	.27593 (.93098)	.32673 (.95806)	.37243 (.99311)	.41182 (1.03535)	.44372 (1.08381)	-.46705 (1.13725)	-.48092 (1.19421)
60	.26359 (.74880)	.34310 (.77644)	.41763 (.81589)	.48559 (.86677)	.54518 (.92849)	.59444 (1.00006)	-.63138 (1.07999)	-.65416 (1.16623)
50	.30732 (.58603)	.40985 (.62133)	.50739 (.67112)	.59806 (.73570)	.67943 (.81512)	.74851 (.90886)	-.80200 (1.01551)	-.83656 (1.13260)
40	.35078 (.42294)	.47485 (.46360)	.59515 (.52100)	.70964 (.59654)	.81523 (.69140)	---	---	---
30	.39231 (.25766)	.53678 (.30074)	.67999 (.36240)	.81990 (.44545)	---	---	---	---
20	.43026 (.08782)	.59421 (.12988)	.76061 (.19177)	---	---	---	---	---
10	.46280 (-.08930)	.64522 (-.05222)	.83496 (.00499)	---	---	---	---	---
0	.48758 (-.27670)	.68704 (-.24918)	---	---	---	---	---	---
-10	.50751* (-.48360)	.72338 (-.47150)	---	---	---	---	---	---
-20	.48812 (-.73406)	.86567* (-.84124)	---	---	---	---	---	---
-30	.38781 (-.96476)	.55209 (-1.10271)	---	---	---	---	---	---
-40	.26583 (-1.14111)	.37784 (-1.24800)	---	---	---	---	---	---
-50	.14798 (-1.28862)	.23054 (-1.37082)	---	---	---	---	---	---
-60	.03499 (-1.42268)	.09524 (-1.48363)	---	---	---	---	---	---
-70	-.07542 (-1.55124)	-.03504 (-1.59227)	---	---	---	---	---	---
-80	-.18569 (-1.67949)	-.16491 (-1.70055)	---	---	---	---	---	---
-90	---	-.29823 (-1.81171)	---	---	---	---	---	---

beyond arbitrary map limits (columns 120°–170°, lat. 0° through –90°)

Lat. W. Long.	90°	80°	70°	60°	50°	40°	30°	20°	10°
90°	-0.14576 (1.24309)	—	—	—	—	—	—	—	—
80	-.15254 (1.07330)	-.12293 (1.07432)	-.09378 (1.08048)	-.06599 (1.09170)	-.04047 (1.10774)	-.01809 (1.12816)	0.00033 (1.15236)	0.01411 (1.17955)	0.02275 (1.20877)
70	-.16395 (.90303)	-.10525 (.90303)	-.04651 (.91292)	.01074 (.93301)	.06470 (.96349)	.11317 (1.00421)	.15365 (1.05436)	.18369 (1.11215)	.20152 (1.17478)
60	-.18043 (.73324)	-.09477 (.73013)	-.00767 (.74005)	.07976 (.76403)	.16594 (.80369)	.24806 (.86133)	.32065 (.93920)	.37468 (1.03623)	.40201 (1.14388)
50	-.20109 (.56481)	-.09192 (.55749)	.01990 (.56421)	.13461 (.58582)	.25295 (.62443)	.37631 (.68480)	.50548 (.77907)	.62083 (.93836)	.64638 (1.13342)
40	-.22411 (.39765)	-.09519 (.38660)	.03637 (.38903)	.17183 (.40460)	.31377 (.43354)	.46682* (.47595)	.64259* (.54614)	—	—
30	-.24741 (.23065)	-.10203 (.21759)	.04468 (.21675)	.19431 (.22664)	.34922 (.24602)	.51120 (.27522)	.68326 (.31537)	—	—
20	-.26899 (.06192)	-.10979 (.04921)	.04816* (.04683)	.20770 (.05280)	.37167 (.06603)	.54280 (.08551)	.72518 (.11131)	—	—
10	-.28689 (-.11090)	-.11634* (-.12083)	.05000 (-.12223)	.21614 (-.11773)	.38494 (-.10970)	.56021 (-.09944)	.74645 (-.08790)	—	—
0	-.29905* (-.29059)	-.11920 (-.29390)	.05292 (-.29122)	.22166 (-.28661)	.39129 (-.28234)	.56601 (-.28009)	.75029 (-.28151)	—	—
-10	-.29984 (-.48267)	-.11376 (-.47202)	.05921 (-.46189)	.22626 (-.45503)	.39254 (-.45295)	.56225 (-.45710)	.73941 (-.46938)	—	—
-20	-.27575 (-.68590)	-.09495 (-.65440)	.07161 (-.63424)	.23171 (-.62366)	.39016 (-.62240)	.55057 (-.63119)	.71601 (-.65175)	—	—
-30	-.21865 (-.88430)	-.05954 (-.83575)	.09194 (-.80677)	.23925 (-.79252)	.38524 (-.79127)	.53215 (-.80304)	.68181 (-.82907)	—	—
-40	-.13981 (-1.06299)	-.00990 (-1.01016)	.12002 (-.97740)	.24931 (-.96122)	.37838 (-.95992)	.50784 (-.97320)	.63813 (-1.00184)	—	—
-50	-.05346 (-1.22345)	.04829 (-1.17590)	.15387 (-1.14498)	.26134 (-1.12947)	.36964 (-1.12858)	.47806 (-1.14214)	.58591 (-1.17057)	beyond arbitrary map limits	—
-60	.03430 (-1.37283)	.11029 (-1.33514)	.19081 (-1.31002)	.27404 (-1.29749)	.35849 (-1.29753)	.44283 (-1.31019)	.52574 (-1.33568)	—	—
-70	.12196 (-1.51739)	.17341 (-1.49156)	.22844 (-1.47435)	.28571 (-1.46615)	.34391 (-1.46721)	.40173 (-1.47764)	.45785 (-1.49748)	—	—
-80	.20970 (-1.66218)	.23631 (-1.64908)	.26481 (-1.64057)	.29445 (-1.63693)	.32443 (-1.63831)	.35394 (-1.64474)	.38215 (-1.65615)	—	—
-90	.29823 (-1.81171)	—	—	—	—	—	—	—	—

*Adjustment to x and y made for discontinuity near axis of conic projections.

18. POLYCONIC PROJECTION

SUMMARY

- Neither conformal nor equal-area.
- Parallels of latitude (except for Equator) are arcs of circles, but are not concentric.
- Central meridian and Equator are straight lines; all other meridians are complex curves.
- Scale is true along each parallel and along the central meridian, but no parallel is "standard."
- Free of distortion only along the central meridian.
- Used almost exclusively in slightly modified form for large-scale mapping in the United States until the 1950's.
- Was apparently originated about 1820 by Hassler.

HISTORY

Shortly before 1820, Ferdinand Rudolph Hassler (fig. 24) began to promote the Polyconic projection, which was to become a standard for much of the official mapping of the United States (Deetz and Adams, 1934, p. 58—60).

Born in Switzerland in 1770, Hassler arrived in the United States in 1805 and was hired 2 years later as the first head of the Survey of the Coast. He was forced to wait until 1811 for funds and equipment, meanwhile teaching to maintain income. After funds were granted, he spent 4 years in Europe securing equipment. Surveying began in 1816, but Congress, dissatisfied with the progress, took the Survey from his control in 1818. The work only foundered. It was returned to Hassler, now superintendent, in 1832. Hassler died in Philadelphia in 1843 as a result of exposure after a fall, trying to save his instruments in a severe wind and hailstorm, but he had firmly established what later became the U.S. Coast and Geodetic Survey (Wraight and Roberts, 1957) and is now the National Ocean Service.

The Polyconic projection, usually called the American Polyconic in Europe, achieved its name because the curvature of the circular arc for each parallel on the map is the same as it would be following the unrolling of a cone which had been wrapped around the globe tangent to the particular parallel of latitude, with the parallel traced onto the cone. Thus, there are many ("poly-") cones involved, rather than the single cone of each regular conic projection. As Hassler himself described the principles, "[t]his distribution of the projection, in an assemblage of sections of surfaces of successive cones, tangents to or cutting a regular succession of parallels, and upon regularly changing central meridians, appeared to me the only one applicable to the coast of the United States" (Hassler, 1825, p. 407—408).

The term "polyconic" is also applied generically by some writers to other projections on which parallels are shown as circular arcs. Most commonly, the term applies to the specific projection described here.

FEATURES

The Polyconic projection (fig. 25) is neither equal-area nor conformal. Along the central meridian, however, it is both distortion free and true to scale. Each parallel is true to scale, but the meridians are lengthened by various amounts to cross each parallel at the correct position along the parallel, so that no parallel is standard in the sense of having conformality (or correct angles), except at the

FIGURE 24.—Ferdinand Rudolph Hassler (1770–1843), first Superintendent of the U.S. Coast Survey and presumed inventor of the Polyconic projection. As a result of his promotion of its use, it became the projection exclusively used for USGS topographic quadrangles for about 70 years.

central meridian. Near the central meridian, which is the case with 7½-minute quadrangles, distortion is extremely small. The Polyconic projection is universal in that tables of rectangular coordinates may be used for any Polyconic projection of the same ellipsoid by merely applying the proper scale and central meridian. U.S. Coast and Geodetic Survey Special Publication No. 5 (1900) replaced tables published in 1884 and was often reprinted because of the universality of the projection (the Clarke 1866 is the reference ellipsoid). Polyconic quadrangle maps prepared to the same scale and for the same central meridian and ellipsoid will fit

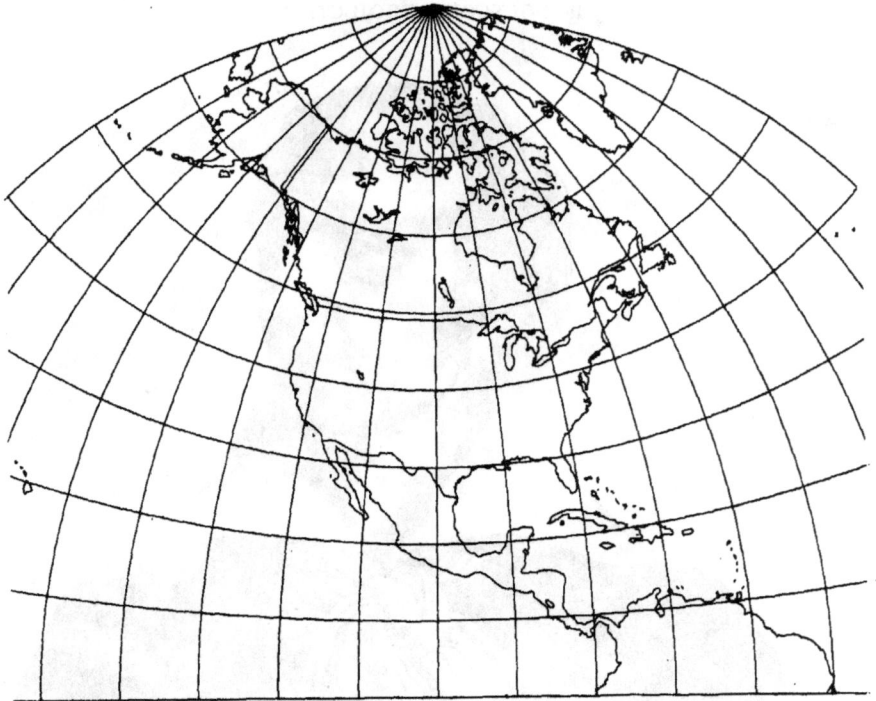

FIGURE 25.—North America on a Polyconic projection grid, central meridian long. 100° W., using a
10° interval. The parallels are arcs of circles which are not concentric, but have radii equal to the
radius of curvature of the parallel at the Earth's surface. The meridians are complex curves formed
by connecting points marked off along the parallels at their true distances. Used by the USGS
for topographic quadrangle maps.

exactly from north to south. Since they are drawn in practice with straight
meridians, they also fit east to west, but discrepancies will accumulate if mosaicking is attempted in both directions.

The parallels are all circular arcs, with the centers of the arcs lying along an
extension of the straight central meridian, but these arcs are not concentric.
Instead, as noted earlier, the radius of each arc is that of the circle developed
from a cone tangent to the sphere or ellipsoid at the latitude. For the sphere, each
parallel has a radius proportional to the cotangent of the latitude. For the ellipsoid,
the radius is slightly different. The Equator is a straight line in either case. Along
the central meridian, the parallels are spaced at their true intervals. For the
sphere, they are therefore equidistant. Each parallel is marked off for meridians
equidistantly and true to scale. The points so marked are connected by the curved
meridians.

USAGE

As geodetic and coastal surveying began in earnest during the 19th century,
the Polyconic projection became a standard, especially for quadrangles. Most
coastal charts produced by the Coast Survey and its successor during the 19th
century were based on one or more variations of the Polyconic projection
(Shalowitz, 1964, p. 138−141). The name of the projection appears on a later
reprint of one of the first published USGS topographic quadrangles, which
appeared in 1886. In 1904, the USGS published tables of rectangular coordinates
extracted from an 1884 Coast and Geodetic Survey report. They were called

"coordinates of curvature," but were actually coordinates for the Polyconic projection, although the latter term was not used (Gannett, 1904, p. 37–48).

As a 1928 USGS bulletin of topographic instructions stated (Beaman, 1928, p. 163):

The topographic engineer needs a projection which is simple in construction, which can be used to represent small areas on any part of the globe, and which, for each small area to which it is applied, preserves shapes, areas, distances, and azimuths in their true relation to the surface of the earth. The polyconic projection meets all these needs and was adopted for the standard topographic map of the United States, in which the 1° quadrangle is the largest unit * * * and the 15' quadrangle is the average unit. * * * Misuse of this projection in attempts to spread it over large areas—that is, to construct a single map of a large area—has developed serious errors and gross exaggeration of details. For example, the polyconic projection is not at all suitable for a single-sheet map of the United States or of a large State, although it has been so employed.

When coordinate plotters and published tables for the State Plane Coordinate System (SPCS) became available in the late 1950's, the USGS ceased using the Polyconic for new maps, in favor of the Transverse Mercator or Lambert Conformal Conic projections used with the SPCS for the area involved. Some of the quadrangles prepared on one or the other of these projections have continued to carry the Polyconic designation, however.

The Polyconic projection was also used for the Progressive Military Grid for military mapping of the United States. There were seven zones, A–G, with central meridians every 8° west from long. 73° W. (zone A), each zone having an origin at lat. 40°30' N. on the central meridian with coordinates $x = 1,000,000$ yards, $y = 2,000,000$ yards (Deetz and Adams, 1934, p. 87–90). Some USGS quadrangles of the 1930's and 1940's display tick marks according to this grid in yards, and many quadrangles then prepared by the Army Map Service and sold by the USGS show a complete grid pattern. This grid was incorporated intact into the World Polyconic Grid (WPG) until both were superseded by the Universal Transverse Mercator grid (Mugnier, 1983).

While quadrangles based on the Polyconic provide low-distortion mapping of the local areas, the inability to mosaic these quadrangles in all directions without gaps makes them less satisfactory for a larger region. Quadrangles based on the SPCS may be mosaicked over an entire zone, at the expense of increased distortion.

For an individual quadrangle 7½ or 15 minutes of latitude or longitude on a side, the distance of the quadrangle from the central meridian of a Transverse Mercator zone or from the standard parallels of a Lambert Conformal Conic zone of the SPCS has much more effect than the type of projection upon the variation in measurement of distances on quadrangles based on the various projections. If the central meridians or standard parallels of the SPCS zones fall on the quadrangle, a change of projection from Polyconic to Transverse Mercator or Lambert Conformal Conic results in a difference of less than 0.001 mm in the measurement of the 700–800 mm diagonals of a 7½-minute quadrangle. If the quadrangle is near the edge of a zone, the discrepancy between measurements of diagonals on two maps of the same quadrangle, one using the Transverse Mercator or Lambert Conformal Conic projection and the other using the Polyconic, can reach about 0.05 mm. These differences are exceeded by variations in expansion and contraction of paper maps, so that these mathematical discrepancies apply only to comparisons of stable-base maps.

Actually, the central meridian of a 7½-minute Polyconic quadrangle may lie along the edge of the map, since 15-minute quadrangles were frequently cut and enlarged to achieve the less extensive coverage. This has a negligible effect upon the map geometry.

Before the Lambert became the projection for the 1:500,000 State base map series, a modified form of the Polyconic was used, but the details are unclear. The

Polyconic was used for the base maps of Alaska until 1972. It has also been used for maps of the United States; but, as stated above, the distortion is excessive at the east and west coasts, and most current maps are drawn to either the Lambert or Albers Conic projections. There are several other modified Polyconic projections, in use or devised, including the Rectangular Polyconic and Bousfield's modification used for northern Canada (Haines, 1981). The best known is that used for the International Map of the World, described on p. 131.

GEOMETRIC CONSTRUCTION

Because of the simplicity of construction using universal tables with which the central meridian and each parallel may be marked off at true distances, the Polyconic projection was favored long after theoretically better projections became known in geodetic circles.

The Polyconic projection must be constructed with curved meridians and parallels if it is used for single-sheet maps of areas with east-west extent of several degrees. Then, however, the inherent distortion is excessive, and a different projection should be considered. For accurate topographic work, the coverage must remain so small that the meridians and parallels may ironically but satisfactorily be drawn as straight-line segments. Official USGS instructions of 1928 declared that

* * * in actual practice on projections of small quadrangles, the parallels are not drawn as arcs of circles, but their intersections with the meridians are plotted from the computed x and y values, and the sections of the parallels between adjacent meridians are drawn as straight lines. In polyconic projections of quadrangles of 1° or smaller meridians may be drawn as straight lines, and in large-scale projections of small quadrangles in low latitudes both meridians and parallels may be drawn as straight lines. For example, the curvature of the parallels of a projection of a 15' quadrangle on a scale of 1:48,000 in latitudes from 0° to 30° is so small that it can not be plotted, and for a 7½' quadrangle on a scale of 1:31,680 or larger the curvature can not be plotted at any latitude (Beaman, 1928, p. 167).

This instruction is essentially repeated in the 1964 edition (USGS, 1964, p. 12−13). The formulas given below are based on curved meridians.

FORMULAS FOR THE SPHERE

The principles stated above lead to the following forward formulas for rectangular coordinates for the spherical form of the Polyconic projection, using radians (see p. 303 for numerical examples):

If ϕ is 0,

$$x = R(\lambda - \lambda_0) \qquad (7-1)$$
$$y = -R\phi_0 \qquad (18-1)$$

If ϕ is not 0,

$$E = (\lambda - \lambda_0) \sin \phi \qquad (18-2)$$
$$x = R \cot \phi \sin E \qquad (18-3)$$
$$y = R[\phi - \phi_0 + \cot \phi (1 - \cos E)] \qquad (18-4)$$

where ϕ_0 is an arbitrary latitude (frequently the Equator) chosen for the origin of the rectangular coordinates at its intersection with λ_0, the central meridian. As with other conics and the Transverse Mercator, the Y axis coincides with the central meridian, y increasing northerly, and the X axis intersects perpendicularly at ϕ_0, x increasing easterly. If $(\lambda - \lambda_0)$ exceeds the range ±180°, 360° must be added or subtracted to place it within the range. For the scale factor h along the meridians (Adams, 1919, p. 144−147):

$$h = (1 - \cos^2 \phi \cos E)/(\sin^2 \phi \cos D) \qquad (18-5)$$

where

$$D = \arctan \left[(E - \sin E)/(\sec^2 \phi - \cos E) \right] \qquad (18-6)$$

If ϕ is 0, this is indeterminate, but h is then $[1 + (\lambda - \lambda_0)^2/2]$. In all cases, the scale factor k along any parallel is 1.0.

The inverse formulas for the sphere are given here in the form of a Newton-Raphson approximation, which converges to any desired accuracy after several iterations, except that if $|\lambda - \lambda_0| > 90°$, a rarely used range, this iteration does not converge, and if $y = -R\phi_0$, it is indeterminate. In the latter case, however,

$$\phi = 0$$
$$\lambda = x/R + \lambda_0 \qquad (7-5)$$

Otherwise, if $y \neq -R\phi_0$, calculations are made in this order:

$$A = \phi_0 + y/R \qquad (18-7)$$
$$B = x^2/R^2 + A^2 \qquad (18-8)$$

Using an initial value of $\phi_n = A$, ϕ_{n+1} is found from equation (18-9),

$$\phi_{n+1} = \phi_n - [A(\phi_n \tan \phi_n + 1) - \phi_n - \tfrac{1}{2}(\phi_n^2 + B) \tan \phi_n]/$$
$$[(\phi_n - A)/\tan \phi_n - 1] \qquad (18-9)$$

The new trial value of ϕ_{n+1} is successively substituted in place of ϕ_n, until ϕ_{n+1} differs from ϕ_n by less than a predetermined convergence limit. Then $\phi = \phi_{n+1}$ as finally determined.

$$\lambda = [\arcsin (x \tan \phi/R)]/\sin \phi + \lambda_0 \qquad (18-10)$$

If $\phi = \pm 90°$, equation (18-10) is indeterminate, but λ may be given any value, such as λ_0.

FORMULAS FOR THE ELLIPSOID

The forward formulas for the ellipsoidal form of the Polyconic projection are only a little more complicated than those for the sphere. These formulas are theoretically exact. They are adapted from formulas given by the Coast and Geodetic Survey (1946, p. 4) (see p. 304 for numerical examples):

If ϕ is zero:

$$x = a(\lambda - \lambda_0) \qquad (7-6)$$
$$y = -M_0 \qquad (18-11)$$

If ϕ is not zero:

$$E = (\lambda - \lambda_0) \sin \phi \qquad (18-2)$$
$$x = N \cot \phi \sin E \qquad (18-12)$$
$$y = M - M_0 + N \cot \phi (1 - \cos E) \qquad (18-13)$$

where

$$M = a[(1 - e^2/4 - 3e^4/64 - 5e^6/256 - \ldots) \phi - (3e^2/8 + 3e^4/32 + 45e^6/1024$$
$$+ \ldots) \sin 2\phi + (15e^4/256 + 45e^6/1024 + \ldots) \sin 4\phi - (35e^6/3072$$
$$+ \ldots) \sin 6\phi + \ldots] \qquad (3-21)$$
$$N = a/(1 - e^2 \sin^2\phi)^{1/2} \qquad (4-20)$$

and M_0 is found from equation (3–21) by using ϕ_0 for ϕ and M_0 for M, with ϕ_0 the latitude of the origin of rectangular coordinates at its intersection with central meridian λ_0. See the spherical formulas for the orientation of axes. The value of $(\lambda - \lambda_0)$ must be adjusted by adding or subtracting 360° if necessary to fall within the range of ±180°. For scale factor h along the meridians ($k = 1.0$ along the parallels):

If ϕ is zero,

$$h = [M' + \tfrac{1}{2}(\lambda - \lambda_0)^2]/(1 - e^2) \tag{18–14}$$

If ϕ is not zero (Adams, 1919, p. 144–146),

$$h = [1 - e^2 + 2(1 - e^2 \sin^2 \phi) \sin^2 \tfrac{1}{2} E/\tan^2 \phi]/[(1 - e^2) \cos D] \tag{18–15}$$

where

$$D = \arctan \{(E - \sin E)/[\sec^2 \phi - \cos E - e^2 \sin^2 \phi/(1 - e^2 \sin^2 \phi)]\} \tag{18–16}$$
$$M' = 1 - e^2/4 - 3e^4/64 - 5e^6/256 - \ldots - 2\,(3e^2/8 + 3e^4/32 + 45e^6/1024$$
$$+ \ldots)\cos 2\phi + 4\,(15e^4/256 + 45e^6/1024 + \ldots)\cos 4\phi - 6$$
$$(35e^6/3072 + \ldots)\cos 6\phi + \ldots. \tag{18–17}$$

For improved computational efficiency using this series, see p. 19.

As with the inverse spherical formulas, the *inverse ellipsoidal formulas* are given in a Newton-Raphson form, converging to any desired degree of accuracy after several iterations. As before, if $|\lambda - \lambda_0| > 90°$, this iteration does not converge, but the projection should not be used in that range in any case. The formulas may be calculated in the following order, given a, e, ϕ_0, λ_0, x, and y. First M_0 is calculated from equation (3–21) above, as in the forward case, with ϕ_0 for ϕ and M_0 for M.

If $y = -M_0$, the iteration is not applicable, but

$$\begin{aligned} \phi &= 0 \\ \lambda &= x/a + \lambda_0 \end{aligned} \tag{7–12}$$

If $y \neq -M_0$, the calculation is as follows:

$$\begin{aligned} A &= (M_0 + y)/a \tag{18–18} \\ B &= x^2/a^2 + A^2 \tag{18–19} \end{aligned}$$

Using an initial value of $\phi_n = A$, the following calculations are made:

$$C = (1 - e^2 \sin^2 \phi_n)^{1/2} \tan \phi_n \tag{18–20}$$

Then M_n and M_n' are found from equations (3–21) and (18–17) above, using ϕ_n for ϕ, M_n for M, and M_n' for M'. Let $M_a = M_n/a$.

$$\phi_{n+1} = \phi_n - [A(CM_a + 1) - M_a - \tfrac{1}{2}(M_a^2 + B)C]/[e^2 \sin 2\phi_n (M_a^2 + B - 2AM_a)/$$
$$4C + (A - M_a)(CM_n' - 2/\sin 2\phi_n) - M_n'] \tag{18–21}$$

Each value of ϕ_{n+1} is substituted in place of ϕ_n, and C, M_n, M_n', and ϕ_{n+1} are recalculated from equations (18–20), (3–21), (18–17), and (18–21), respectively. This process is repeated until ϕ_{n+1} varies from ϕ_n by less than a predetermined convergence value. Then ϕ equals the final ϕ_{n+1}.

$$\lambda = [\arcsin(xC/a)]/\sin \phi + \lambda_0 \tag{18–22}$$

using the C calculated for the last ϕ_n from equation (18–20). If $\phi = \pm 90°$, λ is indeterminate, but may be given any value.

Table 19 lists rectangular coordinates for a band 3° on either side of the central meridian for the ellipsoid extending from lat. 23° to 50° N. Figure 25 shows the graticule applied to a map of North America.

MODIFIED POLYCONIC FOR THE INTERNATIONAL MAP OF THE WORLD

A modified Polyconic projection was devised by Lallemand of France and in 1909 adopted by the International Map Committee (IMC) in London as the basis for the 1:1,000,000-scale International Map of the World (IMW) series. Used for sheets 6° of longitude by 4° of latitude between lats. 60° N. and 60° S., 12° of longitude by 4° of latitude between lats. 60° and 76° N. or S., and 24° by 4° between lats. 76° and 84° N. or S., the projection differs from the ordinary Polyconic in two principal features: All meridians are straight, and there are two meridians (2° east and west of the central meridian on sheets between lats. 60° N. & S.) that are made true to scale. Between lats. 60° & 76° N. and S., the meridians 4° east and west are true to scale, and between 76° & 84°, the true-scale meridians are 8° from the central meridian (United Nations, 1963, p. 22–23; Lallemand, 1911, p. 559).

The top and bottom parallels of each sheet are nonconcentric circular arcs constructed with radii of $N \cot \phi$, where $N = a/(1 - e^2 \sin^2 \phi)^{1/2}$. These radii are the same as the radii on the regular Polyconic for the ellipsoid, and the arcs of these two parallels are marked off true to scale for the straight meridians. The two parallels, however, are spaced from each other according to the true scale along the two standard meridians, not according to the scale along the central meridian, which is slightly reduced. The approximately 440 mm true length of the central meridian at the map scale is thereby reduced by 0.270 to 0.076 mm, depending on the latitude of the sheet. Other parallels of lat. ϕ are circular arcs with radii $N \cot \phi$, intersecting the meridians which are true to scale at the correct points. The parallels strike other meridians at geometrically fixed locations which slightly deviate from the true scale on meridians as well as parallels.

With this modified Polyconic, as with USGS quadrangles based on the rectified Polyconic, adjacent sheets exactly fit together not only north to south, but east to west. There is still a gap when mosaicking in all directions, in that there is a gap between each diagonal sheet and either one or the other adjacent sheet.

In 1962, a U.N. conference on the IMW adopted the Lambert Conformal Conic and Polar Stereographic projections to replace the modified Polyconic (United Nations, 1963, p. 9–10). The USGS has prepared a number of sheets for the IMW series over the years according to the projection officially in use at the time.

FORMULAS FOR THE IMW MODIFIED POLYCONIC

Since the projection was designed solely for this series, the formulas below are based on the ellipsoid. They were derived in 1982 (Snyder, 1982b). The following symbols are used in these formulas:

a = semimajor axis on the given reference ellipsoid
C = distance on the map of latitude ϕ from latitude ϕ_1, measured along the central meridian of longitude λ_0
C_2 = distance on the map of latitude ϕ_2 from latitude ϕ_1, measured along the central meridian of longitude λ_0
e = eccentricity of the given reference ellipsoid
M = distance on the ellipsoid along any meridian from the Equator to ϕ
M_2 = ditto for ϕ_2
M_1 = ditto for ϕ_1

TABLE 19.—*Polyconic Projection: Rectangular coordinates for the Clarke 1866 ellipsoid*

[y coordinates in parentheses under x coordinates. Italic indicates h]

Long. λ Lat. φ	0°	1°	2°	3°
50°	0 (5,540,628) *1.000000*	71,696 (5,541,107) *1.000063*	143,379 (5,542,545) *1.000252*	215,037 (5,544,941) *1.000568*
49	0 (5,429,409) *1.000000*	73,172 (5,429,890) *1.000066*	146,331 (5,431,336) *1.000263*	219,465 (5,433,745) *1.000592*
48	0 (5,318,209) *1.000000*	74,626 (5,318,693) *1.000068*	149,239 (5,320,144) *1.000274*	223,827 (5,322,564) *1.000616*
47	0 (5,207,028) *1.000000*	76,056 (5,207,514) *1.000071*	152,100 (5,208,970) *1.000284*	228,119 (5,211,397) *1.000640*
46	0 (5,095,868) *1.000000*	77,464 (5,096,354) *1.000074*	154,915 (5,097,813) *1.000295*	232,342 (5,100,244) *1.000664*
45	0 (4,984,727) *1.000000*	78,847 (4,985,214) *1.000076*	157,682 (4,986,673) *1.000306*	236,493 (4,989,106) *1.000688*
44	0 (4,873,606) *1.000000*	80,207 (4,874,092) *1.000079*	160,401 (4,875,551) *1.000316*	240,572 (4,877,982) *1.000712*
43°	0 (4,762,505) *1.000000*	81,541 (4,762,990) *1.000082*	163,071 (4,764,446) *1.000327*	244,578 (4,766,872) *1.000736*
42	0 (4,651,423) *1.000000*	82,851 (4,651,907) *1.000084*	165,691 (4,653,358) *1.000338*	248,508 (4,655,777) *1.000760*
41	0 (4,540,361) *1.000000*	84,136 (4,540,843) *1.000087*	168,260 (4,542,288) *1.000348*	252,363 (4,544,696) *1.000784*
40	0 (4,429,319) *1.000000*	85,394 (4,429,798) *1.000090*	170,778 (4,431,235) *1.000359*	256,140 (4,433,630) *1.000808*
39	0 (4,318,296) *1.000000*	86,627 (4,318,772) *1.000092*	173,243 (4,320,199) *1.000369*	259,839 (4,322,577) *1.000831*
38	0 (4,207,292) *1.000000*	87,833 (4,207,764) *1.000095*	175,656 (4,209,180) *1.000380*	263,458 (4,211,539) *1.000855*
37	0 (4,096,308) *1.000000*	89,012 (4,096,775) *1.000098*	178,015 (4,098,178) *1.000390*	266,997 (4,100,515) *1.000878*
36	0 (3,985,342) *1.000000*	90,164 (3,985,805) *1.000100*	180,319 (3,987,192) *1.000400*	270,455 (3,989,504) *1.000901*

TABLE 19.—*Polyconic Projection: Rectangular coordinates for the Clarke 1866 ellipsoid*—Continued

Long. λ Lat. φ	0°	1°	2°	3°
35 _____	0 (3,874,395) *1.000000*	91,289 (3,874,852) *1.000103*	182,568 (3,876,223) *1.000411*	273,830 (3,878,507) *1.000924*
34 _____	0 (3,763,467) *1.000000*	92,385 (3,763,918) *1.000105*	184,762 (3,765,270) *1.000421*	277,121 (3,767,524) *1.000946*
33 _____	0 (3,652,557) *1.000000*	93,454 (3,653,001) *1.000108*	186,899 (3,654,333) *1.000431*	280,328 (3,656,554) *1.000969*
32 _____	0 (3,541,665) *1.000000*	94,494 (3,542,102) *1.000110*	188,980 (3,543,413) *1.000440*	283,449 (3,545,597) *1.000991*
31 _____	0 (3,430,790) *1.000000*	95,505 (3,431,220) *1.000112*	191,002 (3,432,507) *1.000450*	286,484 (3,434,653) *1.001012*
30 _____	0 (3,319,933) *1.000000*	96,487 (3,320,354) *1.000115*	192,967 (3,321,617) *1.000459*	289,432 (3,323,722) *1.001033*
29 _____	0 (3,209,093) *1.000000*	97,440 (3,209,506) *1.000117*	194,872 (3,210,742) *1.000468*	292,291 (3,212,803) *1.001054*
28 _____	0 (3,098,270) *1.000000*	98,363 (3,098,673) *1.000119*	196,719 (3,099,882) *1.000477*	295,062 (3,101,897) *1.001074*
27° _____	0 (2,987,463) *1.000000*	99,256 (2,987,856) *1.000122*	198,505 (2,989,036) *1.000486*	297,742 (2,991,002) *1.001094*
26 _____	0 (2,876,672) *1.000000*	100,119 (2,877,055) *1.000124*	200,231 (2,878,204) *1.000495*	300,332 (2,880,119) *1.001113*
25 _____	0 (2,765,896) *1.000000*	100,951 (2,766,269) *1.000126*	201,896 (2,767,386) *1.000503*	302,831 (2,769,247) *1.001132*
24 _____	0 (2,655,136) *1.000000*	101,753 (2,655,497) *1.000128*	203,500 (2,656,580) *1.000511*	305,237 (2,658,386) *1.001150*
23 _____	0 (2,544,390) *1.000000*	102,523 (2,544,739) *1.000130*	205,042 (2,545,788) *1.000519*	307,551 (2,547,536) *1.001168*

Note: x, y = rectangular coordinates, meters; origin at φ = 0, λ = 0. Y axis increasing north.

h = scale factor along meridian.

k = scale factor along parallel = 1.0.

λ = longitude east of central meridian. For longitude west of central meridian reverse sign of x.

R = radius of circular arc for latitude ϕ as shown on map
R_2 = ditto for ϕ_2
R_1 = ditto for ϕ_1
(x, y) = rectangular coordinates, with the origin at the intersection of ϕ_1 with λ_0, the y axis coinciding with the meridian of longitude λ_0, y increasing northerly, and the x axis perpendicular, x increasing easterly
λ = longitude of any meridian (east longitude is positive)
λ_0 = longitude of central meridian
λ_1 = longitude of true-to-scale meridian east of the central meridian, $2°$ more than λ_0 for most quadrangles
ϕ = any geodetic (or geographic) latitude on the quadrangle map
ϕ_2 = geodetic latitude of the northernmost parallel of a given quadrangle map (north latitude is positive)
ϕ_1 = geodetic latitude of the southernmost parallel of the quadrangle map

Care must be taken to use radians wherever angles are used without trigonometric functions.

The following constants apply to the entire map, given a, e, ϕ_1, ϕ_2, λ_1, and λ_0:

$$x_n = R_n \sin F_n \qquad (18\text{-}23)$$
$$y_1 = R_1 (1-\cos F_1) \qquad (18\text{-}24)$$
$$T_2 = R_2 (1-\cos F_2) \qquad (18\text{-}25)$$

where $n = 1$ and 2, and

$$R_n = a \cot \phi_n/(1-e^2 \sin^2 \phi_n)^{1/2} \qquad (18\text{-}26)$$
$$F_n = (\lambda_1-\lambda_0) \sin \phi_n \qquad (18\text{-}27)$$

with subscripts as required above, but if $\phi_n = 0$, R_n is infinite and equations (18-23) and (18-24) are indeterminate, but $y_1 = 0$, $T_2 = 0$, and

$$x_n = a (\lambda_1-\lambda_0) \qquad (18\text{-}23a)$$

Also for the entire map,

$$y_2 = [(M_2-M_1)^2 - (x_2-x_1)^2]^{1/2} + y_1 \qquad (18\text{-}28)$$
$$C_2 = y_2 - T_2 \qquad (18\text{-}29)$$
$$P = (M_2 y_1 - M_1 y_2)/(M_2-M_1) \qquad (18\text{-}30)$$
$$Q = (y_2-y_1)/(M_2-M_1) \qquad (18\text{-}31)$$
$$P' = (M_2 x_1 - M_1 x_2)/(M_2-M_1) \qquad (18\text{-}32)$$
$$Q' = (x_2-x_1)/(M_2-M_1) \qquad (18\text{-}33)$$

where

$$
\begin{aligned}
M_n = a\,[&(1-e^2/4-3e^4/64-5e^6/256-\ldots)\,\phi_n \\
&- (3e^2/8+3e^4/32+45e^6/1024+\ldots)\sin 2\phi_n \\
&+ (15e^4/256+45e^6/1024+\ldots)\sin 4\phi_n \\
&- (35e^6/3072+\ldots)\sin 6\phi_n + \ldots\,] \qquad (3\text{-}21)
\end{aligned}
$$

with subscripts as required above.

The following values are calculated for each point, given ϕ and λ; to find x and y:

$$x_a = P' + Q' M \qquad (18\text{-}34)$$
$$y_a = P + Q M \qquad (18\text{-}35)$$
$$C = y_a - R \pm (R^2-x_a^2)^{1/2} \qquad (18\text{-}36)$$

where the ± takes the same sign as ϕ. If $\phi = 0$, equation (18−36) is indeterminate, but $C = 0$. M and R are found from (3−21) and (18−26), respectively, omitting subscripts n. Then

$$x_b = R_2 \sin [(\lambda - \lambda_0) \sin \phi_2] \qquad (18-37)$$
$$y_b = C_2 + R_2 \{1 - \cos [(\lambda - \lambda_0) \sin \phi_2]\} \qquad (18-38)$$
$$x_c = R_1 \sin [(\lambda - \lambda_0) \sin \phi_1] \qquad (18-39)$$
$$y_c = R_1 \{1 - \cos [(\lambda - \lambda_0) \sin \phi_1]\} \qquad (18-40)$$

but if $\phi_2 = 0$,

$$x_b = a (\lambda - \lambda_0) \qquad (18-37a)$$
$$y_b = C_2 \qquad (18-38a)$$

or if $\phi_1 = 0$,

$$x_c = a (\lambda - \lambda_0) \qquad (18-39a)$$
$$y_c = 0 \qquad (18-40a)$$

Then

$$D = (x_b - x_c)/(y_b - y_c) \qquad (18-41)$$
$$B = x_c + D (C + R - y_c) \qquad (18-42)$$
$$x = \{B \pm D [R^2 (1 + D^2) - B^2]^{1/2}\}/(1 + D^2) \qquad (18-43)$$
$$y = C + R \pm (R^2 - x^2)^{1/2} \qquad (18-44)$$

where the ± in (18−43) and (18−44) takes the sign opposite that of ϕ. If $\phi = 0$, B and R are infinite, but

$$x = a (\lambda - \lambda_0) \qquad (18-45)$$
$$y = C \qquad (18-46)$$

For the inverse formulas for the IMW Modified Polyconic, given a, e, ϕ_2, ϕ_1, λ_1, λ_0, x and y, to find ϕ and λ:

Step 1: Constants are calculated: x_1, x_2, y_1, M_1, M_2, y_2, C_2, P, Q, P', and Q' from above equations (18−23) through (18−33) and (3−21).

Step 2: A first trial (ϕ, λ), called (ϕ_{t_1}, λ_{t_1}) are calculated:

$$\phi_{t_1} = \phi_2 \qquad (18-47)$$
$$\lambda_{t_1} = [x/(a \cos \phi_{t_1})] + \lambda_0 \qquad (18-48)$$

Step 3: The first test values of (x, y), called (x_{t_1}, y_{t_1}), are calculated from (ϕ_{t_1}, λ_{t_1}), using the latter as (ϕ, λ) in equations (18−34) through (18−46).

Step 4: Test values (x_{t_1}, y_{t_1}) are used with the given (x, y) to adjust (ϕ_{t_1}, λ_{t_1}), to provide second trial values of (ϕ_{t_2}, λ_{t_2}):

$$\phi_{t_2} = [(\phi_{t_1} - \phi_1) (y - y_c)/(y_{t_1} - y_c)] + \phi_1 \qquad (18-49)$$
$$\lambda_{t_2} = [(\lambda_{t_1} - \lambda_0)x/x_{t_1}] + \lambda_0 \qquad (18-50)$$

Step 5: Step 3 is repeated, but using (ϕ_{t_2}, λ_{t_2}) as (ϕ, λ) to obtain (x_{t_2}, y_{t_2}). Step 4 is then repeated, replacing subscripts ($t1$, $t2$) with ($t2$, $t3$), respectively. Steps 3 and 4 are repeated, changing subscripts, until the final (x_{tn}, y_{tn}) vary from (x, y), respectively, by an acceptable total absolute error, such as 1 meter (0.001 mm at map scale).

TABLE 20.—*Modified Polyconic projection for IMW:*
Rectangular coordinates for the International ellipsoid

Latitude	Longitude difference ($\lambda - \lambda_0$)			
	0°	±1°	±2°	±3°
Rectangular coordinates (±x, y) meters				
40°	0.0	85395.9	170781.1	256144.8
	443829.1	444308.8	445745.8	448140.6
39	0.0	86588.8	173167.1	259724.5
	332842.0	333317.3	334743.2	337119.6
38	0.0	87781.4	175552.7	263303.7
	221874.6	222345.9	223759.9	226116.3
37	0.0	88973.9	177937.9	266882.3
	110927.3	111394.4	112795.5	115130.6
36	0.0	90166.1	180322.7	270460.3
	0.0	462.5	1850.0	4162.2
Scale factors (h, k)				
40°	0.999641	0.999730	1.000000	1.000449
	1.000000	1.000000	1.000000	1.000000
39	0.999631	0.999723	1.000000	1.000462
	0.999541	0.999541	0.999540	0.999540
38	0.999620	0.999715	1.000000	1.000474
	0.999394	0.999393	0.999393	0.999392
37	0.999610	0.999707	1.000000	1.000488
	0.999549	0.999549	0.999549	0.999548
36	0.999599	0.999699	1.000000	1.000501
	1.000000	1.000000	1.000000	1.000000

	Rectangular coordinates (±x, y) meters			
	0°	±2°	±4°	±6°
68°	0.0	83632.8	167177.9	250548.0
	445868.7	447222.2	451281.3	458041.7
67	0.0	87188.5	174287.0	261205.5
	334374.6	335774.8	339974.0	346967.9
66	0.0	90743.7	181395.1	271862.0
	222898.0	224344.1	228680.9	235904.0
65	0.0	94298.3	188502.3	282517.5
	111439.6	112930.7	117402.4	124850.3
64	0.0	97852.4	195608.5	293172.1
	0.0	1535.1	6139.0	13807.1
Scale factors (h, k)				
68°	0.999657	0.999743	1.000000	1.000429
	1.000000	1.000000	1.000000	1.000000
67	0.999627	0.999720	1.000000	1.000466
	0.999533	0.999532	0.999531	0.999528
66	0.999596	0.999697	1.000000	1.000504
	0.999394	0.999393	0.999391	0.999387
65	0.999564	0.999673	1.000000	1.000545
	0.999557	0.999556	0.999555	0.999552
64	0.999530	0.999647	1.000000	1.000587
	1.000000	1.000000	1.000000	1.000000

Note: λ_0 is longitude of the central meridian of quadrangle, east being positive.
λ is longitude.
h is scale factor along meridian.
k is scale factor along parallel.
Origin of rectangular coordinates occurs at minimum latitude and central meridian, y increasing northerly, x
    increasing easterly and taking the sign of ($\lambda - \lambda_0$).
Table applies to any quadrangle with the same latitude range.

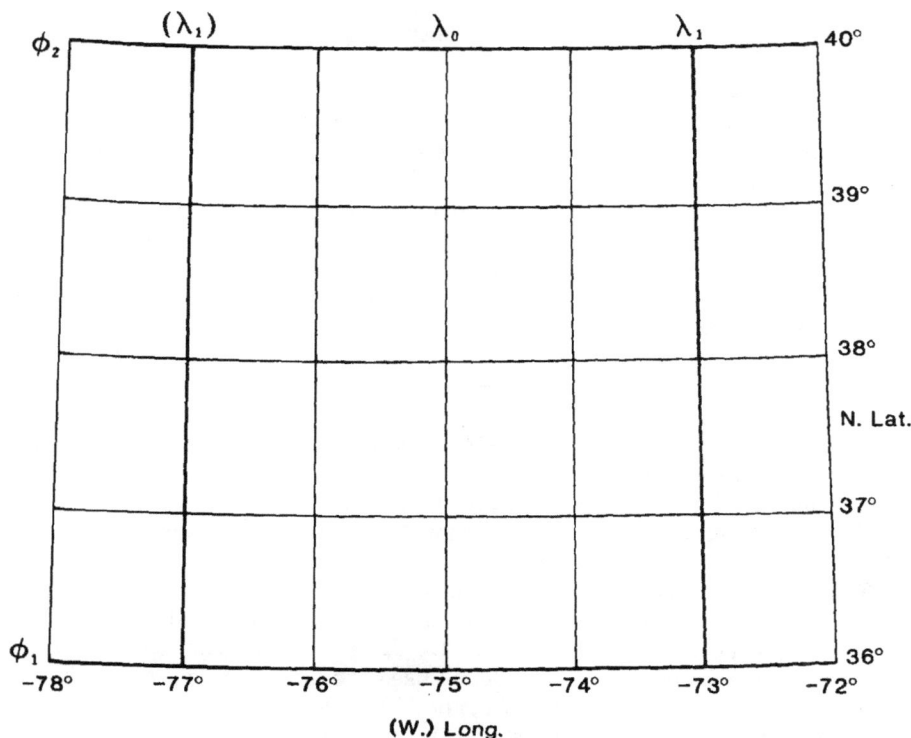

FIGURE 26.—Typical IMW quadrangle graticule—modified Polyconic projection drawn to scale. Parallels are nonconcentric circular arcs; meridians are straight. Lines of true scale are shown heavy. Standard projection for the International Map of the World Series (1:1,000,000-scale) until 1962.

Table 20 provides samples of rectangular coordinates calculated for each degree of typical mid-latitude and far-northern quadrangles. In addition, scale factors h (along the meridian) and k (along the parallel) are shown for the same graticules. The scale factors were calculated by comparing rectangular coordinates 0.01° of latitude apart at constant longitude with the true distances, for h, and a similar change in longitude at constant latitude, for k, rather than analytically. The linear scale error is seen to change less than about 0.06 percent throughout the quadrangle; the scale factor along any given parallel is almost constant, while a given meridian varies up to 0.015 percent in scale. The table is based on the International ellipsoid or spheroid, although the skeletal tables showing rectangular coordinates of parallels ϕ_1 and ϕ_2 and published in earlier technical papers are based on an ellipsoid with a semimajor axis of 6378.24 km and semiminor axis of 6356.56 km. Figure 26 illustrates a typical graticule.

19. BONNE PROJECTION

SUMMARY

- Pseudoconical.
- Central meridian is a straight line. Other meridians are complex curves.
- Parallels are concentric circular arcs, but the poles are points.
- Scale is true along the central meridian and along all parallels.
- No distortion along the central meridian and along the standard parallel.
- Used for atlas maps of continents and for topographic mapping of some countries.
- Sinusoidal projection is equatorial limiting form of Bonne projection.
- Used considerably by Bonne in mid-18th century, but developed by others during the early 16th century.

HISTORY

The name of Rigobert Bonne (1727–1795), a French geographer, is almost universally applied to an equal-area projection which has been used for both large- and small-scale mapping during the past 450 years. During the late 19th and early 20th centuries, the most conspicuous use of the Bonne projection was for maps of continents in atlases.

The Italian Bernardus Sylvanus' world map of 1511 closely approaches the Bonne projection, since its meridians are almost equally spaced along the equidistant and concentric circular parallels. De l'Isle and Coronelli used the Bonne principle for maps of about 1700. Bonne used the projection most notably for a 1752 maritime atlas of the coast of France (Reignier, 1957, p. 164). Continental maps of Europe and Asia appeared on this projection by 1763, and the ellipsoidal version replaced the Cassini projection for French topographic mapping beginning in 1803.

For maps of continents, the Bonne was preceded by its polar limiting form, a cordiform (heart-shaped) world map devised by Johannes Stabius and given wider notice by Johannes Werner about 1514. The Werner projection, as it is usually called, was used in the late 16th century for maps of Asia and Africa by Mercator and Abraham Ortelius, but the "Bonne" projection has less distortion because its projection center is at the center of the region being mapped instead of at the pole. Eventually the Werner projection was made obsolete by the Bonne.

FEATURES AND USAGE

Like the Equidistant Conic with one standard parallel, the Bonne projection (fig. 27) has concentric circular arcs for parallels of latitude. They are equally spaced on the spherical form and spaced in proportion to the true distance along a meridian on the ellipsoidal form. The chosen standard parallel is given its true curvature on the map by making the radius of its circular arc equal to the distance between the parallel and the apex of a cone tangent at the parallel.

Unlike the parallels on the Equidistant Conic and other regular conic projections, but like those on the Polyconic, each parallel is marked off for meridians at the true spacings on either the spherical or ellipsoidal versions, beginning at the straight central meridian. The individual meridians are then shown as complex curves connecting these points. This results in an equal-area projection with true scale along the central meridian and along each parallel, whether spherical or ellipsoidal. The central meridian and the standard parallel are free of local angular and shape distortion as well. The shape distortion increases away from either line, and meridians do not intersect parallels at right angles elsewhere, as they do on regular conic projections.

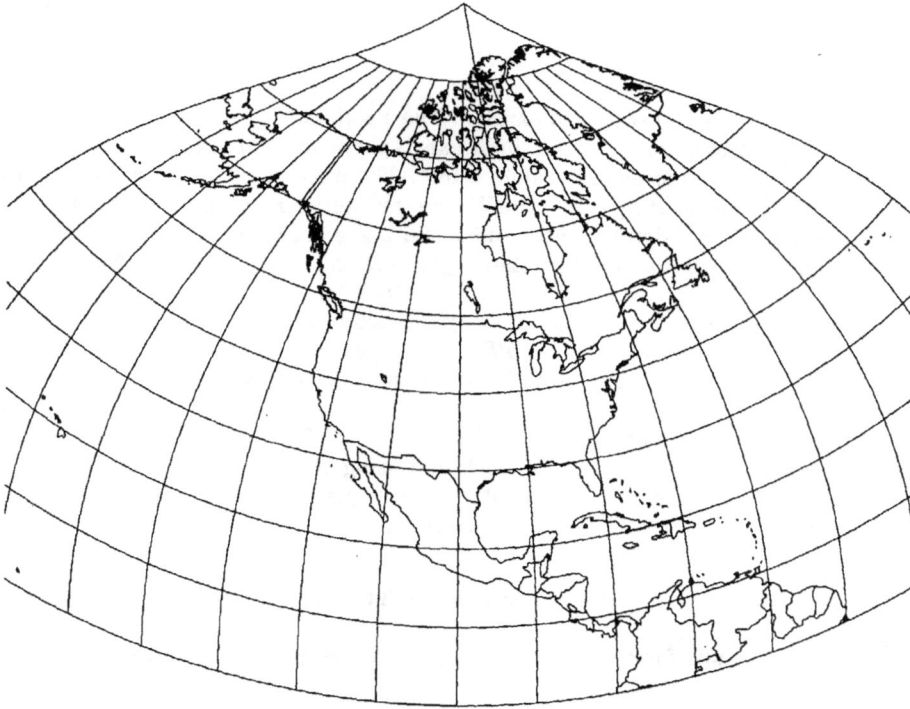

FIGURE 27.—Bonne projection with central parallel at lat. 40° N. Called a pseudoconic projection, this is equal-area and has *no distortion along central meridian or central parallel.* Popular in atlases for maps of continents until mid-20th century.

The combination of curved meridians and concentric circular arcs for parallels has led to the classification of "pseudoconic" for the Bonne projection and for the polar limiting case, the Werner projection, on which the North Pole is the equivalent of the standard parallel. The limiting case with the Equator as the standard parallel is the Sinusoidal, a "pseudocylindrical" projection to be discussed later; the formulas must be changed in this case since the parallels of latitude are straight. Modifications to the Bonne projection, in some cases resulting in non-equal-area projections, were presented by Nell of Germany in 1890 and by Solov'ev of the Soviet Union in the 1940's (Maling, 1960, p. 295−296).

Many atlases of the 19th and early 20th centuries utilized the Bonne projection to show North America, Europe, Asia, and Australia, while the Sinusoidal (as the equatorial Bonne) was used for South America and Africa. The Lambert Azimuthal Equal-Area projection is now generally used by Rand McNally & Co. and Hammond Inc. for maps of continents, while the National Geographic Society prefers its own Chamberlin Trimetric projection for this purpose.

Large-scale use of the Bonne projection for topographic mapping, originally introduced by France, is current chiefly in portions of France, Ireland, Morocco, and some countries in the eastern Mediterranean area (Clifford J. Mugnier, written commun., 1985).

FORMULAS FOR THE SPHERE

The principles stated above lead to the following forward formulas for rectangular coordinates of the spherical form of the Bonne projection, given R, ϕ_1, λ_0, ϕ, and λ, and using radians in equation (19−1),

$$\rho = R \ (\cot \phi_1 + \phi_1 - \phi) \tag{19-1}$$
$$E = \text{R} \ (\lambda - \lambda_0) \ (\cos \phi)/\rho \tag{19-2}$$
$$x = \rho \sin E \tag{19-3}$$
$$y = R \cot \phi_1 - \rho \cos E \tag{19-4}$$

where ϕ_1 is the chosen standard parallel. The Y axis coincides with λ_0, the central meridian, y increasing north, and the X axis is perpendicular at (ϕ_1, λ_0), x increasing east. If $(\lambda - \lambda_0)$ exceeds the range $\pm 180°$, 360° must be added or subtracted to place it within range. If $\phi_1 = 90°$, the Werner projection results, but if ϕ is also 90°, equation (19-2) is indeterminate, and x and y are both zero.

The inverse formulas for the sphere, given R, ϕ_1, λ_0, x, and y, to find (ϕ, λ):

$$\rho = \pm[x^2 + (R \cot \phi_1 - y)^2]^{1/2}, \text{ taking the sign of } \phi_1 \tag{19-5}$$
$$\phi = \cot \phi_1 + \phi_1 - \rho/R \tag{19-6}$$
$$\lambda = \lambda_0 + \rho \ \{ \arctan \ [x/(R \cot \phi_1 - y)]\}/(R \cos \phi) \tag{19-7}$$

using the ϕ determined from (19-6). If $\phi = \pm 90°$, (19-7) is indeterminate, but λ may be given any value, such as λ_0. When using the Fortran ATAN2 function for equation (19-7), and ϕ_1 is negative, the signs of x and $(R \cot \phi_1 - y)$ must be reversed before insertion into the equation.

FORMULAS FOR THE ELLIPSOID

For the forward formulas, given a, e, ϕ_1, λ_0, ϕ, and λ, to find x and y, the following are calculated in order:

$$m = \cos \phi/(1 - e^2 \sin^2 \phi)^{1/2} \tag{14-15}$$
$$\begin{aligned} M = a[&(1 - e^2/4 - 3e^4/64 - 5e^6/256 - \dots) \ \phi \\ &- (3e^2/8 + 3e^4/32 + 45e^6/1024 + \dots) \sin 2\phi \\ &+ (15e^4/256 + 45e^6/1024 + \dots) \sin 4\phi \\ &- (35e^6/3072 + \dots) \sin 6\phi + \dots] \end{aligned} \tag{3-21}$$
$$\rho = am_1/\sin \phi_1 + M_1 - M \tag{19-8}$$
$$E = am(\lambda - \lambda_0)/\rho \tag{19-9}$$
$$x = \rho \sin E \tag{19-10}$$
$$y = am_1/\sin \phi_1 - \rho \cos E \tag{19-11}$$

where ϕ_1 is the chosen central parallel, and m_1 and M_1 are found from (14-15) and (3-21), respectively, by using ϕ_1 instead of ϕ. Axes are the same as those for the spherical form. If both ϕ and ϕ_1 are at the same pole, equation (19-9) is indeterminate, but x and y are both zero.

For *the inverse formulas* for the ellipsoid, given a, e, ϕ_1, λ_0, x and y, to find ϕ and λ, first m_1 and M_1 are calculated as in the forward case from equations (14-15) and (3-21) above. The following are then calculated in order:

$$\rho = \pm[x^2 + (am_1/\sin \phi_1 - y)^2]^{1/2}, \text{ taking the sign of } \phi_1 \tag{19-12}$$
$$M = am_1/\sin \phi_1 + M_1 - \rho \tag{19-13}$$
$$\mu = M/[a(1 - e^2/4 - 3e^4/64 - 5e^6/256 - \dots)] \tag{7-19}$$
$$e_1 = [1 - (1 - e^2)^{1/2}]/[1 + (1 - e^2)^{1/2}] \tag{3-24}$$
$$\begin{aligned} \phi = \mu &+ (3e_1/2 - 27e_1^3/32 + \dots) \sin 2\mu + (21e_1^2/16 \\ &- 55e_1^4/32 + \dots) \sin 4\mu + (151e_1^3/96 - \dots) \sin 6\mu \\ &+ (1097e_1^4/512 - \dots) \sin 8\mu + \dots \end{aligned} \tag{3-26}$$

From (14-15), m is calculated for ϕ, then

$$\lambda = \lambda_0 + \rho \{\arctan[x/(am_1/\sin \phi_1 - y)]\}/(am) \tag{19-14}$$

When using the Fortran ATAN2 function for equation (19-14), and ϕ_1 is negative, the signs of x and $(am_1/\sin \phi_1 - y)$ must be reversed before insertion into the equation. If $\phi = \pm 90°$, (19-14) is indeterminate, but λ may be given any value, such as λ_0.

AZIMUTHAL AND RELATED MAP PROJECTIONS

A third very important group of map projections, some of which have been known for 2,000 years, consists of five major azimuthal (or zenithal) projections and various less-common forms. While cylindrical and conic projections are related to cylinders and cones wrapped around the globe representing the Earth, the azimuthal projections are formed onto a plane which is usually tangent to the globe at either pole, the Equator, or any intermediate point. These variations are called the polar, equatorial (or meridian or meridional), and oblique (or horizon) aspects, respectively. Some azimuthals are true perspective projections; others are not. Although perspective cylindrical and conic projections are much less used than those which are not perspective, the perspective azimuthals are frequently used and have valuable properties. Complications arise when the ellipsoid is involved, but it is used only in special applications that are discussed below.

As stated earlier, azimuthal projections are characterized by the fact that the direction, or azimuth, from the center of the projection to every other point on the map is shown correctly. In addition, on the spherical forms, all great circles passing through the center of the projection are shown as straight lines. Therefore, the shortest route from this center to any other point is shown as a straight line. This fact made some of these projections especially popular for maps as flight and radio transmission became commonplace.

The five principal azimuthals are as follows:
1. **Orthographic.** A true perspective, in which the Earth is projected from an infinite distance onto a plane. The map looks like a globe, thus stressing the roundness of the Earth.
2. **Stereographic.** A true perspective in the spherical form, with the point of perspective on the surface of the sphere at a point exactly opposite the point of tangency for the plane, or opposite the center of the projection, even if the plane is secant. This projection is conformal for sphere or ellipsoid, but the ellipsoidal form is not truly perspective.
3. **Gnomonic.** A true perspective, with the Earth projected from the center onto the tangent plane. All great circles, not merely those passing through the center, are shown as straight lines on the spherical form.
4. **Lambert Azimuthal Equal-Area.** Not a true perspective. Areas are correct, and the overall scale variation is less than that found on the major perspective azimuthals.
5. **Azimuthal Equidistant.** Not a true perspective. Distances from the center of the projection to any other point are shown correctly. Overall scale variation is moderate compared to the perspective azimuthals.

A sixth azimuthal projection of increasing interest in the space age is the general Vertical Perspective (resembling the Orthographic), projecting the Earth from any point in space, such as a satellite, onto a tangent or secant plant. It is used primarily in derivations and pictorial representations.

As a group, the azimuthals have unique esthetic qualities while remaining functional. There is a unity and roundness of the Earth on each (except perhaps the Gnomonic) which is not as apparent on cylindrical and conic projections.

The simplest forms of the azimuthal projections are the polar aspects, in which all meridians are shown as straight lines radiating at their true angles from the center, while parallels of latitude are circles, concentric about the pole. The difference is in the spacing of the parallels. Table 21 lists for the five principal azimuthals the radius of every 10° of latitude on a sphere of radius 1.0 unit, centered on the North Pole. Scale factors and maximum angular deformation are also shown. The distortion is the same for the oblique and equatorial aspects at the same angular distance from the center of the projection, except that h and k are along and perpendicular to, respectively, radii from the center, not necessarily along meridians or parallels.

TABLE 21.—*Comparison of major azimuthal projections: Radius, scale factors, maximum angular distortion for projection of sphere with radius 1.0, North Polar aspect*

Lat.	Orthographic			
	Radius	**h**	**k**	**ω**
90°	0.00000	1.00000	1.0	0.000°
80	.17365	.98481	1.0	.877
70	.34202	.93969	1.0	3.563
60	.50000	.86603	1.0	8.234
50	.64279	.76604	1.0	15.23
40	.76604	.64279	1.0	25.12
30	.86603	.50000	1.0	38.94
20	.93969	.34202	1.0	58.72
10	.98481	.17365	1.0	89.51
0	1.00000	.00000	1.0	180.0
−10	--	--	--	--
−20	--	--	--	--
−30	--	--	--	--
−40	(beyond limits of map)	--	--	--
−50	--	--	--	--
−60	--	--	--	--
−70	--	--	--	--
−80	--	--	--	--
−90	--	--	--	--

Lat.	Stereographic	
	Radius	**k***
90°	0.00000	1.00000
80	.17498	1.00765
70	.35263	1.03109
60	.53590	1.07180
50	.72794	1.13247
40	.93262	1.21744
30	1.15470	1.33333
20	1.40042	1.49029
10	1.67820	1.70409
0	2.00000	2.00000
−10	2.38351	2.42028
−20	2.85630	3.03961
−30	3.46410	4.00000
−40	4.28901	5.59891
−50	5.49495	8.54863
−60	7.46410	14.9282
−70	11.3426	33.1634
−80	22.8601	131.646
−90	∞	∞

There are two principal drawbacks to the azimuthals. First, they are more difficult to construct than the cylindricals and the conics, except for the polar aspects. This drawback was more applicable, however, in the days before computers and plotters, but it is still more difficult to prepare a map having complex curves between plotted coordinates than it is to draw the entire graticule with circles and straight lines. Nevertheless, an increased use of azimuthal projections in atlases and for other published maps may be expected.

Secondly, most azimuthal maps do not have standard parallels or standard meridians. Each map has only one standard point: the center (except for the Stereographic, which may have a standard circle). Thus, the azimuthals are suitable for minimizing distortion in a somewhat circular region such as Antarctica, but not for an area with predominant length in one direction.

TABLE 21.—*Comparison of major azimuthal projections: Radius, scale factors, maximum angular distortion for projection of sphere with radius 1.0, North Polar aspect*—Continued

Lat.	Gnomonic			
	Radius	h	k	ω
90°	0.00000	1.00000	1.00000	0.000°
80	.17633	1.03109	1.01543	.877
70	.36397	1.13247	1.06418	3.563
60	.57735	1.33333	1.15470	8.234
50	.83910	1.70409	1.30541	15.23
40	1.19175	2.42028	1.55572	25.12
30	1.73205	4.00000	2.00000	38.94
20	2.74748	8.54863	2.92380	58.72
10	5.67128	33.1634	5.75877	89.51
0	∞	∞	∞	--
−10	--	--	--	--
−20	--	--	--	--
−30	--	--	--	--
−40	(beyond limits of map)		--	--
−50	--	--	--	--
−60	--	--	--	--
−70	--	--	--	--
−80	--	--	--	--
−90	--	--	--	--

Lat.	Lambert Azimuthal Equal-Area			
	Radius	h	k	ω
90°	0.00000	1.00000	1.00000	0.000°
80	.17431	.99619	1.00382	.437
70	.34730	.98481	1.01543	1.754
60	.51764	.96593	1.03528	3.972
50	.68404	.93969	1.06418	7.123
40	.84524	.90631	1.10338	11.25
30	1.00000	.86603	1.15470	16.43
20	1.14715	.81915	1.22077	22.71
10	1.28558	.76604	1.30541	30.19
0	1.41421	.70711	1.41421	38.94
−10	1.53209	.64279	1.55572	49.07
−20	1.63830	.57358	1.74345	60.65
−30	1.73205	.50000	2.00000	73.74
−40	1.81262	.42262	2.36620	88.36
−50	1.87939	.34202	2.92380	104.5
−60	1.93185	.25882	3.86370	122.0
−70	1.96962	.17365	5.75877	140.6
−80	1.99239	.08716	11.4737	160.1
−90	2.00000	.00000	∞	180.0

TABLE 21.—*Comparison of major azimuthal projections: Radius, scale factors, maximum angular distortion for projection of sphere with radius 1.0, North Polar aspect*

Lat.	Azimuthal Equidistant			
	Radius	h	k	ω
90°	0.00000	1.0	1.00000	0.000°
80	.17453	1.0	1.00510	.291
70	.34907	1.0	1.02060	1.168
60	.52360	1.0	1.04720	2.642
50	.69813	1.0	1.08610	4.731
40	.87266	1.0	1.13918	7.461
30	1.04720	1.0	1.20920	10.87
20	1.22173	1.0	1.30014	15.00
10	1.39626	1.0	1.41780	19.90
0	1.57080	1.0	1.57080	25.66
−10	1.74533	1.0	1.77225	32.35
−20	1.91986	1.0	2.04307	40.09
−30	2.09440	1.0	2.41840	49.03
−40	2.26893	1.0	2.96188	59.36
−50	2.44346	1.0	3.80135	71.39
−60	2.61799	1.0	5.23599	85.57
−70	2.79253	1.0	8.16480	102.8
−80	2.96706	1.0	17.0866	125.6
−90	3.14159	1.0	∞	180.0

Radius = *radius of circle showing given latitude.*

ω = maximum angular deformation.

h = scale factor along meridian of longitude.

k = scale factor along parallel of latitude.

* For Stereographic, h = k and ω = 0.

20. ORTHOGRAPHIC PROJECTION

SUMMARY

- Azimuthal.
- All meridians and parallels are ellipses, circles, or straight lines.
- Neither conformal nor equal-area.
- Closely resembles a globe in appearance, since it is a perspective projection from infinite distance.
- Only one hemisphere can be shown at a time.
- Much distortion near the edge of the hemisphere shown.
- No distortion at the center only.
- Directions from the center are true.
- Radial scale factor decreases as distance increases from the center.
- Scale in the direction of the lines of latitude is true in the polar aspect.
- Used chiefly for pictorial views.
- Used only in the spherical form.
- Known by Egyptians and Greeks 2,000 years ago.

HISTORY

To the layman, the best known perspective azimuthal projection is the Orthographic, although it is the least useful for measurements. While its distortion in shape and area is quite severe near the edges, and only one hemisphere may be shown on a single map, the eye is much more willing to forgive this distortion than to forgive that of the Mercator projection because the Orthographic projection makes the map look very much like a globe appears, especially in the oblique aspect.

The Egyptians were probably aware of the Orthographic projection, and Hipparchus of Greece (2nd century B.C.) used the equatorial aspect for astronomical calculations. Its early name was "analemma," a name also used by Ptolemy, but it was replaced by "orthographic" in 1613 by François d'Aiguillon of Antwerp. While it was also used by Indians and Arabs for astronomical purposes, it is not known to have been used for world maps older than 16th-century works by Albrecht Dürer (1471–1528), the German artist and cartographer, who prepared polar and equatorial versions (Keuning, 1955, p. 6).

FEATURES

The point of perspective for the Orthographic projection is at an infinite distance, so that the meridians and parallels are projected onto the tangent plane with projection lines. All meridians and parallels are shown as ellipses, circles, or straight lines.

As on all polar azimuthal projections, the meridians of the polar Orthographic projection appear as straight lines radiating from the pole at their true angles, while the parallels of latitude are complete circles centered about the pole. On the Orthographic, the parallels are spaced most widely near the pole, and the spacing decreases to zero at the Equator, which is the circle marking the edge of the map (figs. 28, 29A). As a result, the land shapes near the pole are prominent, while lands near the Equator are compressed so that they can hardly be recognized. In spite of the fact that the scale along the meridians varies from the correct value at the pole to zero at the Equator, the scale along every parallel is true.

The equatorial aspect of the Orthographic projection has as its center some point on the Earth's Equator. Here, all the parallels of latitude including the

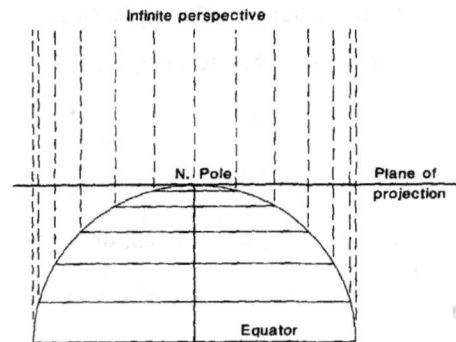

FIGURE 28.—Geometric projection of the parallels of the polar Orthographic projection.

Equator are seen edge-on; thus, they appear as straight parallel lines (fig. 29B). The meridians, which are shaped like circles on the sphere, are projected onto the map at various inclinations to the lines of perspective. The central meridian, seen edge-on, is a straight line. The meridian 90° from the central meridian is shown as a circle marking the limit of the equatorial aspect. This circle is equidistantly marked with parallels of latitude. Other meridians are ellipses of eccentricities ranging from zero (the bounding circle) to 1.0 (the central meridian).

The oblique Orthographic projection, with its center somewhere between the Equator and a pole, gives the classic globelike appearance; and in fact an oblique view, with its center near but not on the Equator or pole, is often preferred to the equatorial or polar aspect for pictorial purposes. On the oblique Orthographic, the only straight line is the central meridian, if it is actually portrayed. All parallels of latitude are ellipses with the same eccentricity (fig. 29C). Some of these ellipses are shown completely and some only partially, while some cannot be shown at all. All other meridians are also ellipses of varying eccentricities. No meridian appears as a circle on the oblique aspect.

The intersection of any given meridian and parallel is shown on an Orthographic projection at the same distance from the central meridian, regardless of whether the aspect is oblique, polar, or equatorial, provided the same central meridian and the same scale are maintained. Scale and distortion, as on all azimuthal projections, change only with the distance from the center. The center of projection has no distortion, but the outer regions are compressed, even though the scale is true along all circles drawn about the center. (These circles are not "standard" lines because the scale is true only in the direction followed by the line.)

USAGE

The Orthographic projection seldom appears in atlases, except as a globe in relief without meridians and parallels. When it does appear, it provides a striking view. Richard Edes Harrison has used the Orthographic for several maps in an atlas of the 1940's partially based on this projection. Frank Debenham (1958) used photographed relief globes extensively in *The Global Atlas*, and Rand McNally has done likewise in their world atlases since 1960. The USGS has used it occasionally as a frontispiece or end map (USGS, 1970; Thompson, 1979), but it also provided a base for definitive maps of voyages of discovery across the North Atlantic (USGS, 1970, p. 133).

It became especially popular during the Second World War when there was stress on the global nature of the conflict. With some space flights of the 1960's,

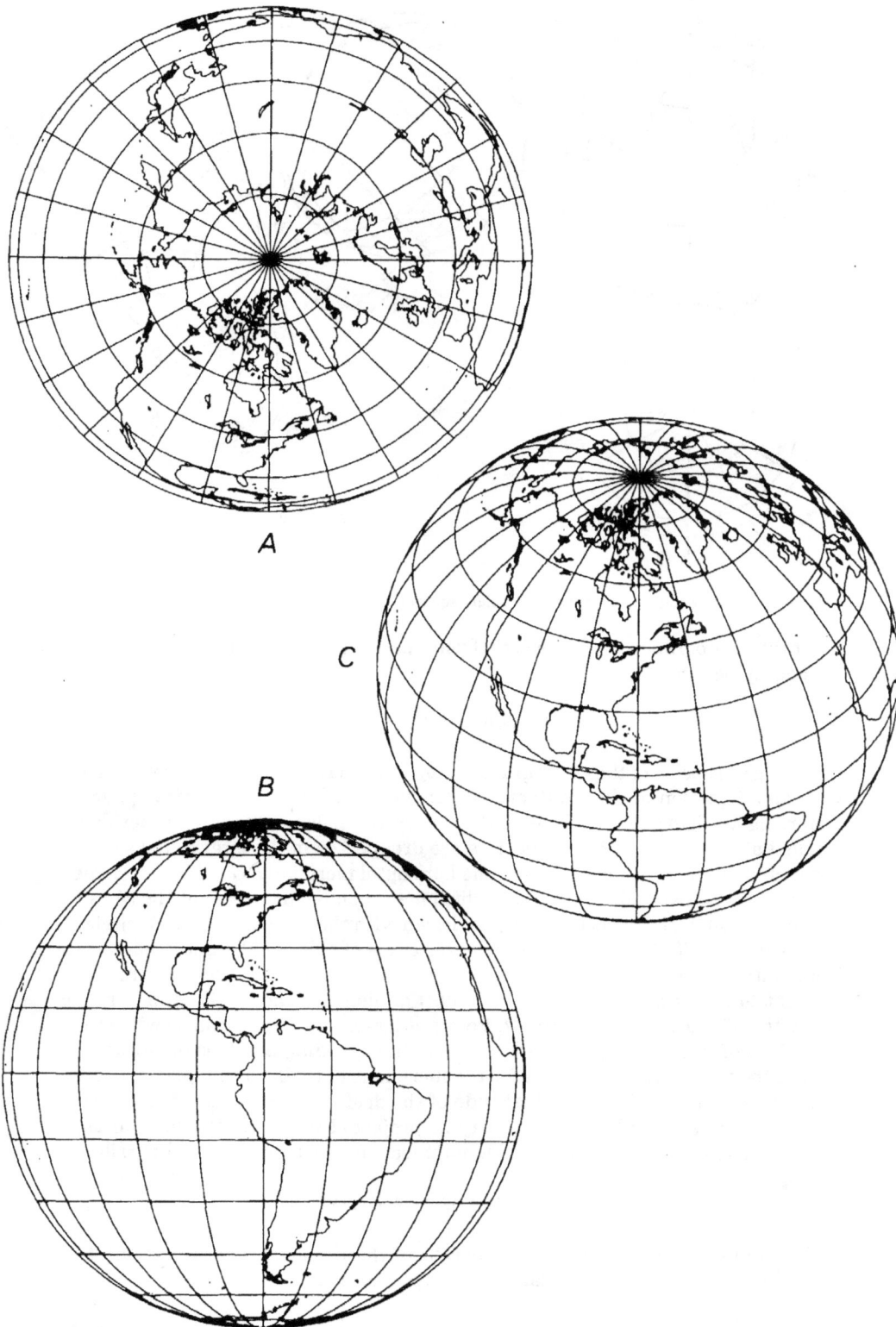

FIGURE 29.—Orthographic projection. (A) Polar aspect. (B) Equatorial aspect, approximately the view of the Moon, Mars, and other outer planets as seen from the Earth. (C) Oblique aspect, centered at lat. 40° N., giving the classic globelike view.

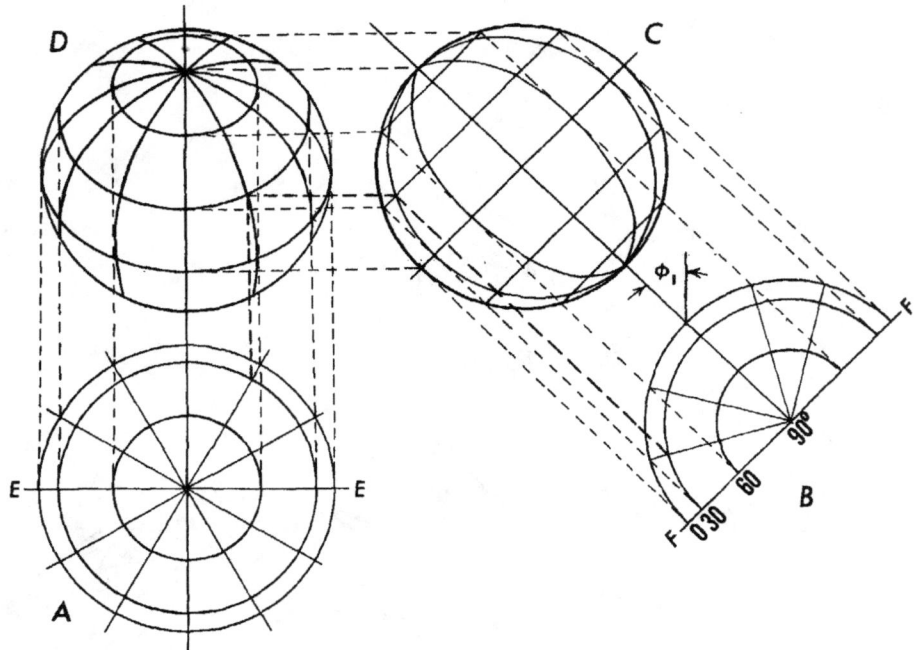

FIGURE 30.—Geometric construction of polar, equatorial, and oblique Orthographic projections.

the first photographs of the Earth from space renewed consciousness of the Orthographic concept.

GEOMETRIC CONSTRUCTION

The three aspects of the Orthographic projection may be graphically constructed with an adaptation of the draftsman's technique shown by Raisz (1962, p. 180). Referring to figure 30, circle A is drawn for the polar aspect, with meridians marked at true angles. Perpendiculars are dropped from the intersections of the outer circle with the meridians onto the horizontal meridian EE. This determines the radii of the parallels of latitude, which may then be drawn about the center.

For the equatorial aspect, circle C is drawn with the same radius as A, circle B is drawn like half of circle A, and the outer circle of C is equidistantly marked to locate intersections of parallels with that circle. Parallels of latitude are drawn as straight lines, with the Equator midway. Parallels are shown tilted merely for use with oblique projection circle D. Points at intersections of parallels with other meridians of B are then projected onto the corresponding parallels of latitude on C, and the new points connected for the meridians of C. By tilting graticule C at an angle ϕ_1 equal to the central latitude of the desired oblique aspect, the corresponding points of circles A and C may be projected vertically and horizontally, respectively, onto circle D to provide intersections for meridians and parallels.

FORMULAS FOR THE SPHERE

To understand the mathematical concept of the Orthographic projection, it is helpful to think in terms of polar coordinates ρ and θ:

$$\rho = R \sin c \qquad (20-1)$$
$$\theta = \pi - Az = 180° - Az \qquad (20-2)$$

'where c is the angular distance of the given point from the center of projection. Az is the azimuth east of north, and θ is the polar coordinate east of south. The distance from the center of a point on an Orthographic map projection is thus proportional to the sine of the angular distance from the center on the sphere. Applying equations (5–3), (5–4), and (5–4a) for great circle distance c and azimuth Az in terms of latitude and longitude, and equations for rectangular coordinates in terms of polar coordinates, the equations for rectangular coordinates for the oblique Orthographic projection reduce to the following, given R, ϕ_1, λ_0, ϕ, and λ (see p. 311 for numerical examples):

$$x = R \cos \phi \sin (\lambda - \lambda_0) \qquad (20\text{–}3)$$
$$y = R[\cos \phi_1 \sin \phi - \sin \phi_1 \cos \phi \cos (\lambda - \lambda_0)] \qquad (20\text{–}4)$$
$$h' = \cos c$$
$$\quad = \sin \phi_1 \sin \phi + \cos \phi_1 \cos \phi \cos (\lambda - \lambda_0) \qquad (20\text{–}5)$$
$$k' = 1.0$$

where ϕ_1 and λ_0 are the latitude and longitude, respectively, of the center point and origin of the projection, h' is the scale factor along a line radiating from the center, and k' is the scale factor in a direction perpendicular to a line radiating from the center. The Y axis coincides with the central meridian λ_0, y increasing northerly. All the parallels are ellipses of eccentricity $\cos \phi_1$. The limit of the map is a circle of radius R.

For the north polar Orthographic, letting $\phi_1 = 90°$, x is still found from (20–3), but

$$y = -R \cos \phi \cos (\lambda - \lambda_0) \qquad (20\text{–}6)$$
$$h = \sin \phi \qquad (20\text{–}7)$$

In polar coordinates,

$$\rho = R \cos \phi \qquad (20\text{–}8)$$
$$\theta = \lambda - \lambda_0 \qquad (20\text{–}9)$$

For the south polar Orthographic, with $\phi_1 = -90°$, x does not change, but

$$y = R \cos \phi \cos (\lambda - \lambda_0) \qquad (20\text{–}10)$$
$$h = -\sin \phi \qquad (20\text{–}11)$$

For polar coordinates, ρ is found from (20–8), but

$$\theta = \pi - \lambda + \lambda_0 \qquad (20\text{–}12)$$

For the equatorial Orthographic, letting $\phi_1 = 0$, x still does not change from (20–3), but

$$y = R \sin \phi \qquad (20\text{–}13)$$

In automatically computing a general set of coordinates for a complete Orthographic map, the distance c from the center should be calculated for each intersection of latitude and longitude to determine whether it exceeds 90° and therefore whether the point is beyond the range of the map. More directly, using equation (5–3),

$$\cos c = \sin \phi_1 \sin \phi + \cos \phi_1 \cos \phi \cos (\lambda - \lambda_0) \qquad (5\text{–}3)$$

if $\cos c$ is zero or positive, the point is to be plotted. If $\cos c$ is negative, the point is not to be plotted.

For the inverse formulas for the sphere, to find ϕ and λ, given R, ϕ_1, λ_0, x, and y:

$$\phi = \arcsin \left[\cos c \sin \phi_1 + (y \sin c \cos \phi_1/\rho)\right] \qquad (20-14)$$

If $\rho = 0$, equations $(20-14)$ through $(20-17)$ are indeterminate, but $\phi = \phi_1$ and $\lambda = \lambda_0$. If ϕ_1 is not $\pm 90°$,

$$\lambda = \lambda_0 + \arctan \left[x \sin c/(\rho \cos \phi_1 \cos c - y \sin \phi_1 \sin c)\right] \qquad (20-15)$$

If ϕ_1 is $90°$,

$$\lambda = \lambda_0 + \arctan \left[x/(-y)\right] \qquad (20-16)$$

If ϕ_1 is $-90°$,

$$\lambda = \lambda_0 + \arctan (x/y) \qquad (20-17)$$

Note that, while the ratio $[x/(-y)]$ in $(20-16)$ is numerically the same as $(-x/y)$, the necessary quadrant adjustment is different when using the Fortran ATAN2 function or its equivalent.

In equations $(20-14)$ and $(20-15)$,

$$\rho = (x^2 + y^2)^{1/2} \qquad (20-18)$$
$$c = \arcsin (\rho/R) \qquad (20-19)$$

Simplification for inverse equations for the polar and equatorial aspects is obtained by giving ϕ_1 values of $\pm 90°$ and $0°$, respectively. They are not given in detail here.

Tables 22 and 23 list rectangular coordinates for the equatorial and oblique aspects, respectively, for a $10°$ graticule with a sphere of radius $R = 1.0$. For the oblique example $\phi_1 = 40°$.

TABLE 22.—*Orthographic projection: Rectangular coordinates for equatorial aspect*

Long. Lat.	y	0°	10°	20°	30°	40°
				x		
90° ___	1.0000	0.0000	0.0000	0.0000	0.0000	0.0000
80 ____	.9848	.0000	.0302	.0594	.0868	.1116
70 ____	.9397	.0000	.0594	.1170	.1710	.2198
60 ____	.8660	.0000	.0868	.1710	.2500	.3214
50 ____	.7660	.0000	.1116	.2198	.3214	.4132
40 ____	.6428	.0000	.1330	.2620	.3830	.4924
30 ____	.5000	.0000	.1504	.2962	.4330	.5567
20 ____	.3420	.0000	.1632	.3214	.4698	.6040
10 ____	.1736	.0000	.1710	.3368	.4924	.6330
0 ____	.0000	.0000	.1736	.3420	.5000	.6428

Long. Lat.	50°	60°	70°	80°	90°
			x		
90° ___	0.0000	0.0000	0.0000	0.0000	0.0000
80 ____	.1330	.1504	.1632	.1710	.1736
70 ____	.2620	.2962	.3214	.3368	.3420
60 ____	.3830	.4330	.4698	.4924	.5000
50 ____	.4924	.5567	.6040	.6330	.6428
40 ____	.5868	.6634	.7198	.7544	.7660
30 ____	.6634	.7500	.8138	.8529	.8660
20 ____	.7198	.8138	.8830	.9254	.9397
10 ____	.7544	.8529	.9254	.9698	.9848
0 ____	.7660	.8660	.9397	.9848	1.0000

Radius of sphere = 1.0
Origin: (x, y) = 0 at (lat., long.) = 0. Y axis increases north. Other quadrants of hemisphere are symmetrical.

TABLE 23.—*Orthographic projection: Rectangular coordinates for oblique aspect centered at lat. 40° N.*

[The circle bounding the hemisphere map has the same coordinates as the λ=90° circle on the equatorial Orthographic projection. The radius of the sphere=1.0. *y* coordinate in parentheses under *x* coordinate]

Lat. \ Long.	0°	10°	20°	30°	40°
90°	0.0000 (.7660)	0.0000 (.7660)	0.0000 (.7660)	0.0000 (.7660)	0.0000 (.7660)
80	.0000 (.6428)	.0302 (.6445)	.0594 (.6495)	.0868 (.6577)	.1116 (.6689)
70	.0000 (.5000)	.0594 (.5033)	.1170 (.5133)	.1710 (.5295)	.2198 (.5514)
60	.0000 (.3420)	.0868 (.3469)	.1710 (.3614)	.2500 (.3851)	.3214 (.4172)
50	.0000 (.1736)	.1116 (.1799)	.2198 (.1986)	.3214 (.2290)	.4132 (.2703)
40	.0000 (.0000)	.1330 (.0075)	.2620 (.0297)	.3830 (.0660)	.4924 (.1152)
30	.0000 (−.1736)	.1504 (−.1652)	.2962 (−.1401)	.4330 (−.0991)	.5567 (−.0434)
20	.0000 (−.3420)	.1632 (−.3328)	.3214 (−.3056)	.4698 (−.2611)	.6040 (−.2007)
10	.0000 (−.5000)	.1710 (−.4904)	.3368 (−.4618)	.4924 (−.4152)	.6330 (−.3519)
0	.0000 (−.6428)	.1736 (−.6330)	.3420 (−.6040)	.5000 (−.5567)	.6428 (−.4924)
−10	.0000 (−.7660)	.1710 (−.7564)	.3368 (−.7279)	.4924 (−.6812)	.6330 (−.6179)
−20	.0000 (−.8660)	.1632 (−.8568)	.3214 (−.8296)	.4698 (−.7851)	.6040 (−.7247)
−30	.0000 (−.9397)	.1504 (−.9312)	.2962 (−.9061)	.4330 (−.8651)	.5567 (−.8095)
−40	.0000 (−.9848)	.1330 (−.9773)	.2620 (−.9551)	.3830 (−.9188)	.4924 (−.8696)
−50	.0000 (−1.0000)	-- --	-- --	-- --	-- --

Origin: (*x*, *y*) = 0 at (lat., long.) = (40°, 0). *Y* axis increases north. Coordinates shown for central meridian (λ = 0) and meridians east of central meridian. For meridians west (negative), reverse signs of meridians and of *x*.

Lat. \ Long.	100°	110°	120°	130°	140°
90°	0.0000 (.7660)	0.0000 (.7660)	0.0000 (.7660)	0.0000 (.7660)	0.0000 (.7660)
80	.1710 (.7738)	.1632 (.7926)	.1504 (.8102)	.1330 (.8262)	.1116 (.8399)
70	.3368 (.7580)	.3214 (.7950)	.2962 (.8298)	.2620 (.8612)	.2198 (.8883)
60	.4924 (.7192)	.4698 (.7733)	.4330 (.8241)	.3830 (.8700)	.3214 (.9096)
50	.6330 (.6586)	.6040 (.7281)	.5567 (.7934)	.4924 (.8524)	.4132 (.9033)
40	.7544 (.5779)	.7198 (.6608)	.6634 (.7386)	.5868 (.8089)	-- --
30	.8529 (.4797)	.8138 (.5734)	-- --	-- --	-- --
20	.9254 (.3669)	-- --	-- --	-- --	-- --

TABLE 23.—*Orthographic projection: Rectangular coordinates for oblique aspect centered at lat. 40° N.*—Continued

Long. Lat.	50°	60°	70°	80°	90°
90°	0.0000 (.7660)	0.0000 (.7660)	0.0000 (.7660)	0.0000 (.7660)	0.0000 (.7660)
80	.1330 (.6827)	.1504 (.6986)	.1632 (.7162)	.1710 (.7350)	.1736 (.7544)
70	.2620 (.5785)	.2962 (.6099)	.3214 (.6447)	.3368 (.6817)	.3420 (.7198)
60	.3830 (.4568)	.4330 (.5027)	.4698 (.5535)	.4924 (.6076)	.5000 (.6634)
50	.4924 (.3212)	.5567 (.3802)	.6040 (.4455)	.6330 (.5151)	.6428 (.5868)
40	.5868 (.1759)	.6634 (.2462)	.7198 (.3240)	.7544 (.4069)	.7660 (.4924)
30	.6634 (.0252)	.7500 (.1047)	.8138 (.1926)	.8529 (.2864)	.8660 (.3830)
20	.7198 (−.1263)	.8138 (−.0400)	.8830 (.0554)	.9254 (.1571)	.9397 (.2620)
10	.7544 (−.2739)	.8529 (−.1835)	.9254 (−.0835)	.9698 (.0231)	.9848 (.1330)
0	.7660 (−.4132)	.8660 (−.3214)	.9397 (−.2198)	.9848 (−.1116)	1.0000 (.0000)
−10	.7544 (−.5399)	.8529 (−.4495)	.9254 (−.3495)	.9698 (−.2429)	--
−20	.7198 (−.6503)	.8138 (−.5640)	.8830 (−.4686)	--	--
−30	.6634 (−.7408)	.7500 (−.6614)	--	--	--
−40	--	--	--	--	--

Long. Lat.	150°	160°	170°	180°
90°	0.0000 (.7660)	0.0000 (.7660)	0.0000 (.7660)	0.0000 (.7660)
80	.0868 (.8511)	.0594 (.8593)	.0302 (.8643)	.0000 (.8660)
70	.1710 (.9102)	.1170 (.9264)	.0594 (.9364)	.0000 (.9397)
60	.2500 (.9417)	.1710 (.9654)	.0868 (.9799)	.0000 (.9848)
50	.3214 (.9446)	.2198 (.9751)	.1116 (.9937)	.0000 (1.0000)
40	--	--	--	--

21. STEREOGRAPHIC PROJECTION

SUMMARY

- Azimuthal.
- Conformal.
- The central meridian and a particular parallel (if shown) are straight lines.
- All meridians on the polar aspect and the Equator on the equatorial aspect are straight lines.
- All other meridians and parallels are shown as arcs of circles.
- A perspective projection for the sphere.
- Directions from the center of the projection are true (except on ellipsoidal oblique and equatorial aspects).
- Scale increases away from the center of the projection.
- Point opposite the center of the projection cannot be plotted.
- Used for polar maps and miscellaneous special maps.
- Apparently invented by Hipparchus (2nd century B.C.).

HISTORY

The Stereographic projection was probably known in its polar form to the Egyptians, while Hipparchus was apparently the first Greek to use it. He is generally considered its inventor. Ptolemy referred to it as "Planisphaerum," a name used into the 16th century. The name "Stereographic" was assigned to it by François d'Aiguillon in 1613. The polar Stereographic was exclusively used for star maps until perhaps 1507, when the earliest-known use for a map of the world was made by Walther Ludd (Gaultier Lud) of St. Dié, Lorraine.

The oblique aspect was used by Theon of Alexandria in the fourth century for maps of the sky, but it was not proposed for geographical maps until Stabius and Werner discussed it together with their cordiform (heart-shaped) projections in the early 16th century. The earliest-known world maps were included in a 1583 atlas by Jacques de Vaulx (c. 1555–97). The two hemispheres were centered on Paris and its opposite point, respectively.

The equatorial Stereographic originated with the Arabs, and was used by the Arab astronomer Ibn-el-Zarkali (1029–87) of Toledo for an astrolabe. It became a basis for world maps in the early 16th century, with the earliest-known examples by Jean Roze (or Rotz), a Norman, in 1542. After Rumold (the son of Gerardus) Mercator's use of the equatorial Stereographic for the world maps of the atlas of 1595, it became very popular among cartographers (Keuning, 1955, p. 7–9; Nordenskiöld, 1889, p. 90, 92–93).

FEATURES

Like the Orthographic, the Stereographic projection is a true perspective in its spherical form. It is the only known true perspective projection of any kind that is also conformal. Its point of projection is on the surface of the sphere at a point just opposite the point of tangency of the plane or the center point of the projection (fig. 31). Thus, if the North Pole is the center of the map, the projection is from the South Pole. All of one hemisphere can be comfortably shown, but it is impossible to show both hemispheres in their entirety from one center. The point on the sphere opposite the center of the map projects at an infinite distance in the plane of the map.

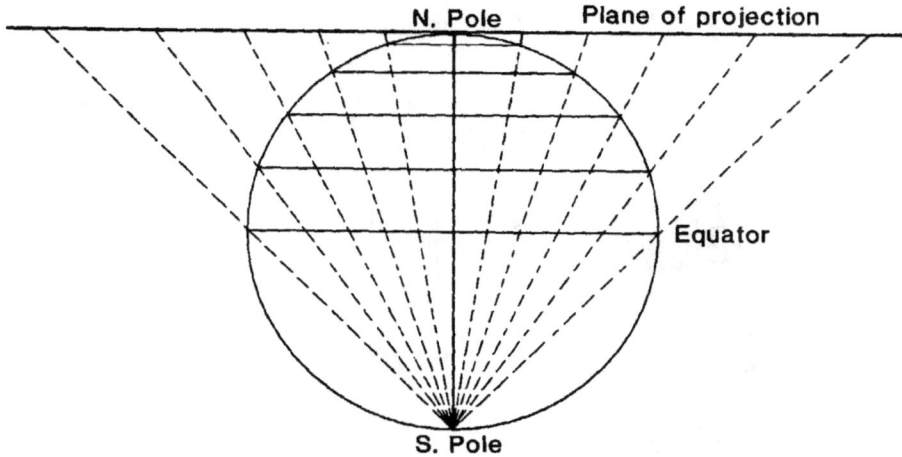

FIGURE 31.—Geometric projection of the polar Stereographic projection.

The polar aspect somewhat resembles other polar azimuthals, with straight radiating meridians and concentric circles for parallels (fig. 32A). The parallels are spaced at increasingly wide distances, the farther the latitude is from the pole (the Orthographic has the opposite feature).

In the equatorial and oblique aspects, the distinctive appearance of the Stereographic becomes more evident: All meridians and parallels, except for two, are shown as circles, and the meridians intersect the parallels at right angles (figs. 32B, C). The central meridian is shown straight, as is the parallel of the same numerical value, but opposite in sign to the central parallel. For example, if lat. 40° N. is the central parallel, then lat. 40° S. is shown as a straight line. For the equatorial aspect with lat. 0° as the central parallel, the Equator, which is of course also its own negative counterpart, is shown straight. (For the polar aspect, this has no meaning since the opposite pole cannot be shown.) Circles for parallels are centered along the central meridian; circles for meridians are centered along the straight parallel. The meridian 90° from the central meridian on the equatorial aspect is shown as a circle bounding the hemisphere. This circle is centered on the projection center and is equidistantly marked for parallels of latitude.

As an azimuthal projection, directions from the center are shown correctly in the spherical form. In the ellipsoidal form, only the polar aspect is truly azimuthal, but it is not perspective, in order to retain conformality. The oblique and equatorial aspects of the ellipsoidal Stereographic, in order to be conformal, are neither azimuthal nor perspective. As with other azimuthal projections, there is no distortion at the center, which may be made the "standard point" true to scale in all directions. Because of the conformality of the projection, a Stereographic map may be given, instead of a "standard point," a "standard circle" (or "standard parallel" in the polar aspect) with an appropriate radius from the center, balancing the scale error throughout the map. (On the ellipsoidal oblique or equatorial aspects, the lines of constant scale are not perfect circles.) This cannot be done with non-conformal azimuthal projections. The Stereographic may also be modified to produce oval and irregular lines of true scale (see p. 203).

USAGE

The oblique aspect of the Stereographic projection has been recently used in the spherical form by the USGS for circular maps of portions of the Moon, Mars, and Mercury, generally centered on a basin. The USGS is currently using the

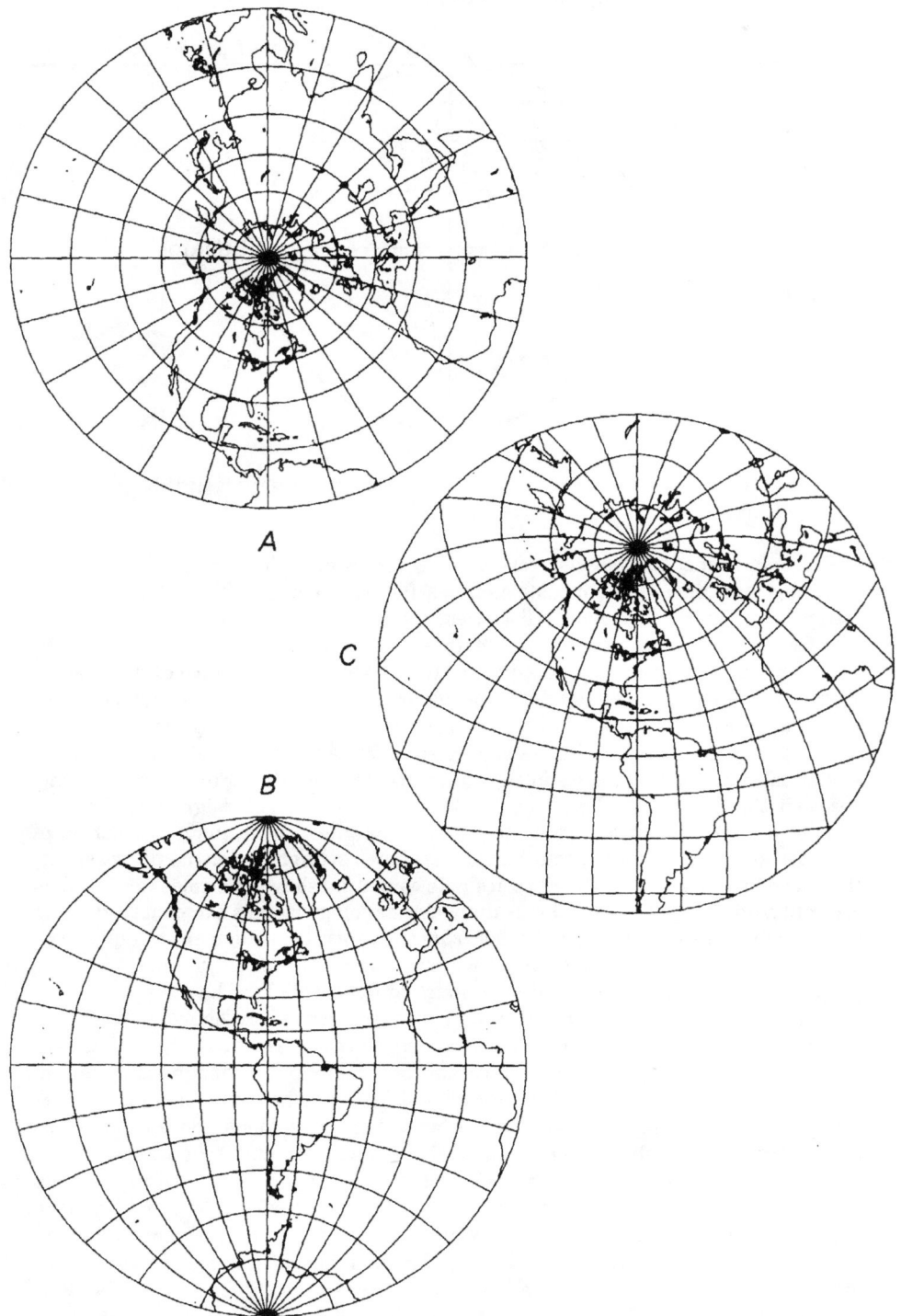

FIGURE 32.—Stereographic projection. (A) Polar aspect; the most common scientific projection for polar
areas of Earth, Moon, and the planets, since it is conformal. (B) Equatorial aspect; often used in the 16th
and 17th centuries for maps of hemispheres. (C) Oblique aspect; centered on lat. 40° N. The Stereographic is the only geometric projection of the sphere which is conformal.

spherical oblique aspect to prepare 1:10,000,000-scale maps of Hydrocarbon Provinces for three continents after a least-squares analysis of over 100 points on each continent to determine optimum parameters for a common conformal projection. For Europe, the central scale factor is 0.976 at a central point of lat. 55°N. and long. 20°E. For Africa, these parameters are 0.941, 5° N., and 20° E. For Asia, they are 0.939, 45° N., and 105° E., respectively.

The USGS has most often used the Stereographic in the polar aspect and ellipsoidal form for maps of Antarctica. For 1:500,000 sketch maps, the standard parallel is 71° S.; for its 1:250,000-scale series between 80° and the South Pole, the standard parallel is 80°14′ S. The Universal Transverse Mercator (UTM) grid employs the UPS (Universal Polar Stereographic) projection from the North Pole to lat. 84° N., and from the South Pole to lat. 80° S. For the UPS, the scale at each pole is reduced to 0.994, resulting in a standard parallel of 81°06′52.3″ N. or S. The UPS central meridian (as defined for λ_0 on p. ix) is the Greenwich meridian, with false eastings and northings of 2,000,000 m at each pole.

In 1962, a United Nations conference changed the polar portion of the International Map of the World (at a scale of 1:1,000,000) from a modified Polyconic to the polar Stereographic. This has consequently affected IMW sheets drawn by the USGS. North of lat. 84° N. or south of lat. 80° S., it is used "with scale matching that of the Modified Polyconic Projection or the Lambert Conformal Conic Projection at Latitudes 84° N. and 80° S." (United Nations, 1963, p. 10). The reference ellipsoid for all these polar Stereographic projections is the International of 1924.

The Astrogeology Center of the Geological Survey at Flagstaff, Ariz., has been using the polar Stereographic for the mapping of polar areas of every planet and satellite for which there is sufficient information in this region (see table 6).

The USGS is preparing a geologic map of the Arctic regions, using as a base an American Geographical Society map of the Arctic at a scale of 1:5,000,000. Drawn to the Stereographic projection, the map is based on a sphere having a radius which gives it the same volume as the International ellipsoid, and lat. 71° N. is made the standard parallel.

FORMULAS FOR THE SPHERE

Mathematically, a point at a given angular distance from the chosen center point on the sphere is plotted on the Stereographic projection at a distance from the center proportional to the trigonometric tangent of half that angular distance, and at its true azimuth, or, if the central scale factor is 1,

$$\rho = 2R \tan \tfrac{1}{2} c \qquad\qquad (21-1)$$
$$\theta = \pi - Az = 180° - Az \qquad\qquad (20-2)$$
$$k = \sec^2 \tfrac{1}{2} c \qquad\qquad (21-1a)$$

where c is the angular distance from the center, Az is the azimuth east of north (see equations (5−3) through (5−4b)), and θ is the polar coordinate east of south. Combining with standard equations, the formulas for rectangular coordinates of the oblique Stereographic projection are found to be as follows, given R, k_0, ϕ_1, λ_0, ϕ, and λ (see p. 312 for numerical examples):

$$x = Rk \cos \phi \sin (\lambda - \lambda_0) \qquad\qquad (21-2)$$
$$y = Rk [\cos \phi_1 \sin \phi - \sin \phi_1 \cos \phi \cos (\lambda - \lambda_0)] \qquad\qquad (21-3)$$

where

$$k = 2k_0/[1 + \sin \phi_1 \sin \phi + \cos \phi_1 \cos \phi \cos (\lambda - \lambda_0)] \qquad\qquad (21-4)$$

and (ϕ_1, λ_0) are the latitude and longitude of the center, which is also the origin. Since this is a conformal projection, k is the scale factor in all directions, based on

TABLE 24.—*Stereographic projection: Rectangular coordinates for equatorial aspect (sphere)*

[One hemisphere; y coordinate in parentheses under x coordinate]

Long. Lat.	0°	10°	20°	30°	40°
90°	0.00000	0.00000	0.00000	0.00000	0.00000
	(2.00000)	(2.00000)	(2.00000)	(2.00000)	(2.00000)
80	.00000	.05150	.10212	.15095	.19703
	(1.67820)	(1.68198)	(1.69331)	(1.71214)	(1.73837)
70	.00000	.08885	.17705	.26386	.34841
	(1.40042)	(1.40586)	(1.42227)	(1.44992)	(1.48921)
60	.00000	.11635	.23269	.34892	.46477
	(1.15470)	(1.16058)	(1.17839)	(1.20868)	(1.25237)
50	.00000	.13670	.27412	.41292	.55371
	(.93262)	(.93819)	(.95515)	(.98421)	(1.02659)
40	.00000	.15164	.30468	.46053	.62062
	(.72794)	(.73277)	(.74749)	(.77285)	(.81016)
30	.00000	.16233	.32661	.49487	.66931
	(.53590)	(.53970)	(.55133)	(.57143)	(.60117)
20	.00000	.16950	.34136	.51808	.70241
	(.35265)	(.35527)	(.36327)	(.37713)	(.39773)
10	.00000	.17363	.34987	.53150	.72164
	(.17498)	(.17631)	(.18037)	(.18744)	(.19796)
0	.00000	.17498	.35265	.53590	.72794
	(.00000)	(.00000)	(.00000)	(.00000)	(.00000)

a central scale factor of k_0, normally 1.0, but which may be reduced. The Y axis coincides with the central meridian λ_0, y increasing northerly and x, easterly.

If $\phi = -\phi_1$, and $\lambda = \lambda_0 \pm 180°$, the point cannot be plotted. Geometrically, it is the point from which projection takes place.

For the north polar Stereographic, with $\phi_1 = 90°$, these simplify to

$$x = 2R\ k_0 \tan (\pi/4 - \phi/2) \sin (\lambda - \lambda_0) \qquad (21-5)$$
$$y = -2R\ k_0 \tan (\pi/4 - \phi/2) \cos (\lambda - \lambda_0) \qquad (21-6)$$
$$k = 2k_0/(1 + \sin \phi) \qquad (21-7)$$
$$\rho = 2R\ k_0 \tan (\pi/4 - \phi/2) \qquad (21-8)$$
$$\theta = \lambda - \lambda_0 \qquad (20-9)$$

For the south polar Stereographic with $\phi_1 = -90°$,

$$x = 2R\ k_0 \tan (\pi/4 + \phi/2) \sin (\lambda - \lambda_0) \qquad (21-9)$$
$$y = 2R\ k_0 \tan (\pi/4 + \phi/2) \cos (\lambda - \lambda_0) \qquad (21-10)$$
$$k = 2k_0/(1 - \sin \phi) \qquad (21-11)$$
$$\rho = 2R\ k_0 \tan (\pi/4 + \phi/2) \qquad (21-12)$$
$$\theta = \pi - \lambda + \lambda_0 \qquad (20-12)$$

For the equatorial aspect, letting $\phi_1 = 0$, x is found from (21-2), but

$$y = R\ k \sin \phi \qquad (21-13)$$
$$k = 2\ k_0/[1 + \cos \phi \cos (\lambda - \lambda_0)] \qquad (21-14)$$

For the inverse formulas for the sphere, given R, k_0, ϕ_1, λ_0, x, and y:

$$\phi = \arcsin\ [\cos c \sin \phi_1 + (y \sin c \cos \phi_1/\rho)] \qquad (20-14)$$

TABLE 24.—*Stereographic projection: Rectangular coordinates for equatorial aspect (sphere)*—Continued

Long. Lat.	50°	60°	70°	80°	90°
90°	0.00000	0.00000	0.00000	0.00000	0.00000
	(2.00000)	(2.00000)	(2.00000)	(2.00000)	(2.00000)
80	.23933	.27674	.30806	.33201	.34730
	(1.77184)	(1.81227)	(1.85920)	(1.91196)	(1.96962)
70	.42957	.50588	.57547	.63588	.68404
	(1.54067)	(1.60493)	(1.68256)	(1.77402)	(1.87939)
60	.57972	.69282	.80246	.90613	1.00000
	(1.31078)	(1.38564)	(1.47911)	(1.59368)	(1.73205)
50	.69688	.84255	.99033	1.13892	1.28558
	(1.08415)	(1.15945)	(1.25597)	(1.37825)	(1.53209)
40	.78641	.95937	1.14080	1.33167	1.53209
	(.86141)	(.92954)	(1.01868)	(1.13464)	(1.28558)
30	.85235	1.04675	1.25567	1.48275	1.73205
	(.64240)	(.69783)	(.77149)	(.86928)	(1.00000)
20	.89755	1.10732	1.33650	1.59119	1.87939
	(.42645)	(.46538)	(.51767)	(.58808)	(.68404)
10	.92394	1.14295	1.38450	1.65643	1.96962
	(.21267)	(.23271)	(.25979)	(.29658)	(.34730)
0	.93262	1.15470	1.40042	1.67820	2.00000
	(.00000)	(.00000)	(.00000)	(.00000)	(.00000)

Radius of sphere = 1.0.

Origin: $(x, y) = 0$ at (lat., long.) = 0. Y axis increases north. Other quadrants of hemisphere are symmetrical.

If $\rho = 0$, equations (20−14) through (20−17) are indeterminate, but $\phi = \phi_1$ and $\lambda = \lambda_0$.

If ϕ_1 is not ±90°:

$$\lambda = \lambda_0 + \arctan\ [x \sin c/(\rho \cos \phi_1 \cos c - y \sin \phi_1 \sin c)] \qquad (20-15)$$

If ϕ_1 is 90°:

$$\lambda = \lambda_0 + \arctan\ [x/(-y)] \qquad (20-16)$$

If ϕ_1 is −90°:

$$\lambda = \lambda_0 + \arctan\ (x/y) \qquad (20-17)$$

In equations (20−14) and (20−15),

$$\rho = (x^2 + y^2)^{1/2} \qquad (20-18)$$
$$c = 2 \arctan\ [\rho/(2Rk_0)] \qquad (21-15)$$

The similarity of formulas for Orthographic, Stereographic, and other azimuthals may be noted. The equations for k' (k for the Stereographic, $k' = 1.0$ for the Orthographic) and the inverse c are the only differences in forward or inverse formulas for the sphere. The formulas are repeated for convenience, unless shown only a few lines earlier.

Table 24 lists rectangular coordinates for the equatorial aspect for a 10° graticule with a sphere of radius $R = 1.0$.

Following are equations for the centers and radii of the circles representing the meridians and parallels of the oblique Stereographic in the spherical form:

Circles for meridians:

Centers:	$x = -2R\ k_0/[\cos\ \phi_1\ \tan\ (\lambda-\lambda_0)]$	(21−16)
	$y = -2R\ k_0\ \tan\ \phi_1$	(21−17)
Radii:	$\rho = 2R\ k_0/[\cos\ \phi_1\ \sin\ (\lambda-\lambda_0)]$	(21−18)

Circles for parallels of latitude:

Centers:	$x = 0$	
	$y = 2R\ k_0\ \cos\ \phi_1/(\sin\ \phi_1+\sin\ \phi)$	(21−19)
Radii:	$\rho = 2R\ k_0\ \cos\ \phi/(\sin\ \phi_1+\sin\ \phi)$	(21−20)

Reduction to the polar and equatorial aspects may be made by letting $\phi_1 = \pm90°$ or $0°$, respectively.

To use a "standard circle" for the spherical Stereographic projection, such that the scale error is a minimum (based on least squares) over the apparent area of the map, the circle has an angular distance c from the center, where

$$c = 2\ \text{arccos}\ (1/\overline{k})^{1/2} \qquad (21-21)$$
$$\overline{k} = \tan^2\ (\beta/2)/(-\ln\ \cos^2\ (\beta/2)\ ) \qquad (21-22)$$

and β is the great circle distance of the circular limit of the region being mapped stereographically. The calculation is only slightly different if minimum error is based on the true area of the map:

$$\overline{k} = -\ln\ \cos^2\ (\beta/2)/\sin^2\ (\beta/2) \qquad (21-23)$$

In either case, c of the standard circle is approximately $\beta/\sqrt{2}$.

FORMULAS FOR THE ELLIPSOID

As noted above, the ellipsoidal forms of the Stereographic projection are nonperspective, in order to preserve conformality. The oblique and equatorial aspects are also slightly nonazimuthal for the same reason. The formulas result from replacing geodetic latitude ϕ in the spherical equations with conformal latitude χ (see equation (3−1)), followed by a small adjustment to the scale at the center of projection (Thomas, 1952, p. 14−15, 128−139). The general forward formulas for the oblique aspect are as follows; given a, e, k_0, ϕ_1, λ_0, ϕ, and λ (see p. 313 for numerical examples):

$$x = A\ \cos\ \chi\ \sin\ (\lambda-\lambda_0) \qquad (21-24)$$
$$y = A\ [\cos\ \chi_1\ \sin\ \chi-\sin\ \chi_1\ \cos\ \chi\ \cos\ (\lambda-\lambda_0)] \qquad (21-25)$$
$$k = A\ \cos\ \chi/(am) \qquad (21-26)$$

where

$$A = 2\ a\ k_0 m_1/\{\cos\ \chi_1\ [1 + \sin\ \chi_1\ \sin\ \chi$$
$$+ \cos\ \chi_1\ \cos\ \chi\ \cos\ (\lambda-\lambda_0)]\} \qquad (21-27)$$
$$\chi = 2\ \arctan\ \{\tan\ (\pi/4 + \phi/2)[(1-e\ \sin\ \phi)/(1+e\ \sin\ \phi)]^{e/2}\}$$
$$-\pi/2 \qquad (3-1)$$

or

$$= 2\ \arctan\ \left[\left(\frac{1+\sin\ \phi}{1-\sin\ \phi}\right)\left(\frac{1-e\ \sin\ \phi}{1+e\ \sin\ \phi}\right)^e\right]^{1/2} - \pi/2 \qquad (3-1a)$$

$$m = \cos\ \phi/(1-e^2\ \sin^2\ \phi)^{1/2} \qquad (14-15)$$

and χ_1 and m_1 are χ and m, respectively, calculated using ϕ_1, the central latitude, in place of ϕ, while k_0 is the scale factor at the center (normally 1.0). The origin of x and y coordinates occurs at the center (ϕ_1, λ_0), the Y axis coinciding with the central meridian λ_0, and y increasing northerly and x, easterly. The scale factor is actually k_0 along a near-circle passing through the origin, except for polar and equatorial aspects, where it occurs only at the central point. The radius of this near-circle is almost 0.4° at midlatitudes, and its center is along the central meridian, approaching the Equator from ϕ_1. The scale factor at the center of the circle is within 0.00001 less than k_0.

In the equatorial aspect, with the substitution of $\phi_1 = 0$ (therefore $\chi_1 = 0$), x is still found from (21–24) and k from (21–26), but

$$y = A \sin \chi \tag{21-28}$$
$$A = 2ak_0/[1 + \cos \chi \cos (\lambda - \lambda_0)] \tag{21-29}$$

For the north polar aspect, substitution of $\phi_1 = 90°$ (therefore $\chi_1 = 90°$) into equations (21–27) and (14–15) leads to an indeterminate A. To avoid this problem, the polar equations may take the form

$$x = \rho \sin (\lambda - \lambda_0) \tag{21-30}$$
$$y = - \rho \cos (\lambda - \lambda_0) \tag{21-31}$$
$$k = \rho/(a\ m) \tag{21-32}$$

where

$$\rho = 2\ ak_0\ t/[(1+e)^{(1+e)}\ (1-e)^{(1-e)}]^{1/2} \tag{21-33}$$
$$t = \tan(\pi/4 - \phi/2)/[(1 - e \sin \phi)/(1 + e \sin \phi)]^{e/2} \tag{15-9}$$

or

$$= \left[\left(\frac{1-\sin \phi}{1+\sin \phi} \right) \left(\frac{1+e \sin \phi}{1-e \sin \phi} \right)^e \right]^{1/2} \tag{15-9a}$$

Equation (21–33) applies only if true scale or known scale factor k_0 is to occur at the pole. For true scale along the circle representing latitude ϕ_c,

$$\rho = am_c\ t/t_c \tag{21-34}$$

Then the scale at the pole is

$$k_p = (1/2)\ m_c\ [(1+e)^{(1+e)}\ (1-e)^{(1-e)}]^{1/2}/(a\ t_c) \tag{21-35}$$

In equations (21–34) and (21–35), m_c and t_c are found from equations (14–15) and (15–9), respectively, substituting ϕ_c in place of ϕ.

For the south polar aspect, the equations for the north polar aspect may be used, but the signs of x, y, ϕ_c, ϕ, λ, and λ_0 must be reversed to be used in the equations.

For the inverse formulas for the ellipsoid, the oblique and equatorial aspects (where ϕ_1 is not ±90°) may be solved as follows, given a, e, k_0, ϕ_1, λ_0, x, and y:

$$\phi = 2 \arctan \{\tan (\pi/4 + \chi/2)[(1 + e \sin \phi)/(1 - e \sin \phi)]^{e/2}\}$$
$$- \pi/2 \tag{3-4}$$
$$\lambda = \lambda_0 + \arctan [x \sin c_e/(\rho \cos \chi_1 \cos c_e - y \sin \chi_1 \sin c_e)] \tag{21-36}$$

where

$$\chi = \arcsin [\cos c_e \sin \chi_1 + (y \sin c_e \cos \chi_1/\rho)] \tag{21-37}$$

but if $\rho = 0$, $\chi = \chi_1$ and $\lambda = \lambda_0$.

$$\rho = (x^2 + y^2)^{1/2} \qquad\qquad\qquad\qquad\qquad (20-18)$$
$$c_e = 2 \arctan [\rho \cos \chi_1/(2\ a\ k_0\ m_1)] \qquad\qquad (21-38)$$

and m_1 is found from equation (14−15) above, using ϕ_1 in place of ϕ. Equation (3−4) involves iteration, using χ as the first trial ϕ in the right-hand side, solving for a new trial ϕ on the left side, substituting into the right side, etc., until ϕ changes by less than a preset convergence (such as 10^{-9} radians). Conformal latitude χ_1 is found from (3−1), using ϕ_1 for ϕ. The factor c_e is not the true angular distance, as it is in the spherical case, but it is a convenient expression similar in nature to c, used to find ϕ and λ.

To avoid the iteration of (3−4), this series may be used instead:

$$\phi = \chi + (e^2/2 + 5e^4/24 + e^6/12 + 13e^8/360 + \dots)\sin 2\chi$$
$$+ (7e^4/48 + 29e^6/240 + 811e^8/11520 + \dots)$$
$$\sin 4\chi + (7e^6/120 + 81e^8/1120 + \dots)$$
$$\sin 6\chi + (4279e^8/161280 + \dots)\sin 8\chi + \dots \qquad (3-5)$$

For improved computational efficiency using this series, see p. 19.

The inverse equations for the north polar ellipsoidal Stereographic are as follows; given a, e, ϕ_c, k_0 (if $\phi_c = 90°$), λ_0, x, and y:

$$\phi = \pi/2 - 2 \arctan \{t[(1 - e \sin \phi)/(1 + e \sin \phi)]^{e/2}\} \qquad (7-9)$$
$$\lambda = \lambda_0 + \arctan [x/(-y)] \qquad\qquad\qquad (20-16)$$

Equation (7−9) for ϕ also involves iteration. For the first trial, $(\pi/2 - 2 \arctan t)$ is substituted for ϕ in the right side, and the procedure for solving equation (3−4) just above is followed:

If ϕ_c (the latitude of true scale) is 90°,

$$t = \rho[(1 + e)^{(1+e)} (1 - e)^{(1-e)}]^{1/2}/(2a\ k_0) \qquad (21-39)$$

If ϕ_c is not 90°,

$$t = \rho t_c/(a\ m_c) \qquad\qquad\qquad\qquad (21-40)$$

In either case,

$$\rho = (x^2 + y^2)^{1/2} \qquad\qquad\qquad\qquad (20-18)$$

and t_c and m_c are found from equations (15−9) and (14−15), respectively, listed with the forward equations, using ϕ_c in place of ϕ. Scale factor k is found from equation (21−26) or (21−32) above, for the ϕ found from equation (3−4), (3−5), or (7−9), depending on the aspect.

To avoid iteration, series (3−5) above may be used in place of (7−9), where

$$\chi = \pi/2 - 2 \arctan t \qquad\qquad\qquad (7-13)$$

Inverse equations for the south polar aspect are the same as those for the north polar aspect, but the signs of x, y, λ_0, ϕ_c, ϕ, and λ must be reversed.

Polar coordinates for the ellipsoidal form of the polar Stereographic are given in table 25, using the International ellipsoid and a central scale factor of 1.0.

To convert coordinates measured on an existing Stereographic map (or other azimuthal map projection), the user may choose any meridian for λ_0 on the polar aspect, but only the original meridian and parallel may be used for λ_0 and ϕ_1, respectively, on other aspects.

TABLE 25.—*Ellipsoidal polar Stereographic projection: Polar coordinates*

[International ellipsoid; central scale factor =1.0]

Latitude	Radius, meters	k, scale factor
90°	0.0	1.000000
89	111,702.7	1.000076
88	223,421.7	1.000305
87	335,173.4	1.000686
86	446,974.1	1.001219
85	558,840.1	1.001906
84	670,788.1	1.002746
83	782,834.3	1.003741
82	894,995.4	1.004889
81	1,007,287.9	1.006193
80	1,119,728.7	1.007653
79	1,232,334.4	1.009270
78	1,345,122.0	1.011045
77	1,458,108.4	1.012979
76	1,571,310.9	1.015073
75	1,684,746.8	1.017328
74	1,798,433.4	1.019746
73	1,912,388.4	1.022329
72	2,026,629.5	1.025077
71	2,141,174.8	1.027993
70	2,256,042.3	1.031078
69	2,371,250.5	1.034335
68	2,486,818.0	1.037765
67	2,602,763.6	1.041370
66	2,719,106.4	1.045154
65	2,835,865.8	1.049117
64	2,953,061.4	1.053264
63	3,070,713.2	1.057595
62	3,188,841.4	1.062115
61	3,307,466.7	1.066826
60	3,426,609.9	1.071732

22. GNOMONIC PROJECTION

SUMMARY

- Azimuthal and perspective.
- All meridians and the Equator are straight lines.
- All parallels except the Equator and poles are ellipses, parabolas, or hyperbolas.
- Neither conformal nor equal-area.
- All great circles are shown as straight lines.
- Less than one hemisphere may be shown around a given center.
- No distortion at the center only.
- Distortion and scale rapidly increase away from the center.
- Directions from the center are true.
- Used only in the spherical form.
- Known by Greeks 2,000 years ago.

HISTORY

The Gnomonic is the perspective projection of the globe from the center onto a plane tangent to the surface. It was used by Thales (636?−546?B.C.) of Miletus for star maps. Called "horologium" (sundial or clock) in early times, it was given the name "gnomonic" in the 19th century. It has also been called the Gnomic and the Central projection. The name Gnomonic is derived from the fact that the meridians radiate from the pole (or are spaced, on the equatorial aspect) just as the corresponding hour markings on a sundial for the same central latitude. The gnomon of the sundial is the elevated straightedge pointed toward the pole and casting its shadow on the various hour markings as the sun moves across the sky.

FEATURES AND USAGE

The outstanding (and only useful) feature of the Gnomonic projection results from the fact that each great-circle arc, the shortest distance between any two points on the surface of a sphere, lies in a plane passing through the center of the globe. Therefore, all great-circle arcs project as straight lines on this projection. The scale is badly distorted along such a plotted great circle, but the route is precise for the sphere.

Because the projection is from the center of the globe (fig. 33), it is impossible to show even a full hemisphere with the Gnomonic. Thus, if either pole is the point of tangency and center (the polar aspect), the Equator cannot be shown. Except at the center, the distortion of shape, area, and scale on the Gnomonic projection is so great that it has seldom been used for atlas maps. Historical exceptions are several sets of star maps from the late 18th century and terrestrial maps of 1803. These maps were plotted with the sphere projected onto the six faces of a tangent cube. The globe has also been projected from the mid-16th to

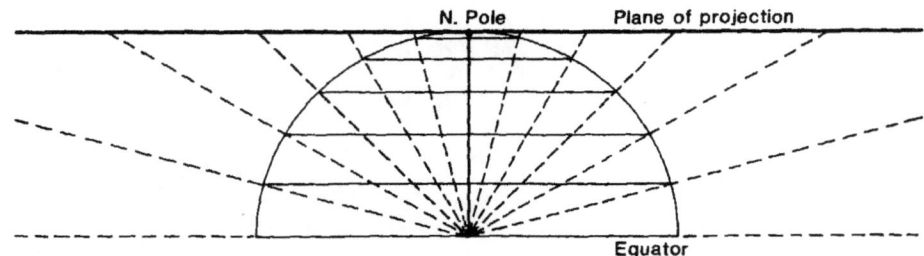

FIGURE 33.—Geometric projection of the parallels of the polar Gnomonic projection.

the mid-20th centuries, using the Gnomonic projection as well as others, onto the faces of other polyhedra. Generally, the projection is used for plotting great-circle paths, although the USGS has not used the projection for published maps.

The meridians of the polar Gnomonic projection appear straight, as on other polar azimuthal projections, and parallels of latitude are circles centered about the pole (fig. 34A). The parallels are closest near the pole, and their spacings increase away from the pole much more rapidly than they do on the polar Stereographic. The radii are proportional to the trigonometric tangent of the arc distance from the pole.

On the equatorial aspect, meridians are straight parallel lines perpendicular to the Equator, which is also straight (fig. 34B). The meridians are closest near the central meridian, and the spacing is rapidly increased away from it, the distance from center in proportion to the tangent of the difference in longitude. The parallels other than the Equator are all hyperbolic arcs, symmetrical about the Equator.

Since meridians are great-circle paths, they are also plotted straight on the oblique aspect of the Gnomonic, but they intersect at the pole (fig. 34C). They are not spaced at equal angles. The Equator is a straight line perpendicular to the central meridian. If the central latitude is north of the Equator, its colatitude (90° minus the latitude) is shown as a parabolic arc, more northern latitudes are ellipses, and more southern latitudes are hyperbolas. If the central latitude is south of the Equator, opposite signs apply.

Various graphical constructions have been published, but they are not described here because of the ease of plotting or calculating coordinates by computer, and because they do not add significantly to the understanding of this projection.

FORMULAS FOR THE SPHERE

A point at a given angular distance from the chosen center point on the sphere is plotted on the Gnomonic projection at a distance from the center proportional to the trigonometric tangent of that angular distance, and at its true azimuth, or

$$\rho = R \tan c \tag{22-1}$$
$$\theta = \pi - Az = 180° - Az \tag{20-2}$$
$$h' = 1/\cos^2 c \tag{22-2}$$
$$k' = 1/\cos c \tag{22-3}$$

where c is the angular distance of the given point from the center of projection. Az is the azimuth east of north, and θ is the polar coordinate east of south. The term k' is the scale factor in a direction perpendicular to the radius from the center of the map, not along the parallel except on the polar aspect. The scale factor h' is measured in the direction of the radius. Combining with standard equations, the formulas for rectangular coordinates of the oblique Gnomonic projection are as follows, given R, ϕ_1, λ_0, ϕ, and λ, to find x and y (see p. 319 for numerical examples):

$$x = Rk' \cos \phi \sin (\lambda - \lambda_0) \tag{22-4}$$
$$y = Rk' [\cos \phi_1 \sin \phi - \sin \phi_1 \cos \phi \cos (\lambda - \lambda_0)] \tag{22-5}$$

where k' is found from (22-3) above,

$$\cos c = \sin \phi_1 \sin \phi + \cos \phi_1 \cos \phi \cos (\lambda - \lambda_0) \tag{5-3}$$

and (ϕ_1, λ_0) are latitude and longitude of the projection center and origin. The Y axis coincides with the central meridian λ_0, y increasing northerly. The meridians

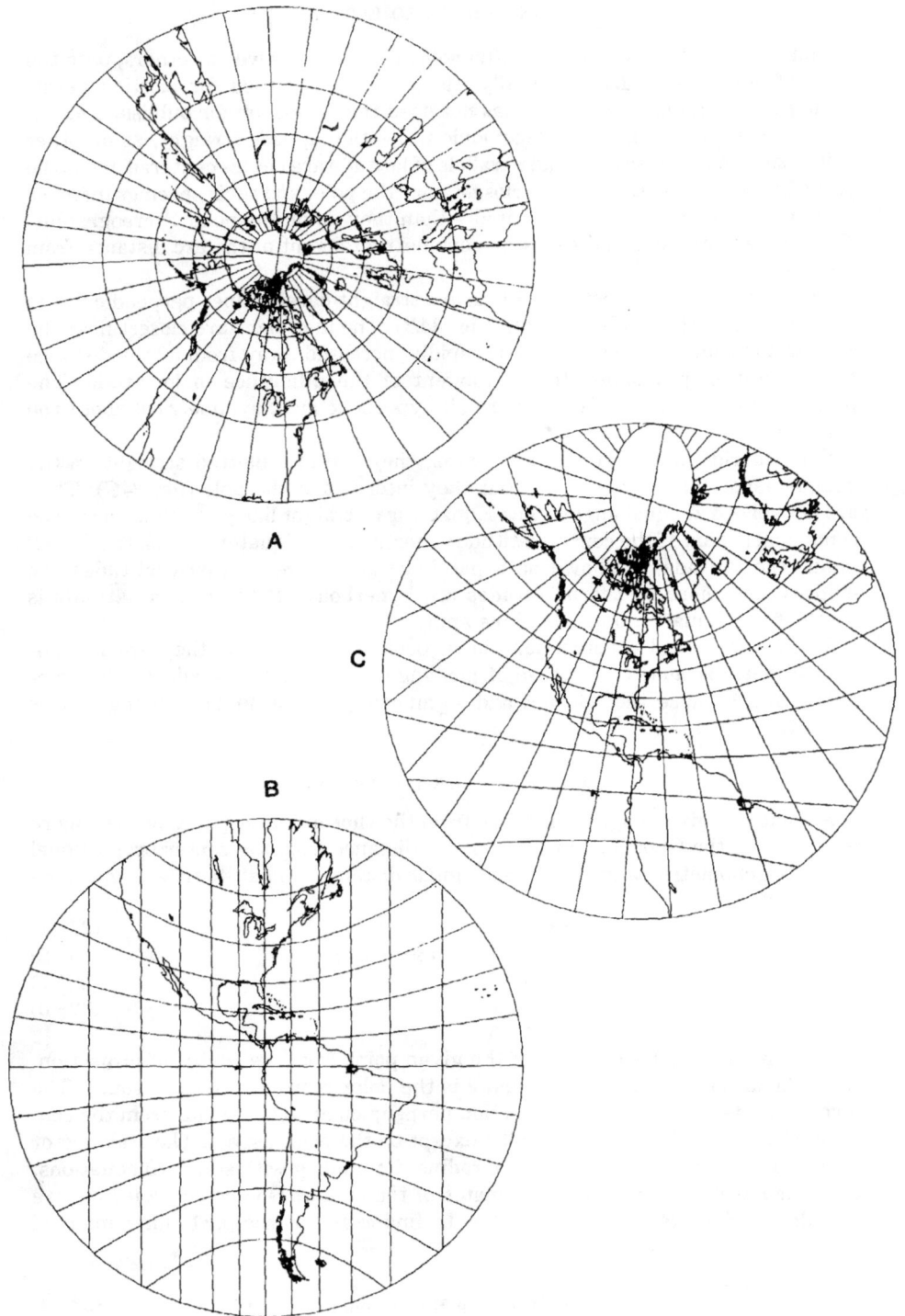

FIGURE 34.—Gnomonic projection, range 60° from center. (A) Polar aspect. (B) Equatorial aspect. (C) Oblique aspect, centered at lat. 40° N. All great-circle paths are straight lines on these maps.

are straight lines, but the parallels are conic sections for which the eccentricity = $(\cos \phi_1/\sin \phi)$. (If the eccentricity is zero, for $\phi_1 = \pm 90°$, they are circles. If the eccentricity is less than 1, they are ellipses; if equal to 1, a parabola; if greater than 1, a hyperbolic arc.)

For the north polar Gnomonic, letting $\phi_1 = 90°$,

$$x = R \cot \phi \sin (\lambda - \lambda_0) \tag{22-6}$$
$$y = -R \cot \phi \cos (\lambda - \lambda_0) \tag{22-7}$$

In polar coordinates,

$$\rho = R \cot \phi \tag{22-8}$$
$$\theta = \lambda - \lambda_0 \tag{22-9}$$

For the south polar Gnomonic, with $\phi_1 = -90°$,

$$x = -R \cot \phi \sin (\lambda - \lambda_0) \tag{22-10}$$
$$y = R \cot \phi \cos (\lambda - \lambda_0) \tag{22-11}$$

In polar coordinates,

$$\rho = -R \cot \phi \tag{22-12}$$
$$\theta = \pi - \lambda + \lambda_0 \tag{22-13}$$

For the equatorial Gnomonic, letting $\phi_1 = 0$,

$$x = R \tan (\lambda - \lambda_0) \tag{22-14}$$
$$y = R \tan \phi/\cos (\lambda - \lambda_0) \tag{22-15}$$

In automatically computing a general set of coordinates for a Gnomonic map, equation (5-3) above should be used to reject points equal to or greater than 90° from the center. That is, if $\cos c$ is zero or negative, the point is to be rejected. If $\cos c$ is positive, it may or may not be plotted depending on the desired limits of the map.

For the inverse formulas for the sphere, to find ϕ and λ, given R, ϕ_1, λ_0, x, and y:

$$\phi = \arcsin [\cos c \sin \phi_1 + (y \sin c \cos \phi_1/\rho)] \tag{20-14}$$

If $\rho = 0$, equations (20-14) through (20-17) are indeterminate, but $\phi = \phi_1$, and $\lambda = \lambda_0$. If ϕ_1 is not $\pm 90°$,

$$\lambda = \lambda_0 + \arctan [x \sin c/(\rho \cos \phi_1 \cos c - y \sin \phi_1 \sin c)] \tag{20-15}$$

If ϕ_1 is 90°,

$$\lambda = \lambda_0 + \arctan [x/(-y)] \tag{20-16}$$

If ϕ_1 is $-90°$,

$$\lambda = \lambda_0 + \arctan (x/y) \tag{20-17}$$

In equations (20-14) and (20-15),

$$\rho = (x^2 + y^2)^{1/2} \tag{20-18}$$
$$c = \arctan (\rho/R) \tag{22-16}$$

Table 26 lists rectangular coordinates for the equatorial aspect for a 10° graticule with a sphere of radius R = 1.0.

TABLE 26.—*Gnomonic projection: Rectangular coordinates for equatorial aspect*

Long.	0°	10°	20°	30°	40°	50°	60°	70°	80°
x	0.0000	0.1763	0.3640	0.5774	0.8391	1.1918	1.7321	2.7475	5.6713
Lat.					*y*				
80°	5.6713	5.7588	6.0353	6.5486	7.4033	8.8229	11.3426	16.5817	32.6596
70	2.7475	2.7899	2.9238	3.1725	3.5866	4.2743	5.4950	8.0331	15.8221
60	1.7321	1.7588	1.8432	2.0000	2.2610	2.6946	3.4641	5.0642	9.9745
50	1.1918	1.2101	1.2682	1.3761	1.5557	1.8540	2.3835	3.4845	6.8630
40	0.8391	0.8520	0.8930	0.9689	1.0954	1.3054	1.6782	2.4534	4.8322
30	.5774	.5863	.6144	.6667	0.7537	0.8982	1.1547	1.6881	3.3248
20	.3640	.3696	.3873	.4203	.4751	.5662	0.7279	1.0642	2.0960
10	.1763	.1790	.1876	.2036	.2302	.2743	.3527	0.5155	1.0154
0	.0000	.0000	.0000	.0000	.0000	.0000	.0000	.0000	0.0000

Radius of sphere = 1.0.

Origin: (x,y) = 0 at (lat., long.) = 0. Y axis increases north. Other quadrants of hemisphere are symmetrical. 90th meridian or pole cannot be shown.

23. GENERAL PERSPECTIVE PROJECTION

SUMMARY

- Often used to show the Earth or other planets and satellites as seen from space.
- Orthographic, Stereographic, and Gnomonic projections are special forms of the Vertical Perspective.
- Vertical Perspective projections are azimuthal; Tilted Perspectives are not.
- Central meridian and a particular parallel (if shown) are straight lines.
- Other meridians and parallels are usually arcs of circles or ellipses, but some may be parabolas or hyperbolas.
- Neither conformal (unless Stereographic) nor equal-area.
- If the point of perspective is above the sphere or ellipsoid, less than one hemisphere may be shown, unless the view is from infinity (Orthographic). If below center of globe or beyond the far surface, more than one hemisphere may be shown.
- No distortion at the center if a Vertical Perspective is projected onto a tangent plane. Considerable distortion near the projection limit.
- Directions from the center are true on the Vertical Perspective for the sphere and for the polar ellipsoidal form.
- Known by Greeks and Egyptians 2,000 years ago in limiting forms.

HISTORY AND USAGE

Whenever the Earth is photographed from space, the camera records the view as a perspective projection. If the camera precisely faces the center of the Earth, the projection is Vertical Perspective. Otherwise, a Tilted Perspective projection is obtained. Perspective views have also served other purposes.

With the complication of plotting coordinates for general perspective projections, there was little known interest in them until the 18th century, except for the well-known special cases of the Orthographic, Stereographic, and Gnomonic projections, which are perspective from infinity, the opposite surface, and the center of the sphere, respectively.

In 1701, the French mathematician Philippe De la Hire (1640−1718) found that if the point of perspective is placed 1.71 times the radius of the globe from the center in a direction opposite that of the plane of projection, the Equator on the polar Vertical Perspective projection has exactly twice the radius of the 45th parallel. The other parallels are not quite proportionally spaced, but this represented a use of geometric projection to achieve low distortion. Several other scientists, such as Antoine Parent in 1702 and various mathematicians of the late 19th century, extended this approach to obtain low-distortion projections which meet other criteria.

Of special interest was British geodesist A.R. Clarke's use of least squares to obtain in 1862 the Vertical Perspective projection with minimum error for the portion of the Earth bounded by a given spherical circle. He determined parameters for several continental areas, and he also presented the "Twilight" projection, with a bounding circle 108° from the center and centered to show much of the land mass of the Earth in one map. All these low- and minimum-error perspective projections were based on "far-side" points of perspective, and they were projected onto a secant plane to reduce overall error (Close and Clarke, 1911, p. 655−656; Snyder, 1985a).

Space exploration beginning in 1957 led to a renewed interest in the perspective projection, although Richard Edes Harrison had used several perspective views in a World War II atlas of 1944. Now the concern was for the pictorial view from space, not for minimal distortion. Albert L. Nowicki of the U.S. Army Map Service presented the AMS Lunar Projection, which is a far-side Vertical Perspec-

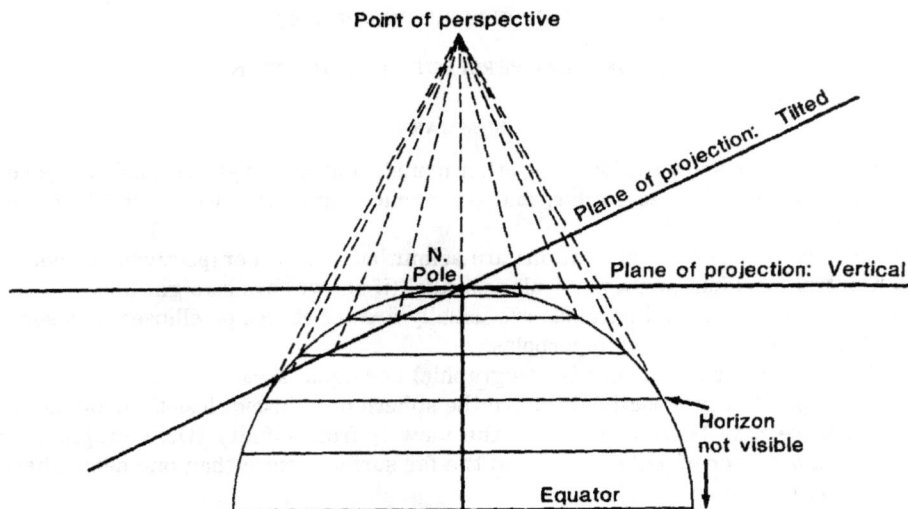

FIGURE 35.—Geometric projection of the parallels of the polar Perspective projections, Vertical and Tilted. Distance of point of perspective from center of Earth may be varied, as may the angle of tilt. For "far-side" projection, "point of perspective" would be shown below Equator and usually below South Pole on this drawing.

tive based on a perspective center about 1.54 times the radius from the center, to show somewhat more than one hemisphere of the Moon. This recognized the fact that more than half the Moon is seen from the Earth over a period of time. Nowicki called this a "modified Stereographic" projection (Nowicki, 1962). This name has been applied elsewhere to "far-side" Vertical Perspectives, none of which are conformal; it is applied later in this book to complex-algebra modifications of the Stereographic which are conformal but not perspective.

The Tilted Perspective projection is more complicated to compute, but since it has been the projection used in effect for most space photographs, such as those from the manned Gemini and Apollo space missions, it has been analyzed in recent literature.

Weather maps issued by the U.S. National Weather Service have regularly been based on a Vertical Perspective projection as seen from geosynchronous satellites near the Equatorial plane and 42,000 km from the Earth's center. The USGS has not used the Perspective projection to date for published maps.

FEATURES

The general Perspective projection (excepting the three common forms) should be considered primarily as a basis for a view of the Earth from space. The various historical studies described above and leading to low-error azimuthal projections have little practical value, since nonperspective azimuthal projections, like the Azimuthal Equidistant, may be used instead.

It is therefore of little interest to compute distortion at various locations on the map. There is no distortion at the center of projection with the Vertical Perspective onto a tangent plane (figs. 35 and 36), but there is shape, area, and scale distortion almost everywhere else on perspective maps (except that the Stereographic is conformal). The rapidity with which distortion increases varies with the location of the point of perspective and with the tilt of the plane to the line connecting this point with the center of the Earth (figs. 35 and 37). For the Vertical Perspective, this plane is perpendicular to this line.

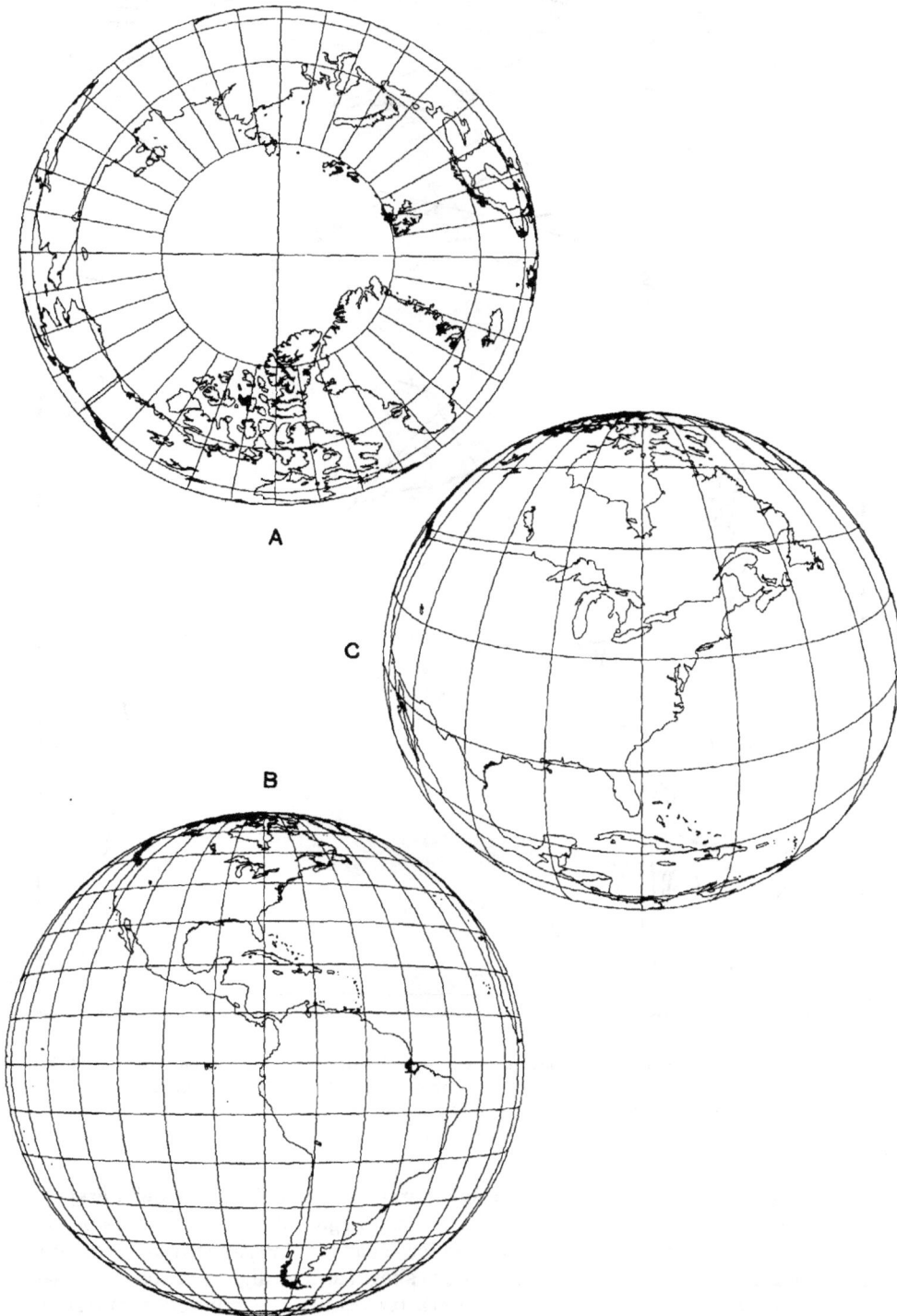

FIGURE 36.—Vertical Perspective projection. (A) Polar aspect, from 2,000 km above the Earth's surface. (B) Equatorial aspect, from geosynchronous satellite, 35,800 km above the Earth's surface. (C) Oblique aspect, centered at lat. 40° N., from 2,000 km above the Earth's surface.

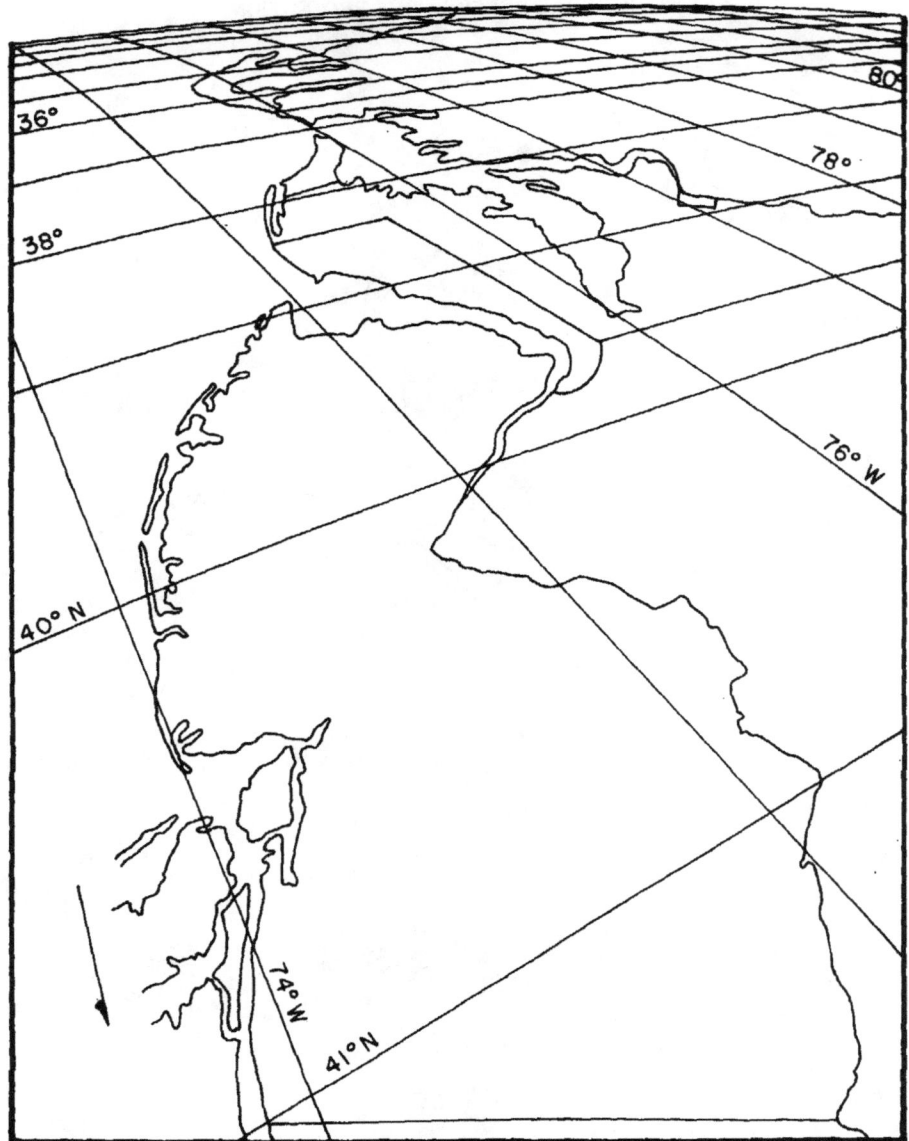

FIGURE 37.—Tilted Perspective projection. Eastern seaboard viewed from a point about 160 km above Newburgh, N.Y. Parameters using symbols in text: ϕ_1 = 41° 30′ N. lat., λ_0 = 74° 00′ W. long., ω = 55°, γ = 210°, P = 1.025. 1° graticule.

While the equations listed below are generally suitable for "far-side" Perspective projections (from below the surface), using negative distances to the points of perspective, the features are described for "near-side" Perspectives. For many perspective maps, one parallel of latitude is shown as a straight line (on the equatorial Orthographic aspect, all are straight). Its location is computed from formulas given below. The central meridian is also straight, as are all meridians on vertical polar aspects. Parallels of latitude on vertical polar aspects are concentric circles. Nearly all other meridians and parallels are elliptical arcs, except that certain angles of tilt may cause some meridians and parallels to be shown as parabolas or hyperbolas.

The horizon or limit of the map is outlined by one of the conic sections, depending on the angle of tilt and the location of the point of perspective. For the sphere, if there is no tilt, the outline is a circle. It is an ellipse, parabola, or hyperbola if the cosine of the tilt angle is greater than, equal to, or less than, respectively, the radius of the sphere divided by the distance from its center to the point of perspective.

For pictorial and small-scale usage, the spherical equations are adequate. For special large-scale applications, such as Landsat returned-beam-vidicon (RBV) and Space Shuttle Large-Format-Camera images and photographs, the ellipsoidal equations are necessary. The formulas are given below for several possible alternatives.

FORMULAS FOR THE SPHERE

VERTICAL PERSPECTIVE PROJECTION

A point at a given angular distance c from the center, and at an azimuth Az east of north is plotted in accordance with the following polar coordinates (θ is measured east of south):

$$\rho = R(P-1) \sin c/(P-\cos c) \tag{23-1}$$
$$\theta = \pi - Az = 180° - Az \tag{20-2}$$
$$h' = (P-1)(P \cos c - 1)/(P-\cos c)^2 \tag{23-2}$$
$$k' = (P-1)/(P-\cos c) \tag{23-3}$$

P is the distance of the point of perspective from the center of the Earth, divided by the radius R of the Earth as a sphere. It is positive in the direction of the center of the projection (for the "view from space") and negative in the opposite direction (for a far-side perspective such as those by Clarke and Nowicki (above), or the Stereographic, for which $P = -1$). In terms of the height H of the point of perspective above the surface, $P = H/R + 1$, or $H = R(P-1)$. The term k' is the scale factor in a direction perpendicular to the radius from the center of the map, not along the parallel, except in the polar aspect. The scale factor h' is measured in the direction of the radius.

Combining with standard equations, the formulas for rectangular coordinates of the oblique Vertical Perspective projection are as follows, given R, P, ϕ_1, λ_0, ϕ, and λ, to find x and y (see p. 320 for numerical examples):

$$x = R k' \cos \phi \sin (\lambda - \lambda_0) \tag{22-4}$$
$$y = R k' [\cos \phi_1 \sin \phi - \sin \phi_1 \cos \phi \cos (\lambda - \lambda_0)] \tag{22-5}$$

where k' is found from (23-3) above,

$$\cos c = \sin \phi_1 \sin \phi + \cos \phi_1 \cos \phi \cos (\lambda - \lambda_0) \tag{5-3}$$

and (ϕ_1, λ_0) are latitude and longitude of the projection center and origin. The Y axis coincides with the central meridian λ_0, y increasing northerly. The map limit is a circle of radius $R[(P-1)/(P+1)]^{1/2}$. Meridians and parallels are generally elliptical arcs, but the central meridian and the latitude whose sine equals $P \sin \phi_1$ are straight lines. For automatic plotting, equation (5-3) should be used to reject points for which $\cos c$ is less than $1/P$. These are beyond the range of the map, regardless of whether P is positive or negative.

Because of the number of other equations below, the simplified equations for polar and equatorial aspects are not given here. They may be obtained by entering the appropriate values of ϕ_1 in equations (22-4), (22-5), and (5-3). Table 27 shows rectangular coordinates for a hemisphere as seen from a geosynchronous satellite.

TABLE 27.—*Vertical Perspective projection: Rectangular coordinates
for equatorial aspect from geosynchronous satellite*

[y coordinate in parentheses under x coordinate]

Long. Lat.	0°	10°	20°	30°	40°	50°	60°	70°	
80°	0.0000	0.0263	0.0517	—	—	—	—	—	
	(.8586)	(.8582)	(.8572)	—	—	—	—	—	
70	.0000	.0531	.1044	0.1520	0.1943	0.2301	0.2581	—	
	(.8412)	(.8405)	(.8385)	(.8351)	(.8306)	(.8251)	(.8189)	—	
60	.0000	.0796	.1563	.2271	.2896	.3418	.3820	0.4094	
	(.7953)	(.7943)	(.7914)	(.7867)	(.7804)	(.7727)	(.7641)	(.7547)	
50	.0000	.1048	.2054	.2979	.3789	.4458	.4967	.5304	
	(.7203)	(.7191)	(.7156)	(.7100)	(.7026)	(.6936)	(.6835)	(.6727)	
40	.0000	.1275	.2496	.3614	.4587	.5382	.5978	.6363	
	(.6171)	(.6159)	(.6123)	(.6065)	(.5988)	(.5895)	(.5792)	(.5682)	
30	.0000	.1465	.2867	.4146	.5252	.6149	.6813	.7232	
	(.4884)	(.4872)	(.4840)	(.4787)	(.4717)	(.4634)	(.4542)	(.4444)	
20	.0000	.1610	.3148	.4548	.5753	.6725	.7436	.7879	0.
	(.3384)	(.3375)	(.3350)	(.3311)	(.3258)	(.3195)	(.3125)	(.3052)	(.
10	.0000	.1701	.3324	.4798	.6065	.7082	.7822	.8277	.
	(.1732)	(.1727)	(.1714)	(.1692)	(.1664)	(.1630)	(.1593)	(.1553)	(.
0	.0000	.1732	.3384	.4884	.6171	.7203	.7953	.8412	.
	(.0000)	(.0000)	(.0000)	(.0000)	(.0000)	(.0000)	(.0000)	(.0000)	(.

Radius of sphere = 1.0. Radius of bounding circle = 0.8588. Point of perspective is P = 6.62 radii from center (35,800 km above Earth's surface). See fig.
Origin: (x, y) = 0 at (lat., long.) = 0. Y axis increases north. Other quadrants of hemisphere are symmetrical. Dashes indicate invisible graticule intersections.
and 90th meridians are also invisible.

For the inverse formulas for the Vertical Perspective projection of the sphere, given R, P, ϕ_1, λ_0, x, and y, to find ϕ and λ:

$$\phi = \arcsin [\cos c \sin \phi_1 + (y \sin c \cos \phi_1/\rho)] \qquad (20-14)$$

If $\rho = 0$, equations (20−14) through (20−17) are indeterminate, but $\phi = \phi_1$ and $\lambda = \lambda_0$.

$$\lambda = \lambda_0 + \arctan [x \sin c/(\rho \cos \phi_1 \cos c - y \sin \phi_1 \sin c)] \qquad (20-15)$$

In equations (20−14) and (20−15),

$$\rho = (x^2 + y^2)^{1/2} \qquad (20-18)$$
$$c = \arcsin \{[P - (1 - \rho^2(P+1)/(R^2(P-1)))^{1/2}]/$$
$$[R(P-1)/\rho + \rho/(R(P-1))]\} \qquad (23-4)$$

In (23−4), if P is negative and ρ is greater than $R(P-1)/P$, c must be subtracted from 180° to place it in the proper quadrant.

TILTED PERSPECTIVE PROJECTION

The following equations are used in conjunction with the equations above for the Vertical Perspective. While they may be combined, it is easier to follow and more practical to program separately these equations to follow (for forward) or precede (for inverse) those above. For the forward equations, given R, P, ϕ_1, λ_0, ω, γ, ϕ, and λ, (x,y) is first calculated from equations (5−3), (23−3), (22−4), and (22−5) in order, then

$$x_t = (x \cos \gamma - y \sin \gamma) \cos \omega/A \qquad (23-5)$$
$$y_t = (y \cos \gamma + x \sin \gamma)/A \qquad (23-6)$$

where

$$A = \{(y \cos \gamma + x \sin \gamma) \sin \omega/H\} + \cos \omega \qquad (23-7)$$
$$H = R(P-1) \qquad (23-8)$$

γ is the azimuth east of north of the Y axis, the most upward-tilted axis of the plane of projection relative to the tangent plane, and ω is the upward angle of tilt, or the angle between the Y_t axis and the tangent plane. The X_t axis lies at the intersection of the tangent and tilted planes. The rectangular coordinates (x_t, y_t) lie in the tilted plane, with the origin at (ϕ_1, λ_0) and the Y_t axis oriented at azimuth γ rather than due north (see fig. 38).

Restated in terms of a camera in space, the camera is placed at a distance RP from the center of the Earth, perpendicularly over point (ϕ_1, λ_0). The camera is horizontally turned to face γ clockwise from north, and then tilted (90°−ω) downward from horizontal, "horizontal" meaning parallel to a plane tangent to the sphere at (ϕ_1, λ_0). The photograph is then taken, placing points (ϕ, λ) in positions (x_t, y_t), based on a scale reduction in R. The straight meridian and parallel of the Vertical Perspective are also straight on the Tilted form.

If the tilted plane is perpendicular to the line connecting the point of perspective and the horizon, $\omega = \arcsin (1/P)$. Points for which $\cos c$ (equation (5−3)) is less than $(1/P)$ are beyond the map limits, as on the Vertical Perspective, but the map limit is now a conic section of eccentricity equal to $\sin \omega/(1 - 1/P^2)^{1/2}$. This limit may be plotted by inserting the (x,y) coordinates of the circle representing the Vertical Perspective map limit into equations (23−5) through (23−7) for final plotting coordinates (x_t, y_t), after stating the original equations for the circle in parametric form,

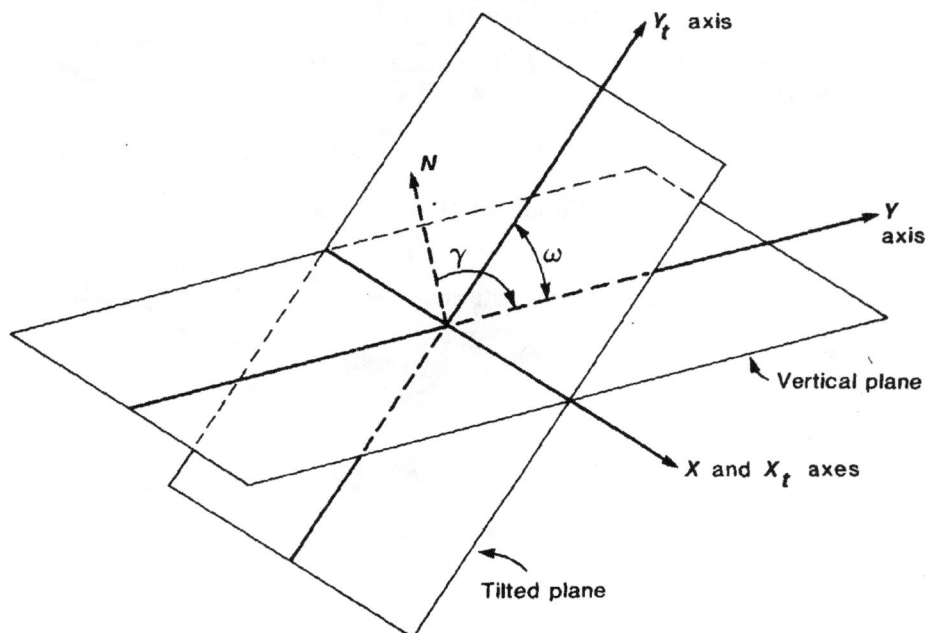

FIGURE 38.—Coordinate system for Tilted Perspective projection. The north (N) arrow lies in the vertical plane for the equatorial or oblique aspect. See figure 35 for projection of points onto these planes.

$$x = R[(P-1)/(P+1)]^{1/2} \sin \theta \qquad\qquad (23-9)$$
$$y = R[(P-1)/(P+1)]^{1/2} \cos \theta \qquad\qquad (23-10)$$

in which θ is given successive values from 0° to 360°.

For the inverse equations for the Tilted Perspective projection of the sphere, given R, P, ϕ_1, λ_0, ω, γ, x_t and y_t, first H is calculated from (23−8), and (x,y) are calculated from these equations:

$$M = Hx_t/(H-y_t \sin \omega) \qquad\qquad (23-11)$$
$$Q = Hy_t \cos \omega/(H-y_t \sin \omega) \qquad\qquad (23-12)$$
$$x = M \cos \gamma + Q \sin \gamma \qquad\qquad (23-13)$$
$$y = Q \cos \gamma - M \sin \gamma \qquad\qquad (23-14)$$

Then these values of (x,y) are inserted in equations (20−14) through (20−18) and (23−4) for inversing the Vertical Perspective, to obtain (ϕ, λ).

It is also possible to use projective constants K_1-K_{11} for the sphere as well as the ellipsoid in equations below, but this is not often done for the sphere. If desired, the formulas below may be used for the sphere if the eccentricity is made zero.

FORMULAS FOR THE ELLIPSOID

VERTICAL PERSPECTIVE PROJECTION

Because of the increased number of equations, they are given in the order of use. Given a, e^2, P, ϕ_1, λ_0, h_0, ϕ, λ, and h, to find x and y (For numerical examples see p. 323):

$$N = a/(1-e^2 \sin^2 \phi)^{1/2} \qquad\qquad (4-20)$$
$$N_1 = a/(1-e^2 \sin^2 \phi_1)^{1/2} \qquad\qquad (8-23)$$
$$C = [(N+h)/a] \cos \phi \qquad\qquad (23-15)$$
$$S = \{[N(1-e^2) + h]/a\} \sin \phi \qquad\qquad (23-16)$$
$$\phi_g = \phi_1 - \arcsin [N_1 e^2 \sin \phi_1 \cos \phi_1/(Pa)] \qquad\qquad (23-17)$$
$$H = Pa \cos \phi_g/\cos \phi_1 - N_1 - h_0 \qquad\qquad (23-18)$$
$$K = H/[P \cos (\phi_1-\phi_g) - S \sin \phi_1 - C \cos \phi_1 \cos (\lambda-\lambda_0)] \qquad\qquad (23-19)$$
$$x = KC \sin (\lambda-\lambda_0) \qquad\qquad (23-19a)$$
$$y = K[P \sin (\phi_1-\phi_g) + S \cos \phi_1 - C \sin \phi_1 \cos (\lambda-\lambda_0)] \qquad\qquad (23-20)$$

where P = the distance of the point of perspective from the center of the Earth, divided by a, the semimajor axis.

H = the height of the point of perspective in a direction perpendicular to the surface of the ellipsoid at nadir point (ϕ_1, λ_0), but measured from the height h_0 of the nadir above the ellipsoid, not above sea level.

ϕ_g = the geocentric latitude of the point of perspective, measured as the angle between the direct line from the center to this point, and the equatorial plane, not as the geocentric latitude corresponding to ϕ_1.

h = the height of (ϕ, λ) above the ellipsoid. The use of h makes these formulas more general, but for most plotting of graticules it would be zero.

If H is given rather than P, the latter may be computed as follows:

$$P = (\cos \phi_1/\cos \phi_g) (H+N_1+h_0)/a \qquad\qquad (23-21)$$

Since ϕ_g is calculated from P in equation (23-17), iteration is involved, with ϕ_1 as the first trial value of ϕ_g. The comments following the forward formulas for the sphere apply approximately here. The straight parallel is the latitude ϕ whose sine equals $Pa \sin \phi_g/[N(1-e^2)+h]$, if h is a constant, such as zero. This is an iterative calculation with successive substitution of ϕ, starting with ϕ_1 as a trial. The central meridian λ_0 is also straight.

For the inverse formulas for the Vertical Perspective projection of the ellipsoid, given a, e^2, P, ϕ_1, λ_0, h_0, h, x, and y, to find ϕ, λ:

Equations (23-17) and (23-18) are used to compute ϕ_g and H (or (23-21) to compute P if H is given), then

$$B = P \cos (\phi_1-\phi_g) \qquad\qquad (23-22)$$
$$D = P \sin (\phi_1-\phi_g) \qquad\qquad (23-23)$$
$$L = 1-e^2 \cos^2 \phi_1 \qquad\qquad (23-24)$$
$$G = 1-e^2 \sin^2 \phi_1 \qquad\qquad (23-25)$$
$$J = 2e^2 \sin \phi_1 \cos \phi_1 \qquad\qquad (23-26)$$
$$u = - 2BLH - 2DGy + BJy + DHJ \qquad\qquad (23-27)$$
$$v = LH^2 + Gy^2 - HJy + (1-e^2)x^2 \qquad\qquad (23-28)$$

If h is zero, $E = 1$ in the next equation (23-29). If h is not zero, use a first trial $E = 1$.
Then,

$$t = P^2 (1-e^2 \cos^2 \phi_g) - E (1-e^2) \qquad\qquad (23-29)$$
$$K' = [- u + (u^2-4tv)^{1/2}]/2t \qquad\qquad (23-30)$$
$$X = a[(B-H/K') \cos \phi_1 - (y/K'-D) \sin \phi_1] \qquad\qquad (23-31)$$
$$Y = ax/K' \qquad\qquad (23-32)$$
$$S = (y/K'-D) \cos \phi_1 + (B-H/K') \sin \phi_1 \qquad\qquad (23-33)$$
$$\lambda = \lambda_0 + \arctan (Y/X) \qquad\qquad (23-34)$$

If h is not zero, ϕ may be initially estimated at $\arcsin S$ to calculate a trial ϕ from equation (23-35) and then E from (23-36). Equations (23-29) through (23-36) are iterated using the latest values of ϕ, E, and h (based on the height above the ellipsoid at the trial ϕ, λ) until ϕ changes by a negligible amount.

$$\phi = \arcsin \{S/[(1-e^2)/(1-e^2 \sin^2 \phi)^{1/2} + h/a]\} \qquad (23-35)$$
$$E = [1/(1-e^2 \sin^2 \phi)^{1/2} + h/a]^2 - e^2 \sin^2 \phi \, [1/(1-e^2 \sin^2 \phi) - h^2/(a^2-a^2 e^2)] \quad (23-36)$$

If h is zero, no iteration or previous estimate for ϕ is necessary, and ϕ may be found as follows:

$$\phi = \arctan \{S/[(1-e^2) \, (1-e^2-S^2)]^{1/2}\} \qquad (23-37)$$

TILTED PERSPECTIVE PROJECTION USING "CAMERA" PARAMETERS

Given a, e^2, P, ϕ_1, λ_0, h_0, ω, γ, ϕ, λ, and h, to find x_t and y_t, first (x, y) are calculated from $(23-15)$ through $(23-20)$, then (x_t, y_t) from $(23-5)$ through $(23-7)$, but $(23-8)$ is not used. Definitions following each of these sets of formulas apply, but the limits (horizons) of the map do not precisely follow the spherical formulas given. The ellipsoidal form is unnecessarily complicated to extend to the map limits in any case.

For the corresponding inverse formulas, given a, e^2, P, ϕ_1, λ_0, h_0, ω, γ, h, x_t and y_t, to find ϕ and λ, first (x, y) are calculated using $(23-11)$ through $(23-14)$, then (ϕ, λ) are calculated from $(23-17)$, $(23-18)$, and $(23-22)$ through $(23-37)$.

TILTED PERSPECTIVE PROJECTION USING PROJECTIVE EQUATIONS

When a photograph or other plane image is obtained from space, projective equations with 11 constants may be used to find the rectangular coordinates of any point of known latitude, longitude, and height above the ellipsoid, in the plane of the image, instead of directly using the orientation of the camera or sensor. The 3-dimensional rectangular coordinates of a point on or off the Earth's surface can be found from the following equations, taking the semimajor axis a of the Earth as 1.0:

$$X = C \cos \lambda \qquad (23-38)$$
$$Y = C \sin \lambda \qquad (23-39)$$
$$Z = S \qquad (23-40)$$

where C and S are found from equations $(23-15)$ and $(23-16)$ respectively, the X and Y axis lie in the Earth's equatorial plane, with the X axis intersecting the prime meridian ($\lambda = 0$), and the Z axis coincides with the Earth's polar axis. The values of X, Y, and Z increase from the origin at the center of the Earth toward $\lambda = 0$, $\lambda = 90°$, and the North Pole, respectively, but they are dimensionless in the above equations.

The projective equations are as follows,

$$x'_t = (K_1 X + K_2 Y + K_3 Z + K_4)/(K_5 X + K_6 Y + K_7 Z + 1) \qquad (23-41)$$
$$y_t' = (K_8 X + K_9 Y + K_{10} Z + K_{11})/(K_5 X + K_6 Y + K_7 Z + 1) \qquad (23-42)$$

where (x_t', y_t') are coordinates in the tilted plane, but relative to any pair of perpendicular axes and any origin, rather than those of (x_t, y_t) as described following $(23-8)$. Constants in the denominators are dimensionless, but those in the numerators are in the same units as (x_t', y_t').

The 11 constants K_n may be determined either from points on the "space photograph" or from the parameters of the "camera." Although least squares and other corrections are used in determining these constants in analytical photogrammetry for highest precision, the approach given here is confined to the use of measurements which are assumed to the precise. The reader is referred to other texts for the least-squares approach.

To determine K_1-K_{11} from six widely spaced identified points on the image, equations (23−41) and (23−42) may be transposed as follows:

$$XK_1 + YK_2 + ZK_3 + K_4 - x_t'XK_5 - x_t'YK_6 - x_t'ZK_7$$
$$+ 0K_8 + 0K_9 + 0K_{10} + 0K_{11} = x_t' \qquad (23-43)$$

$$0K_1 + 0K_2 + 0K_3 + 0K_4 - y_t'XK_5 - y_t'YK_6 - y_t'ZK_7$$
$$+ XK_8 + YK_9 + ZK_{10} + K_{11} = y_t' \qquad (23-44)$$

Using a separate pair of these two equations for each of the six points, and omitting one of the twelve equations, the equations are suitable for solution as eleven simultaneous equations with eleven unknowns (K_1-K_{11}), using standard programs. A less satisfactory but usable procedure involving only seven simultaneous equations is detailed in Snyder (1981c, p. 158).

To determine K_1-K_{11} from parameters of the projection, first H is found from (23−18), then

$$
\begin{aligned}
U &= P[\sin(\phi_1-\phi_g) \cos \gamma \sin \omega + \cos (\phi_1-\phi_g) \cos \omega] & (23-45) \\
F &= (\sin \phi_1 \sin \lambda_0 \cos \gamma - \cos \lambda_0 \sin \gamma)/U & (23-46) \\
V &= (\sin \phi_1 \sin \lambda_0 \sin \gamma + \cos \lambda_0 \cos \gamma) \cos \omega/U & (23-47) \\
M &= (\sin \phi_1 \cos \lambda_0 \sin \gamma - \sin \lambda_0 \cos \gamma) \cos \omega/U & (23-48) \\
N &= (\sin \phi_1 \cos \lambda_0 \cos \gamma + \sin \lambda_0 \sin \gamma)/U & (23-49) \\
W &= (-\sin \gamma \cos \omega \cos \theta - \cos \gamma \sin \theta)/U & (23-50) \\
T &= (-\sin \gamma \cos \omega \sin \theta + \cos \gamma \cos \theta)/U & (23-51) \\
K_5 &= -N \sin \omega - \cos \phi_1 \cos \lambda_0 \cos \omega/U & (23-52) \\
K_6 &= -F \sin \omega - \cos \phi_1 \sin \lambda_0 \cos \omega/U & (23-53) \\
K_7 &= (\cos \phi_1 \cos \gamma \sin \omega - \sin \phi_1 \cos \omega)/U & (23-54) \\
K_1 &= H (M \cos \theta + N \sin \theta) + K_5 x_0 & (23-55) \\
K_2 &= H(V \cos \theta + F \sin \theta) + K_6 x_0 & (23-56) \\
K_3 &= HW \cos \phi_1 + K_7 x_0 & (23-57) \\
K_4 &= HWP \sin (\phi_1-\phi_g) + x_0 & (23-58) \\
K_8 &= H (M \sin \theta - N \cos \theta) + K_5 y_0 & (23-59) \\
K_9 &= H (V \sin \theta - F \cos \theta) + K_6 y_0 & (23-60) \\
K_{10} &= HT \cos \phi_1 + K_7 y_0 & (23-61) \\
K_{11} &= HTP \sin (\phi_1-\phi_g) + y_0 & (23-62)
\end{aligned}
$$

where, to review the meanings of previously defined symbols,

(ϕ_1, λ_0) = latitude and longitude of the projection center and origin

ϕ_g = geocentric latitude of the point of perspective, found from equation (23−17)

γ = azimuth east of north of the Y_t axis, or the most upward-tilted axis of the plane of projection

ω = upward angle of tilt

P = distance from the center of the Earth to the point of perspective, divided by a, the semimajor axis.

New symbols are as follows:

θ = clockwise angle through which the (X_t, Y_t) axes are rotated for the arbitrary axes (X_t', Y_t') used for the constants K_1-K_{11}. This may be made zero to retain the (X_t, Y_t) axes.

(x_0, y_0) = offsets of the (X_t, Y_t) axes to establish a different origin for the (X_t', Y_t') axes. They may also be set at zero to retain the (X_t, Y_t) axes.

The two sets of axes are related as follows:

$$x_t' = x_t \cos \theta - y_t \sin \theta + x_0 \qquad (23-62a)$$
$$y_t' = y_t \cos \theta + x_t \sin \theta + y_0 \qquad (23-62b)$$

For inverse computations using projective constants, given $K_1 - K_{11}$, x_t', and y_t', to find ϕ and λ, the following are calculated in order:

$$A_1 = x_t' K_5 - K_1 \qquad (23-63)$$
$$A_2 = x_t' K_6 - K_2 \qquad (23-64)$$
$$A_3 = x_t' K_7 - K_3 \qquad (23-65)$$
$$A_4 = K_4 - x_t' \qquad (23-66)$$
$$A_5 = y_t' K_5 - K_8 \qquad (23-67)$$
$$A_6 = y_t' K_6 - K_9 \qquad (23-68)$$
$$A_7 = y_t' K_7 - K_{10} \qquad (23-69)$$
$$A_8 = K_{11} - y_t' \qquad (23-70)$$
$$A_9 = A_1 A_8 - A_4 A_5 \qquad (23-71)$$
$$A_{10} = A_1 A_7 - A_3 A_5 \qquad (23-72)$$
$$A_{11} = A_2 A_5 - A_1 A_6 \qquad (23-73)$$
$$A_{12} = A_2 A_7 - A_3 A_6 \qquad (23-74)$$
$$A_{13} = A_2 A_8 - A_4 A_6 \qquad (23-75)$$
$$A_{14} = A_{10}{}^2 + A_{11}{}^2/(1-e^2) + A_{12}{}^2 \qquad (23-76)$$
$$A_{15} = A_9 A_{10} + A_{12} A_{13} \qquad (23-77)$$
$$A_{16} = A_9{}^2 - E A_{11}{}^2 + A_{13}{}^2 \qquad (23-78)$$

where E is found from (23−36) if h is not zero, or $E = 1$ if h is zero. Then

$$S = (A_{15}/A_{14}) \pm [(A_{15}/A_{14})^2 - A_{16}/A_{14}]^{1/2} \qquad (23-79)$$

and ϕ is found from (23−35) if h is not zero, or (23−37) if h is zero, taking one sign in (23−79) for the latitude desired, and the opposite sign for the latitude hidden from view at the same coordinates. The same sign applies throughout the map, once it is determined for a point for which the latitude is obviously right or wrong.

$$\lambda = \arctan [(A_9 - A_{10}S)/(A_{12}S - A_{13})] \qquad (23-80)$$

In this case the ATAN2 function is not used, but 180° must be added to or subtracted from λ if the denominator has the same sign as A_{11}.

If h is not zero, E is initially assumed to be 1. After trial values of ϕ and λ are determined above, an h suitable for that point may be used with the new ϕ in calculating E; then A_{16}, S, ϕ and λ are recalculated. Iteration continues until the change in the calculated ϕ is negligible.

If h is zero, since $E = 1$ and (23−37) is explicit in ϕ, no iteration is required.

Finally, to compute "camera" parameters from given constants $K_1 - K_{11}$ (Bender, ca. 1970, p. 26−27), given a, e^2, and an assumed h_0, first the following three simultaneous equations are solved for X_0, Y_0, and Z_0, the space coordinates of the point of perspective divided by a (see description of axes following (23−40)):

$$\begin{aligned} K_1 X_0 + K_2 Y_0 + K_3 Z_0 &= -K_4 \\ K_8 X_0 + K_9 Y_0 + K_{10} Z_0 &= -K_{11} \qquad (23-81) \\ K_5 X_0 + K_6 Y_0 + K_7 Z_0 &= -1 \end{aligned}$$

Then the coordinates (x_p, y_p) of the principal point of the "space photograph" are found as the point where a perpendicular dropped from the point of perspective strikes the plane of the map:

$$x_p = (K_1K_5 + K_2K_6 + K_3K_7)/(K_5^2 + K_6^2 + K_7^2) \qquad (23-82)$$
$$y_p = (K_5K_8 + K_6K_9 + K_7K_{10})/(K_5^2 + K_6^2 + K_7^2) \qquad (23-83)$$

The parameters reviewed after equation (23−62) are then found as follows (except that ϕ_g is an intermediate latitude described after (23−20)):

$$\lambda_0 = \arctan(Y_0/X_0) \qquad (23-84)$$
$$P = (X_0^2 + Y_0^2 + Z_0^2)^{1/2} \qquad (23-85)$$
$$\phi_g = \arcsin(Z_0/P) \qquad (23-86)$$
$$\phi_1 = \phi_g + \arcsin\{e^2 \sin\phi_1 \cos\phi_1/[P(1-e^2\sin^2\phi_1)^{1/2}]\} \qquad (23-87)$$

which is solved for ϕ_1, with ϕ_g as the first approximation for ϕ_1, and iterating with successive substitution.

$$H = a[P\cos\phi_g/\cos\phi_1 - 1/(1-e^2\sin^2\phi_1)^{1/2} - h_0/a] \qquad (23-88)$$

using for h_0 the height at (ϕ_1, λ_0). The forward equations (23−15), (23−16), and (23−38) through (23−40) are now used to calculate X, Y, and Z for (ϕ_1, λ_0, h_0). Substituting these values and K_1-K_{11} into (23−41) and (23−42), x_0 is found as x_t', and y_0 as y_t'. Then

$$\omega = \arcsin\{[(x_0-x_p)^2 + (y_p-y_0)^2]^{1/2}/H\} \qquad (23-89)$$
$$\theta = \arctan[(x_0-x_p)/(y_p-y_0)] \qquad (23-90)$$

Then, (x_t', y_t') are calculated for $(\phi_1 + 0.02°, \lambda_0)$ from (23−41) and (23−42) and the necessary preceding equations, in order to obtain the direction of the Y_t axis, and from this value of (x_t', y_t') are calculated

$$x_t = (x_t'-x_0)\cos\theta + (y_t'-y_0)\sin\theta \qquad (23-91)$$
$$y_t = (y_t'-y_0)\cos\theta - (x_t'-x_0)\sin\theta \qquad (23-92)$$
$$\gamma = -\arctan[x_t/(y_t\cos\omega)] \qquad (23-93)$$

24. LAMBERT AZIMUTHAL EQUAL-AREA PROJECTION

SUMMARY

- Azimuthal.
- Equal-Area.
- All meridians in the polar aspect, the central meridian in other aspects, and the Equator in the equatorial aspect are straight lines.
- The outer meridian of a hemisphere in the equatorial aspect (for the sphere) and the parallels in the polar aspect (sphere or ellipsoid) are circles.
- All other meridians and parallels are complex curves.
- Not a perspective projection.
- Scale decreases radially as the distance increases from the center, the only point without distortion.
- Scale increases in the direction perpendicular to radii as the distance increases from the center.
- Directions from the center are true for the sphere and the polar ellipsoidal forms.
- Point opposite the center is shown as a circle surrounding the map (for the sphere).
- Used for maps of continents and hemispheres.
- Presented by Lambert in 1772.

HISTORY

The last major projection presented by Johann Heinrich Lambert in his 1772 *Beiträge* was his azimuthal equal-area projection (Lambert, 1772, p. 75–78). His name is usually applied to the projection in modern references, but it is occasionally called merely the Azimuthal (or Zenithal) Equal-Area projection. Not only is it equal-area, with, of course, the azimuthal property showing true directions from the center of the projection, but its scale at a given distance from the center varies less from the scale at the center than the scale of any of the other major azimuthals (see table 21).

Lambert discussed the polar and equatorial aspects of the Azimuthal Equal-Area projection, but the oblique aspect is just as popular now. The polar aspect was apparently independently derived by Lorgna in Italy in 1789, and the latter was called the originator in a publication a century later (USC&GS, 1882, p. 290). G. A. Ginzburg proposed two modifications of the general Lambert Azimuthal projection in 1949 to reduce the angular distortion at the expense of creating a slight distortion in area (Maling, 1960, p. 206). A common modification was devised by Ernst Hammer in 1892 and is called the Hammer or Hammer-Aitoff projection. It consists of halving the vertical coordinates of the equatorial aspect of one hemisphere and doubling the values of the meridians from center. It retains equality of area, but it is no longer azimuthal.

FEATURES

The Lambert Azimuthal Equal-Area projection is not a perspective projection. It may be called a "synthetic" azimuthal in that it was derived for the specific purpose of maintaining equal area. The ellipsoidal form maintains equal area, but it is not quite azimuthal except in the polar aspect, so the name for the general ellipsoidal form is a slight misnomer, although it looks like the spherical azimuthal form and has most of its other characteristics.

The polar aspect (fig. 39A), like that of the Orthographic and Stereographic, has circles for parallels of latitude, all centered about the North or South Pole, and straight equally spaced radii of these circles for meridians. The difference is,

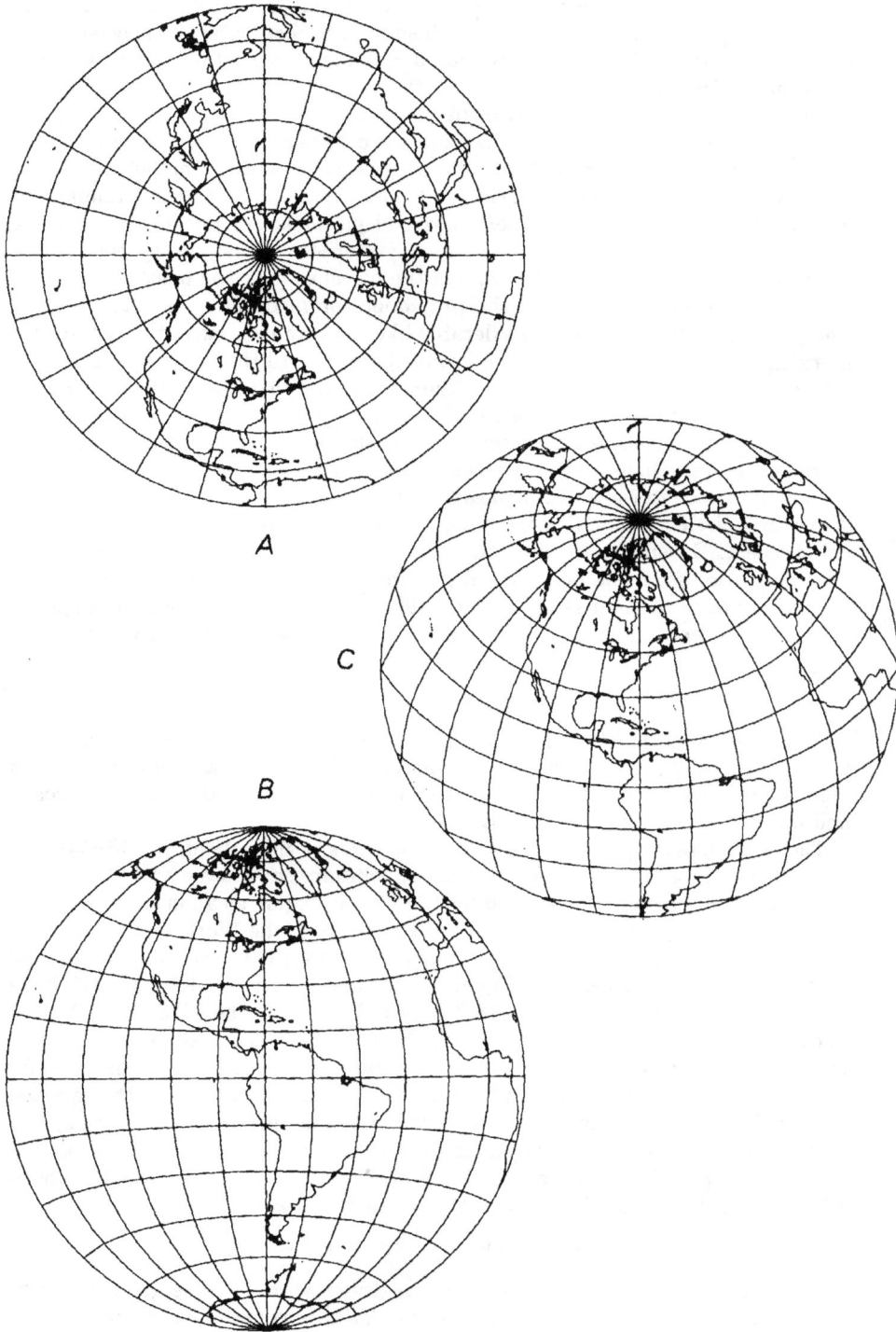

FIGURE 39.—Lambert Azimuthal Equal-Area projection. (A) Polar aspect showing one hemisphere; the entire globe may be included in a circle of 1.41 times the diameter of the Equator. (B) Equatorial aspect; frequently used in atlases for maps of the Eastern and Western hemispheres. (C) Oblique aspect; centered on lat. 40° N.

once again, in the spacing of the parallels. For the Lambert, the spacing between the parallels gradually decreases with increasing distance from the pole. The opposite pole, not visible on either the Orthographic or Stereographic, may be shown on the Lambert as a large circle surrounding the map, almost half again as far as the Equator from the center. Normally, the projection is not shown beyond one hemisphere (or beyond the Equator in the polar aspect).

The equatorial aspect (fig. 39*B*) has, like the other azimuthals, a straight Equator and straight central meridian, with a circle representing the 90th meridian east and west of the central meridian. Unlike those for the Orthographic and Stereographic, the remaining meridians and parallels are uncommon complex curves. The chief visual distinguishing characteristic is that the spacing of the meridians near the 90th meridian and of the parallels near the poles is about 0.7 of the spacing at the center of the projection, or moderately less to the eye. The parallels of latitude look considerably like circular arcs, except near the 90th meridians, where they exhibit a noticeable turn toward the nearest pole.

The oblique aspect (fig. 39*C*) of the Lambert Azimuthal Equal-Area resembles the Orthographic to some extent, until it is seen that crowding is far less pronounced as the distance from the center increases. Aside from the straight central meridian, all meridians and parallels are complex curves, not ellipses.

In both the equatorial and oblique aspects, the point opposite the center may be shown as a circle surrounding the map, corresponding to the opposite pole in the polar aspect. Except for the advantage of showing the entire Earth in an equal-area projection from one point, the distortion is so great beyond the inner hemisphere that for world maps the Earth should be shown as two separate hemispherical maps, the second map centered on the point opposite the center of the first map.

USAGE

The spherical form in all three aspects of the Lambert Azimuthal Equal-Area projection has appeared in recent commercial atlases for Eastern and Western Hemispheres (replacing the long-used Globular projection) and for maps of oceans and most of the continents and polar regions.

The polar aspect appears in the *National Atlas* (USGS, 1970, p. 148–149) for maps delineating north and south polar expeditions, at a scale of 1:39,000,000. It is used at a scale of 1:20,000,000 for the Arctic Region as an inset on the 1978 USGS Map of Prospective Hydrocarbon Provinces of the World.

The USGS has prepared six base maps of the Pacific Ocean on the spherical form of the Lambert Azimuthal Equal-Area. Four sections, at 1:10,000,000, have centers at 35° N., 150° E.; 35° N., 135° W.; 35° S., 135° E.; and 40° S., 100° W. The Pacific-Antarctic region, at a scale of 1:8,300,000, is centered at 20° S. and 165° W., while a Pacific Basin map at 1:17,100,000 is centered at the Equator and 160° W. (The last two maps were originally erroneously labeled with scales that are too small.) The base maps have been used for individual geographic, geologic, tectonic, minerals, and energy maps. The USGS has also cooperated with the National Geographic Society in revising maps of the entire Moon drawn to the spherical form of the equatorial Lambert Azimuthal Equal-Area.

GEOMETRIC CONSTRUCTION

The polar aspect (for the sphere) may be drawn with a simple geometric construction: In figure 40, if angle *AOR* is the latitude φ and *P* is the pole at the center, *PA* is the radius of that latitude on the polar map. The oblique and equatorial aspects have no direct geometric construction. They may be prepared indirectly by using other azimuthal projections (Harrison, 1943), but it is now simpler to plot automatically or manually from rectangular coordinates which are generated by a relatively simple computer program. The formulas are given below.

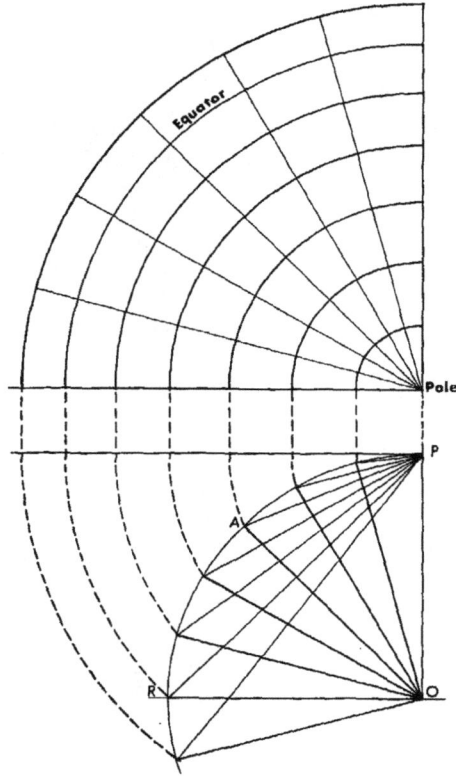

FIGURE 40.—Geometric construction of polar Lambert Azimuthal Equal-Area projection.

FORMULAS FOR THE SPHERE

On the Lambert Azimuthal Equal-Area projection for the sphere, a point at a given angular distance from the center of projection is plotted at a distance from the center proportional to the sine of half that angular distance, and at its true azimuth, or

$$\rho = 2 R \sin (c/2) \qquad (24-1)$$
$$\theta = \pi - Az = 180° - Az \qquad (20-2)$$
$$h' = \cos (c/2) \qquad (24-1a)$$
$$k' = \sec (c/2) \qquad (24-1b)$$

where c is the angular distance from the center, Az is the azimuth east of north (see equations (5–3) through (5–4b)), and θ is the polar coordinate east of south. The term k' is the scale factor in a direction perpendicular to the radius from the center of the map, not along the parallel, except in the polar aspect. The scale factor h' in the direction of the radius equals $1/k'$. After combining with standard equations, the formulas for rectangular coordinates for the oblique Lambert Azimuthal Equal-Area projection may be written as follows, given R, ϕ_1, λ_0, ϕ, and λ:

$$x = R k' \cos \phi \sin (\lambda - \lambda_0) \qquad (22-4)$$
$$y = R k' [\cos \phi_1 \sin \phi - \sin \phi_1 \cos \phi \cos (\lambda - \lambda_0)] \qquad (22-5)$$

where

$$k' = \{2/[1 + \sin \phi_1 \sin \phi + \cos \phi_1 \cos \phi \cos (\lambda - \lambda_0)]\}^{1/2} \qquad (24-2)$$

and (ϕ_1, λ_0) are latitude and longitude of the projection center and origin. The Y axis coincides with the central meridian λ_0, y increasing northerly. For the point opposite the center, at latitude $-\phi_1$ and longitude $\lambda_0 \pm 180°$, these formulas give indeterminates. This point, if the map is to cover the entire sphere, is plotted as a circle of radius $2R$.

For the north polar Lambert Azimuthal Equal-Area, with $\phi_1 = 90°$, since k' is k for the polar aspect, these formulas simplify to the following (see p. 332 for numerical examples):

$$x = 2R \sin (\pi/4 - \phi/2) \sin (\lambda - \lambda_0) \qquad (24-3)$$
$$y = -2R \sin (\pi/4 - \phi/2) \cos (\lambda - \lambda_0) \qquad (24-4)$$
$$k = \sec (\pi/4 - \phi/2) \qquad (24-5)$$
$$h = 1/k = \cos (\pi/4 - \phi/2) \qquad (24-6)$$

or, using polar coordinates,

$$\rho = 2R \sin (\pi/4 - \phi/2) \qquad (24-7)$$
$$\theta = \lambda - \lambda_0 \qquad (20-9)$$

For the south polar aspect, with $\phi_1 = -90°$,

$$x = 2R \cos (\pi/4 - \phi/2) \sin (\lambda - \lambda_0) \qquad (24-8)$$
$$y = 2R \cos (\pi/4 - \phi/2) \cos (\lambda - \lambda_0) \qquad (24-9)$$
$$k = 1/\sin (\pi/4 - \phi/2) \qquad (24-10)$$
$$h = \sin (\pi/4 - \phi/2) \qquad (24-11)$$

or

$$\rho = 2R \cos (\pi/4 - \phi/2) \qquad (24-12)$$
$$\theta = \pi - \lambda + \lambda_0 \qquad (20-12)$$

For the equatorial aspect, letting $\phi_1 = 0$, x is found from (22–4), but

$$y = Rk' \sin \phi \qquad (24-13)$$

and

$$k' = \{2/[1 + \cos \phi \cos (\lambda - \lambda_0)]\}^{1/2} \qquad (24-14)$$

The maximum angular deformation ω for any of these aspects, derived from equations (4–7) through (4–9), and from the fact that $h' = 1/k'$ for equal-area maps:

$$\sin (\omega/2) = (k'^2 - 1)/(1 + k'^2) \qquad (24-15)$$

For the inverse formulas for the sphere, given R, ϕ_1, λ_0, x, and y:

$$\phi = \arcsin [\cos c \sin \phi_1 + (y \sin c \cos \phi_1/\rho)] \qquad (20-14)$$

If $\rho = 0$, equations (20–14) through (20–17) are indeterminate, but $\phi = \phi_1$ and $\lambda = \lambda_0$.

If ϕ_1 is not $\pm 90°$:

$$\lambda = \lambda_0 + \arctan [x \sin c/(\rho \cos \phi_1 \cos c - y \sin \phi_1 \sin c)] \qquad (20-15)$$

If ϕ_1 is 90°:

$$\lambda = \lambda_0 + \arctan [x/(-y)] \tag{20-16}$$

If ϕ_1 is −90°:

$$\lambda = \lambda_0 + \arctan (x/y) \tag{20-17}$$

In equations (20−14) and (20−15),

$$\rho = (x^2 + y^2)^{1/2} \tag{20-18}$$
$$c = 2 \arcsin [\rho/(2R)] \tag{24-16}$$

It may again be noted that several of the above forward and inverse equations apply to the other azimuthals.

Table 28 lists rectangular coordinates for the equatorial aspect for a 10° graticule with a sphere of radius $R = 1.0$.

FORMULAS FOR THE ELLIPSOID

As noted above, the ellipsoidal oblique aspect of the Lambert Azimuthal Equal-Area projection is slightly nonazimuthal in order to preserve equality of area. To date, the USGS has not used the ellipsoidal form in any aspect. The formulas are analogous to the spherical equations, but involve replacing the geodetic latitude ϕ with authalic latitude β (see equation (3−11)). In order to achieve correct scale in all directions at the center of projection, that is, to make the center a "standard point," a slight adjustment using D is also necessary. The general forward formulas for the oblique aspect are as follows, given a, e, ϕ_1, λ_0, ϕ, and λ (see p. 333 for numerical examples):

$$x = B\, D \cos \beta \sin (\lambda - \lambda_0) \tag{24-17}$$
$$y = (B/D) [\cos \beta_1 \sin \beta - \sin \beta_1 \cos \beta \cos (\lambda - \lambda_0)] \tag{24-18}$$

where

$$B = R_q \{2/[1 + \sin \beta_1 \sin \beta + \cos \beta_1 \cos \beta \cos (\lambda - \lambda_0)]\}^{1/2} \tag{24-19}$$
$$D = a\, m_1/(R_q \cos \beta_1) \tag{24-20}$$
$$R_q = a\, (q_p/2)^{1/2} \tag{3-13}$$
$$\beta = \arcsin (q/q_p) \tag{3-11}$$
$$q = (1 - e^2) \{\sin \phi/(1 - e^2 \sin^2 \phi) - [1/(2\,e)] \ln$$
$$[(1 - e \sin \phi)/(1 + e \sin \phi)]\} \tag{3-12}$$
$$m = \cos \phi/(1 - e^2 \sin^2 \phi)^{1/2} \tag{14-15}$$

and β_1 is found from (3−11), using q_1 for q, while q_1 and q_p are found from (3−12) using ϕ_1 and 90°, respectively, for ϕ, and m_1 is found from (14−15), calculated for ϕ_1. The origin occurs at (ϕ_1, λ_0), the Y axis coinciding with the central meridian λ_0, and y increasing northerly. For the equatorial aspect, the equations simplify as follows:

$$x = a \cos \beta \sin (\lambda - \lambda_0) \{2/[1 + \cos \beta \cos (\lambda - \lambda_0)]\}^{1/2} \tag{24-21}$$
$$y = (R_q^2/a) \sin \beta \{2/[1 + \cos \beta \cos (\lambda - \lambda_0)]\}^{1/2} \tag{24-22}$$

For the polar aspects, D is indeterminate using equations above, but the following equations may be used instead. For the north polar aspect, $\phi_1 = 90°$,

TABLE 28.—*Lambert Azimuthal Equal-Area projection: Rectangular coordinates for equatorial aspect (sphere)*

[One hemisphere; *y* coordinate in parentheses under *x* coordinate]

Long. Lat.	0°	10°	20°	30°	40°
90° _____	0.00000 (1.41421)	0.00000 (1.41421)	0.00000 (1.41421)	0.00000 (1.41421)	0.00000 (1.41421)
80 _____	.00000 (1.28558)	.03941 (1.28702)	.07788 (1.29135)	.11448 (1.29851)	.14830 (1.30842)
70 _____	.00000 (1.14715)	.07264 (1.14938)	.14391 (1.15607)	.21242 (1.16725)	.27676 (1.18296)
60 _____	.00000 (1.00000)	.10051 (1.00254)	.19948 (1.01021)	.29535 (1.02311)	.38649 (1.04143)
50 _____	.00000 (.84524)	.12353 (.84776)	.24549 (.85539)	.36430 (.86830)	.47831 (.88680)
40 _____	.00000 (.68404)	.14203 (.68631)	.28254 (.69317)	.41999 (.70483)	.55281 (.72164)
30 _____	.00000 (.51764)	.15624 (.51947)	.31103 (.52504)	.46291 (.53452)	.61040 (.54826)
20 _____	.00000 (.34730)	.16631 (.34858)	.33123 (.35248)	.49337 (.35915)	.65136 (.36883)
10 _____	.00000 (.17431)	.17231 (.17497)	.34329 (.17698)	.51158 (.18041)	.67588 (.18540)
0 _____	.00000 (.00000)	.17431 (.00000)	.34730 (.00000)	.51764 (.00000)	.68404 (.00000)

Radius of sphere = 1.0.
Origin: $(x, y) = 0$ at (lat., long.) = 0. Y axis increases north. Other quadrants of hemisphere are symmetrical.

$$x = \rho \sin (\lambda - \lambda_0) \qquad (21-30)$$
$$y = -\rho \cos (\lambda - \lambda_0) \qquad (21-31)$$
$$k = \rho/(a\ m) \qquad (21-32)$$

where

$$\rho = a(q_p - q)^{1/2} \qquad (24-23)$$

and q_p and q are found from (3−12) as before and m from (14−15) above. Since the meridians and parallels intersect at right angles, and this is an equal-area projection, $h = 1/k$.

For the south polar aspect, ($\phi_1 = -90°$), equations (21−30) and (21−32) remain the same, but

$$y = \rho \cos (\lambda - \lambda_0) \qquad (24-24)$$

and

$$\rho = a(q_p + q)^{1/2} \qquad (24-25)$$

For the inverse formulas for the ellipsoid, the oblique and equatorial aspects (where ϕ_1 is not ±90°) may be solved as follows, given a, e, ϕ_1, λ_0, x, and y.

$$\phi = \phi + \frac{(1-e^2 \sin^2 \phi)^2}{2 \cos \phi} \left[\frac{q}{1-e^2} - \frac{\sin \phi}{1-e^2 \sin^2 \phi} + \frac{1}{2e}\ \ln \left(\frac{1-e \sin \phi}{1+e \sin \phi} \right) \right] \qquad (3-16)$$

$$\lambda = \lambda_0 + \arctan [x \sin c_e/(D \rho \cos \beta_1 \cos c_e - D^2 y \sin \beta_1 \sin c_e)] \qquad (24-26)$$

where

$$q = q_p [\cos c_e \sin \beta_1 + (Dy \sin c_e \cos \beta_1/\rho)] \qquad (24-27)$$

TABLE 28.—*Lambert Azimuthal Equal-Area projection: Rectangular coordinates for equatorial aspect (sphere)*—Continued

Long. Lat.	50°	60°	70°	80°	90°
90°	0.00000	0.00000	0.00000	0.00000	0.00000
	(1.41421)	(1.41421)	(1.41421)	(1.41421)	(1.41421)
80	.17843	.20400	.22420	.23828	.24558
	(1.32096)	(1.33594)	(1.35313)	(1.37219)	(1.39273)
70	.33548	.38709	.43006	.46280	.48369
	(1.20323)	(1.22806)	(1.25741)	(1.29114)	(1.32893)
60	.47122	.54772	.61403	.66797	.70711
	(1.06544)	(1.09545)	(1.13179)	(1.17481)	(1.22474)
50	.58579	.68485	.77342	.84909	.90904
	(.91132)	(.94244)	(.98088)	(1.02752)	(1.08335)
40	.67933	.79778	.90620	1.00231	1.08335
	(.74411)	(.77298)	(.80919)	(.85401)	(.90904)
30	.75197	.88604	1.01087	1.12454	1.22474
	(.56674)	(.59069)	(.62108)	(.65927)	(.70711)
20	.80380	.94928	1.08635	1.21347	1.32893
	(.38191)	(.39896)	(.42078)	(.44848)	(.48369)
10	.83488	.98731	1.13192	1.26747	1.39273
	(.19217)	(.20102)	(.21240)	(.22694)	(.24558)
0	.84524	1.00000	1.14715	1.28558	1.41421
	(.00000)	(.00000)	(.00000)	(.00000)	(.00000)

but if $\rho = 0$, then $q = q_p \sin \beta_1$, and $\lambda - \lambda_0$.

$$\rho = [(x/D)^2 + (Dy)^2]^{1/2} \qquad (24-28)$$
$$c_e = 2 \arcsin (\rho/2 \, R_q) \qquad (24-29)$$

and D, R_q, q_p, and β_1 are found from equations (24−20), (3−13), (3−12), (3−11), and (14−15), as in the forward equations above. The factor c_e is not the true angular distance, as c is in the spherical case, but it is a convenient number similar in nature to c, used to find ϕ and λ. Equation (3−16) requires iteration by successive substitution, using arcsin $(q/2)$ as the first trial ϕ on the right side, calculating ϕ on the left side, substituting this new ϕ on the right side, etc., until the change in ϕ is negligible. If, in equation (24−27),

$$q = \pm\{1 - [(1-e^2)/(2\,e)] \ln [(1-e)/(1+e)]\} \qquad (14-20)$$

the iteration does not converge, but $\phi = \pm 90°$, taking the sign of q.

To avoid the iteration, equations (3−16), (24−27), and (14−20) may be replaced with the series

$$\phi = \beta + (e^2/3 + 31e^4/180 + 517e^6/5040 + \ldots) \sin 2\beta$$
$$+ (23e^4/360 + 251e^6/3780 + \ldots) \sin 4\beta + (761e^6/45360 + \ldots)$$
$$\sin 6\beta + \ldots \qquad (3-18)$$

where β, the authalic latitude, is found thus:

$$\beta = \arcsin [\cos c_e \sin \beta_1 + (Dy \sin c_e \cos \beta_1/\rho)] \qquad (24-30)$$

Equations (24−26), (24−28), and (24−29) still apply. In (24−30), if $\rho = 0$, $\beta = \beta_1$. For improved computational efficiency using this series, see p. 19.

TABLE 29.—*Ellipsoidal polar Lambert Azimuthal Equal-Area projection (International ellipsoid)*

Latitude	Radius, meters	h	k
90°	0.0	1.000000	1.000000
89	111,698.4	.999962	1.000038
88	223,387.7	.999848	1.000152
87	335,058.5	.999657	1.000343
86	446,701.8	.999391	1.000610
85	558,308.3	.999048	1.000953
84	669,868.8	.998630	1.001372
83	781,374.2	.998135	1.001869
82	892,815.4	.997564	1.002442
81	1,004,183.1	.996918	1.003092
80	1,115,468.3	.996195	1.003820
79	1,226,661.9	.995397	1.004625
78	1,337,754.7	.994522	1.005508
77	1,448,737.6	.993573	1.006469
76	1,559,601.7	.992547	1.007509
75	1,670,337.9	.991446	1.008628
74	1,780,937.2	.990270	1.009826
73	1,891,390.6	.989018	1.011104
72	2,001,689.2	.987691	1.012462
71	2,111,824.0	.986289	1.013902
70	2,221,786.2	.984812	1.015422

h = scale factor along meridian.
k = scale factor along parallel.

The inverse formulas for the polar aspects involve relatively simple transformations of above equations (21−30), (21−31), and (24−23), except that ϕ is found from the iterative equation (3−16), listed just above, in which q is calculated as follows:

$$q = \pm[q_p - (\rho/a)^2] \qquad (24-31)$$

taking the sign of ϕ_1. The series (3−18) may be used instead for ϕ, where

$$\beta = \pm \arcsin \{1-\rho^2/[a^2[1-((1-e^2)/(2\,e))\ln((1-e)/(1+e))]]\} \qquad (24-32)$$

taking the sign of ϕ_1. In any case,

$$\rho = (x^2 + y^2)^{1/2} \qquad (20-18)$$

while

$$\lambda = \lambda_0 + \arctan[x/(-y)] \qquad (20-16)$$

for the north polar case, and

$$\lambda = \lambda_0 + \arctan(x/y) \qquad (20-17)$$

for the south polar case.

Table 29 lists polar coordinates for the ellipsoidal polar aspect of the Lambert Azimuthal Equal-Area, using the International ellipsoid.

To convert coordinates measured on an existing Lambert Azimuthal Equal-Area map (or other azimuthal map projection), the user may choose any meridian for λ_0 on the polar aspect, but only the original meridian and parallel may be used for λ_0 and ϕ_1, respectively, on other aspects.

25. AZIMUTHAL EQUIDISTANT PROJECTION

SUMMARY

- Azimuthal.
- Distances measured from the center are true.
- Distances not measured along radii from the center are not correct.
- The center of projection is the only point without distortion.
- Directions from the center are true (except on some oblique and equatorial ellipsoidal forms).
- Neither equal-area nor conformal.
- All meridians on the polar aspect, the central meridian on other aspects, and the Equator on the equatorial aspect are straight lines.
- Parallels on the polar projection are circles spaced at true intervals (equidistant for the sphere).
- The outer meridian of a hemisphere on the equatorial aspect (for the sphere) is a circle.
- All other meridians and parallels are complex curves.
- Not a perspective projection.
- Point opposite the center is shown as a circle (for the sphere) surrounding the map.
- Used in the polar aspect for world maps and maps of polar hemispheres.
- Used in the oblique aspect for atlas maps of continents and world maps for aviation and radio use.
- Known for many centuries in the polar aspect.

HISTORY

While the Orthographic is probably the most familiar azimuthal projection, the Azimuthal Equidistant, especially in its polar form, has found its way into many atlases with the coming of the air age for maps of the Northern and Southern Hemispheres or for world maps. The simplicity of the polar aspect for the sphere, with equally spaced meridians and equidistant concentric circles for parallels of latitude, has made it easier to understand than most other projections. The primary feature, showing distances and directions correctly from one point on the Earth's surface, is also easily accepted. In addition, its linear scale distortion is moderate and falls between that of equal-area and conformal projections.

Like the Orthographic, Stereographic, and Gnomonic projections, the Azimuthal Equidistant was apparently used centuries before the 15th-century surge in scientific mapmaking. It is believed that Egyptians used the polar aspect for star charts, but the oldest existing celestial map on the projection was prepared in 1426 by Conrad of Dyffenbach. It was also used in principle for small areas by mariners from earliest times in order to chart coasts, using distances and directions obtained at sea.

The first clear examples of the use of the Azimuthal Equidistant for polar maps of the Earth are those included by Gerardus Mercator as insets on his 1569 world map, which introduced his famous cylindrical projection. As Northern and Southern Hemispheres, the projection appeared in a manuscript of about 1510 by the Swiss Henricus Loritus, usually called Glareanus (1488–1563), and by several others in the next few decades (Keuning, 1955, p. 4–5). Guillaume Postel is given credit in France for its origin, although he did not use it until 1581. Antonio Cagnoli even gave the projection his name as originator in 1799 (Deetz and Adams, 1934, p. 163; Steers, 1970, p. 234). Philippe Hatt developed ellipsoidal versions of the oblique aspect which are used by the French and the Greeks for coastal or topographic mapping.

Two projections with similar names are called the Two-Point Azimuthal and the Two-Point Equidistant projections. Both were developed about 1920 independently by Maurer (1919) of Germany and Close (1921) of England. The first projection (rarely used) is geometrically a tilting of the Gnomonic projection to provide true azimuths from either of two chosen points instead of from just one. Like the Gnomonic, it shows all great circle arcs as straight lines and is limited to one hemisphere. The Two-Point Equidistant has received moderate use and interest, and shows true distances, but not true azimuths, from either of two chosen points to any other point on the map, which may be extended to show the entire world (Close, 1934).

The Chamberlin Trimetric projection is an approximate "three-point equidistant" projection, constructed so that distances from three chosen points to any other point on the map are approximately correct. The latter distances cannot be exactly true, but the projection is a compromise which the National Geographic Society uses as a standard projection for maps of most continents. This projection was geometrically constructed by the Society, of which Wellman Chamberlin (1908–76) was chief cartographer for many years.

An ellipsoidal adaptation of the Two-Point Equidistant was made by Jay K. Donald of American Telephone and Telegraph Company in 1956 to develop a grid still used by the Bell Telephone system for establishing the distance component of long distance rates. Still another approach is Bomford's modification of the Azimuthal Equidistant, in which the usual circles of constant scale factor perpendicular to the radius from the center are made ovals to give a better average scale factor on a map with a rectangular border (Lewis and Campbell, 1951, p. 7, 12–15).

FEATURES

The Azimuthal Equidistant projection, like the Lambert Azimuthal Equal-Area, is not a perspective projection, but in the spherical form, and in some of the ellipsoidal forms, it has the azimuthal characteristic that all directions or azimuths are correct when measured from the center of the projection. As its special feature, all distances are at true scale when measured between this center and any other point on the map.

The polar aspect (fig. 41A), like other polar azimuthals, has circles for parallels of latitude, all centered about the North or South Pole, and equally spaced radii of these circles for meridians. The parallels are, however, spaced equidistantly on the spherical form (or according to actual parallel spacings on the ellipsoid). A world map can extend to the opposite pole, but distortion becomes infinite. Even though the map is finite, the point for the opposite pole is shown as a circle twice the radius of the mapped Equator, thus giving an infinite scale factor along that circle. Likewise, the countries of the outer hemisphere are visibly increasingly distorted as the distance from the center increases, while the inner hemisphere has little enough distortion to appear rather satisfactory to the eye, although the east-west scale along the Equator is almost 60 percent greater than the scale at the center.

As on other azimuthals, there is no distortion at the center of the projection and, as on azimuthals other than the Stereographic, the scale cannot be reduced at the center to provide a standard circle of no distortion elsewhere. It is possible to use an average scale over the map involved to minimize variations in scale error in any direction, but this defeats the main purpose of the projection, that of providing true distance from the center. Therefore, the scale at the projection center should be used for any Azimuthal Equidistant map.

The equatorial aspect (fig. 41B) is least used of the three Azimuthal Equidistant aspects, primarily because there are no cities along the Equator from which

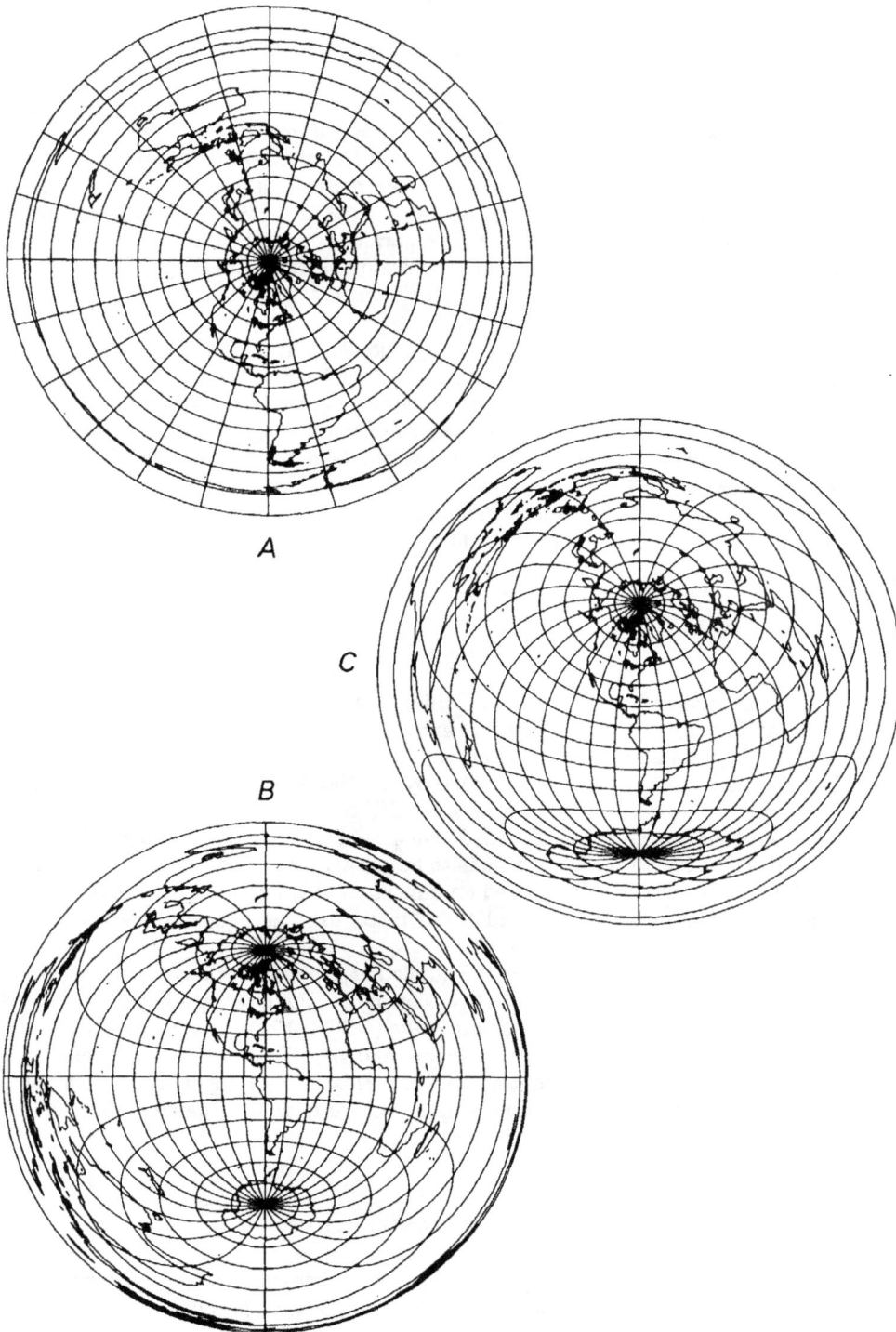

FIGURE 41.—Azimuthal Equidistant projection. (A) Polar aspect extending to the South Pole; commonly used in atlases for polar maps. (B) Equatorial aspect. (C) Oblique aspect centered on lat. 40° N. Distance from the center is true to scale.

distances in all directions have been of much interest to map users. Its potential use as a map of the Eastern or Western Hemisphere was usually supplanted first by the equatorial Stereographic projection, later by the Globular projection (both graticules drawn entirely with arcs of circles and straight lines), and now by the equatorial Lambert Azimuthal Equal-Area.

For the equatorial Azimuthal Equidistant projection of the sphere, the only straight lines are the central meridian and the Equator. The outer circle for one hemisphere (the meridian 90° east and west of the central meridian) is equidistantly marked off for the parallels, as it is on other azimuthals. The other meridians and parallels are complex curves constructed to maintain the correct distances and azimuths from the center. The parallels cross the central meridian at their true equidistant spacings, and the meridians cross the Equator equidistantly. The map can be extended, like the polar aspect, to include a much-distorted second hemisphere on the same center.

The oblique Azimuthal Equidistant projection (fig. 41C) rather resembles the oblique Lambert Azimuthal Equal-Area when confined to the inner hemisphere centered on any chosen point between Equator and pole. Except for the straight central meridian, the graticule consists of complex curves, positioned to maintain true distance and azimuth from the center. When the outer hemisphere is included, the difference between the Equidistant and the Lambert becomes more pronounced, and while distortion is as extreme as in other aspects, the distances and directions of the features from the center now outweigh the distortion for many applications.

USAGE

The polar aspect of the Azimuthal Equidistant has regularly appeared in commercial atlases issued during the past century as the most common projection for maps of the north and south polar areas. It is used for polar insets on Van der Grinten-projection world maps published by the National Geographic Society and used as base maps (including the insets) by USGS. The polar Azimuthal Equidistant projection is also normally used when a hemisphere or complete sphere centered on the North or South Pole is to be shown. The oblique aspect has been used for maps of the world centered on important cities or sites and occasionally for maps of continents. Nearly all these maps use the spherical form of the projection.

The USGS has used the Azimuthal Equidistant projection in both spherical and ellipsoidal form. An oblique spherical version of the Earth centered at lat. 40° N., long. 100° W., appears in the *National Atlas* (USGS, 1970, p. 329). At a scale of 1:175,000,000, it does not show meridians and parallels, but shows circles at 1,000-mile intervals from the center. The ellipsoidal oblique aspect is used for the plane coordinate projection system in approximate form for Guam and in nearly rigorous form for islands in Micronesia.

GEOMETRIC CONSTRUCTION

The polar Azimuthal Equidistant is among the easiest projections to construct geometrically, since the parallels of latitude are equally spaced in the spherical case and the meridians are drawn at their true angles. There are no direct geometric constructions for the oblique and equatorial aspects. Like the Lambert Azimuthal Equal-Area, they may be prepared indirectly by using other azimuthal projections (Harrison, 1943), but automatic computer plotting or manual plotting of calculated rectangular coordinates is the most suitable means now available.

FORMULAS FOR THE SPHERE

On the Azimuthal Equidistant projection for the sphere, a given point is plotted at a distance from the center of the map proportional to the distance on the sphere and at its true azimuth, or

$$\rho = R \, c \tag{25-1}$$
$$\theta = \pi - Az = 180° - Az \tag{20-2}$$

where c is the angular distance from the center, Az is the azimuth east of north (see equations (5−3) through (5−4b)), and θ is the polar coordinate east of south. For k' and h', see equation (25−2) and the statement below. Combining various equations, the rectangular coordinates for the oblique Azimuthal Equidistant projection are found as follows, given R, ϕ_1, λ_0, ϕ, and λ (see p. 337 for numerical examples):

$$x = R \, k' \cos \phi \sin (\lambda - \lambda_0) \tag{22-4}$$
$$y = R \, k' \, [\cos \phi_1 \sin \phi - \sin \phi_1 \cos \phi \cos (\lambda - \lambda_0)] \tag{22-5}$$

where

$$k' = c/\sin c \tag{25-2}$$
$$\cos c = \sin \phi_1 \sin \phi + \cos \phi_1 \cos \phi \cos (\lambda - \lambda_0) \tag{5-3}$$

and (ϕ_1, λ_0) are latitude and longitude of the center of projection and origin. The Y axis coincides with the central meridian λ_0, and y increases northerly. If $\cos c = 1$, equation (25−2) is indeterminate, but $k' = 1$, and $x = y = 0$. If $\cos c = -1$, the point opposite the center $(-\phi_1, \lambda_0 \pm 180°)$ is indicated; it is plotted as a circle of radius πR. The term k' is the scale factor in a direction perpendicular to the radius from the center of the map, not along the parallel, except in the polar aspect. The scale factor h' in the direction of the radius is 1.0.

For the north polar aspect, with $\phi_1 = 90°$,

$$x = R(\pi/2 - \phi) \sin (\lambda - \lambda_0) \tag{25-3}$$
$$y = - R \, (\pi/2 - \phi) \cos (\lambda - \lambda_0) \tag{25-4}$$
$$k = (\pi/2 - \phi)/\cos \phi \tag{25-5}$$
$$h = 1.0$$
$$\rho = R(\pi/2 - \phi) \tag{25-6}$$
$$\theta = \lambda - \lambda_0 \tag{20-9}$$

For the south polar aspect, with $\phi_1 = -90°$,

$$x = R(\pi/2 + \phi) \sin (\lambda - \lambda_0) \tag{25-7}$$
$$y = R \, (\pi/2 + \phi) \cos (\lambda - \lambda_0) \tag{25-8}$$
$$k = (\pi/2 + \phi)/\cos \phi \tag{25-9}$$
$$h = 1.0$$
$$\rho = R(\pi/2 + \phi) \tag{25-10}$$
$$\theta = \pi - \lambda + \lambda_0 \tag{20-12}$$

For the equatorial aspect, with $\phi_1 = 0$, x is found from (22−4) and k' from (25−2), but

$$y = R \, k' \sin \phi \tag{25-11}$$

TABLE 30.—*Azimuthal Equidistant projection: Rectangular coordinates for equatorial aspect (sphere)*

[One hemisphere; $R=1$. y coordinates in parentheses under x coordinates]

Long. Lat.	0°	10°	20°	30°	40°
90° _____	0.00000	0.00000	0.00000	0.00000	0.00000
	(1.57080)	(1.57080)	(1.57080)	(1.57080)	(1.57080)
80 _____	.00000	.04281	.08469	.12469	.16188
	(1.39626)	(1.39829)	(1.40434)	(1.41435)	(1.42823)
70 _____	.00000	.07741	.15362	.22740	.29744
	(1.22173)	(1.22481)	(1.23407)	(1.24956)	(1.27137)
60 _____	.00000	.10534	.20955	.31145	.40976
	(1.04720)	(1.05068)	(1.06119)	(1.07891)	(1.10415)
50 _____	.00000	.12765	.25441	.37931	.50127
	(.87266)	(.87609)	(.88647)	(.90408)	(.92938)
40 _____	.00000	.14511	.28959	.43276	.57386
	(.69813)	(.70119)	(.71046)	(.72626)	(.74912)
30 _____	.00000	.15822	.31607	.47314	.62896
	(.52360)	(.52606)	(.53355)	(.54634)	(.56493)
20 _____	.00000	.16736	.33454	.50137	.66762
	(.34907)	(.35079)	(.35601)	(.36497)	(.37803)
10 _____	.00000	.17275	.34546	.51807	.69054
	(.17453)	(.17541)	(.17810)	(.18270)	(.18943)
0 _____	.00000	.17453	.34907	.52360	.69813
	(.00000)	(.00000)	(.00000)	(.00000)	(.00000)

and

$$\cos c = \cos \phi \cos (\lambda - \lambda_0) \qquad (25-12)$$

The maximum angular deformation ω for any of these aspects, using equations (4−7) through (4−9), since $h' = 1.0$:

$$\sin \tfrac{1}{2}\omega = (k'-1)/(k'+1) \qquad (25-13)$$
$$= (c-\sin c)/(c+\sin c) \qquad (25-14)$$

For the inverse formulas for the sphere, given R, ϕ_1, λ_0, x, and y:

$$\phi = \arcsin [\cos c \sin \phi_1 + (y \sin c \cos \phi_1/\rho)] \qquad (20-14)$$

If $\rho = 0$, equations (20−14) through (20−17) are indeterminate, but $\phi = \phi_1$ and $\lambda = \lambda_0$.

If ϕ_1 is not ±90°:

$$\lambda = \lambda_0 + \arctan [x \sin c/(\rho \cos \phi_1 \cos c - y \sin \phi_1 \sin c)] \qquad (20-15)$$

If ϕ_1 is 90°:

$$\lambda = \lambda_0 + \arctan [x/(-y)] \qquad (20-16)$$

If ϕ_1 is −90°:

$$\lambda = \lambda_0 + \arctan (x/y) \qquad (20-17)$$

In equations (20−14) and (20−15),

$$\rho = (x^2 + y^2)^{1/2} \qquad (20-18)$$
$$c = \rho/R \qquad (25-15)$$

TABLE 30.—*Azimuthal Equidistant projection: Rectangular coordinates for equatorial aspect (sphere)*—Continued

Long. Lat.	50°	60°	70°	80°	90°
90°	0.00000	0.00000	0.00000	0.00000	0.00000
	(1.57080)	(1.57080)	(1.57080)	(1.57080)	(1.57080)
80	.19529	.22399	.24706	.26358	.27277
	(1.44581)	(1.46686)	(1.49104)	(1.51792)	(1.54693)
70	.36234	.42056	.47039	.50997	.53724
	(1.29957)	(1.33423)	(1.37533)	(1.42273)	(1.47607)
60	.50301	.58948	.66711	.73343	.78540
	(1.13733)	(1.17896)	(1.22963)	(1.28993)	(1.36035)
50	.61904	.73106	.83535	.92935	1.00969
	(.96306)	(1.00602)	(1.05942)	(1.12464)	(1.20330)
40	.71195	.84583	.97392	1.09409	1.20330
	(.77984)	(.81953)	(.86967)	(.93221)	(1.00969)
30	.78296	.93436	1.08215	1.22487	1.36035
	(.59010)	(.62291)	(.66488)	(.71809)	(.78540)
20	.83301	.99719	1.15965	1.31964	1.47607
	(.39579)	(.41910)	(.44916)	(.48772)	(.53724)
10	.86278	1.03472	1.20620	1.37704	1.54693
	(.19859)	(.21067)	(.22634)	(.24656)	(.27277)
0	.87266	1.04720	1.22173	1.39626	1.57080
	(.00000)	(.00000)	(.00000)	(.00000)	(.00000)

Radius of sphere = 1.0.
Origin: $(x, y) = 0$ at (lat., long.) = 0. Y axis increases north. Other quadrants of hemisphere are symmetrical.

Except for (25−15), the above inverse formulas are the same as those for the other azimuthals, and (25−2) is the only change from previous azimuthals among the general (oblique) formulas (22−4) through (5−3) for the forward calculations as listed above.

Table 30 shows rectangular coordinates for the equatorial aspect for a 10° graticule with a sphere of radius $R = 1.0$.

FORMULAS FOR THE ELLIPSOID

The formulas for the polar aspect of the ellipsoidal Azimuthal Equidistant projection are relatively simple and are theoretically accurate for a map of the entire world. However, such a use is unnecessary because the errors of the sphere versus the ellipsoid become insignificant when compared to the basic errors of projection. The polar form is truly azimuthal as well as equidistant. Given a, e, ϕ_1, λ_0, ϕ, and λ, for the north polar aspect, $\phi_1 = 90°$ (see p. 338 for numerical examples):

$$x = \rho \sin (\lambda - \lambda_0) \qquad (21-30)$$
$$y = -\rho \cos (\lambda - \lambda_0) \qquad (21-31)$$
$$k = \rho/(a\,m) \qquad (21-32)$$

where

$$\rho = M_p - M \qquad (25-16)$$
$$M = a\,[(1 - e^2/4 - 3e^4/64 - 5e^6/256 - \ldots)\phi - (3e^2/8 + 3e^4/32$$
$$+ 45e^6/1024 + \ldots)\sin 2\phi + (15e^4/256 + 45e^6/1024 + \ldots)$$
$$\sin 4\phi - (35e^6/3072 + \ldots)\sin 6\phi + \ldots] \qquad (3-21)$$

with M_p the value of M for a ϕ of 90°,
and $m = \cos \phi/(1 - e^2 \sin^2 \phi)^{1/2}$ \hfill (14−15)

TABLE 31.—*Ellipsoidal Azimuthal Equidistant projection (International ellipsoid)—Polar Aspect*

Latitude	Radius, meters	h	k
90°	0.0	1.0	1.000000
89	111,699.8	1.0	1.000051
88	223,399.0	1.0	1.000203
87	335,096.8	1.0	1.000457
86	446,792.5	1.0	1.000813
85	558,485.4	1.0	1.001270
84	670,175.0	1.0	1.001830
83	781,860.4	1.0	1.002492
82	893,541.0	1.0	1.003256
81	1,005,216.2	1.0	1.004124
80	1,116,885.2	1.0	1.005095
79	1,228,547.5	1.0	1.006169
78	1,340,202.4	1.0	1.007348
77	1,451,849.2	1.0	1.008631
76	1,563,487.4	1.0	1.010019
75	1,675,116.3	1.0	1.011513
74	1,786,735.3	1.0	1.013113
73	1,898,343.8	1.0	1.014821
72	2,009,941.3	1.0	1.016636
71	2,121,527.1	1.0	1.018560
70	2,233,100.9	1.0	1.020594

h = scale factor along meridian.
k = scale factor along parallel.

For improved computational efficiency using this series, see p. 19.

For the south polar aspect, the equations for the north polar aspect apply, except that equations (21−31) and (25−16) become

$$y = \rho \cos (\lambda - \lambda_0) \qquad (24-23)$$
$$\rho = M_p + M \qquad (25-17)$$

The origin falls at the pole in either case, and the Y axis follows the central meridian λ_0. For the north polar aspect, λ_0 is shown below the pole, and y increases along λ_0 toward the pole. For the south polar aspect, λ_0 is shown above the pole, and y increases along λ_0 away from the pole.

Table 31 lists polar coordinates for the ellipsoidal aspect of the Azimuthal Equidistant, using the International ellipsoid.

For the oblique and equatorial aspects of the ellipsoidal Azimuthal Equidistant, both nearly rigorous and approximate sets of formulas have been derived. For mapping of Guam, the National Geodetic Survey and the USGS use an approximation to the ellipsoidal oblique Azimuthal Equidistant called the "Guam projection." It is described by Claire (1968, p. 52−53) as follows (changing his symbols to match those in this publication):

The plane coordinates of the geodetic stations on Guam were obtained by first computing the geodetic distances [c] and azimuths [Az] of all points from the origin by inverse computations. The coordinates were then computed by the equations: [$x = c \sin Az$ and $y = c \cos Az$]. This really gives a true azimuthal equidistant projection. The equations given here are simpler, however, than those for a geodetic inverse computation, and the resulting coordinates computed using them will not be significantly different from those computed rigidly by inverse computation. This is the reason it is called an approximate azimuthal equidistant projection.

The formulas for the Guam projection are equivalent to the following:

$$x = a (\lambda - \lambda_0) \cos \phi / (1 - e^2 \sin^2 \phi)^{1/2} \qquad (25-18)$$
$$y = M - M_1 + x^2 \tan \phi (1 - e^2 \sin^2 \phi)^{1/2}/(2a) \qquad (25-19)$$

where M and M_1 are found from equation (3−21) for ϕ and ϕ_1. Actually, the original formulas are given in terms of seconds of rectifying latitude and geodetic latitude and longitude, but they may be written as above. The x coordinate is thus taken as the distance along the parallel, and y is the distance along the central meridian λ_0 with adjustment for curvature of the parallel. The origin occurs at (ϕ_1, λ_0), the Y axis coincides with the central meridian, and y increases northerly.

For Guam, $\phi_1 = 13°28'20.87887''$ N. lat. and $\lambda_0 = 144°44'55.50254''$ E. long., with 50,000 m added to both x and y to eliminate negative numbers. The Clarke 1866 ellipsoid is used. The above formulas provide coordinates within 5 mm at full scale of those using ellipsoidal Polyconic formulas (p. 129) for the region of Guam.

A more complicated and more accurate approach to the oblique ellipsoidal Azimuthal Equidistant projection is used for plane coordinates of various individual islands of Micronesia. In this form, the true distance and azimuth of any point on the island or in nearby waters are measured from the origin chosen for the island and along the normal section or plane containing the perpendicular to the surface of the ellipsoid at the origin. This is not exactly the same as the shortest or geodesic distance between the points, but the difference is negligible (Bomford, 1971, p. 125). This distance and azimuth are plotted on the map. The projection is, therefore, equidistant and azimuthal with respect to the center and appears to satisfy all the requirements for an ellipsoidal Azimuthal Equidistant projection, although it is described as a "modified" form. The origin is assigned large-enough values of x and y to prevent negative readings.

The formulas for calculation of this distance and azimuth have been published in various forms, depending on the maximum distance involved. The projection system for Micronesia makes use of "Clarke's best formula" and Robbins' inverse of this. These are considered suitable for lines up to 800 km in length. The formulas below, rearranged slightly from Robbins' formulas as given in Bomford (1971, p. 136−137), are extended to produce rectangular coordinates. No iteration is required. They are listed in the order of use, given a central point at lat. ϕ_1, long. λ_0, coordinates x_0 and y_0 of the central point, the Y axis along the central meridian λ_0, y increasing northerly, ellipsoidal parameters a and e, and ϕ and λ.

To find x and y:

$$N_1 = a/(1-e^2 \sin^2 \phi_1)^{1/2} \qquad (4-20a)$$
$$N = a/(1-e^2 \sin^2 \phi)^{1/2} \qquad (4-20)$$
$$\psi = \arctan\,[(1-e^2)\tan\phi + e^2 N_1 \sin\phi_1/(N\cos\phi)] \qquad (25-20)$$
$$Az = \arctan\,\{\sin(\lambda-\lambda_0)/[\cos\phi_1 \tan\psi - \sin\phi_1 \cos(\lambda-\lambda_0)]\} \qquad (25-21)$$

The ATAN2 Fortran function should be used with equation (25−21), but it is not applicable to (25−20).

If $\sin Az = 0$,

$$s = \pm \arcsin(\cos\phi_1 \sin\psi - \sin\phi_1 \cos\psi) \qquad (25-22)$$

taking the sign of $\cos Az$.

If $\sin Az \neq 0$,

$$s = \arcsin\,[\sin(\lambda-\lambda_0)\cos\psi/\sin Az] \qquad (25-22a)$$

In either case,

$$G = e \sin\phi_1/(1-e^2)^{1/2} \qquad (25-23)$$
$$H = e \cos\phi_1 \cos Az/(1-e^2)^{1/2} \qquad (25-24)$$
$$c = N_1 s\{1-s^2 H^2(1-H^2)/6 + (s^3/8)GH(1-2H^2) + (s^4/120)[H^2(4-7H^2)-3G^2(1-7H^2)] - (s^5/48)GH\} \qquad (25-25)$$

TABLE 32.—*Plane coordinate systems for Micronesia: Clarke 1866 ellipsoid*

Group	Islands	Station at Origin	ϕ_1 Lat N.	λ_0 Long E.	Meters x_0	y_0
Caroline Islands _____	Yap	Yap Secor	9°32'48.898"	138°10'07.084"	39,987.92	60,022.98
	Palau	Arakabesan Is.	7°21'04.3996"	134°27'01.6015"	50,000.00	150,000.00
	Ponape	Distad (USE)	6°57'54.2725"	158°12'33.4772"	80,122.82	80,747.24
	Truk Atoll	Truk Secor RM 1	7°27'22.3600"	151°50'17.8530"	60,000.00	70,000.00
Mariana Islands _____	Saipan	Saipan	15°11'05.6830"	145°44'29.9720"	28,657.52	67,199.99
	Rota	Astro	14°07'58.8608"	145°08'03.2275"	5,000.00	5,000.00
Marshall Islands _____	Majuro Atoll	Dalap	7°05'14.0"	171°22'34.5"	85,000.00	40,000.00

x_0, y_0 = rectangular coordinates of center of projection.
ϕ_1, λ_0 = geodetic coordinates of center of projection.

$$x = c \cos Az + x_0 \qquad (25-26)$$
$$y = c \cos Az + y_0 \qquad (25-27)$$

where c is the geodesic distance, and Az is azimuth east of north.

Table 32 shows the parameters for the various islands mapped with this projection.

Inverse formulas for the polar ellipsoidal aspect, given a, e, ϕ_1, λ_0, x, and y:

$$\phi = \mu + (3e_1/2 - 27\,e_1{}^3/32 + \ldots)\sin 2\mu + (21\,e_1{}^2/16 - 55\,e_1{}^4/32 + \ldots)\sin$$
$$4\mu + (151\,e_1{}^3/96 - \ldots)\sin 6\mu + (1097\,e_1{}^4/512 - \ldots)\sin 8\mu + \ldots \qquad (3-26)$$

where

$$e_1 = [1 - (1-e^2)^{1/2}]/[1 + (1-e^2)^{1/2}] \qquad (3-24)$$
$$\mu = M/[a\,(1-e^2/4-3e^4/64-5e^6/256-\ldots)] \qquad (7-19)$$
$$M = M_p - \rho \text{ for the north polar case,} \qquad (25-28)$$

and

$$M = \rho - M_p \text{ for the south polar case.} \qquad (25-29)$$

For improved computational efficiency using series (3−26) see p. 19. Equation (3−21), listed with the forward equations, is used to find M_p for $\phi = 90°$. For either case,

$$\rho = (x^2 + y^2)^{1/2} \qquad (20-18)$$

For longitude, for the north polar case,

$$\lambda = \lambda_0 + \arctan[x/(-y)] \qquad (20-16)$$

For the south polar case,

$$\lambda = \lambda_0 + \arctan(x/y) \qquad (20-17)$$

Inverse formulas for the Guam projection (Claire, 1968, p. 53) involve an iteration of two equations, which may be rearranged and rewritten in the following form consistent with the above formulas. Given a, e, ϕ_1, λ_0, x, and y, M_1 is calculated for ϕ_1 from (3−21), given with forward equations. (If false northings and eastings are included in x and y, they must be subtracted first.)

Then, first assuming $\phi = \phi_1$,

$$M = M_1 + y - x^2 \tan \phi \,(1-e^2 \sin^2 \phi)^{1/2}/(2\,a) \qquad (25-30)$$

Using this M, μ is calculated from (7−19) and inserted into the right side of (3−26) to solve for a new ϕ on the left side. This is inserted into (25−30), a new M is found, and it is resubstituted into (7−19), μ into (3−26), etc., until ϕ on the left side of (3−26) changes by less than a chosen convergence figure, for a final ϕ. Then

$$\lambda = \lambda_0 + x\,(1-e^2 \sin^2 \phi)^{1/2}/(a \cos \phi) \qquad (25-31)$$

The inverse Guam formulas arbitrarily stop at three iterations, which are sufficient for the small area.

For the Micronesia version of the ellipsoidal Azimuthal Equidistant projection, the inverse formulas given below are "Clarke's best formula," as given in Bomford (1971, p. 133) and do not involve iteration. They have also been rearranged to begin with rectangular coordinates, but they are also suitable for finding latitude and longitude accurately for a point at any given distance c (up to about 800 km) and azimuth Az (east of north) from the center, if equations (25−32) and (25−33) are deleted. In order of use, given a, e, central point at lat. ϕ_1, long. λ_0, rectangular coordinates of center x_0, y_0, and x and y for another point, to find ϕ and λ:

$$c = [(x-x_0)^2 + (y - y_0)^2]^{1/2} \tag{25−32}$$

$$Az = \arctan[(x-x_0)/(y-y_0)] \tag{25−33}$$

$$N_1 = a/(1-e^2 \sin^2 \phi_1)^{1/2} \tag{4−20a}$$

$$A = -e^2 \cos^2 \phi_1 \cos^2 Az/(1-e^2) \tag{25−34}$$

$$B = 3e^2 (1-A) \sin \phi_1 \cos \phi_1 \cos Az/(1-e^2) \tag{25−35}$$

$$D = c/N_1 \tag{25−36}$$

$$E = D - A(1+A)D^3/6 - B(1+3A)D^4/24 \tag{25−37}$$

$$F = 1 - AE^2/2 - BE^3/6 \tag{25−38}$$

$$\psi = \arcsin(\sin \phi_1 \cos E + \cos \phi_1 \sin E \cos Az) \tag{25−39}$$

$$\lambda = \lambda_0 + \arcsin(\sin Az \sin E/\cos \psi) \tag{25−40}$$

$$\phi = \arctan[(1-e^2 F \sin \phi_1/\sin \psi) \tan \psi/(1-e^2)] \tag{25−41}$$

The ATAN2 function of Fortran, or equivalent, should be used in equation (25−33), but not (25−41).

To convert coordinates measured on an existing Azimuthal Equidistant map (or other azimuthal map projection), the user may choose any meridian for λ_0 on the polar aspect, but only the original meridian and parallel may be used for λ_0 and ϕ_1, respectively, on other aspects.

26. MODIFIED-STEREOGRAPHIC CONFORMAL PROJECTIONS

SUMMARY

- Modified azimuthal.
- Conformal.
- All meridians and parallels are normally complex curves, although some may be straight under some conditions.
- Scale is true along irregular lines, but map is usually designed to minimize scale variation throughout a selected region.
- Map is normally not symmetrical about any axis or point.
- Used for maps of continents in the Eastern Hemisphere, for the Pacific Ocean, and for maps of Alaska and the 50 United States.
- Specific forms devised by Miller, Lee, and Snyder, 1950–84.

HISTORY AND USAGE

Two short mathematical formulas, taken as a pair, absolutely define the conformal transformation of one surface onto another surface. These formulas (see p. 27) are called the Cauchy-Riemann equations, after two 19th-century mathematicians who added rigorous analysis to principles developed in the middle of the 18th century by physicist D'Alembert. Much later, Driencourt and Laborde (1932, vol. 14, p. 242) presented a fairly simple series (equation (26–4) below without the R), involving complex algebra (with imaginary numbers), that fully satisfies the Cauchy-Riemann equations and permits the formation of an endless number of new conformal map projections when certain constants are changed.

The advantage of this series is that lines of constant scale may be made to follow one of a variety of patterns, instead of following the great or small circles of the common conformal projections. The disadvantage is that the length of the series and the computations become increasingly lengthy as the irregularity of the lines of constant scale increases, but this problem has decreased with the development of computers.

Laborde (1928; Reignier, 1957, p. 130) applied this transformation to the mapping of Madagascar, starting with the Oblique Mercator projection and applying the complex equation up to the third-order or cubic terms. Miller (1953) used the same order of complex equation, but began with an oblique Stereographic projection. His resulting map of Europe and Africa has oval lines of constant scale (fig. 42); this projection is called the Miller Oblated (or Prolated) Stereographic. He subsequently (Miller, 1955) prepared similar projections for Asia and Australasia, each precisely conformal, but he linked them with nonconformal "fill-in" projections to provide a continuous map (in several sheets) of the land masses of the Eastern Hemisphere.

Lee (1974) designed a map of the Pacific Ocean, also using an oblique Stereographic with a third-order complex polynomial. The third-order polynomials used by Laborde, Miller, and Lee make relatively moderate computational demands, because several of the coefficients are zero, and the complex algebra can be readily simplified to equations without imaginary numbers. Recently Reilly (1973) and the writer (Snyder, 1984a, 1985a) have used much higher-order complex equations, but modern computers can readily handle them. Reilly used sixth-order coefficients with the regular Mercator for the new official New Zealand Map Grid, while the writer, beginning with oblique Stereographic projections, used sixth-order coefficients for a map of Alaska and tenth-order for a map of the 50 United States (figs. 43, 44). For these sixth- and tenth-order equations, only one coefficient is zero, but the other coefficients were computed using least squares. The projection for Alaska was used in 1985 by Alvaro F. Espinosa of the USGS to

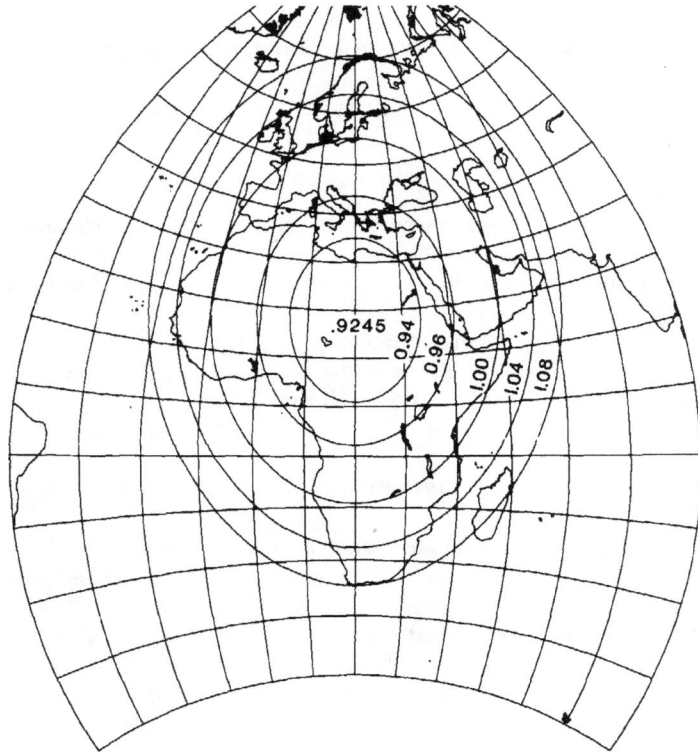

FIGURE 42.—Miller Oblated Stereographic projection of Europe and Africa, showing oval lines of constant scale. Center of projection lat. 18° N., long. 20° E.

depict earthquake information for that State. The "Modified Transverse Mercator" projection is still being used by the USGS for most maps of Alaska.

In addition, the writer (Snyder, 1984b) used oblique Stereographics as bases with third- to fifth-order equations, most coefficients remaining zero, to surround maps with lines of constant scale which are nearly regular polygons or rectangles (fig. 45). This minimizes error within a map as conventionally published.

FEATURES

The common feature linking the endless possibilities of map projections discussed in this chapter is the fact that they are perfectly conformal regardless of the order of the complex-algebra transformation, and regardless of the initial projection, provided it is also conformal.

Chebyshev (1856) stated that a region may be best shown conformally if the sum of the squares of the scale errors (scale factors minus 1) over the region is a minimum. He further declared that this results if the region is bounded by a line of constant scale. This was proven later. Thus the Stereographic is suitable for regions approximately circular in shape, but regions bounded by ovals, regular polygons, or rectangles may be mapped with nearly minimum error by suitably altering the Stereographic with the complex-algebra transformation.

If the region is irregular, such as Alaska, the region of interest may be divided into small elements, and the coefficients may be calculated using least squares to minimize the scale variation for the region shown. The resulting coefficients for

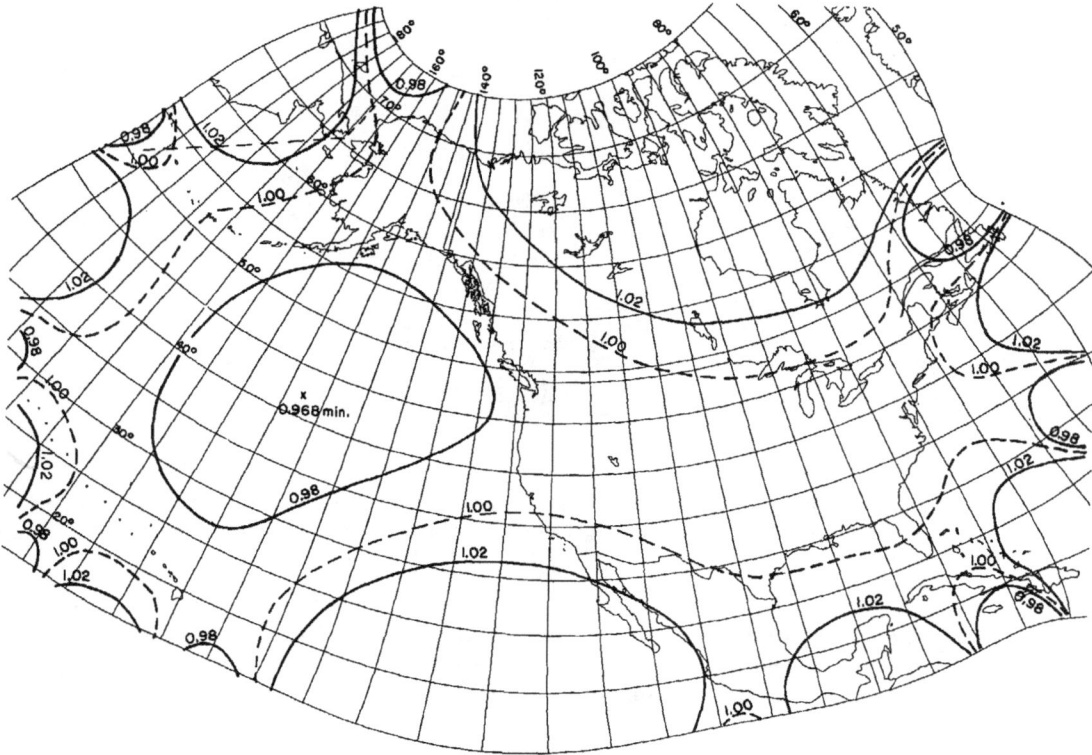

FIGURE 43.—GS-50 projection, with lines of constant scale factor superimposed. All 50 States, including islands and passages between Alaska, Hawaii, and the conterminous 48 States are shown with scale factors ranging only from 1.02 to 0.98.

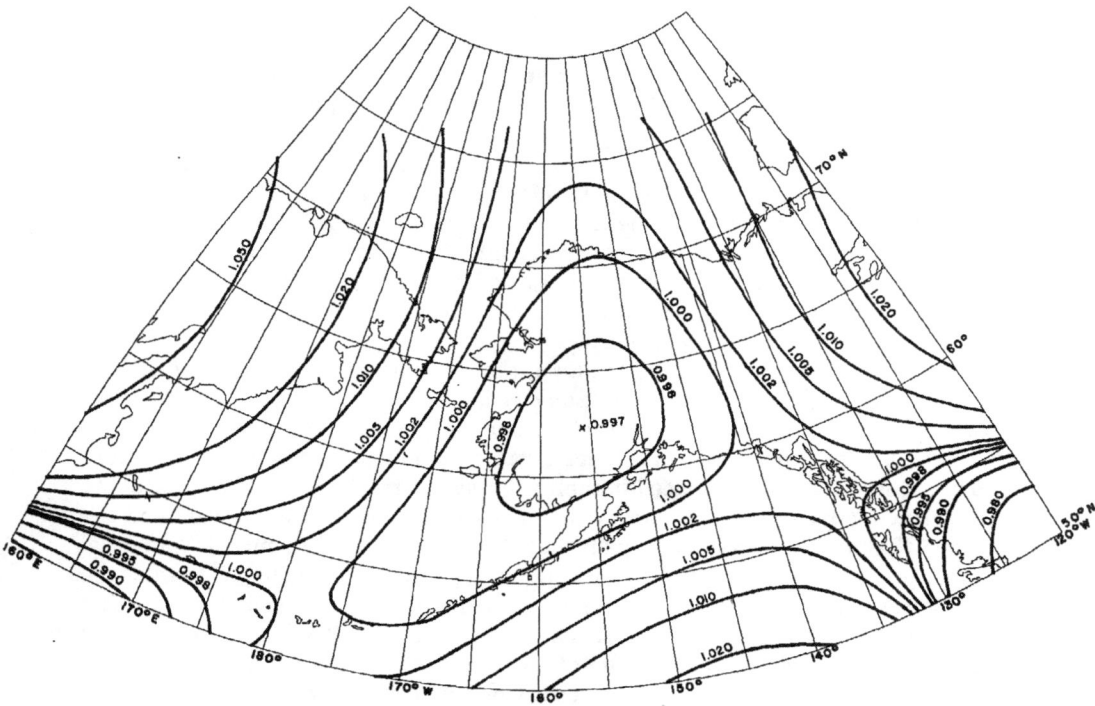

FIGURE 44.—Modified-Stereographic Conformal projection of Alaska, with lines of constant scale superimposed. Scale factors for Alaska range from 0.997 to 1.003, one-fourth the range for a corresponding conic projection.

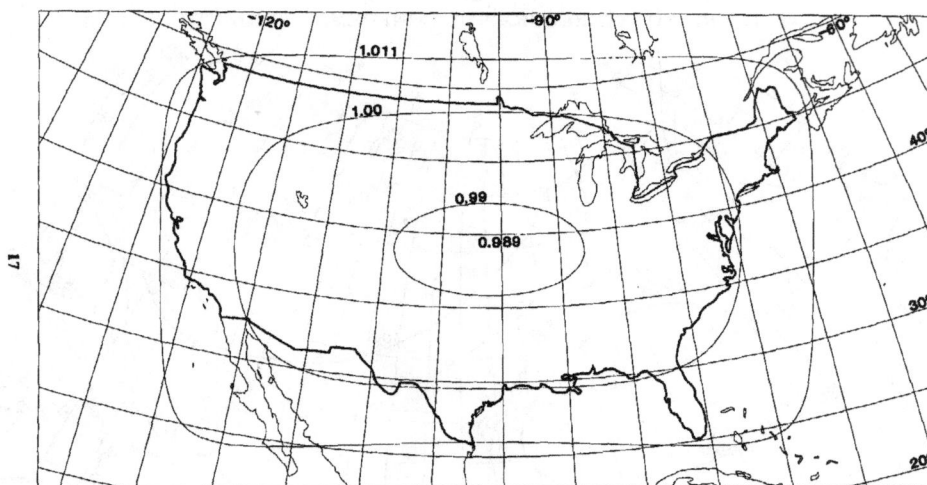

FIGURE 45.—Modified-Stereographic Conformal projection of 48 United States, bounded by a near-rectangle of constant scale. Three lines of constant scale are superimposed. Region bounded by near-rectangle has minimum error.

the selected projections are given below, but the formulas for least-squares summation are not included here because they are lengthy and are only needed to devise new projections. For them the reader may refer to Snyder (1984a, 1984b, 1985a).

The reduction of scale variation by using this complex-algebra transformation makes the ellipsoidal form even more important. This form is also simpler in these cases than for the Transverse Mercator and some other projections, because the lines of true scale normally do not follow a selected meridian, parallel, or other easily identifiable line in any case. Therefore, use of the conformal latitude in place of the geographic latitude is sufficient for the ellipsoidal form. This merely slightly shifts the lines of constant scale from one set of arbitrary locations to another. The coefficients have somewhat different values, however.

The meridians and parallels of the Modified-Stereographic projections are generally curved, and there is usually no symmetry about any point or line. There are limitations to these transformations. Most of them can only be used within a limited range, depending on the number and values of coefficients. As the distance from the projection center increases, the meridians, parallels, and shorelines begin to exhibit loops, overlapping, and other undesirable curves. A world map using the GS50 (50-State) projection is almost illegible, with meridians and parallels intertwined like wild vines.

Within the intended range of the map, the Modified-Stereographic projections can reduce the range of scale variation considerably when compared with standard conformal projections. The tenth-order complex-algebra modification used for the 50-State projection has a scale range of only ±2 percent (or 4 percent overall) for all 50 States placed in their relative geographical positions, including all islands, adjacent waters, water channels connecting Alaska, Hawaii, and the other 48 States, and nearby Canada and Mexico (fig. 43). A Lambert Conformal Conic projection previously used with standard parallels 37° and 65° N. to show the 50 States has a scale range of +12 to −3 percent (or 15 percent overall). The sixth-order modification for the Alaska map, called the Modified-Stereographic Conformal projection, has a range of ±0.3 percent (or 0.6 percent overall) for Alaska itself, while a Lambert Conformal Conic with standard parallels 55° and 65° N. ranges from +2.0 to −0.4 percent, or 2.4 percent overall.

The bounding of regions by ovals, near-regular polygons, or near-rectangles of constant scale results in improvement of scale variation by amounts depending on the size and shape of the boundary. The improvement in mean scale error is about 15 to 20 percent using a near-square instead of the circle of the base Stereographic projection. Using a Modified-Stereographic bounded by a near-rectangle instead of an oblique Mercator projection provides a mean improvement of up to 30 percent in some cases, but only 5 to 10 percent in cases involving a long narrow region. For fig. 45, the range of scale is ±1.1 percent (or 2.2 percent overall) within the 48 States, while the Lambert Conformal Conic normally used has a range of +2.4 to −0.6 percent (or 3.0 percent overall).

The improvement for the region in question is made at the expense of scale preservation outside the region. The regular conic projections maintain the same scale range around the entire world between the same latitude limits, even though most of that region is not shown on the regional maps described above.

FORMULAS FOR THE SPHERE

The Modified-Stereographic conformal projections which have a scale range of more than 5 percent, such as regions bounded by rectangles 80° by 40° in spherical degrees, may satisfactorily be computed for the sphere instead of the ellipsoid. As stated above, development of coefficients is not shown here. For the calculation of final rectangular coordinates, given R, ϕ_1, λ_0, A_1 through A_m, B_1 through B_m, ϕ, and λ, and to find x and y (see p. 344 for numerical examples):

$$k' = 2/[1 + \sin \phi_1 \sin \phi + \cos \phi_1 \cos \phi \cos (\lambda - \lambda_0)] \tag{26-1}$$

$$x' = k' \cos \phi \sin (\lambda - \lambda_0) \tag{26-2}$$

$$y' = k' [\cos \phi_1 \sin \phi - \sin \phi_1 \cos \phi \cos (\lambda - \lambda_0)] \tag{26-3}$$

$$x + iy = R \sum_{j=1}^{m} (A_j + iB_j) (x' + iy')^j \tag{26-4}$$

$$k = \left| \sum_{j=1}^{m} j (A_j + iB_j) (x' + iy')^{j-1} \right| k' \tag{26-5}$$

where k' is the scale factor on the base Stereographic map, (x', y') are rectangular coordinates for a globe of radius 1 on the base map, (x, y) are rectangular coordinates on the final map, k is the scale factor on the final map, (ϕ_1, λ_0) are the central latitude and longitude of the projection, (ϕ, λ) are the latitude and longitude of the point to be plotted, R is the radius of the sphere, (A_j, B_j) are the coefficients for $j = 1$ to $j = m$, the order of the equation, and i^2 is -1. Equations (26−1) through (26−3) are similar to the forward equations listed under the regular Stereographic projection, but there are slight differences. The formulas for this projection as published in Snyder (1984a, 1985a) introduce R (and a for the ellipsoid) at the wrong points, although answers are correct.

For the practical computation of equations (26−4) and (26−5), Knuth's (1969) algorithm is recommended instead of them. Let

$$r = 2x'; s' = (x')^2 + (y')^2; g_0 = 0; g_f = A_f + iB_f; a_1 = g_m;$$
$$b_1 = g_{m-1}; c_1 = mg_m; d_1 = (m-1) g_{m-1}; a_j = b_{j-1} + ra_{j-1};$$
$$b_j = g_{m-j} - s'a_{j-1}; c_j = d_{j-1} + rc_{j-1}; d_j = (m-j)g_{m-j} - s'c_{j-1} \tag{26-6}$$

After j is given the value of successive integers from 2 to m for a_j and b_j and 2 to $(m-1)$ for c_j and d_j, then

$$x + iy = R\left[(x' + iy')\,a_m + b_m\right] \qquad (26{-}7)$$

$$F_2 + iF_1 = (x' + iy')\,c_{m-1} + d_{m-1} \qquad (26{-}8)$$

$$k = (F_2^2 + F_1^2)^{1/2}\,k' \qquad (26{-}9)$$

For the Modified-Stereographic Conformal projections with ovals, near-regular polygons, or near-rectangles as bounding lines of constant scale, since there are only two or three non-zero coefficients, plus a possible rotation, equations $(26{-}4)$ and $(26{-}5)$ may be simplified to avoid a need for the use of i or Knuth's algorithm. The above formulas are more general, however, once they are programmed. For the simplified forms, the reader is referred to Miller (1953) and Snyder (1984b). If k is not being calculated in the above formulas, the four equations of $(26{-}6)$ which include c or d, as well as $(26{-}8)$ and $(26{-}9)$, may be omitted. For constants, see table 33.

For *inverse equations*, given R, ϕ_1, λ_0, A_1 through A_m, B_1 through B_m, x, and y, to find ϕ and λ, first a Newton-Raphson iteration may be used as follows to find (x', y'):

$$\Delta\,(x' + iy') = -\,f\,(x' + iy')/(F_2 + iF_1) \qquad (26{-}10)$$

where

$$f\,(x' + iy') = \sum_{j=1}^{m}\,(A_j + iB_j)(x' + iy')^j - (x + iy)/R \qquad (26{-}11)$$

$$F_2 + iF_1 = \sum_{j=1}^{m} j\,(A_j + iB_j)\,(x' + iy')^{j-1} \qquad (26{-}12)$$

and the first trial value of x' is (x/R) and of y' is (y/R). The Knuth algorithm is equally suitable here, using all of the equations in $(26{-}6)$, assigning j values which are described following those equations, and replacing equations $(26{-}11)$ and $(26{-}12)$ with $(26{-}13)$ and $(26{-}8)$, respectively.

$$f(x' + iy') = (x' + iy')a_m + b_m - (x + iy)/R \qquad (26{-}13)$$

After the trial values of (x', y') are adjusted with $(26{-}10)$ until the change in each is negligible (3–4 iterations are normally enough), the final (x', y') is converted to (ϕ, λ) without iteration as follows:

$$\rho = [(x')^2 + (y')^2]^{1/2} \qquad (26{-}14)$$

$$c = 2\,\arctan\,(\rho/2) \qquad (26{-}15)$$

$$\phi = \arcsin\,[\cos c \sin \phi_1 + (y' \sin c \cos \phi_1/\rho)] \qquad (26{-}16)$$

$$\lambda = \lambda_0 + \arctan\,[x' \sin c/(\rho \cos \phi_1 \cos c - y' \sin \phi_1 \sin c)] \qquad (26{-}17)$$

If $\rho = 0$, equations $(26{-}16)$ and $(26{-}17)$ are indeterminate, but $\phi = \phi_1$ and $\lambda = \lambda_0$.

FORMULAS FOR THE ELLIPSOID

For higher precision maps taking greater advantage of the reduced scale variation available with Modified-Stereographic Conformal projections, the ellipsoidal formulas should be used. Given a, e, ϕ_1, λ_0, A_1 through A_m, B_1 through B_m, ϕ, and λ, to find x and y (special numerical examples are not given, but examples of the ellipsoidal Stereographic, p. 313, and of the spherical Modified-Stereographic p. 344, are sufficiently similar):

TABLE 33.—*Modified-Stereographic Conformal projections: Coefficients for specific forms*

Taking the Earth as a sphere:

Miller Oblated Stereographic projection for Europe and Africa (fig. 42):

R = 6,371,221 m at full scale
m = 3
ϕ_1 = 18°N. lat.
λ_0 = 20°E. long. as constructed (18° in Miller (1953))
A_1 = 0.924500
A_3 = 0.019430
$A_2, B_1, B_2, B_3 = 0$

Lee Oblated Stereographic projection for the Pacific Ocean:

R = Not stated
m = 3
ϕ_1 = 10°S. lat.
λ_0 = 165°W. long.
A_1 =   0.721316
A_3 = −0.00881625
B_3 = −0.00617325
$A_2, B_1, B_2 = 0$

GS50 projection for the 50 States (fig. 43; ellipsoidal formulas and constants should normally be used):

R = 6,370,997 m at full scale
m = 10
ϕ_1 = 45°N. lat.
λ_0 = 120°W. long.

A_1 = 0.9842990	B_1 = 0
A_2 = 0.0211642	B_2 = 0.0037608
A_3 = −0.1036018	B_3 = −0.0575102
A_4 = −0.0329095	B_4 = −0.0320119
A_5 = 0.0499471	B_5 = 0.1223335
A_6 = 0.0260460	B_6 = 0.0899805
A_7 = 0.0007388	B_7 = −0.1435792
A_8 = 0.0075848	B_8 = −0.1334108
A_9 = −0.0216473	B_9 = 0.0776645
A_{10} = −0.0225161	B_{10} = 0.0853673

Modified-Stereographic Conformal projection for Alaska (fig. 44; ellipsoidal formulas and constants should normally be used):

R = 6,370,997 m at full scale
m = 6
ϕ_1 = 64°N. lat.
λ_0 = 152°W. long.

A_1 = 0.9972523	B_1 = 0
A_2 = 0.0052513	B_2 = −0.0041175
A_3 = 0.0074606	B_3 = 0.0048125
A_4 = −0.0153783	B_4 = −0.1968253
A_5 = 0.0636871	B_5 = −0.1408027
A_6 = 0.3660976	B_6 = −0.2937382

Modified-Stereographic Conformal projection for United States bounded by near-rectangle (fig. 45):

R = 6,370,997 m at full scale
m = 5
ϕ_1 = 39°N. lat.
λ_0 = 96°W. long.
A_1 =   0.98879
A_3 = −0.050909
A_5 =   0.075528
$A_2, A_4, B_1, B_2, B_3, B_4, B_5 = 0$

TABLE 33.—*Modified-Stereographic Conformal projections: Coefficients for specific forms*—Continued

Taking the Earth as an ellipsoid:

GS50 projection for the 50 States:

$$a = 6,378,206.4 \text{ m at full scale (Clarke 1866 ellipsoid)}$$
$$e^2 = 0.00676866$$
$$m = 10$$
$$\phi_1 = 45°\text{N. lat.}$$
$$\lambda_0 = 120°\text{W long.}$$

$A_1 = 0.9827497$	$B_1 = 0$
$A_2 = 0.0210669$	$B_2 = 0.0053804$
$A_3 = -0.1031415$	$B_3 = -0.0571664$
$A_4 = -0.0323337$	$B_4 = -0.0322847$
$A_5 = 0.0502303$	$B_5 = 0.1211983$
$A_6 = 0.0251805$	$B_6 = 0.0895678$
$A_7 = -0.0012315$	$B_7 = -0.1416121$
$A_8 = 0.0072202$	$B_8 = -0.1317091$
$A_9 = -0.0194029$	$B_9 = 0.0759677$
$A_{10} = -0.0210072$	$B_{10} = 0.0834037$

Modified-Stereographic Conformal projection for Alaska:

$$a = 6,378,206.4 \text{ m at full scale (Clarke 1866 ellipsoid)}$$
$$e^2 = 0.00676866$$
$$m = 6$$
$$\phi_1 = 64°\text{N. lat.}$$
$$\lambda_0 = 152°\text{W long.}$$

$A_1 = 0.9945303$	$B_1 = 0$
$A_2 = 0.0052083$	$B_2 = -0.0027404$
$A_3 = 0.0072721$	$B_3 = 0.0048181$
$A_4 = -0.0151089$	$B_4 = -0.1932526$
$A_5 = 0.0642675$	$B_5 = -0.1381226$
$A_6 = 0.3582802$	$B_6 = -0.2884586$

$$\chi = 2 \arctan \{\tan (\pi/4 + \phi/2)[(1 - e \sin \phi)/(1 + e \sin \phi)]^{e/2}\} - \pi/2 \qquad (3-1)$$
$$m = \cos \phi/(1 - e^2 \sin^2 \phi)^{1/2} \qquad (14-15)$$
$$s = 2/[1 + \sin \chi_1 \sin \chi + \cos \chi_1 \cos \chi \cos (\lambda - \lambda_0)] \qquad (26-18)$$
$$k' = s \cos \chi/m \qquad (26-19)$$
$$x' = s \cos \chi \sin (\lambda - \lambda_0) \qquad (26-20)$$
$$y' = s [\cos \chi_1 \sin \chi - \sin \chi_1 \cos \chi \cos (\lambda - \lambda_0)] \qquad (26-21)$$

where χ_1 is found as χ (the conformal latitude) from equation (3–1) by substituting ϕ_1 for ϕ. The (x', y') thus found are converted to (x, y) with unchanged equations (26–4) and (26–5), or (26–6) through (26–9) as listed under spherical formulas with accompanying explanations, except that R in (26–4) or (26–7) is replaced with a, the semimajor axis of the ellipsoid of eccentricity e, and the constants used must be those for the ellipsoidal projection.

For *inverse equations*, given a, e, ϕ_1, λ_0, A_1 through A_m, B_1 through B_m, x, and y, to find ϕ and λ, the Newton-Raphson iteration of spherical equations (26–10) through (26–13) is used unchanged to find (x', y') except that R is replaced with a, and ellipsoidal constants must be used. After convergence, the final (x', y') is converted to (ϕ, λ) without iteration. Equations (26–14), (26–15), and (3–1) are used to calculate ρ, c, and χ_1 as before.
Then,

$$\chi = \arcsin \left[\cos c \, \sin \chi_1 + (y' \sin c \, \cos \chi_1 / \rho) \right] \qquad (26\text{--}22)$$

$$\phi = 2 \arctan \left\{ \tan (\pi/4 + \chi/2)[(1 + e \sin \phi)/(1 - e \sin \phi)]^{e/2} \right\} - \pi/2 \qquad (3\text{--}4)$$

$$\lambda = \lambda_0 + \arctan \left[x' \sin c / (\rho \cos \chi_1 \cos c - y' \sin \chi_1 \sin c) \right] \qquad (26\text{--}23)$$

If $\rho = 0$, equations (26–22) and (26–23) are indeterminate, but $\chi = \chi_0$ and $\lambda = \lambda_0$. Equation (3–4), which should not use the ATAN2 function or equivalent, involves iteration by successive substitution, using χ as the first trial ϕ on the right side of the equation, calculating ϕ on the left, using the new value of ϕ on the right side, and so forth, until the change in ϕ is negligible. Tables 34 and 35 list representative rectangular coordinates for the ellipsoidal forms of the 50-State and Alaska projections, to be used in the above formulas.

TABLE 34.—*GS50 projection for 50 States: Rectangular coordinates for Clarke 1866 ellipsoid*

[y coordinate in parentheses below x coordinate; k (scale factor) in italics. Equatorial radius of ellipsoid, a = 1 unit; eccentricity is based on Clarke 1866 ellipsoid. Origin 45°N. lat., 120°W. long., Y axis north from origin]

	Longitude					
Latitude	165°	180°	−165°	−150°	−135°	−12
75°	−0.29450	−0.26954	−0.22462	−0.16629	−0.09888	−0.0
	(0.68122)	(0.62252)	(0.57777)	(0.54832)	(0.53514)	(0.5
	0.96940	*0.93350*	*0.94680*	*0.98600*	*1.04351*	*1.1*
60	−0.56708	−0.47652	−0.37432	−0.25945	−0.13450	−0.0
	(0.55579)	(0.44931)	(0.36467)	(0.30448)	(0.26964)	(0.2
	1.11056	*1.03320*	*0.99720*	*0.98684*	*0.99638*	*1.0*
45	−0.78438	−0.65970	−0.51358	−0.35313	−0.18060	0.0
	(0.40816)	(0.25882)	(0.14804)	(0.06723)	(0.01707)	(0.0
	1.10999	*1.01071*	*0.97599*	*0.96761*	*0.97461*	*0.9*
30	−0.99437	−0.82970	−0.64556	−0.44699	−0.23176	−0.0
	(0.18093)	(0.05909)	(−0.06996)	(−0.16831)	(−0.23587)	(−0.2
	0.92437	*0.99489*	*0.98110*	*0.97955*	*1.01526*	*1.0*
15	−1.26654	−0.99879	−0.77655	−0.54614	−0.30348	0.0
	(0.37724)	(−0.17525)	(−0.29355)	(−0.40445)	(−0.50718)	(−0.5
	5.35283	*1.26758*	*1.02533*	*0.96750*	*1.17269*	*1.2*

	Longitude				
Latitude	−105°	−90°	−75°	−60°	−45°
75°	0.05019	0.12669	0.20199	0.27474	0.34149
	(0.56135)	(0.60290)	(0.66601)	(0.75568)	(0.88349)
	1.21874	*1.35483*	*1.55468*	*1.87521*	*2.42649*
60	0.13189	0.26642	0.39828	0.52663	0.66111
	(0.27713)	(0.31778)	(0.37908)	(0.45182)	(0.50581)
	1.05301	*1.09134*	*1.12313*	*1.11492*	*1.17353*
45	0.18215	0.35975	0.52792	0.68068	0.78758
	(0.01665)	(0.06457)	(0.14091)	(0.24688)	(0.42634)
	0.99055	*0.99558*	*0.99806*	*1.02418*	*1.44787*
30	0.22878	0.44683	0.65324	0.83776	1.04409
	(−0.23806)	(−0.17878)	(−0.08678)	(0.03834)	(0.00223)
	1.00249	*0.99481*	*1.00384*	*0.87806*	*2.72764*
15	0.28360	0.53662	0.76638	1.12680	0.56142
	(−0.49621)	(−0.43117)	(−0.31713)	(0.21682)	(1.25008)
	1.00240	*1.10194*	*0.84755*	*2.88781*	*16.99865*

TABLE 35.—*Modified-Stereographic Conformal projection for Alaska: Rectangular coordinates for Clarke 1866 ellipsoid*

[y coordinate in parentheses below x coordinate; k (scale factor) in italics. Equatorial radius of ellipsoid, a = 1 unit; eccentricity is based on Clarke 1866 ellipsoid. Origin: 64° Lat., −152 Y axis north from origin]

	Longitude						
Latitude	170°	180°	−170°	−160°	−150°	−140°	−
75°	−0.16211	−0.12311	−0.08081	−0.03641	0.00892	0.05402	0.0
	(0.24589)	(0.22161)	(0.20445)	(0.19469)	(0.19248)	(0.19786)	(0.2
	1.01917	*1.01147*	*1.00600*	*1.00306*	*1.00264*	*1.00459*	*1.0*
70	−0.21520	−0.16271	−0.10647	−0.04782	0.01192	0.07140	0.1
	(0.17360)	(0.14228)	(0.12028)	(0.10779)	(0.10494)	(0.11178)	(0.1
	1.03059	*1.01497*	*1.00535*	*1.00062*	*0.99993*	*1.00304*	*1.0*
65	−0.26675	−0.20094	−0.13124	−0.05888	0.01475	0.08813	0.1
	(0.09941)	(0.06222)	(0.03605)	(0.02112)	(0.01767)	(0.02578)	(0.
	1.03421	*1.01364*	*1.00273*	*0.99805*	*0.99768*	*1.00108*	*1.0*
60	−0.31591	−0.23765	−0.15521	−0.06968	−0.01744	0.10427	0.1
	(0.02389)	(−0.01813)	(−0.04808)	(−0.06536)	(−0.06946)	(−0.06013)	(−0.0
	1.02672	*1.00804*	*0.99991*	*0.99758*	*0.99834*	*1.00020*	*1.0*
55	−0.36252	−0.27315	−0.17873	−0.08047	0.01999	0.12022	0.2
	(−0.05185)	(−0.09835)	(−0.13210)	(−0.15191)	(−0.15683)	(−0.14611)	(−0.1
	1.00925	*1.00166*	*0.99931*	*1.00127*	*1.00536*	*1.00467*	*0.9*
50	−0.40740	−0.30816	−0.20222	−0.09163	0.02232	0.13669	0.2
	(−0.12654)	(−0.17828)	(−0.21616)	(−0.23888)	(−0.24516)	(−0.23284)	(−0.2
	0.98940	*1.00073*	*1.00245*	*1.00955*	*1.02260*	*1.02237*	*0.9*

SPACE MAP PROJECTIONS

One of the most recent developments in map projections has been that involving a time factor, relating a mapping satellite revolving in an orbit about a rotating Earth. With the advent of automated continuous mapping in the near future, the static projections previously available are not sufficient to provide the accuracy merited by the imagery, without frequent readjustment of projection parameters and discontinuity at each adjustment. Projections appropriate for such satellite mapping are much more complicated mathematically, but, once derived, can be handled by computer.

Several such space map projections have been conceived, and all but one have been mathematically developed. The Space Oblique Mercator projection, suitable for mapping imagery from Landsat and other vertically scanning satellites, is described below, and is followed by a discussion of Satellite-Tracking projections. The Space Oblique Conformal Conic is a still more complex projection, currently only in conception, but for which mathematical development will be required if satellite side-looking imagery has been developed to an extent sufficient to encourage its use.

27. SPACE OBLIQUE MERCATOR PROJECTION

SUMMARY

- Modified cylindrical projection with map surface defined by satellite orbit.
- Designed especially for continuous mapping of satellite imagery.
- Basically conformal, especially in region of satellite scanning.
- Groundtrack of satellite, a curved line on the globe, is shown as a curved line on the map and is continuously true to scale as orbiting continues.
- All meridians and parallels are curved lines, except meridian at each polar approach.
- Recommended only for a relatively narrow band along the groundtrack.
- Developed 1973–79 by Colvocoresses, Snyder, and Junkins.

HISTORY

The launching of an Earth-sensing satellite by the National Aeronautics and Space Administration in 1972 led to a new era of mapping on a continuous basis from space. This satellite, first called ERTS-1 and renamed Landsat 1 in 1975, was followed by two others, all of which circled the Earth in a nearly circular orbit inclined about 99° to the Equator and scanning a swath about 185 km (officially 100 nautical miles) wide from an altitude of about 919 km. The fourth and fifth Landsat satellites involved circular orbits inclined about 98° and scanning from an altitude of about 705 km.

Continuous mapping of this band required a new map projection. Although conformal mapping was desired, the normal choice, the Oblique Mercator projection, is unsatisfactory for two reasons. First, the Earth is rotating at the same time the satellite is moving in an orbit which lies in a plane almost at a right angle to the plane of the Equator, with the double-motion effect producing a curved groundtrack, rather than one formed by the intersection of the Earth's surface with a plane passing through the center of the Earth. Second, the only available Oblique Mercator projections for the ellipsoid are for limited coverage near the chosen central point.

What was needed was a map projection on which the groundtrack remained true-to-scale throughout its course. This course did not, in the case of Landsat 1, 2, or 3, return to the same point for 251 revolutions. (For Landsat 4 and 5, the cycle is 233 revolutions.) It was also felt necessary that conformality be closely maintained within the range of the swath mapped by the satellite.

Alden P. Colvocoresses of the Geological Survey was the first to realize not only that such a projection was needed, but also that it was mathematically feasible. He defined it geometrically (Colvocoresses, 1974) and immediately began to appeal for the development of formulas. The following formulas resulted from the writer's response to Colvocoresses' appeal made at a geodetic conference at The Ohio State University in 1976. While the formulas were derived (1977–79) for Landsat, they are applicable to any satellite orbiting the Earth in a circular or elliptical orbit and at any inclination. Less complete formulas were also developed in 1977 by John L. Junkins, then of the University of Virginia. The following formulas are limited to nearly circular orbits. A complete derivation for orbits of any ellipticity is given by Snyder (1981b) and another summary by Snyder (1978b).

FEATURES AND USAGE

The Space Oblique Mercator (SOM) projection visually differs from the Oblique Mercator projection in that the central line (the groundtrack of the orbiting satellite) is slightly curved, rather than straight. For Landsat, this groundtrack appears as a nearly sinusoidal curve crossing the X axis at an angle of about 8°. The scanlines, perpendicular to the orbit in space, are slightly skewed with respect to the perpendicular to the groundtrack when plotted on the sphere or ellipsoid. Due to Earth rotation, the scanlines on the Earth (or map) intersect the groundtrack at about 86° near the Equator, but at 90° when the groundtrack makes its closest approach to either pole. With the curved groundtrack, the scanlines generally are skewed with respect to the X and Y axes, inclined about 4° to the Y axis at the Equator, and not at all at the polar approaches.

The orbit for Landsat 1, 2, and 3 intersected the plane of the Equator at an inclination of about 99°, measured as the angle between the direction of satellite revolution and the direction of Earth rotation. Thus the groundtrack reached limits of about lat. 81° N. and S. (180° minus 99°). The 185-km swath scanned by Landsat, about 0.83° on either side of the groundtrack, led to complete coverage of the Earth from about lat. 82° N. to 82° S. in the course of the 251 revolutions. With a nominal altitude of about 919 km, the time of one revolution was 103.267 minutes, and the orbit was designed to complete the 251 revolutions in exactly 18 days. Landsat 4 and 5, launched in 1982 and 1984, respectively, scanned the same width, but with an orbit of different radius and inclination, as stated above.

As on the normal Oblique Mercator, all meridians and parallels are curved lines, except for the meridian crossed by the groundtrack at each polar approach. While the straight meridians are 180° apart on the normal Oblique Mercator, they are about 167° apart on the SOM for Landsat 1, 2, and 3, since the Earth advanced about 26° in longitude for each revolution of the satellite.

As developed, the SOM is not perfectly conformal for either the sphere or ellipsoid, although the error is negligible within the scanning range for either. Along the groundtrack, scale in the direction of the groundtrack is correct for sphere or ellipsoid, while conformality is correct for the sphere and within 0.0005 percent of correct for the ellipsoid. At 1° away from the groundtrack, the Tissot Indicatrix (the ellipse of distortion) is flattened a maximum of 0.001 percent for the sphere and a maximum of 0.006 percent for the ellipsoid (this would be zero if conformal). The scale 1° away from the groundtrack averages 0.015 percent greater than that at the groundtrack, a value which is fundamental to projection. As a result of the slight nonconformality, the scale 1° away from the groundtrack on the ellipsoid then varies from 0.012 to 0.018 percent more than the scale along the groundtrack.

A prototype version of the SOM was used by NASA with a geometric analogy proposed by Colvocoresses (1974) while he was seeking the more rigorous mathematical development. This consisted basically of moving an obliquely tangent cylinder back and forth on the sphere so that a circle around it which would normally be tangent shifted to follow the groundtrack. This is suitable near the Equator but leads to errors of about 0.1 percent near the poles, even for the sphere. In 1977, John B. Rowland of the USGS applied the Hotine Oblique Mercator (described previously) to Landsat 1, 2, and 3 orbits in five stationary zones, with smaller but significant errors (up to twice the scale variation of the SOM) resulting from the fact that the groundtrack cannot follow the straight central line of the HOM. In addition, there are discontinuities at the zone changes. This was done to fill the void resulting from the lack of SOM formulas.

For Landsat 4 and 5, the final SOM equations replaced the HOM for mapping. Figures 46 and 47 show the SOM extended to two orbits with a 30° graticule and for one-fourth of an orbit with a 10° graticule, respectively. The progressive advance of meridians may be seen in figure 46. Both views are for Landsat 4 and 5 constants.

FORMULAS FOR THE SPHERE

Both iteration and numerical integration are involved in the formulas as presented for sphere or ellipsoid. The iteration is quite rapid (three to five iterations

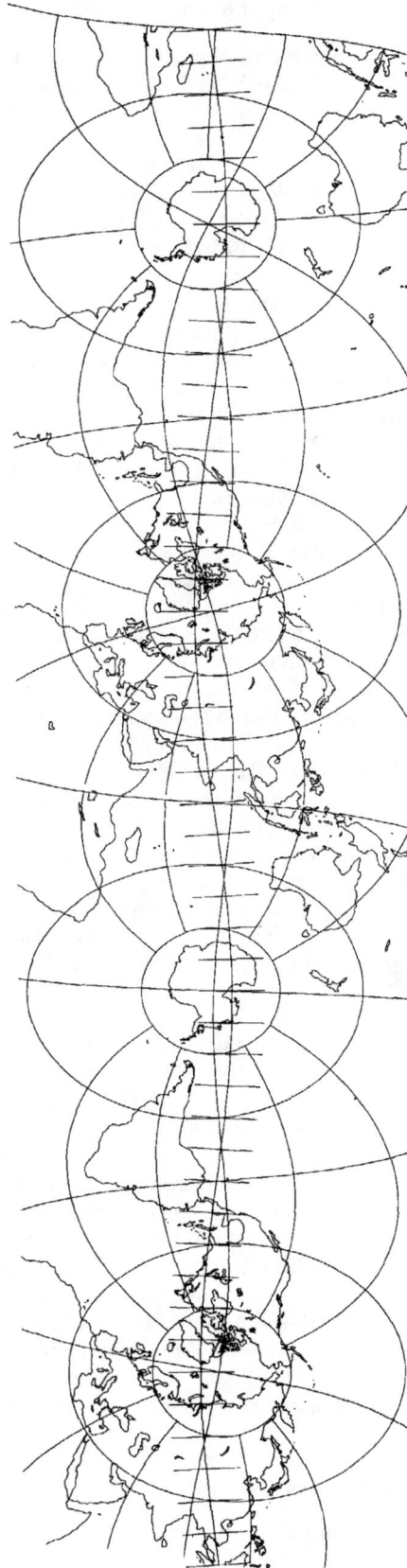

FIGURE 46.—Two orbits of the Space Oblique Mercator projection, shown for Landsat 5, paths 15 (left) and 31. Designed for a narrow band along groundtrack, which remains true to scale. Note the rotation of the Earth with successive orbits. Scan lines, extended 15° from groundtrack, are short lines nearly perpendicular to it.

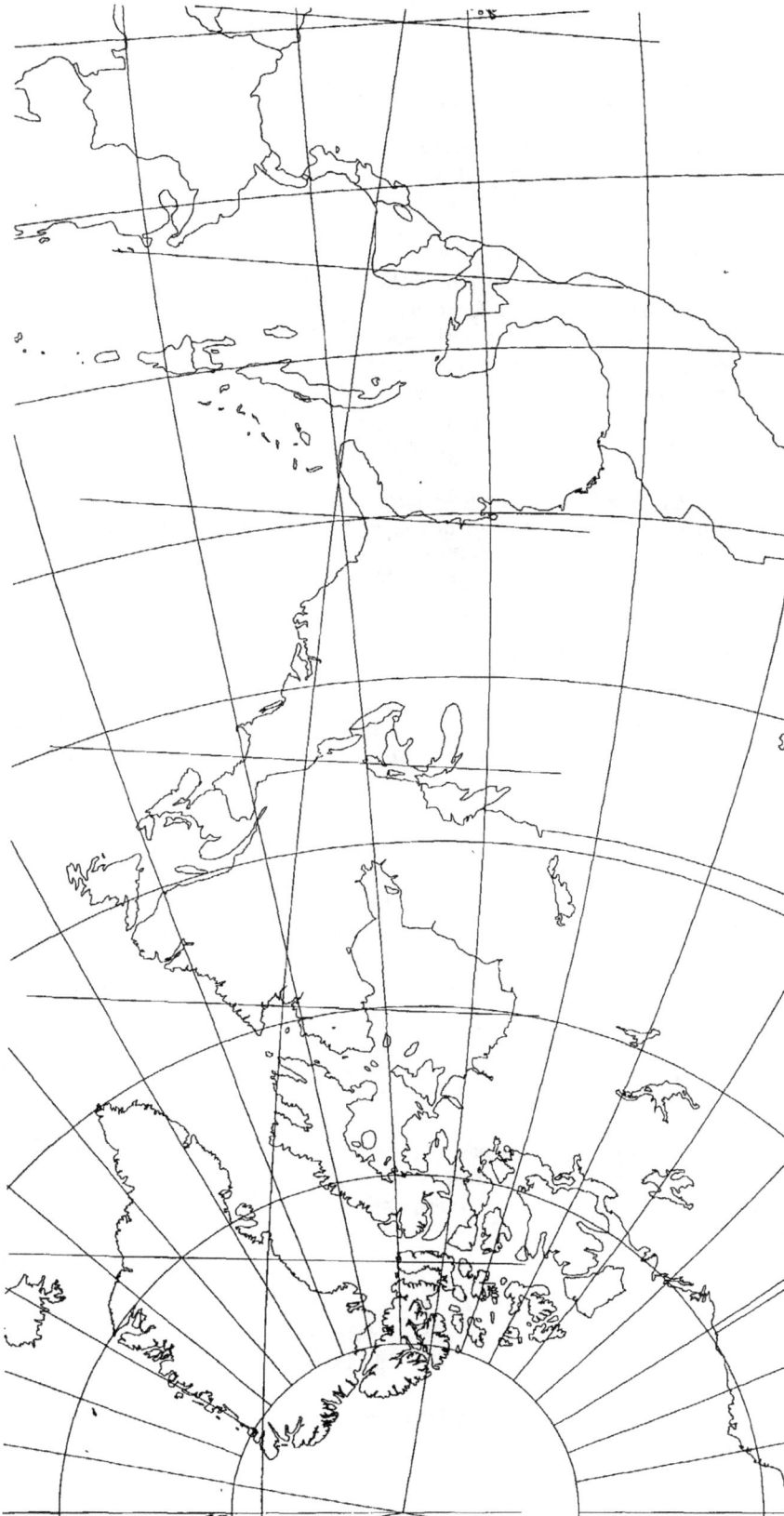

FIGURE 47.—One quadrant of the Space Oblique Mercator projection for Landsat 5, path 15. An "enlargement" of part of figure 46 beginning at the North Pole.

required for ten-place accuracy), and the numerical integration is greatly simplified by the use of rapidly converging Fourier series. The coefficients for the Fourier series may be calculated once for a given satellite orbit. [Some formulas below are slightly simplified from those first published (Snyder, 1978b).]

For the forward equations, for the sphere and circular orbit, to find (x, y) for a given (ϕ, λ), it is necessary to be given R, i, P_2, P_1, λ_0, ϕ, and λ, where

R = radius of the globe at the scale of the map.
i = angle of inclination between the plane of the Earth's Equator and the plane of the satellite orbit, measured counterclockwise from the Equator to the orbital plane at the ascending node (99.092° for Landsat 1, 2, 3; 98.20° for Landsat 4, 5).
P_2 = time required for revolution of the satellite (103.267 min for Landsat 1, 2, 3; 98.884 min. for Landsat 4, 5).
P_1 = length of Earth's rotation with respect to the precessed ascending node. For Landsat, the satellite orbit is Sun-synchronous; that is, it is always the same with respect to the Sun, equating P_1 to the solar day (1,440 min). The ascending node is the point on the satellite orbit at which the satellite crosses the Earth's equatorial plane in a northerly direction.
λ_0 = geodetic longitude of the ascending node at time $t=0$.
(ϕ,λ) = geodetic latitude and longitude of point to be plotted on map.
t = time elapsed since the satellite crossed the ascending node for the orbit considered to be the initial one. This may be the current orbit or any earlier one, as long as the proper λ_0 is used.

First, various constants applying to the entire map for all the satellite's orbits should be calculated a single time (see p. 347 for numerical examples):

$$B = (2/\pi) \int_0^{\pi/2} [(H-S^2)/(1 + S^2)^{1/2}]d\lambda' \tag{27-1}$$
$$A_n = [4/(\pi n)]\int_0^{\pi/2}[(H-S^2)/(1 + S^2)^{1/2}] \cos n\lambda' \, d\lambda' \tag{27-2}$$

for $n=2$ and 4 only.

$$C_n = [4(H + 1)/(\pi n)]\int_0^{\pi/2} [S/(1 + S^2)^{1/2}] \cos n\lambda' \, d\lambda' \tag{27-3}$$

for $n=1$ and 3 only.

For calculating A_n, B, and C_n, numerical integration using Simpson's rule is recommended, with 9° intervals in λ' (sufficient for ten-place accuracy). The terms shown are sufficient for seven-place accuracy, ample for the sphere. For H and S in equations (27−1) through (27−3):

$$H = 1-(P_2/P_1) \cos i \tag{27-4}$$
$$S = (P_2/P_1) \sin i \cos \lambda' \tag{27-5}$$

To find x and y, with the X axis passing through each ascending and descending node (wherever the groundtrack crosses the Equator), x increasing in the direction of satellite motion, and the Y axis passing through the ascending node for time $t=0$:

$$x/R = B\lambda' + A_2 \sin 2\lambda' + A_4 \sin 4\lambda' + \ldots$$
$$-[S/(1 + S^2)^{1/2}] \ln \tan (\pi/4 + \phi'/2) \tag{27-6}$$
$$y/R = C_1 \sin \lambda' + C_3 \sin 3\lambda' + \ldots$$
$$+ [1/(1 + S^2)^{1/2}] \ln \tan (\pi/4 + \phi'/2) \tag{27-7}$$

where B, A_n, and C_n are constants calculated above, S is calculated from (27−5) for each point, and

$$\lambda' = \arctan \left(\cos i \tan \lambda_t + \sin i \tan \phi / \cos \lambda_t \right) \tag{27-8}$$
$$\lambda_t = \lambda - \lambda_0 + (P_2/P_1) \lambda' \tag{27-9}$$
$$\phi' = \arcsin \left(\cos i \sin \phi - \sin i \cos \phi \sin \lambda_t \right) \tag{27-10}$$
$$\lambda_0 = 128.87° - (360°/251)p \quad \text{(Landsat 1, 2, 3 only)} \tag{27-11}$$
$$\quad\quad = 129.30° - (360°/233)p \quad \text{(Landsat 4, 5 only)} \tag{27-11a}$$

p = path number of Landsat orbit for which the ascending node occurs at time $t = 0$. This ascending node is prior to the start of the path, so that the path extends from ¼ orbit past this node to ⁵/₄ orbit past it.

λ' = "transformed longitude," the angular distance along the groundtrack, measured from the initial ascending node ($t = 0$), and directly proportional to t for a circular orbit, or $\lambda' = 360° \, t/P_2$.

λ_t = a "satellite-apparent" longitude, the longitude relative to λ_0 seen by the satellite if the Earth were stationary.

ϕ' = "transformed latitude," the angular distance from the groundtrack, positive to the left of the satellite as it proceeds along the orbit.

Finding λ' from equations (27–8) and (27–9) involves iteration performed in the following manner: After selecting ϕ and λ, the λ' of the nearest polar approach, λ_p', is used as the first trial λ' on the right side of (27–9); λ_t is calculated and substituted into (27–8) to find a new λ'. A quadrant adjustment (see below) is applied to λ', since the computer normally calculates arctan as an angle between −90° and 90°, and this λ' is used as the next trial λ' in (27–9), etc., until λ' changes by less than a chosen convergence factor. The value of λ_p' may be determined as follows, for any number of revolutions:

$$\lambda_p' = 90° \times (4 N + 2 \pm 1) \tag{27-12}$$

where N is the number of orbits completed at the last ascending node before the satellite passes the nearest pole, and the $\pm$ takes minus in the Northern Hemisphere and plus in the Southern (either for the Equator). Thus, if only the first path number past the ascending node is involved, λ_p' is 90° for the first quadrant (North Pole to Equator), 270° for the second and third quadrants (Equator to South Pole to Equator), and 450° for the fourth quadrant (Equator to North Pole). For quadrant adjustment to λ' calculated from (27–8), the Fortran ATAN2 or its equivalent should not be used. Instead, λ' should be increased by λ_p' minus the following factor: 90° times sin λ_p' times ± 1 (taking the sign of cos λ_{tp}, where $\lambda_{tp} = \lambda - \lambda_0 + (P_2/P_1)\lambda_p'$). If cos λ_{tp} is zero, the final λ' is λ_p'. Thus, the adder to the arctan is 0° for the quadrant between the ascending node and the start of the path, and 180°, 180°, 360°, and 360°, respectively, for the four quadrants of the first path.

The closed forms of equations (27–6) and (27–7) are as follows:

$$x/R = \int_0^{\lambda'} [(H - S^2)/(1 + S^2)^{1/2}]d\lambda' - [S/(1 + S^2)^{1/2}]$$
$$\ln \tan (\pi/4 + \phi'/2) \tag{27-6a}$$
$$y/R = (H + 1) \int_0^{\lambda'} [S/(1 + S^2)^{1/2}]d\lambda' + [1/(1 + S^2)^{1/2}]$$
$$\ln \tan (\pi/4 + \phi'/2) \tag{27-7a}$$

Since these involve numerical integration for each point, the series forms, limiting numerical integration to once per satellite, are distinctly preferable. These are Fourier series, and equations (27–2) and (27–3) normally require integration from 0 to 2π, without the multiplier 4, but the symmetry of the circular orbit permits the simplification as shown for the nonzero coefficients.

For inverse formulas for the sphere, given R, i, P_2, P_1, λ_0, x, and y, with ϕ and λ required: Constants B, A_n, C_n, and λ_0 must be calculated from (27–1) through (27–3) and (27–11) just as they were for the forward equations. Then,

$$\lambda = \arctan\ [(\cos i \sin \lambda' - \sin i \tan \phi')/\cos \lambda'] - (P_2/P_1)\lambda' + \lambda_0 \qquad (27\text{--}13)$$

$$\phi = \arcsin\ (\cos i \sin \phi' + \sin i \cos \phi' \sin \lambda') \qquad (27\text{--}14)$$

·where the ATAN2 function of Fortran is useful for (27−13), except that it may be necessary to add or subtract 360° to place λ between long. 180° E. (+) and 180° W. (−), and

$$\lambda' = [x/R + Sy/R - A_2 \sin 2\lambda' - A_4 \sin 4\lambda' - S(C_1 \sin \lambda' + C_3 \sin 3\lambda')]/B \qquad (27\text{--}15)$$

$$\ln \tan(\pi/4 + \phi'/2) = (1 + S^2)^{1/2}(y/R - C_1 \sin \lambda' - C_3 \sin 3\lambda') \qquad (27\text{--}16)$$

Equation (27−15) is iterated by trying almost any λ' (preferably $x/(BR)$) in the right side, solving for λ' on the left and using the new λ' for the next trial, etc., until there is no significant change between successive trial values. Equation (27−16) uses the final λ' calculated from (27−15).

The closed form of equation (27−15) given below involves repeated numerical integration as well as iteration, making its use almost prohibitive:

$$(x + Sy)/R = \int_0^\lambda [(H - S^2)/(1 + S^2)^{1/2}]d\lambda' + S(H + 1)\int_0^{\lambda'} [S/(1 + S^2)^{1/2}]d\lambda' \qquad (27\text{--}15a)$$

The following closed form of (27−16) requires the use of the last integral calculated from (27−15a):

$$\ln \tan(\pi/4 + \phi'/2) = (1 + S^2)^{1/2}\{(y/R) - (H + 1)\int_0^{\lambda'} [S/(1 + S^2)^{1/2}]d\lambda'\} \qquad (27\text{--}16a)$$

The original published forms of these equations include several other Fourier coefficient calculations which slightly save computer time when continuous mapping is involved. The resulting equations are more complicated, so they are omitted here for simplicity. The above equations are as accurate and only slightly less efficient.

The values of coefficients for Landsat 1, 2, and 3 ($P_2/P_1 = 18/251$; $i = 99.092°$) are listed here as examples:

$$
\begin{aligned}
A_2 &= -0.0018820 \\
A_4 &= 0.0000007 \\
B &= 1.0075654142 \text{ for } \lambda' \text{ in radians} \\
&= 0.0175853339 \text{ for } \lambda' \text{ in degrees} \\
C_1 &= 0.1421597 \\
C_3 &= -0.0000296
\end{aligned}
$$

It is also of interest to determine values of φ, λ, or λ' along the groundtrack, given any one of the three (as well as P_2, P_1, i, and λ_0). Given φ,

$$\lambda' = \arcsin\ (\sin \phi/\sin i) \qquad (27\text{--}17)$$

$$\lambda = \arctan\ [(\cos i \sin \lambda')/\cos \lambda'] - (P_2/P_1)\lambda' + \lambda_0 \qquad (27\text{--}18)$$

If φ is given for a descending part of the orbit (daylight on Landsat), subtract λ' from the λ' of the nearest descending node (180°, 540°, . . .). If the orbit is ascending, add λ' to the λ' of the nearest ascending node (0°, 360°, . . .). For a given path, only 180° and 360°, respectively, are involved.

Given λ,

$$\lambda' = \arctan\ (\tan \lambda_t/\cos i) \qquad (27\text{--}19)$$

$$\lambda_t = \lambda - \lambda_0 + (P_2/P_1)\lambda' \qquad (27\text{--}9)$$

$$\phi = \arcsin\ (\sin i \sin \lambda') \qquad (27\text{--}20)$$

Table 36.—*Scale factors for the spherical Space Oblique Mercator projection using Landsat 1, 2, and 3 constants*

λ'	$\phi' = 1°$				$\phi' = -1°$			
	h	k	ω°	m_ϕ	h	k	ω°	m_ϕ
0° ____	1.000154	1.000151	0.0006	1.000152	1.000154	1.000151	0.0006	1.000152
5 _____	1.000153	1.000151	.0006	1.000151	1.000154	1.000151	.0006	1.000152
10 _____	1.000153	1.000151	.0006	1.000151	1.000155	1.000151	.0006	1.000153
15 _____	1.000153	1.000151	.0006	1.000151	1.000155	1.000151	.0006	1.000153
20 _____	1.000152	1.000151	.0006	1.000150	1.000156	1.000151	.0006	1.000154
25 _____	1.000152	1.000151	.0006	1.000150	1.000156	1.000151	.0006	1.000154
30 _____	1.000152	1.000151	.0005	1.000149	1.000156	1.000151	.0005	1.000154
35 _____	1.000152	1.000150	.0005	1.000149	1.000156	1.000151	.0005	1.000154
40 _____	1.000152	1.000150	.0005	1.000150	1.000156	1.000151	.0005	1.000154
45 _____	1.000152	1.000150	.0004	1.000150	1.000156	1.000151	.0005	1.000154
50 _____	1.000152	1.000150	.0004	1.000150	1.000156	1.000151	.0004	1.000154
55 _____	1.000152	1.000150	.0004	1.000150	1.000155	1.000151	.0004	1.000154
60 _____	1.000153	1.000151	.0003	1.000151	1.000155	1.000151	.0003	1.000154
65 _____	1.000153	1.000151	.0003	1.000151	1.000155	1.000151	.0003	1.000153
70 _____	1.000153	1.000151	.0002	1.000152	1.000154	1.000151	.0002	1.000153
75 _____	1.000153	1.000151	.0002	1.000152	1.000154	1.000151	.0002	1.000153
80 _____	1.000153	1.000151	.0001	1.000152	1.000153	1.000152	.0001	1.000153
85 _____	1.000153	1.000152	.0001	1.000152	1.000153	1.000152	.0001	1.000152
90 _____	1.000152	1.000151	.0001	1.000152	1.000152	1.000152	.0000	1.000152

Notes: λ' = angular position along groundtrack, from ascending node.

   φ' = angular distance away from groundtrack, positive in direction away from North Pole.

   h = scale factor along meridian of longitude.

   k = scale factor along parallel of latitude.

   ω = maximum angular deformation.

   m_ϕ = scale factor along line of constant φ'.

   m_λ = scale factor along line of constant λ'.

      = sec φ', or 1.000152 at φ' = 1°.

   If φ' = 0°, h, k, and m_ϕ = 1.0, while ω = 0.

Given λ', equations (27−18) and (27−20) may be used for λ and φ, respectively. Equations (27−6) and (27−7), with φ' = 0, convert these values to x and y. Equations (27−19) and (27−9) require joint iteration, using the same procedure as that for the pair of equations (27−8) and (27−9) given earlier. The λ calculated from equation (27−18) should have the same quadrant adjustment as that described for (27−13).

The formulas for scale factors h and k and maximum angular deformation ω are so lengthy that they are not given here. They are available in Snyder (1981b). Table 36 lists these values as calculated for the spherical SOM using Landsat constants. Although calculated for Landsat 1, 2, and 3, they are almost identical for 4 and 5.

FORMULAS FOR THE ELLIPSOID AND CIRCULAR ORBIT

Since the SOM is intended to be used only for the mapping of relatively narrow strips, it is highly recommended that the ellipsoidal form be used to take advantage of the high accuracy of scale available, especially as the imagery is further developed and used for more precise measurement. In addition to the normal modifications to the above spherical formulas for ellipsoidal equivalents, an additional element is introduced by the fact that Landsat is designed to scan vertically, rather than in a geocentric direction. Therefore, "pseudotransformed" latitude φ" and longitude λ" have been introduced. They relate to a geocentric groundtrack for an orbit in a plane through the center of the Earth. The regular transformed coordinates φ' and λ' are related to the actual vertical groundtrack. The two groundtracks are only a maximum of 0.008° apart, although a lengthwise displacement of 0.028° for a given position may occur.

If the eccentricity of the ellipsoid is made zero, the formulas reduce to the spherical formulas above. These formulas vary slightly, but not significantly, from those published in Snyder (1978b, 1981b). In practice, the coordinates for each picture element (pixel) should not be calculated because of computer time required. Linear interpolation between occasional calculated points can be developed with adequate accuracy.

For the forward formulas, given a, e, i, P_2, P_1, λ_0, R_0, ϕ, and λ, find x and y. As with the spherical formulas, the X axis passes through each ascending and descending node, x increasing in the direction of satellite motion, and the Y axis intersects perpendicularly at the ascending node for the time $t = 0$. Defining terms,

a, e = semimajor axis and eccentricity of ellipsoid, respectively (as for other ellipsoidal projections).

R_0 = radius of the circular orbit of the satellite.

i, P_2, P_1, λ_0, ϕ, λ are as defined for the spherical SOM formulas. For constants applying to the entire map (see p. 354 for numerical examples):

$$B = (2/\pi)\int_0^{\pi/2}[(HJ-S^2)/(J^2 + S^2)^{1/2}]d\lambda'' \tag{27-21}$$

$$A_n = [4/(\pi n)]\int_0^{\pi/2}[(HJ-S^2)/(J^2 + S^2)^{1/2}]\cos n\lambda''d\lambda'' \tag{27-22}$$

$$C_n = [4/(\pi n)]\int_0^{\pi/2}[S(H+J)/(J^2 + S^2)^{1/2}]\cos n\lambda''d\lambda'' \tag{27-23}$$

$$J = (1-e^2)^3 \tag{27-24}$$

$$W = [(1-e^2\cos^2 i)^2/(1-e^2)^2]-1 \tag{27-25}$$

$$Q = e^2\sin^2 i/(1-e^2) \tag{27-26}$$

$$T = e^2\sin^2 i(2-e^2)/(1-e^2)^2 \tag{27-27}$$

$$j_n = (4/\pi)\int_0^{\pi/2}\phi''\sin n\lambda'd\lambda' \tag{27-28}$$

$$m_n = (4/\pi)\int_0^{\pi/2}(\lambda''-\lambda')\sin n\lambda'd\lambda' \tag{27-29}$$

where ϕ'' and λ'' are determined in these last two equations for the groundtrack as functions of λ', from equations (27–43), (27–34), (27–35), and (27–36).

To calculate A_n, B, and C_n, the following functions, varying with λ'', are used:

$$S = (P_2/P_1)\sin i\cos\lambda''\{(1 + T\sin^2\lambda'')/[(1 + W\sin^2\lambda'')(1 + Q\sin^2\lambda'')]\}^{1/2} \tag{27-30}$$

$$H = \left[\frac{1 + Q\sin^2\lambda''}{1 + W\sin^2\lambda''}\right]^{1/2}\left[\frac{1 + W\sin^2\lambda''}{(1 + Q\sin^2\lambda'')^2} - (P_2/P_1)\cos i\right] \tag{27-31}$$

These constants may be determined from numerical integration, using Simpson's rule with 9° intervals. For circular orbits, A_n if n is odd, C_n if n is even, j_n if n is even, and m_n if n is odd are all zero. The above integration to $\pi/2$ is suitable, due to symmetry, only for non-zero coefficients. Integration to 2π would be necessary to show that other coefficients are zero.

To find x and y from ϕ and λ:

$$x/a = B\lambda'' + A_2\sin 2\lambda'' + A_4\sin 4\lambda'' + \ldots - [S/(J^2 + S^2)^{1/2}]\ln\tan(\pi/4 + \phi''/2) \tag{27-32}$$

$$y/a = C_1\sin\lambda'' + C_3\sin 3\lambda'' + \ldots + [J/(J^2 + S^2)^{1/2}]\ln\tan(\pi/4 + \phi''/2) \tag{27-33}$$

where

$$\lambda'' = \arctan[\cos i\tan\lambda_t + (1-e^2)\sin i\tan\phi/\cos\lambda_t] \tag{27-34}$$

$$\lambda_t = \lambda-\lambda_0 + (P_2/P_1)\lambda'' \tag{27-35}$$

$$\phi'' = \arcsin \{[(1-e^2) \cos i \sin \phi - \sin i \cos \phi \sin \lambda_t]/$$
$$(1-e^2 \sin^2 \phi)^{1/2}\} \qquad (27-36)$$
$$\lambda_0 = 128.87° - (360°/251)p \text{ (Landsat 1, 2, 3 only)} \qquad (27-37)$$
$$= 129.30° - (360°/233)p \text{ (Landsat 4, 5 only)} \qquad (27-38)$$

Equations (27−34) and (27−35) are iterated together as were (27−8) and (27−9). Equation (27−30) is used to find S for the given λ'' in equations (27−32) and (27−33). For improved computational efficiency using these and subsequent series, see p. 19.

The closed forms of equations (27−32) and (27−33) are given below, but the repeated numerical integration necessitates replacement by the series forms.

$$x/a = \int_0^\lambda [(HJ-S^2)/(J^2 + S^2)^{1/2}] \, d\lambda'' - [S/(J^2+S^2)^{1/2}]$$
$$\ln \tan (\pi/4 + \phi''/2) \qquad (27-32a)$$
$$y/a = \int_0^\lambda [S(H+J)/(J^2+S^2)^{1/2}] \, d\lambda'' + [J/(J^2+S^2)^{1/2}]$$
$$\ln \tan (\pi/4 + \phi''/2) \qquad (27-33a)$$

While the above equations are sufficient for plotting a graticule according to the SOM projection, it is also desirable to relate these points to the true vertical groundtrack. To find ϕ'' and λ'' in terms of ϕ' and λ', the shift between these two sets of coordinates is so small it is equivalent to an adjustment, requiring only small Fourier coefficients, and very lengthy calculations if they are not used. The use of Fourier series is therefore highly recommended, although the one-time calculation of coefficients is more difficult than the foregoing calculation of A_n, B, and C_n.

$$\phi'' = \phi' + j_1 \sin \lambda' + j_3 \sin 3\lambda' + \ldots \qquad (27-39)$$
$$\lambda'' = \lambda' + m_2 \sin 2\lambda' + m_4 \sin 4\lambda' + \ldots \qquad (27-40)$$

For a circular orbit, λ' is $2\pi t/P_2$, where t is the time from the initial ascending node.

The equations for functions of the satellite groundtrack, both forward and inverse, are given here, since some are used in calculating j_n and m_n as well. In any case a, e, i, P_2, P_1, λ_0, and R_0 must be given. For λ' and λ, if ϕ is given:

$$\phi_g = \phi - \arcsin \{ae^2 \sin \phi \cos \phi/[R_0 (1-e^2 \sin^2 \phi)^{1/2}]\} \qquad (27-41)$$
$$\lambda' = \arcsin (\sin \phi_g/\sin i) \qquad (27-42)$$

where ϕ_g is the geocentric latitude of the point geocentrically under the satellite, not the geocentric latitude corresponding to the vertical groundtrack latitude ϕ.

$$\lambda = \arctan [(\cos i \sin \lambda')/\cos \lambda'] - (P_2/P_1)\lambda' + \lambda_0 \qquad (27-43)$$

If λ of a point along the groundtrack is given, to find λ' and ϕ,

$$\lambda' = \arctan (\tan \lambda_t/\cos i) \qquad (27-19)$$
$$\lambda_t = \lambda - \lambda_0 + (P_2/P_1) \lambda' \qquad (27-35)$$

These two equations are iterated as a group, but the first trial λ' and the quadrant adjustments should follow the procedures listed for equation (27−8).

$$\phi = \arcsin (\sin i \sin \lambda') + \arcsin \{ae^2 \sin \phi \cos \phi/$$
$$[R_0 (1-e^2 \sin^2 \phi)^{1/2}]\} \qquad (27-44)$$

Iteration is involved in (27−44), beginning with a trial ϕ of arcsin (sin i sin λ').

If λ' of a point along the groundtrack is given, ϕ is found from (27-44), and (27-43) provides λ. Only (27-44) requires iteration for these calculations.

Inverse formulas for the ellipsoidal form of the SOM projection, with circular orbit, follow: Given a, e, i, P_2, P_1, λ_0, R_0, x, and y, to find ϕ and λ, the general Fourier and other constants are first determined as described at the beginning of the forward equations. Then

$$\lambda = \lambda_t - (P_2/P_1)\ \lambda'' + \lambda_0 \tag{27-45}$$

where

$$\lambda_t = \arctan(V/\cos \lambda'') \tag{27-46}$$
$$V = \{[1 - \sin^2 \phi''/(1 - e^2)]\cos i \sin \lambda'' - \sin i \sin \phi''\ [(1 + Q \sin^2 \lambda'')$$
$$(1 - \sin^2 \phi'') - U \sin^2 \phi'']^{1/2}\}/[1 - \sin^2 \phi''\ (1 + U)] \tag{27-47}$$
$$U = e^2 \cos^2 i/(1 - e^2) \tag{27-48}$$

while λ'' and ϕ'' are found from (27-51) and (27-52) below.

$$\phi = \arctan\ \{(\tan \lambda'' \cos \lambda_t - \cos i \sin \lambda_t)/[(1 - e^2)\sin i]\} \tag{27-49}$$

If $i = 0$, equation (27-49) is indeterminate, but

$$\phi = \arcsin\ \{\sin \phi''/[(1 - e^2)^2 + e^2 \sin^2 \phi'']^{1/2}\} \tag{27-50}$$

No iteration is involved in equations (27-45) through (27-50), and the ATAN2 function of Fortran should be used with (27-46), but not (27-49), adding or subtracting 360° to or from λ if necessary in (27-45) to place it between longs. 180° E. and W.

Iteration is required to find λ'' from x and y:

$$\lambda'' = [x/a + (S/J)(y/a) - A_2 \sin 2\lambda'' - A_4 \sin 4\lambda''$$
$$- (S/J)(C_1 \sin \lambda'' + C_3 \sin 3\lambda'')]/B \tag{27-51}$$

using equation (27-30) and various constants. Iteration involves substitution of a trial $\lambda'' = x/a\ B$ in the right side, finding a new λ'' on the left side, etc.

For ϕ'', the λ'' just calculated is used in the following equation:

$$\ln \tan(\pi/4 + \phi''/2) = (1 + S^2/J^2)^{1/2}\ (y/a - C_1 \sin \lambda'' - C_3 \sin 3\lambda'') \tag{27-52}$$

The closed forms of equations (27-51) and (27-52) involve both iteration and repeated numerical integration and are impractical:

$$x/a + (S/J)(y/a) = \int_0^{\lambda''}[(HJ - S^2)/(J^2 + S^2)^{1/2}]d\lambda''$$
$$+ (S/J)\int_0^{\lambda''}[S(H + J)/(J^2 + S^2)^{1/2}]d\lambda'' \tag{27-51a}$$
$$\ln \tan(\pi/4 + \phi''/2) = [1 + (S/J)^2]^{1/2}\{y/a - \int_0^{\lambda''}[S(H + J)/$$
$$(J^2 + S^2)^{1/2}]d\lambda''\} \tag{27-52a}$$

For ϕ' and λ' in terms of ϕ'' and λ'', the same Fourier series developed for equations (27-39) and (27-40) may be used with reversal of signs, since the correction is so small. That is,

$$\phi' = \phi'' - j_1 \sin \lambda'' - j_3 \sin 3\lambda'' - \ldots \tag{27-53}$$
$$\lambda' = \lambda'' - m_2 \sin 2\lambda'' - m_4 \sin 4\lambda'' - \ldots \tag{27-54}$$

Equations (27-53) and (27-54) are, of course, not the exact inverses of (27-39) and (27-40), although the correct coefficients may be derived by an analogous numerical integration in terms of λ'', rather than λ'. The inverse values of ϕ' and λ' from (27–53) and (27–54) are within 0.000003° and 0.000009°, respectively, of the true inverses of (27–39) and (27–40) for the Landsat orbits.

The following values of Fourier coefficients for the ellipsoidal SOM are listed for Landsat orbits, using the Clarke 1866 ellipsoid ($a = 6,378,206.4$ m and $e^2 = 0.00676866$) and a circular orbit:

	Landsat 1, 2, 3	Landsat 4, 5	
$B =$	1.005798138	1.004560314	for λ'' in radians
$=$	0.0175544891	0.017532885	for λ'' in degrees
$A_2 =$	-0.0010979201	-0.0009425101	
$A_4 =$	-0.0000012928	-0.0000012678	
$A_6 =$	-0.0000000021	-0.0000000021	
$C_1 =$	0.1434409899	0.1375926735	
$C_3 =$	0.0000285091	0.0000299489	
$C_5 =$	-0.0000000011	0.0000000004	
$R =$	7,294,690	7,081,000	meters
$i =$	99.092°	98.20°	
$P_2/P_1 =$	18/251	16/233	
$j_1 =$	0.00855567	0.00619893	for ϕ'' and ϕ' in degrees
$j_3 =$	0.00081784	0.00061698	"
$j_5 =$	-0.00000263	-0.00000308	"
$m_2 =$	-0.02384005	-0.01901574	for λ'' and λ' in degrees
$m_4 =$	0.00010606	0.00011587	"
$m_6 =$	0.00000019	0.00000024	"

Additional Fourier constants have been developed in the published literature for other functions of circular orbits. They add to the complication of the equations, but not to the accuracy, and only slightly to continuous mapping efficiency. A further simplification from published formulas is the elimination of a function F, which nearly cancels out in the range involved in imaging.

As in the spherical form of SOM, the formulas for scale factors h and k and maximum angular deformation ω are too lengthy to include here, although they are given by Snyder (1981b). Table 37 presents these values for Landsat constants for the scanning range required. Values for Landsat 4 and 5 are nearly identical with those shown for 1, 2, and 3.

FORMULAS FOR THE ELLIPSOID AND NONCIRCULAR ORBIT

The following formulas accommodate a slight ellipticity in the satellite orbit. They provide a true-to-scale groundtrack for an orbit of any eccentricity, if the orbital motion follows Kepler's laws for two-bodied systems, but the areas scanned by the satellite as shown on the map are distorted beyond the accuracy desired if the eccentricity of the orbit exceeds about 0.05, well above the maximum reported eccentricity of Landsat orbits (about 0.002). For greater eccentricities, more complex formulas (Snyder, 1981b) are recommended. If the orbital eccentricity is made zero, these formulas readily reduce to those for a circular orbit.

For the forward formulas, given a, e, i, P_2, P_1, λ_0, a', e', γ, ϕ, and λ, find x and y. Again, the X axis passes through each ascending and descending node, x increasing in the direction of satellite motion, and the Y axis intersects perpendicularly at the ascending node for the time $t = 0$. Defining terms,

a', e' = semimajor axis and eccentricity of satellite orbit, respectively.
γ = longitude of the perigee relative to the ascending node.
a and e are as defined for the ellipsoidal/circular formulas, and i, P_2, P_1, λ_0, ϕ,

TABLE 37.—*Scale factors for the ellipsoidal Space Oblique Mercator projection using Landsat 1, 2, and 3 constants*

λ''	ϕ''	h	k	$\omega°$	$\sin \frac{1}{2} \omega$
0°	1°	1.000154	1.000151	0.0006	0.000005
	0	1.000000	1.000000	.0000	.000000
	−1	1.000154	1.000151	.0006	.000005
15	1	1.000161	1.000151	.0022	.000019
	0	1.000000	1.000000	.0001	.000000
	−1	1.000147	1.000151	.0011	.000010
30	1	1.000167	1.000150	.0033	.000029
	0	1.000000	1.000000	.0001	.000001
	−1	1.000142	1.000150	.0025	.000021
45	1	1.000172	1.000150	.0036	.000031
	0	.999999	1.000000	.0001	.000001
	−1	1.000138	1.000150	.0031	.000027
60	1	1.000174	1.000150	.0031	.000027
	0	.999999	1.000000	.0002	.000001
	−1	1.000136	1.000150	.0028	.000025
75	1	1.000174	1.000152	.0019	.000016
	0	.999999	1.000000	.0001	.000000
	−1	1.000135	1.000150	.0019	.000016
90	1	1.000170	1.000156	.0008	.000007
	0	.999999	1.000000	.0000	.000000
	−1	1.000133	1.000151	.0010	.000009

Notes: λ' = angular position along geocentric groundtrack, from ascending node.

ϕ'' = angular distance away from geocentric groundtrack, positive in direction away from North Pole.

h = scale factor along meridian of longitude.

k = scale factor along parallel of latitude.

ω = maximum angular deformation.

$\sin \frac{1}{2} \omega$ = maximum variation of scale factors from true conformal values.

and λ are as defined for the spherical SOM formulas. For constants applying to the entire map (a numerical example is not given for the non-circular orbit):

$$B_1 = [1/(2\pi)]\int_0^{2\pi}[(HJ-S^2)/(J^2+S^2)^{1/2}]d\lambda'' \qquad (27-55)$$

$$B_2 = [1/(2\pi)]\int_0^{2\pi}[S(H+J)/(J^2+S^2)^{1/2}]d\lambda'' \qquad (27-56)$$

$$A_n = [1/(\pi n)]\int_0^{2\pi}[(HJ-S^2)/(J^2+S^2)^{1/2}]\cos n\lambda''d\lambda'' \qquad (27-57)$$

$$A'_n = [1/(\pi n)]\int_0^{2\pi}[(HJ-S^2)/(J^2+S^2)^{1/2}]\sin n\lambda''d\lambda'' \qquad (27-58)$$

$$C_n = [1/(\pi n)]\int_0^{2\pi}[S(H+J)/(J^2+S^2)^{1/2}]\cos n\lambda''d\lambda'' \qquad (27-59)$$

$$C'_n = [1/(\pi n)]\int_0^{2\pi}[S(H+J)/(J^2+S^2)^{1/2}]\sin n\lambda''d\lambda'' \qquad (27-60)$$

$$J = (1-e^2)^3 \qquad (27-24)$$

$$W = [(1-e^2\cos^2 i)^2/(1-e^2)^2]-1 \qquad (27-25)$$

$$Q = e^2\sin^2 i/(1-e^2) \qquad (27-26)$$

$$T = e^2\sin^2 i\,(2-e^2)/(1-e^2)^2 \qquad (27-27)$$

$$H_1 = B_1/(B_1^2+B_2^2)^{1/2} \qquad (27-61)$$

$$S_1 = B_2/(B_1^2+B_2^2)^{1/2} \qquad (27-62)$$

$$j_n = (1/\pi)\int_0^{2\pi}\phi''\sin n\lambda'd\lambda' \qquad (27-63)$$

$$j'_n = (1/\pi)\int_0^{2\pi}\phi''\cos n\lambda'd\lambda' \qquad (27-64)$$

$$m_n = (1/\pi)\int_0^{2\pi}(\lambda''-\lambda')\sin n\lambda'd\lambda' \qquad (27-65)$$

$$m'_n = (1/\pi)\int_0^{2\pi}(\lambda''-\lambda')\cos n\lambda'd\lambda' \qquad (27-66)$$

where ϕ'' and λ'' are determined in these last four equations for the groundtrack as functions of λ', from equations (27−69a), (27−87), (27−86), (27−85), (27−88), and (27−34), (27−74) through (27−76), and (27−36).

To calculate A_n, A'_n, B_n, C_n, and C'_n, the following functions, varying with λ'', are used:

$$S = (P_2/P_1)L'\sin i\cos\lambda''[(1 + T\sin^2\lambda'')/[(1 + W\sin^2\lambda'')$$
$$(1+Q\sin^2\lambda'')]]^{1/2} \qquad (27-67)$$

$$H = \left[\frac{1 + Q \sin^2 \lambda''}{1 + W \sin^2 \lambda''}\right]^{1/2} \left[\frac{1 + W \sin^2 \lambda''}{(1 + Q \sin^2 \lambda'')^2} - (P_2/P_1)L' \cos i\right] \qquad (27-67a)$$

$$L' = (1 - e' \cos E')^2/(1 - e'^2)^{1/2} \qquad (27-68)$$
$$E' = 2 \arctan \{\tan [(\lambda'' - \gamma)/2] [(1 - e')/(1 + e')]^{1/2}\} \qquad (27-69)$$

These constants may be determined from numerical integration, using Simpson's rule with 9° intervals. Unlike the case for circular orbits, integration must occur through the 360° or 2π cycle, as indicated. Many more terms are needed than for circular orbits.

To find x and y from ϕ and λ:

$$x/a = x'H_1 + y'S_1 \qquad (27-70)$$
$$y/a = y'H_1 - x'S_1 \qquad (27-71)$$

where

$$x' = B_1\lambda'' + \sum_{n=1}^{n} A_n \sin n\lambda'' - \sum_{n=1}^{n} A'_n \cos n\lambda'' + \sum_{n=1}^{n} A'_n - [S/(J^2 + S^2)^{1/2}]$$
$$\ln \tan (\pi/4 + \phi''/2) \qquad (27-72)$$

$$y' = B_2\lambda'' + \sum_{n=1}^{n} C_n \sin n\lambda'' - \sum_{n=1}^{n} C'_n \cos n\lambda'' + \sum_{n=1}^{n} C'_n + [J/(J^2 + S^2)^{1/2}]$$
$$\ln \tan (\pi/4 + \phi''/2) \qquad (27-73)$$

$$\lambda'' = \arctan [\cos i \tan \lambda_t + (1 - e^2) \sin i \tan \phi/\cos \lambda_t] \qquad (27-34)$$
$$\lambda_t = \lambda - \lambda_0 + (P_2/P_1)(L + \gamma) \qquad (27-74)$$
$$L = E' - e' \sin E' \qquad (27-75)$$
$$E' = 2 \arctan \{\tan [(\lambda'' - \gamma)/2] [(1 - e')/(1 + e')]^{1/2}\} \qquad (27-76)$$
$$\phi'' = \arcsin \{[(1 - e^2) \cos i \sin \phi - \sin i \cos \phi \sin \lambda_t]/$$
$$(1 - e^2 \sin^2 \phi)^{1/2}\} \qquad (27-36)$$

Function E' is called the "eccentric anomaly" along the orbit, and L is the "mean anomaly" or mean longitude of the satellite measured from perigee and directly proportional to time.

Equations (27-34), (27-74) through (27-76), and (27-36) are solved by special iteration as described for equations (27-8) and (27-9) in the spherical formulas, except that λ'' replaces λ', and each trial λ'' is placed in (27-76), from which E' is calculated, then L from (27-75), λ_t from (27-74), and another trial λ'' from (27-34). This cycle is repeated until λ'' changes by less than the selected convergence. The last value of λ_t found is then used in (27-36) to find ϕ''.

Equation (27-67) is used to find S for the given λ'' in equations (27-72) and (27-73).

The closed forms of equations (27-72) and (27-73) are (27-32a) and (27-33a), respectively, in which the repeated numerical integration necessitates replacement by the series forms.

As in the case of the circular orbit, it is also desirable to relate these points to the true vertical groundtrack. To find ϕ'' and λ'' in terms of ϕ' and λ', the following series are employed:

$$\phi'' = \phi' + \sum_{n=1}^{n} j_n \sin n\lambda' + \sum_{n=1}^{n} j_n' \cos n\lambda' - \sum_{n=1}^{n} j'_n \qquad (27-77)$$

$$\lambda'' = \lambda' + \sum_{n=1}^{n} m_n \sin n\lambda' + \sum_{n=1}^{n} m'_n \cos n\lambda' - \sum_{n=1}^{n} m'_n \qquad (27-78)$$

For λ' in terms of time t from the initial ascending node,

$$\lambda' = \gamma + 2 \arctan \{[\tan (E'/2)] [(1 + e')/(1 - e')]^{1/2}\} \qquad (27-79)$$
$$E' = e' \sin E' + L_0 + 2\pi t/P_2 \qquad (27-80)$$

$$L_0 = E'_0 - e' \sin E'_0 \qquad (27\text{-}81)$$
$$E'_0 = -2 \arctan \{[\tan (\gamma/2)] [(1-e')/(1+e')]^{1/2}\} \qquad (27\text{-}82)$$

Equation (27–80) requires iteration, converging rapidly by substituting an initial trial $E' = L_0 + 2\pi t/P_2$ in the right side, finding a new E' on the left, substituting it on the right, etc., until sufficient convergence occurs.

The equations for functions of the satellite groundtrack, both forward and inverse, are given here, since some are used in calculating j_n and m_n as well. In any case a, e, i, P_2, P_1, λ_0, a', e', and γ must be given. For λ' and λ, if ϕ is given:

$$\lambda' = \arcsin (\sin \phi_g/\sin i) \qquad (27\text{-}83)$$
$$\phi_g = \phi - \arcsin \{ae^2 \sin \phi \cos \phi/[R_0(1-e^2 \sin^2 \phi)^{1/2}]\} \qquad (27\text{-}84)$$
$$R_0 = a' (1-e' \cos E') \qquad (27\text{-}85)$$
$$E' = 2 \arctan \{\tan [(\lambda'-\gamma)/2] [(1-e')/(1+e')]^{1/2}\} \qquad (27\text{-}69a)$$

where ϕ_g is the geocentric latitude of the point geocentrically under the satellite, not the geocentric latitude corresponding to the vertical groundtrack latitude ϕ, and R_0 is the radius vector to the satellite from the center of the Earth.

These equations are solved as a group by iteration, inserting a trial $\lambda' = \arcsin$ $(\sin \phi/\sin i)$ in (27–69a), solving (27–85), (27–84), and (27–83) for a new λ', etc. Each trial λ' must be adjusted for quadrant. If the satellite is traveling north, add 360° times the number of orbits completed at the nearest ascending node (0, 1, 2, etc.). If traveling south, subtract λ' from 360° times the number of orbits completed at the nearest descending node (1/2, 3/2, 5/2, etc.). For λ,

$$\lambda = \arctan[(\cos i \sin \lambda')/\cos \lambda'] - (P_2/P_1)(L+\gamma) + \lambda_0 \qquad (27\text{-}86)$$
$$L = E' - e' \sin E' \qquad (27\text{-}87)$$

using the λ' and E' finally found just above.

If λ of a point along the groundtrack is given, to find λ' and ϕ,

$$\lambda' = \arctan (\tan \lambda_t/\cos i) \qquad (27\text{-}19)$$
$$\lambda_t = \lambda - \lambda_0 + (P_2/P_1)(L+\gamma) \qquad (27\text{-}74)$$

and L is found from (27–87) and (27–69a) above. The four equations are iterated as a group, as above, but the first trial λ' and the quadrant adjustments should follow the procedures listed for equation (27–8).

$$\phi = \arcsin (\sin i \sin \lambda') + \arcsin \{ae^2 \sin \phi \cos \phi/ [R_0(1-e^2 \sin^2 \phi)^{1/2}]\} \qquad (27\text{-}88)$$

where R_0 is determined from (27–85) and (27–69a), using the λ' determined just above. Iteration is involved in (27–88), beginning with a trial ϕ of arcsin (sin i sin λ').

If λ' of a point along the groundtrack is given, ϕ is found from (27–88), (27–85), and (27–69a), while λ is found from (27–86), (27–87), and (27–69a). Only (27–88) requires iteration for these calculations.

Inverse formulas for the ellipsoidal form of the SOM projection, with an orbit of 0.05 eccentricity or less, follow: Given a, e, i, P_2, P_1, λ_0, a', e', γ, x, and y, to find ϕ and λ, the general Fourier and other constants are first determined as described at the beginning of the forward equations for noncircular orbits. Then

$$\lambda = \lambda_t - (P_2/P_1) (L+\gamma) + \lambda_0 \qquad (27\text{-}89)$$

where

$$\lambda_t = \arctan (V/\cos \lambda'') \tag{27-46}$$

$$V = \{[1-\sin^2 \phi''/(1-e^2)] \cos i \sin \lambda'' - \sin i \sin \phi'' [(1+Q \sin^2 \lambda'')$$
$$(1-\sin^2 \phi'') - U \sin^2 \phi'']^{1/2}\}/[1-\sin^2 \phi'' (1+U)] \tag{27-47}$$

$$U = e^2 \cos^2 i/(1-e^2) \tag{27-48}$$

while L is found from (27−87), E' from (27−76), and λ'' and ϕ'' from (27−90) and (27−91) below.

$$\phi = \arctan \{(\tan \lambda'' \cos \lambda_t - \cos i \sin \lambda_t)/[(1-e^2) \sin i]\} \tag{27-49}$$

If $i = 0$, equation (27−49) is indeterminate, but

$$\phi = \arcsin \{\sin \phi''/[(1-e^2)^2 + e^2 \sin^2 \phi'']^{1/2}\} \tag{27-50}$$

No iteration is involved in the above equations, and the ATAN2 function of Fortran should be used with (27−46), but not (27−49), adding or subtracting 360° to or from λ if necessary in (27−89) to place it between longs. 180°E. and W.

Iteration is required to find λ'' from x and y:

$$\lambda'' = \{x' + (S/J) y' - \sum_{n=1}^{n}[A_n + (S/J)C_n] \sin n \lambda'' + \sum_{n=1}^{n}[A'_n + (S/J)C'_n]$$
$$\cos n\lambda'' - \sum_{n=1}^{n}[A'_n + (S/J)C'_n]\}/[B_1 + (S/J)B_2] \tag{27-90}$$

using equations (27−67), (27−92), (27−93), and various constants. Iteration involves substitution of a trial $\lambda'' = x'/B_1$ in the right side, finding a new λ'' on the left side, etc.

For ϕ'', the λ'' just calculated is used in the following equation:

$$\ln \tan (\pi/4 + \phi''/2) = (1+S^2/J^2)^{1/2}(y' - B_2 \lambda'' - \sum_{n=1}^{n} C_n \sin n \lambda'' + \sum_{n=1}^{n}$$
$$C'_n \cos n \lambda'' - \sum_{n=1}^{n} C'_n) \tag{27-91}$$

where

$$x' = (x/a) H_1 - (y/a)S_1 \tag{27-92}$$
$$y' = (y/a) H_1 + (x/a)S_1 \tag{27-93}$$

The closed forms of equations (27−90) and (27−91) involve both iteration and repeated numerical integration and are impractical:

$$x' + (S/J)y' = \int_0^{\lambda''}[(HJ-S^2)/(J^2+S^2)^{1/2}]d\lambda''$$
$$+ (S/J)\int_0^{\lambda''}[S(H+J)/(J^2+S^2)^{1/2}]d\lambda'' \tag{27-90a}$$

$$\ln \tan (\pi/4 + \phi''/2) = [1 + (S/J)^2]^{1/2}\{y' - \int_0^{\lambda''}[S(H+J)/$$
$$(J^2+S^2)^{1/2}]d\lambda''\} \tag{27-91a}$$

For ϕ' and λ' in terms of ϕ'' and λ'', the same Fourier series developed for equations (27−77) and (27−78) may be used with reversal of signs, since the correction is so small. That is,

$$\phi' = \phi'' - \sum_{n=1}^{n} j_n \sin n \lambda'' - \sum_{n=1}^{n} j'_n \cos n \lambda'' + \sum_{n=1}^{n} j'_n \tag{27-94}$$

$$\lambda' = \lambda'' - \sum_{n=1}^{n} m_n \sin n \lambda'' - \sum_{n=1}^{n} m'_n \cos n \lambda'' + \sum_{n=1}^{n} m'_n \tag{27-95}$$

As with the circular orbit, equations (27−94) and (27−95) are not the exact inverses of (27−77) and (27−78), although the correct coefficients may be derived by an analogous numerical integration in terms of λ'', rather than λ'.

28. SATELLITE-TRACKING PROJECTIONS

SUMMARY

- All groundtracks for satellites orbiting the Earth with the same orbital parameters are shown as straight lines on the map.
- Cylindrical or conical form available.
- Neither conformal nor equal-area.
- All meridians are equally spaced straight lines, parallel on cylindrical form and converging to a common point on conical form.
- All parallels are straight and parallel on cylindrical form and are concentric circular arcs on conical form. Parallels are unequally spaced.
- Conformality occurs along two chosen parallels. Scale is correct along one of these parallels on the conical form and along both on the cylindrical form.
- Developed 1977 by Snyder.

HISTORY, FEATURES, AND USAGE

The Landsat mapping system which inspired the development of the Space Oblique Mercator (SOM) projection also inspired the development of a simpler type of projection with a different purpose. While the SOM is used for low-distortion mapping of the strips scanned by the satellite, the Satellite-Tracking projections are designed solely to show the groundtracks for these or other satellites as straight lines, thus facilitating their plotting on a map. As a result, the other features of such maps are minimal, although they may be designed to reduce overall distortion in particular regions.

The writer developed the formulas in 1977 after essentially completing the mathematical development of the formulas for the SOM. The formulas for the Satellite-Tracking projections, with derivations, were published later (Snyder, 1981a). Arnold (1984) further analyzed the distortion. These formulas are confined to circular orbits and the spherical Earth. Because of the small-scale maps resulting, the ellipsoidal forms are hardly justified.

Charts of groundtracks have to date continued to employ the Lambert Conformal Conic projection, on which the groundtracks are slightly curved. The writer is not aware of any use of the new projection, except that a Chinese map of about 1982 claims this feature.

The projections were developed in two basic forms, the cylindrical and the conic, with variations of features within the latter category. The cylindrical form (fig. 48) has straight parallel equidistant meridians and straight parallels of latitude which are perpendicular to the meridians. The parallels of latitude are increasingly spaced away from the Equator, and for Landsat orbits the spacing changes more rapidly than it does on the Mercator projection. The Equator or two parallels of latitude equidistant from the Equator may be made standard, without shape or scale distortion, as on several other cylindrical projections.

The groundtracks for the various orbits are plotted on the cylindrical form as diagonal equidistant straight lines. The descending orbital groundtracks (north to south) are parallel to each other, and the ascending groundtracks (south to north) are parallel to each other but with a direction in mirror image to that of the descending lines. The ascending and descending groundtracks meet at the northern and southern tracking limits, lats. 80.9° N. and S. for Landsat 1, 2, and 3. The map projection does not extend closer to the poles, although the mapmaker can arbitrarily extend the map using any convenient projection. The extension does not affect the purpose of the projection.

The groundtracks are not shown at constant scale, just as the straight great-circle paths on the Gnomonic and straight rhumb lines on the Mercator projection are not at constant scale. The complete tracks appear to be a sequence of zig-zag lines, although for Landsat normally only the descending (daylight) groundtracks should be shown to reduce confusion, since interest is normally confined to them.

While the cylindrical form of the Satellite-Tracking projections is of more interest if much of the world is to be shown, the conic form applies to most continents and countries, just as do the usual cylindrical and conic projections. On each conic Satellite-Tracking projection, the meridians are equally spaced straight lines converging at a common point, and the parallels are unequally spaced circular arcs centered on the same point. There are three types of distortion patterns available with the conic form:

1. For the normal map (fig. 49) of a continent or country, there can be conformality or no shape distortion along two chosen parallels, but correct scale at only one of them. The groundtracks break at the closest tracking limit, but the map cannot be extended to the other tracking limit in many cases, since it extends infinitely before reaching that latitude.

2. If one of the parallels with conformality is made a tracking limit, the groundtracks do not break at this tracking limit, since there can be no distortion there (fig. 50).

3. If both parallels with conformality are made the same, the projection has just one standard parallel. If this parallel is made the tracking limit, the conic projection becomes the closest approximation to an azimuthal projection (fig. 51). For Landsat orbits, the cone constant of such a limiting projection is about 0.96, so the developed cone is about 4 percent less than a full circle, and the projection somewhat resembles a polar Gnomonic projection. With orbits of lower inclination, the approach to azimuthal becomes less.

For each of the conics, the straight groundtracks are equidistant, they have constant inclinations to each meridian being crossed at a given latitude on a given map, and they are not at constant scale. They are also all tangent to a circle slightly smaller than the latitude circle for the tracking limit in case 1 above, and tangent to the tracking limit itself in cases 2 and 3. As in the case of the cylindrical form, any extension of the map from the tracking limit to a pole is cosmetic and arbitrary, since the groundtracks do not pass through this region.

FORMULAS FOR THE SPHERE

Forward formulas (see p. 360 for numerical examples):

For the *Cylindrical Satellite-Tracking projection*, R, i, P_2, P_1, λ_0, ϕ_1, ϕ, and λ must be given, where

R =radius of the globe at the scale of the map.

i =angle of inclination between the plane of the Earth's Equator and the plane of the satellite orbit, measured counterclockwise from the Equator to the orbital plane at the ascending node (99.092° for Landsat 1, 2, 3; 98.20° for Landsat 4, 5).

P_2 =time required for revolution of the satellite (103.267 min for Landsat 1, 2, 3; 98.884 min. for Landsat 4, 5).

P_1 =length of Earth's rotation with respect to the precessed ascending node. For Landsat, the satellite orbit is Sun-synchronous; that is, it is always the same with respect to the Sun, equating P_1 to the solar day (1,440 min). The ascending node is the point on the satellite orbit at which the satellite crosses the Earth's equatorial plane in a northerly direction.

λ_0 =central meridian.

ϕ_1 =standard parallel (N. and S.).

(ϕ, λ) =geodetic latitude and longitude of point to be plotted on map.

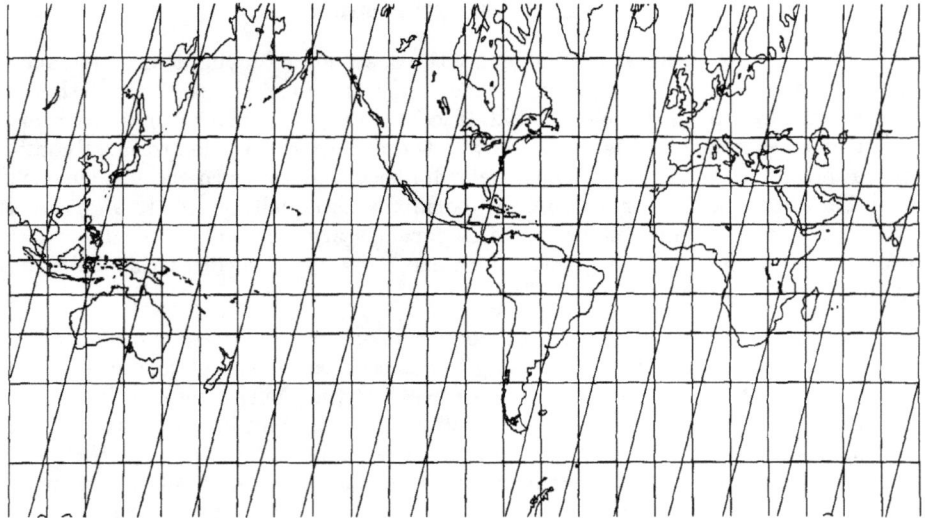

FIGURE 48.—Cylindrical Satellite-Tracking projection (standard parallels 30° N. and S.). Landsat 1, 2, 3 orbits. Groundtracks (paths 15, 30, 45, etc.) are shown as straight diagonal lines. They continue broken at tracking limits (not shown).

$$F_1' = [(P_2/P_1) \cos^2\phi_1 - \cos i]/(\cos^2\phi_1 - \cos^2 i)^{1/2} \qquad (28-1)$$

$$F' = [(P_2/P_1) \cos^2\phi - \cos i]/(\cos^2\phi - \cos^2 i)^{1/2} \qquad (28-1a)$$

$$\lambda' = -\arcsin(\sin\phi/\sin i) \qquad (28-2)$$

$$\lambda_t = \arctan(\tan\lambda' \cos i) \qquad (28-3)$$

$$L = \lambda_t - (P_2/P_1)\lambda' \qquad (28-4)$$

$$x = R(\lambda - \lambda_0)\cos\phi_1 \qquad (28-5)$$

$$y = R\,L\,\cos\phi_1/F_1' \qquad (28-6)$$

$$k = \cos\phi_1/\cos\phi \qquad (28-7)$$

$$h = k\,F'/F_1' \qquad (28-8)$$

Geometrically, F' is the tangent of the angle on the globe between the groundtrack and the meridian at latitude ϕ, and F_1' is the tangent of this angle both on the globe and on the map at latitude ϕ_1. Scale factors h and k apply along the meridian and parallel, respectively. If the latitude is closer to either pole than the corresponding tracking limit, equation (28-2) cannot be solved, and the point cannot be mapped using these formulas. The X axis lies along the Equator, x increasing easterly, and the Y axis lies along the central meridian, y increasing northerly. If $(\lambda - \lambda_0)$ lies outside the range $\pm 180°$, 360° should be added or subtracted so it will fall inside the range.

For the *Conic Satellite-Tracking projection* with two parallels having conformality, R, i, P_2, P_1, λ_0, ϕ_0, ϕ_1, ϕ_2, ϕ, and λ must be given, where the symbols are defined above, except that ϕ_2 is the other parallel of conformality, but without true scale, and ϕ_0 is the latitude crossing the central meridian at the desired origin of rectangular coordinates. For constants which apply to the entire map,

$$F_n = \arctan\{[(P_2/P_1)\cos^2\phi_n - \cos i]/(\cos^2\phi_n - \cos^2 i)^{1/2}\} \qquad (28-9)$$

$$\lambda'_n = -\arcsin(\sin\phi_n/\sin i) \qquad (28-2a)$$

$$\lambda_{tn} = \arctan(\tan\lambda_n' \cos i) \qquad (28-3a)$$

$$L_n = \lambda_{tn} - (P_2/P_1)\lambda_n' \qquad (28-4a)$$

$$n = (F_2 - F_1)/(L_2 - L_1) \qquad (28-10)$$

$$s_0 = F_1 - n\,L_1 \qquad (28-11)$$

$$\rho_0 = R\cos\phi_1 \sin F_1/[n\sin(nL_0 + s_0)] \qquad (28-12)$$

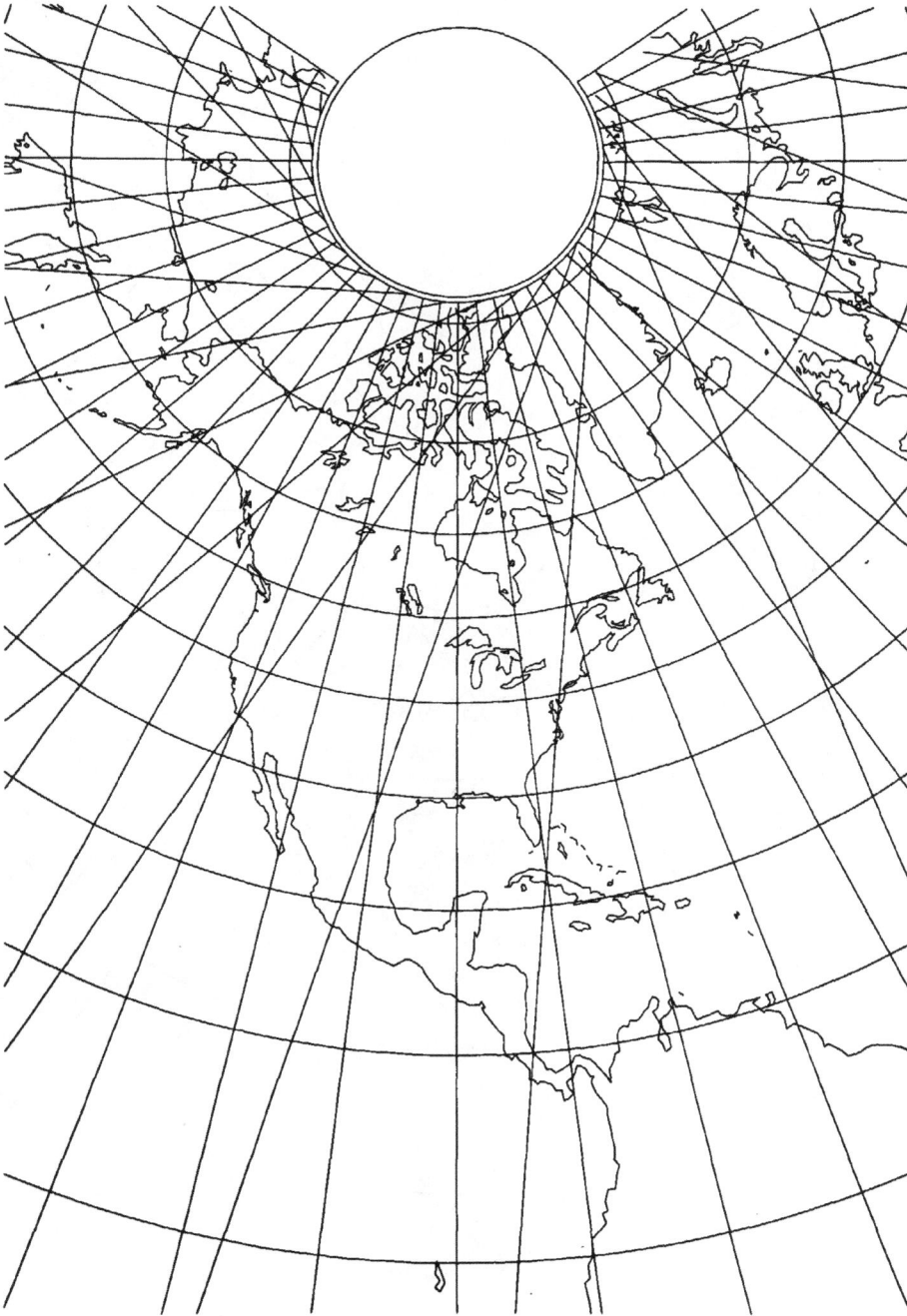

FIGURE 49.—Conic Satellite-Tracking projection (conformality at lats. 45° and 70° N.). Landsat 1, 2, 3 orbits. Groundtracks (paths 15, 30, 45, etc.) are shown as diagonal straight lines. They continue broken (not shown) at tracking limit, the smallest incomplete circle. The complete circle is the circle of tangency.

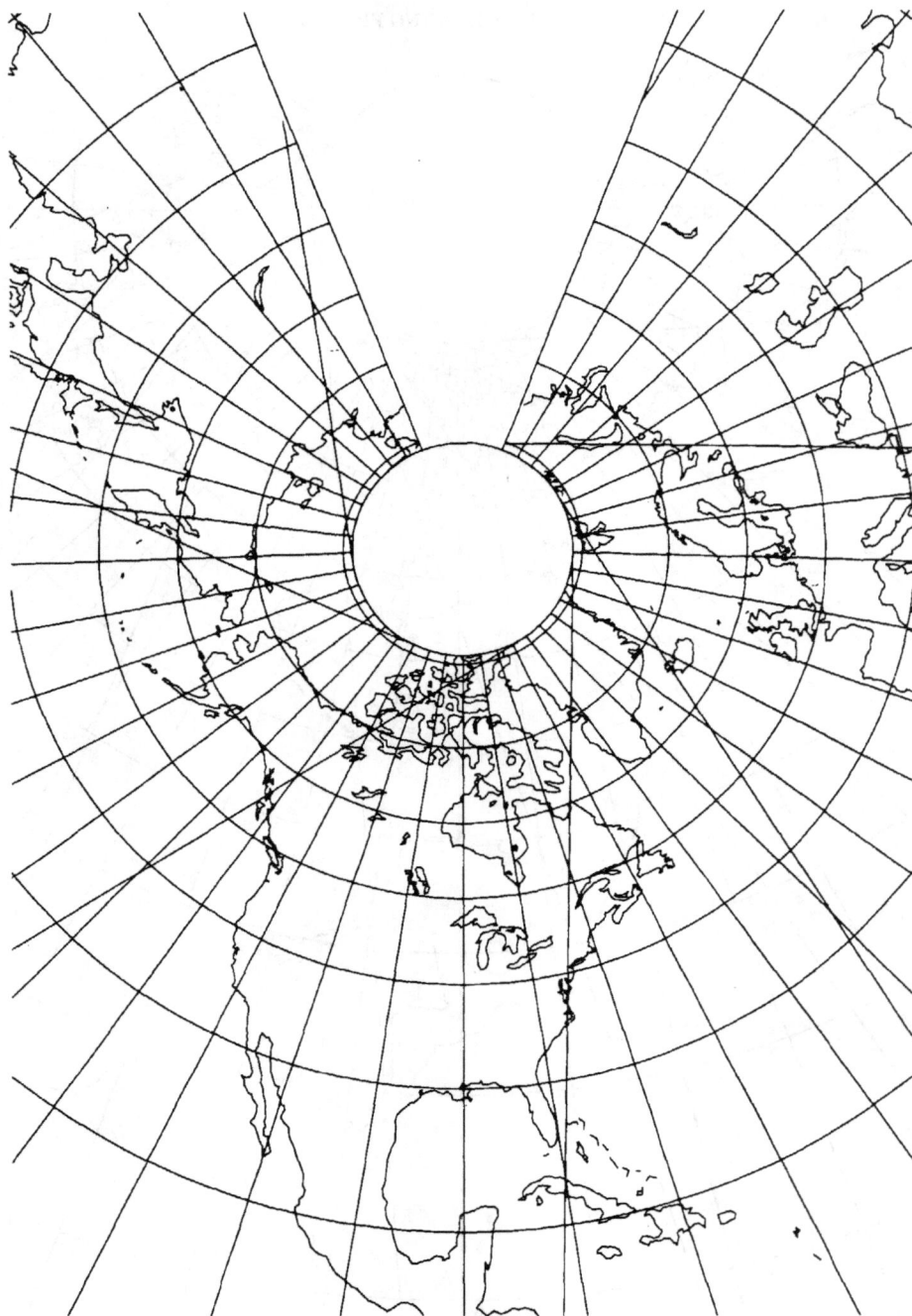

FIGURE 50.—Conic Satellite-Tracking projection (conformality at lats. 45° and 80.9° N.). Landsat 1, 2, 3 orbits. Diagonal groundtracks (paths 15, 60, 105, etc.) are straight, unbroken even at the tracking limit, which is the same as the circle of tangency (inner circle).

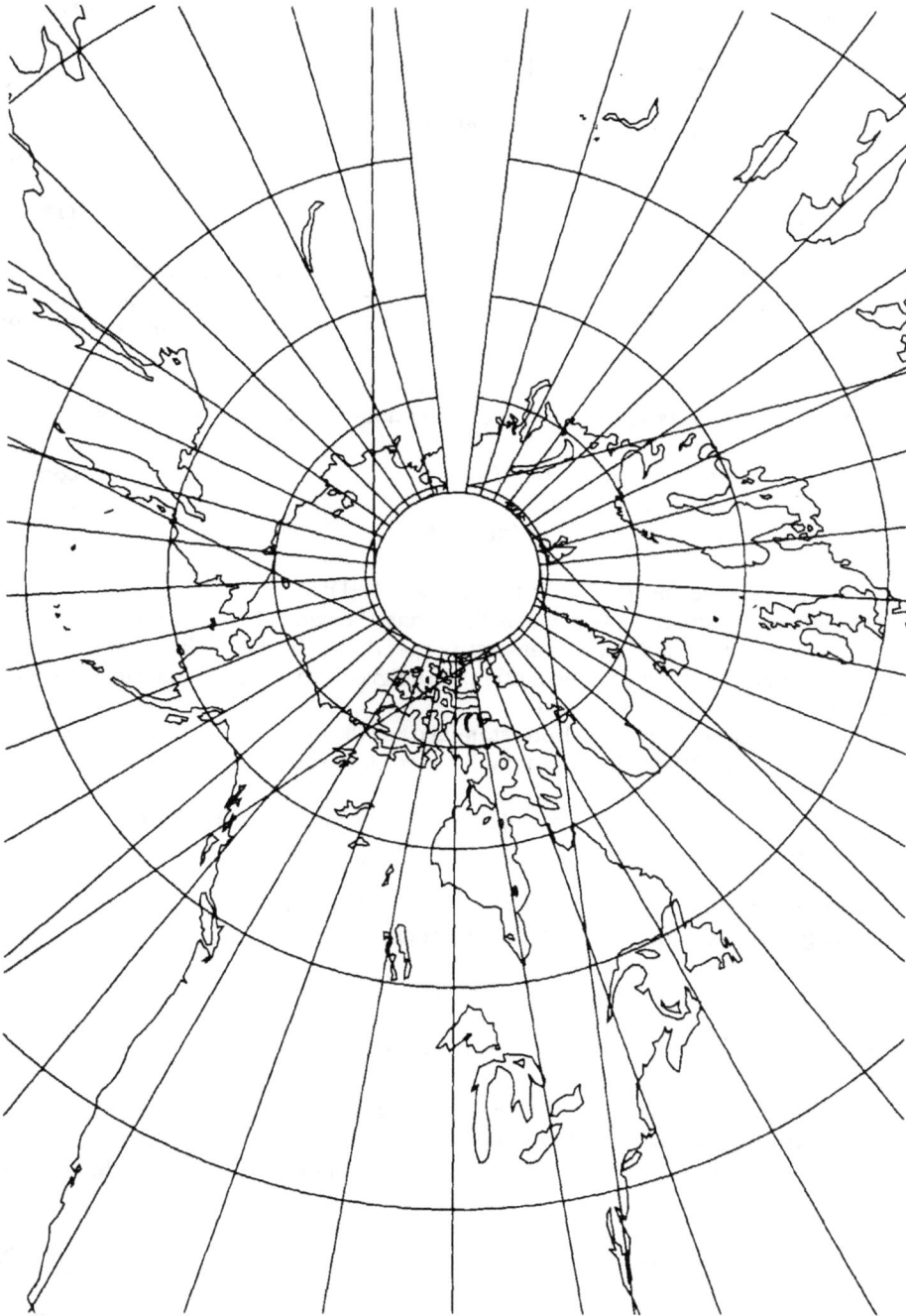

FIGURE 51.—Conic Satellite-Tracking projection (standard parallel 80.9° N.). Landsat 1, 2, 3 orbits. Groundtracks are as described on Fig. 50. The nearest approach to an azimuthal projection for these orbits. Inner circle is tracking limit and circle of tangency.

in which subscript n in equations (28–9) and (28–2a) through (28–4a) is made 0, 1, or 2 as required for (28–10) through (28–12), and subscript n is omitted for calculating F and L for formulas below.

For plotting each point (ϕ, λ),

$$\rho \quad =R \cos \phi_1 \sin F_1/[n \sin (nL + s_0)] \qquad (28–13)$$
$$\theta \quad =n (\lambda-\lambda_0) \qquad (14–4)$$
$$x \quad =\rho \sin \theta \qquad (14–1)$$
$$y \quad =\rho_0 - \rho \cos \theta \qquad (14–2)$$

If n is positive and L is equal to or less than $(-s_0/n)$, or if n is negative and L is equal to or greater than $(-s_0/n)$, the point cannot or should not be plotted. The limiting latitude ϕ for $L=(-s_0/n)$ may be found using (28–20) through (28–22) below.

In addition, ρ_s, the radius of the circle to which groundtracks are tangent on the map, and scale factors h and k, defined above, are found as follows:

$$\rho_s \quad =R \cos \phi_1 (\sin F_1)/n \qquad (28–14)$$
$$k \quad =\rho n/(R \cos \phi) \qquad (28–15)$$
$$h \quad =k \tan F/\tan (nL + s_0) \qquad (28–16)$$

Radius ρ_s may be inserted into equations (14–1) and (14–2) in place of ρ for rectangular coordinates. The Y axis lies along the central meridian λ_0, y increasing northerly, and the X axis intersects perpendicularly at ϕ_0, x increasing easterly. Geometrically, F_1 is the inclination of the groundtrack to the meridian at latitude ϕ_1, and n is the cone constant.

For the conic projection with one standard parallel, $\phi_1 = \phi_2$, but equation (28–10) is indeterminate. The following may be used in its place:

$$n = \sin \phi_1 [(P_2/P_1)(2 \cos^2 i-\cos^2 \phi_1) - \cos i]/\{[(P_2/P_1) \cos^2 \phi_1-\cos i]$$
$$[(P_2/P_1)[(P_2/P_1) \cos^2 \phi_1 - 2 \cos i] + 1]\} \qquad (28–17)$$

For the conic projection with one standard parallel ϕ_1 which is equal to the upper tracking limit, equation (28–17) may be considerably simplified to the following:

$$n = \sin i/[(P_2/P_1) \cos i - 1]^2 \qquad (28–18)$$

Other equations for the conic form remain the same.

Inverse Formulas (see p. 362 for numerical examples):

For the *cylindrical* form, the same constants must be given as those listed for the forward formulas (R, i, P_2, P_1, λ_0, and ϕ_1), and F_1' must be calculated from equation (28–1). For a given (x, y), to find (ϕ, λ):

$$L \quad =y F_1'/(R \cos \phi_1) \qquad (28–19)$$
$$\lambda_t \quad =L + (P_2/P_1) \lambda' \qquad (28–20)$$
$$\lambda' \quad =\arctan (\tan \lambda_t/\cos i) \qquad (28–21)$$
$$\phi \quad = - \arcsin (\sin \lambda' \sin i) \qquad (28–22)$$
$$\lambda \quad = \lambda_0 + x/(R \cos \phi_1) \qquad (28–23)$$

Equations (28–20) and (28–21) must be iterated as a pair, using ($-90°$) as the first trial λ' in equation (28–20), solving for λ_t, inserting it into (28–21), finding a new λ' without using the equivalent of the Fortran ATAN2 function, and using it in (28–20), until λ' changes by a negligible amount. This final λ' is used in (28–22) to find ϕ.

A generally faster solution of (28-20) and (28-21) involves the use of a Newton-Raphson iteration in place of those two equations, although equations are longer:

$$A = \tan[L + (P_2/P_1)\lambda']/\cos i \qquad (28-24)$$

$$\Delta\lambda' = -(\lambda' - \arctan A)/[1 - (A^2 + 1/\cos^2 i)(P_2/P_1)\cos i/(A^2 + 1)] \qquad (28-25)$$

The first trial λ' is again $(-90°)$ in equation (28-24) and (28-25). The adjustment $\Delta\lambda'$ is added to each successive trial until reasonable convergence occurs.

For any of the *conic* forms, the initial constants R, i, P_2, P_1, λ_0, ϕ_0, and ϕ_1 alone or both ϕ_1 and ϕ_2 must be given. In addition, all constants in equations (28-9) through (28-12), (28-2a) through (28-4a), and (28-17) or (28-18) if necessary must be calculated. For a given (x, y), to find (ϕ, λ),

$$\rho = \pm[x^2 + (\rho_0 - y)^2]^{1/2}, \text{ taking the sign of } n \qquad (14-10)$$

$$\theta = \arctan[x/(\rho_0 - y)] \qquad (14-11)$$

$$L = [\arcsin(R\cos\phi_1\sin F_1/(\rho n)) - s_0]/n \qquad (28-26)$$

From L, λ' and then ϕ are found using equations (28-20) through (28-22), or (28-24), (28-25), and (28-22), with iteration as described above. Then

$$\lambda = \lambda_0 + \theta/n \qquad (14-9)$$

Sample coordinates for several of the Satellite-Tracking projections are shown in tables 38 through 40.

TABLE 38.—*Cylindrical Satellite-Tracking projection: Rectangular coordinates*

Landsat 1, 2, 3 orbits: $i = 99.092°$
$P_2 = 103.267$ min.
$P_1 = 1440.0$ min.
Globe radius: $R = 1.0$

ϕ_1 F_1 x	0° 13.09724° 0.017453λ°			±30° 13.96868° 0.015115λ°			±45° 15.71115° 0.012341λ°	
$\pm\phi$	$\pm y$	h	k	$\pm y$	h	k	$\pm y$	h
TL*	7.23571	∞	6.32830	5.86095	∞	5.48047	4.23171	∞
80°	5.35080	55.0714	5.75877	4.33417	44.6081	4.98724	3.12934	32.2078
70	2.34465	6.89443	2.92380	1.89918	5.58452	2.53209	1.37124	4.03212
60	1.53690	3.18846	2.00000	1.24489	2.58266	1.73205	0.89883	1.86473
50	1.09849	2.01389	1.55572	0.88979	1.63126	1.34730	0.64244	1.17780
40	0.79741	1.49787	1.30541	0.64591	1.21328	1.13052	0.46636	0.87601
30	0.56135	1.23456	1.15470	0.45470	1.00000	1.00000	0.32830	0.72202
20	0.35952	1.09298	1.06418	0.29121	0.88532	0.92160	0.21026	0.63921
10	0.17579	1.02179	1.01543	0.14239	0.82766	0.87939	0.10281	0.59758
0°	0.00000	1.00000	1.00000	0.00000	0.81000	0.86603	0.00000	0.58484

* Tracking limit, 80.908° = (180° − i)
See text for other symbols.

TABLE 39.—*Conic Satellite-Tracking projections with two conformal parallels: Polar coordinates*

Landsat 1, 2, 3 orbits (i, P_2, P_1 same as Table 38)
Globe radius: $R = 1.0$

ϕ_1 ϕ_2 n F_1 ρ_s	30° 60° 0.49073 13.96868° 0.42600			45° 70° 0.69478 15.71115° 0.27559			45° 80.908° 0.88475 15.71115° 0.21642	
ϕ	ρ	h	k	ρ	h	k	ρ	h
TL*	0.50439	∞	1.56635	0.28663	∞	1.26024	0.21642	1.21172
80°	0.59934	3.72928	1.69373	0.33014	1.93850	1.32093	0.23380	1.08325
70	0.98470	1.61528	1.41283	0.57297	1.16394	1.16394	0.40484	0.90832
60	1.22500	1.20228	1.20228	0.75975	1.00596	1.05572	0.55875	0.87290
50	1.41806	1.03521	1.08260	0.93154	0.97914	1.00689	0.71504	0.93344
45	1.50659	0.99771	1.04556	1.01774	1.00000	1.00000	0.79921	1.00000
40	1.59281	0.98135	1.02035	1.10669	1.04212	1.00374	0.89042	1.09569
30	1.76478	1.00000	1.00000	1.30060	1.19708	1.04342	1.10616	1.40901
20	1.94551	1.08181	1.01599	1.53188	1.47984	¯1.13263	1.39852	2.00877
10	2.14662	1.23677	1.06965	1.82978	1.98371	1.29091	1.84527	3.28641
0	2.38332	1.49781	1.16956	2.25035	2.94795	1.56351	2.66270	6.72124
−10	2.67991	1.94172	1.33539	2.92503	5.10490	2.06361	4.79153	22.2902
−20°	3.08210	2.75586	1.60953	4.26519	11.6380	3.15356	29.3945	898.207
ML**	−60.65° (ρ = ∞)			−38.52° (ρ = ∞)			−21.86° (ρ = ∞)	

* Tracking limit, 80.908° = (180° − i)
** Minimum latitude, at infinite radius
See text for other symbols.

TABLE 40.—*Near-Azimuthal Conic Satellite-Tracking projection: Polar coordinates*

Landsat 1, 2, 3 orbits (i, P_2, P_1 same as Table 38)
Globe radius: $R = 1.0$
$\phi_1 = 80.908°$
$n = 0.96543$
$F_1 = \pm90°$
$\rho_s = 0.16368$

ϕ	ρ	h	k
TL*	0.16368	1.00000	1.00000
80°	0.17953	1.00076	0.99813
70	0.35986	1.09115	1.01579
60	0.57095	1.36647	1.10243
50	0.85650	1.99000	1.28641
40	1.31643	3.53452	1.65907
30	2.28682	8.83705	2.54931
20°	6.22402	58.0828	6.39449
ML**	13.70° (ρ = ∞)		

* Tracking limit, 80.908° = (180° − i)
** Minimum latitude, of infinite radius
See text for other symbols.

PSEUDOCYLINDRICAL AND MISCELLANEOUS
MAP PROJECTIONS

29. VAN DER GRINTEN PROJECTION

SUMMARY

- Neither equal-area nor conformal. Not pseudocylindrical.
- Shows entire globe enclosed in a circle.
- Central meridian and Equator are straight lines.
- All other meridians and parallels are arcs of circles.
- A curved modification of the Mercator projection, with great distortion in the polar areas.
- Equator is true to scale.
- Used for world maps.
- Used only in the spherical form.
- Presented by van der Grinten in 1904.

HISTORY, FEATURES, AND USAGE

In a 1904 issue of a German geographical journal, Alphons J. van der Grinten (1852–?) of Chicago presented four projections showing the entire Earth. Aside from having a straight Equator and central meridian, three of the projections consist of arcs of circles for meridians and parallels; the other projection has straight-line parallels. The projections are neither conformal nor equal-area (van der Grinten, 1904; 1905). They were patented in the United States by van der Grinten in 1904.

The best-known Van der Grinten projection, his first (fig. 52), shows the world in a circle and was invented in 1898. It is designed for use in the spherical form only. There are no special features to preserve in an ellipsoidal form. It has been used by the National Geographic Society for their standard world map since 1943, printed at various scales and with the central meridian either through America or along the Greenwich meridian; this use has prompted others to employ the projection. The U.S. Department of Agriculture adopted the projection as the base map for economic data in the 1940's, and this led to frequent use in geography textbooks (Wong, 1965, p. 117). The USGS has used one of the National Geographic maps as a base for a four-sheet set of maps of World Subsea Mineral Resources, 1970, one at a scale of 1:60,000,000 and three at 1:39,283,200 (a scale used by the National Geographic), and for three smaller maps in the *National Atlas* (USGS, 1970, p. 150–151, 332–335). All the USGS maps have a central meridian of long. 85° W., passing through the United States.

Van der Grinten emphasized that this projection blends the Mercator appearance with the curves of the Mollweide, an equal-area projection described later. He included a simple graphical construction and limited formulas showing the mathematical coordinates along the central meridian, the Equator, and the outer (180th) meridian. The meridians are equally spaced along the Equator, but the spacing between the parallels increases with latitude, so that the 75th parallels are shown about halfway between the Equator and the respective poles. Because of the polar exaggerations, most published maps using the Van der Grinten projection do not extend farther into the polar regions than the northern shores of Greenland and the outer rim of Antarctica.

The National Geographic Society prepared the base map graphically. General mathematical formulas have been published in recent years and are only useful with computers, since they are fairly complex for such a simply drawn projection (O'Keefe and Greenberg, 1977; Snyder, 1979b).

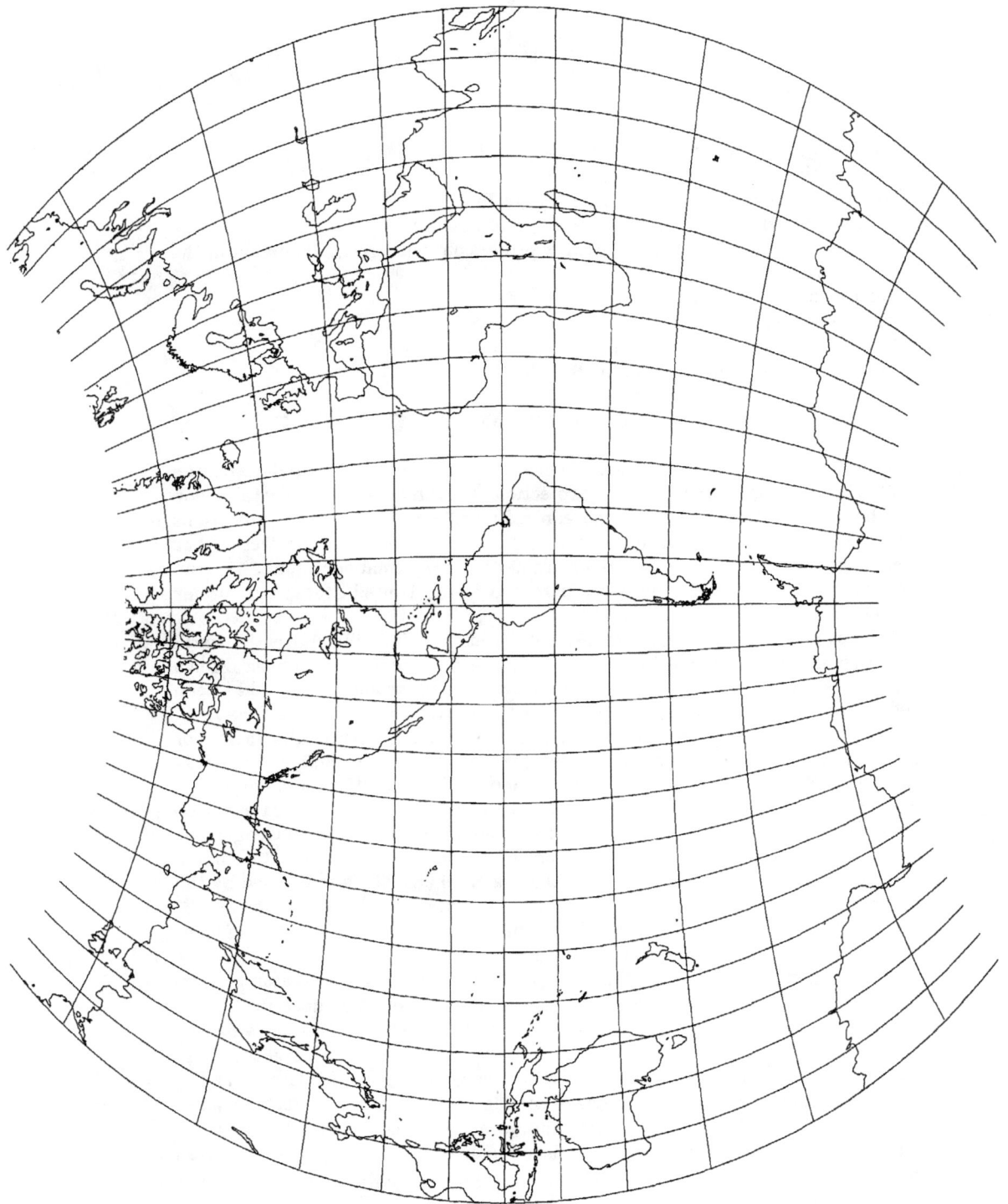

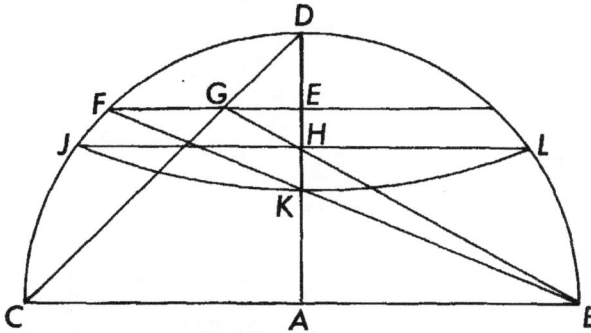

FIGURE 53.—Geometric construction of the Van der Grinten projection.

GEOMETRIC CONSTRUCTION

The meridians are circular arcs equally spaced on the Equator and joined at the poles. For parallels, referring to figure 53, semicircle CDB is drawn centered at A. Diagonal CD is drawn. Point E is marked so that the ratio of EA to AD is the same as the ratio of the latitude to 90°. Line FE is drawn parallel to CB, and FB and GB are connected. At H, the intersection of GB and AD, JHL is drawn parallel to CB. A circular arc, representing the parallel of latitude, is then drawn through JKL.

FORMULAS FOR THE SPHERE

The general formulas published are in two forms. Both sets give identical results, but the 1979 formulas are somewhat shorter and are given here with some rearrangement and addition of new inverse equations. For the forward calculations, given R, λ_0, ϕ, and λ (giving true scale along the Equator), to find x and y (see p. 363 for numerical examples):

$$x = \pm \pi R \left[A(G-P^2) + [A^2(G-P^2)^2 - (P^2+A^2)(G^2-P^2)]^{1/2}\right]/(P^2+A^2) \qquad (29-1)$$

taking the sign of $(\lambda-\lambda_0)$. Note that $(\lambda-\lambda_0)$ must fall between $+180°$ and $-180°$; if necessary, 360° must be added or subtracted. The X axis lies along the Equator, x increasing easterly, while the Y axis coincides with the central meridian λ_0.

$$y = \pm \pi R\{PQ - A[(A^2+1)(P^2+A^2)-Q^2]^{1/2}\}/(P^2+A^2) \qquad (29-2)$$

taking the sign of ϕ,

where

$$\begin{aligned}
A &= \tfrac{1}{2}|\pi/(\lambda-\lambda_0)-(\lambda-\lambda_0)/\pi| & (29-3) \\
G &= \cos\theta/(\sin\theta+\cos\theta-1) & (29-4) \\
P &= G(2/\sin\theta-1) & (29-5) \\
\theta &= \arcsin|2\phi/\pi| & (29-6) \\
Q &= A^2+G & (29-6a)
\end{aligned}$$

But if $\phi=0$ or $\pm90°$, or $\lambda=\lambda_0$, these equations are indeterminate. In that case, if $\phi=0$,

$$x = R(\lambda - \lambda_0) \tag{29-7}$$

and

$$y = 0$$

or if $\lambda = \lambda_0$, or $\phi = \pm 90°$

$$x = 0$$

and

$$y = \pm \pi R \, \tan(\theta/2) \tag{29-8}$$

taking the sign of ϕ. It may be noted that absolute values (symbol | |) are used in several cases. The origin is at the center ($\phi = 0$, $\lambda = \lambda_0$).

For the inverse equations, given R, λ_0, x, and y, to find ϕ and λ: Because of the complications involved, the equations are given in the order of use. This is closely based upon a recent, noniterative algorithm by Rubincam (1981):

$$X = x/(\pi R) \tag{29-9}$$
$$Y = y/(\pi R) \tag{29-10}$$
$$c_1 = -|Y|(1 + X^2 + Y^2) \tag{29-11}$$
$$c_2 = c_1 - 2Y^2 + X^2 \tag{29-12}$$
$$c_3 = -2c_1 + 1 + 2Y^2 + (X^2 + Y^2)^2 \tag{29-13}$$
$$d = Y^2/c_3 + (2c_2^3/c_3^3 - 9c_1c_2/c_3^2)/27 \tag{29-14}$$
$$a_1 = (c_1 - c_2^2/3c_3)/c_3 \tag{29-15}$$
$$m_1 = 2(-a_1/3)^{1/2} \tag{29-16}$$
$$\theta_1 = (1/3) \arccos(3d/a_1m_1) \tag{29-17}$$
$$\phi = \pm \pi[-m_1 \cos(\theta_1 + \pi/3) - c_2/3c_3] \tag{29-18}$$

taking the sign of y.

$$\lambda = \pi|X^2 + Y^2 - 1 + [1 + 2(X^2 - Y^2) + (X^2 + Y^2)^2]^{1/2}|/2X + \lambda_0 \tag{29-19}$$

but if $X = 0$, equation (29–19) is indeterminate. Then

$$\lambda = \lambda_0 \tag{29-20}$$

The formulas for scale factors are quite lengthy and are not included here. Rectangular coordinates are given in table 41 for a map of the world with unit radius of the outer circle, or $R = 1/\pi$. The longitude is measured from the central meridian. Only one quadrant of the map is given, but the map is symmetrical about both X and Y axes.

30. SINUSOIDAL PROJECTION

SUMMARY

- Pseudocylindrical projection.
- Equal-area.
- Central meridian is a straight line; all other meridians are shown as equally spaced sinusoidal curves.
- Parallels are equally spaced straight lines, parallel to each other. Poles are points.
- Scale is true along central meridian and all parallels.
- Used for world maps with single central meridian or in interrupted form with several central meridians.
- Used for maps of South America and Africa.
- Used since the mid-16th century.

HISTORY

There is an almost endless number of possible projections with horizontal straight lines for parallels of latitude and curved lines for meridians. They are sometimes called pseudocylindrical because of their partial similarity to cylindrical projections. Scores of such projections have been presented, purporting various special advantages, although several are strikingly similar to other members of the group (Snyder, 1977). While there were rudimentary projections with straight parallels used as early as the 2nd century B.C. by Hipparchus, the first such projection still used for scientific mapping of the sphere is the Sinusoidal.

This projection (fig. 54), used for world maps as well as maps of continents and other regions, especially those bordering the Equator, has been given many names after various presumed originators, but it is most frequently called by the name used here. Among the first to show the Sinusoidal projection was Jean Cossin of Dieppe, who used it for a world map of 1570. In addition, it was used by Jodocus Hondius for maps of South America and Africa in some of his editions of Mercator's atlases of 1606–1609. This is probably the basis for one of the names of the projection: The Mercator Equal-Area. Nicolas Sanson (1600–67) of France used it in about 1650 for maps of continents, while John Flamsteed (1646–1719) of England later used it for star maps. Thus, the name "Sanson-Flamsteed" has often been applied to the Sinusoidal projection, even though they were not the originators (Keuning, 1955, p. 24; Deetz and Adams, 1934, p. 161).

While maps of North America are no longer drawn to the Sinusoidal, South America and Africa are still shown on this projection in recent Rand McNally atlases.

FEATURES AND USAGE

The simplicity of construction, either graphically or mathematically, combined with the useful features obtained, make the Sinusoidal projection not only popular to use, but a popular object of study for interruptions, transformations, and combination with other projections.

On the normal Sinusoidal projection, the parallels of latitude are equally spaced straight parallel lines, and the central meridian is a straight line crossing the parallels perpendicularly. The Equator is marked off from the central meridian equidistantly for meridians at the same scale as the latitude markings on the central meridian, so the Equator for a complete world map is twice as long as the central meridian. The other parallels of latitude are also marked off for meridians in proportion to the true distances from the central meridian. The meridians

TABLE 41.—*Van der Grinten projection: Rectangular coordinates*

[y coordinate in parentheses under x coordinate]

Long. Lat.	0°	10°	20°	30°	40°
90°........	0.00000 (1.00000)	0.00000 (1.00000)	0.00000 (1.00000)	0.00000 (1.00000)	0.00000 (1.00000)
80........	.00000 (.60961)	.03491 (.61020)	.06982 (.61196)	.10473 (.61490)	.13963 (.61902)
70........	.00000 (.47759)	.04289 (.47806)	.08581 (.47948)	.12878 (.48184)	.17184 (.48517)
60........'	.00000 (.38197)	.04746 (.38231)	.09495 (.38336)	.14252 (.38511)	.19020 (.38756)
50........	.00000 (.30334)	.05045 (.30358)	.10094 (.30430)	.15149 (.30551)	.20215 (.30721)
40........	.00000 (.23444)	.05251 (.23459)	.10504 (.23505)	.15764 (.23582)	.21031 (.23690)
30........	.00000 (.17157)	.05392 (.17166)	.10787 (.17192)	.16185 (.17235)	.21588 (.17295)
20........	.00000 (.11252)	.05485 (.11256)	.10972 (.11267)	.16460 (.11286)	.21951 (.11313)
10........	.00000 (.05573)	.05538 (.05574)	.11077 (.05577)	.16616 (.05581)	.22156 (.05588)
0........	.00000 (.00000)	.05556 (.00000)	.11111 (-.00000)	.16667 (.00000)	.22222 (.00000)

Long. Lat.	100°	110°	120°	130°	140°
90°........	0.00000 (1.00000)	0.00000 (1.00000)	0.00000 (1.00000)	0.00000 (1.00000)	0.00000 (1.00000)
80........	.34699 (.66917)	.38069 (.68174)	.41394 (.69548)	.44668 (.71035)	.47882 (.72631)
70........	.43163 (.52588)	.47493 (.53621)	.51810 (.54756)	.56110 (.55992)	.60385 (.57328)
60........	.47903 (.41762)	.52754 (.42525)	.57608 (.43366)	.62463 (.44282)	.67313 (.45275)
50........	.50899 (.32792)	.56059 (.33317)	.61228 (.33894)	.66404 (.34524)	.71585 (.35207)
40........	.52871 (.25001)	.58218 (.25333)	.63575 (.25697)	.68939 (.26094)	.74310 (.26523)
30........	.54168 (.18026)	.59626 (.18209)	.65091 (.18411)	.70562 (.18631)	.76038 (.18869)
20........	.54979 (.11635)	.60499 (.11716)	.66022 (.11804)	.71548 (.11901)	.77077 (.12005)
10........	.55419 (.05668)	.60967 (.05688)	.66516 (.05710)	.72066 (.05734)	.77617 (.05760)
0........	.55555 (.00000)	.61111 (.00000)	.66667 (.00000)	.72222 (.00000)	.77778 (.00000)

TABLE 41.—*Van der Grinten projection: Rectangular coordinates*—Continued

Long. Lat.	50°	60°	70°	80°	90°
90°........	0.00000	0.00000	0.00000	0.00000	0.00000
	(1.00000)	(1.00000)	(1.00000)	(1.00000)	(1.00000)
80.........	.17450	.20932	.24403	.27859	.31293
	(.62435)	(.63088)	(.63863)	(.64760)	(.65778)
70.........	.21498	.25821	.30152	.34488	.38827
	(.48946)	(.49473)	(.50100)	(.50828)	(.51657)
60.........	.23800	.28594	.33403	.38225	.43059
	(.39073)	(.39462)	(.39925)	(.40462)	(.41074)
50........	.25293	.30385	.35492	.40614	.45750
	(.30940)	(.31208)	(.31527)	(.31897)	(.32319)
40........	.26308	.31596	.36897	.42210	.47535
	(.23829)	(.24000)	(.24202)	(.24436)	(.24703)
30........	.26998	.32415	.37841	.43275	.48718
	(.17373)	(.17468)	(.17581)	(.17711)	(.17860)
20........	.27445	.32944	.38446	.43953	.49464
	(.11347)	(.11389)	(.11439)	(.11497)	(.11562)
10........	.27697	.33239	.38782	.44327	.49872
	(.05597)	(.05607)	(.05620)	(.05634)	(.05650)
0........	.27778	.33333	.38889	.44444	.50000
	(.00000)	(.00000)	(.00000)	(.00000)	(.00000)

Long. Lat.	150°	160°	170°	180°
90°.........	0.00000	0.00000	0.00000	0.00000
	(1.00000)	(1.00000)	(1.00000)	(1.00000)
80..........	.51028	.54101	.57093	.60000
	(.74331)	(.76130)	(.78021)	(.80000)
70..........	.64631	.68843	.73013	.77139
	(.58762)	(.60293)	(.61919)	(.63636)
60..........	.72156	.76988	.81804	.86603
	(.46344)	(.47488)	(.48707)	(.50000)
50..........	.76768	.81951	.87132	.92308
	(.35942)	(.36729)	(.37569)	(.38462)
40..........	.79686	.85066	.90448	.95831
	(.26986)	(.27482)	(.28010)	(.28571)
30..........	.81518	.87003	.92490	.97980
	(.19125)	(.19398)	(.19690)	(.20000)
20..........	.82609	.88143	.93678	.99216
	(.12117)	(.12237)	(.12365)	(.12500)
10..........	.83168	.88721	.94274	.99827
	(.05788)	(.05817)	(.05849)	(.05882)
0..........	.83333	.88889	.94444	1.00000
	(.00000)	(.00000)	(.00000)	(.00000)

Radius of map = 1.0. Radius of sphere = $1/x$.

Origin: (x, y) 0 at (lat, long) = 0. Y axis increases north. One quadrant given. Other quadrants of world map are symmetrical.

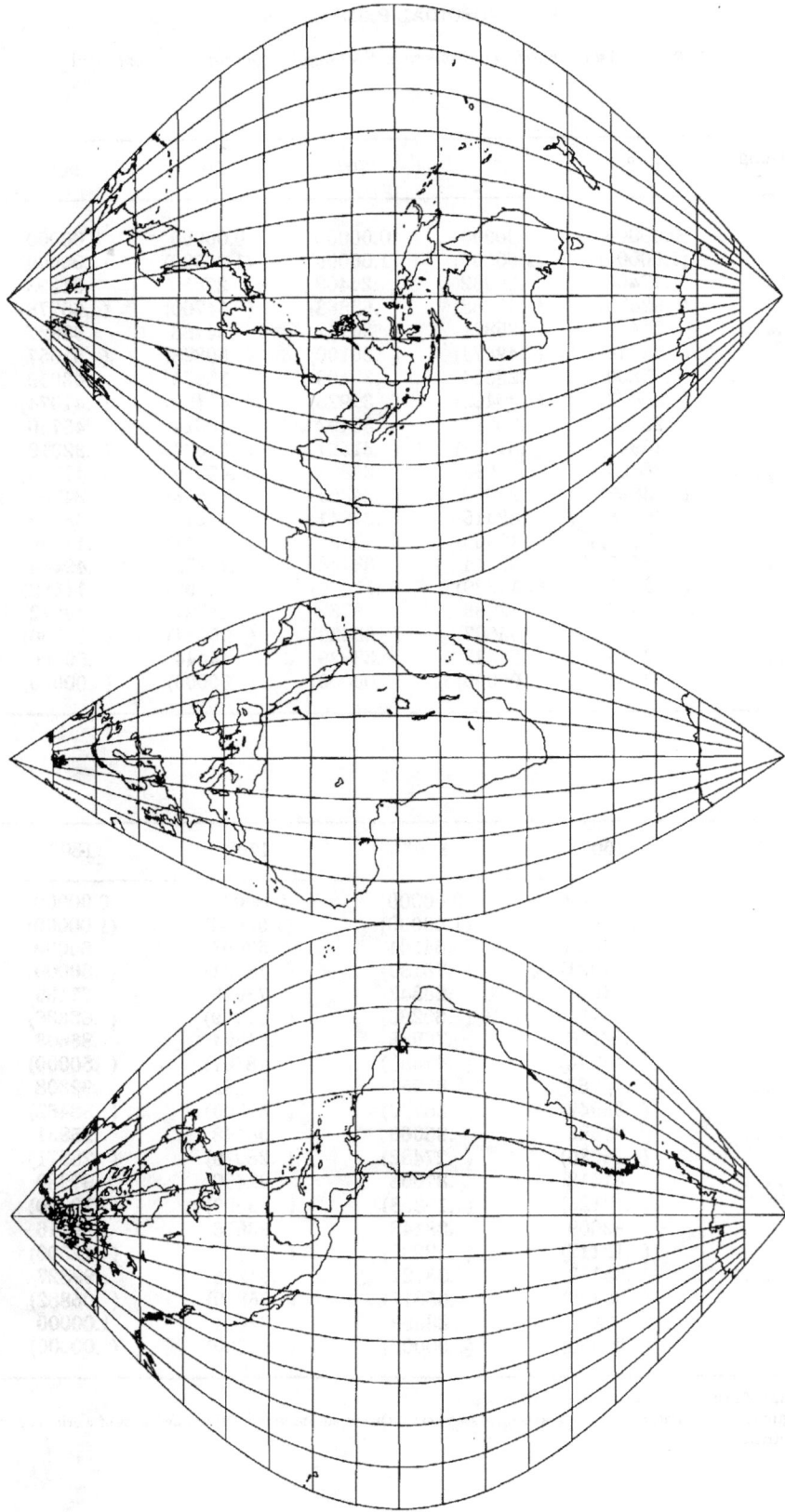

FIGURE 54.—Interrupted Sinusoidal projection as used by the USGS. The oldest pseudocylindrical projection, it shows areas correctly.

connect these markings from pole to pole. Since the spacings on the parallels are proportional to the cosine of the latitude, and since parallels are equally spaced, the meridians form curves which may be called cosine, sine, or sinusoidal curves; hence, the name.

Areas are shown correctly. There is no distortion along the Equator and central meridian, but distortion becomes pronounced near the outer meridians, especially in the polar regions.

Because of this distortion, J. Paul Goode (1862−1932) of the University of Chicago developed an interrupted form of the Sinusoidal in 1916 with several meridians chosen as central meridians without distortion and a limited expanse east and west for each section. The central meridians may be different for Northern and Southern Hemispheres and may be selected to minimize distortion of continents or of oceans instead. Ultimately, Goode combined the portion of the interrupted Sinusoidal projection between about lats. 40° N. and S. with the portions of the Mollweide or Homolographic projection (described later) not in this zone, to produce the Homolosine projection used in Rand McNally's *Goode's Atlas* (Goode, 1925).

In 1927, the Sinusoidal was shown interrupted in three symmetrical segments in the *Nordisk Världs Atlas*, Stockholm, serving as the base for the Sinusoidal as shown in Deetz and Adams (1934, p. 161). It is this interrupted form which served in turn as the base for a three-sheet set by the USGS in 1978 at a scale of 1:20,000,000, entitled Map of Prospective Hydrocarbon Provinces of the World. With interruptions occurring at longs. 160° W., 20° W., and 60° E., and the three central meridians equidistant from these limits, the sheets show (1) North and South America; (2) Europe, West Asia, and Africa; and (3) East Asia, Australia, and the Pacific; respectively. The maps extend pole to pole, but no data are shown for Antarctica. An inset of the Arctic region at the same scale is drawn to the polar Lambert Azimuthal Equal-Area projection. A similar map is being prepared by the USGS showing sedimentary basins of the world.

The Sinusoidal projection is normally used in the spherical form, adequate for the usual small-scale usage, but the ellipsoidal form has been used for topographic mapping in Ecuador (C. J. Mugnier, pers. comm., 1985).

FORMULAS FOR THE SPHERE

The formulas for the Sinusoidal projection are perhaps the simplest of those for any projection described in this bulletin, except for the Equidistant Cylindrical. For the forward case, given R, λ_0, ϕ, and λ, to find x and y (see p. 365 for numerical examples):

$$x = R(\lambda - \lambda_0) \cos \phi \qquad (30-1)$$
$$y = R\phi \qquad (30-2)$$
$$h = [1 + (\lambda - \lambda_0)^2 \sin^2 \phi]^{1/2} \qquad (30-3)$$
$$k = 1.0$$
$$\theta' = \arcsin (1/h) \qquad (30-4)$$
$$\omega = 2 \arctan |\tfrac{1}{2}(\lambda - \lambda_0) \sin \phi| \qquad (30-5)$$

where θ' is the angle of intersection of a given meridian and parallel (see equation (4−14)), and h, k, and ω are other distortion factors as previously used. The X axis coincides with the Equator, with x increasing easterly, while the Y axis follows the central meridian λ_0, y increasing northerly. It is necessary to adjust $(\lambda - \lambda_0)$, if it falls outside the range $\pm 180°$, by adding or subtracting 360°. For the interrupted form, values of x are calculated for each section with respect to its own central meridian λ_0.

In equations (30−1) through (30−5), radians must be used, or ϕ and λ in degrees must be multiplied by $\pi/180°$.

For the *inverse* formulas, given R, λ_0, x, and y, to find ϕ and λ:

$$\phi = y/R \tag{30−6}$$
$$\lambda = \lambda_0 + x/(R \cos \phi) \tag{30−7}$$

but if $\phi = \pm\pi/2$, equation (30−7) is indeterminate, and λ may be given an arbitrary value such as λ_0.

FORMULAS FOR THE ELLIPSOID

The ellipsoidal form may be made by spacing parallels along the central meridian(s) true to scale for the ellipsoid and meridians along each parallel also true to scale. The projection remains equal-area, while the parallels are not quite equally spaced, and the meridians are no longer perfect sinusoids. Specifically, given a, e, λ_0, ϕ, and λ, to find x and y (see p. 366 for numerical examples):

$$x = a (\lambda-\lambda_0) \cos \phi/(1-e^2 \sin^2 \phi)^{1/2} \tag{30−8}$$
$$y = M \tag{30−9}$$

where

$$\begin{aligned}
M = a [&(1-e^2/4-3e^4/64-5e^6/256- \ldots) \phi \\
&-(3e^2/8 + 3e^4/32 + 45e^6/1024 + \ldots) \sin 2 \phi \\
&+(15e^4/256 + 45e^6/1024 + \ldots) \sin 4 \phi \\
&-(35e^6/3072 + \ldots) \sin 6 \phi + \ldots]
\end{aligned} \tag{3−21}$$

Axes are the same as those for the spherical form above.

For the *inverse* formulas, given a, e, λ_0, x, and y, to find ϕ and λ:

$$\begin{aligned}
\phi = \mu + &(3e_1/2-27e_1^3/32 + \ldots) \sin 2\mu + (21e_1^2/16 \\
&-55e_1^4/32 + \ldots) \sin 4\mu + (151e_1^3/96 - \ldots) \sin 6\mu \\
&+(1097e_1^4/512 - \ldots) \sin 8\mu + \ldots
\end{aligned} \tag{3−26}$$

where

$$e_1 = [1-(1-e^2)^{1/2}]/[1 + (1-e^2)^{1/2}] \tag{3−24}$$
$$\mu = M/[a(1-e^2/4-3e^4/64-5e^6/256- \ldots)] \tag{7−19}$$

and

$$M = y \tag{30−10}$$

Then

$$\lambda = \lambda_0 + x (1-e^2 \sin^2 \phi)^{1/2}/(a \cos \phi) \tag{30−11}$$

but if $\phi = \pm\pi/2$, equation (30−11) is indeterminate, and λ may be given an arbitrary value such as λ_0.

31. MOLLWEIDE PROJECTION

SUMMARY

- Pseudocylindrical.
- Equal-area.
- Central meridian is a straight line; 90th meridians are circular arcs; all other meridians are equally spaced elliptical arcs.
- Parallels are unequally spaced straight lines, parallel to each other. Poles are points.
- Scale is true along latitudes 40°44′ N. and S.
- Used for world maps with single central meridian or in interrupted form with several central meridians.
- Inspiration for several other projections.
- Presented by Mollweide in 1805.

HISTORY AND USAGE

The second oldest pseudocylindrical projection which is still in use (after the Sinusoidal) was presented by Carl B. Mollweide (1774−1825) of Halle, Germany, in 1805 (Mollweide, 1805). It is an equal-area projection of the Earth within an ellipse. It has had a profound effect on world map projections in the 20th century, especially as an inspiration for other important projections. It lay dormant until J. Babinet reintroduced it in 1857 under the name "homalographic." It has been called Babinet's Equal-Surface or the Elliptical projection, but it is most often called the Mollweide, Homalographic, or Homolographic.

J. Paul Goode, after interrupting the Sinusoidal projection, made similar interruptions of the Mollweide in 1916 to minimize distortion of continents or oceans. Ultimately he combined them to produce the Homolosine projection.

Other projections directly inspired by the Mollweide have been the Van der Grinten, described earlier, and the Boggs Eumorphic, in which the y coordinates of the Sinusoidal and Mollweide are arithmetically averaged, and the x coordinates are derived to maintain equality of area (Boggs, 1929). J. Fairgrieve in 1928 (Steers, 1970, p. 172) was the first of several to use the oblique aspect, and John Bartholomew applied the name "Atlantis" to a transverse Mollweide centered on the Atlantic Ocean and used as the frontispiece in *The Times Atlas* of 1958. Allen K. Philbrick (1953) combined the Sinusoidal and Mollweide in a manner different from the Goode Homolosine, using both normal and oblique aspects. Less direct inspiration by the Mollweide has led to several other projections, especially pseudocylindrical, some of which have lines for poles.

Some other projections showing the world in an ellipse, especially the Hammer and the Briesemeister, originate from the Lambert Azimuthal Equal-Area projection, not the Mollweide. Another projection occasionally seen is identical with the Mollweide, except that the parallels are equally spaced, and therefore the projection is not equal-area. It was first used in a rudimentary form in the 16th century.

FEATURES

Unlike the Sinusoidal projection, which has been satisfactorily used for continental maps, the Mollweide projection (fig. 55) is normally used as a world map, and occasionally for a very large region such as the Pacific Ocean. This is because only two points on the Mollweide are completely free of distortion unless the projection is interrupted. These are the points at latitudes 40°44′12″ N. and S. on the central meridian(s).

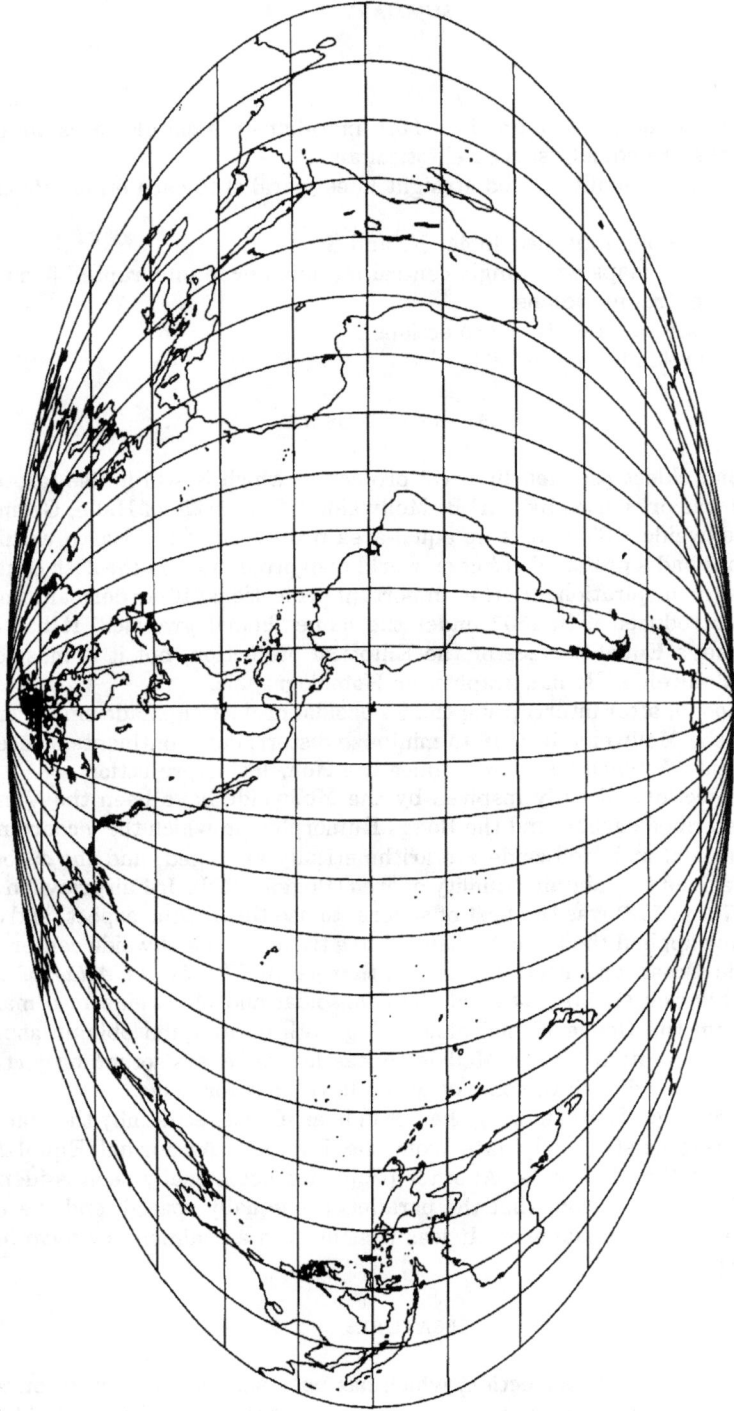

FIGURE 55.—Mollweide projection. An equal-area projection of the world bounded with an ellipse, and the basis of several other projections.

The world is shown in an ellipse with the Equator, its major axis, twice as long as the central meridian, its minor axis. The meridians 90° east and west of the central meridian form a complete circle. All other meridians are elliptical arcs which, with their opposite numbers on the other side of the central meridian, form complete ellipses which meet at the two poles. The central meridian is the major axis of meridian ellipses less than 90° from it, and a portion of the Equator is the minor axis. Meridians greater than 90° have the reverse arrangement for their axes. Meridians are equally spaced along the Equator and along all other parallels. The 90th meridians form a circle.

The parallels of latitude are straight parallel lines, but they are not equally spaced. Their spacing may be determined from the facts that the projection is equal-area and that the 90th meridians are circular. As a result, the regions along the Equator are stretched 23 percent in a north-south direction relative to east-west dimensions. This stretching decreases along the central meridian to zero at the 40°44' latitudes, and becomes compression nearer the poles. The distortion near the outer meridians is considerable at high latitudes, but less than that on the Sinusoidal. The distortion along the Equator led Bromley (1965) to propose flattening the projection in a north-south direction and expanding east-west, to provide an Equator free of distortion, but the Equator thereby becomes 2.47 times as long as the central meridian.

Because the Mollweide projection is normally used at a small scale, there is little justification for an ellipsoidal form.

FORMULAS FOR THE SPHERE

The forward formulas for the Mollweide require iteration, but they are otherwise relatively simple. Given R, λ_0, ϕ, and λ, to find x and y (see p. 367 for numerical examples):

$$x = (8^{1/2}/\pi) R (\lambda - \lambda_0) \cos \theta \qquad (31-1)$$
$$y = 2^{1/2} R \sin \theta \qquad (31-2)$$

where

$$2\theta + \sin 2\theta = \pi \sin \phi \qquad (31-3)$$

The X axis coincides with the Equator, x increasing easterly, and the Y axis coincides with the central meridian, y increasing northerly. Angle θ is not a polar coordinate here; it is a parametric angle, geometrically the angle as seen from the center of the map between the Equator and the position of latitude ϕ on the 90th meridian circle.

Equation (31-3) may be solved with rapid convergence (but slow at the poles) using a Newton-Raphson iteration consisting of the following instead of (31-3):

$$\Delta\theta' = -(\theta' + \sin \theta' - \pi \sin \phi)/(1 + \cos \theta') \qquad (31-4)$$

With ϕ as the first trial θ', $\Delta\theta'$ is calculated from (31-4), this value is added to the preceding trial θ' to obtain the next trial θ', and the calculation is repeated with (31-4) until $\Delta\theta'$ is less than a predetermined convergence value. Then, using the final θ', θ is calculated as follows:

$$\theta = \theta'/2 \qquad (31-5)$$

Note that all these formulas are in terms of radians.

For the *inverse* formulas, given R, λ_0, x, and y, to find ϕ and λ, no iteration is required:

$$\theta = \arcsin \left[y/(2^{1/2} R) \right] \tag{31-6}$$
$$\phi = \arcsin \left[(2\theta + \sin 2\theta)/\pi \right] \tag{31-7}$$
$$\lambda = \lambda_0 + \pi x/(8^{1/2} R \cos \theta) \tag{31-8}$$

If ϕ is $\pm 90°$, equation (31-8) is indeterminate, but λ may be made λ_0. Table 42 lists the rectangular coordinates of the 90th meridian for a sphere of radius $(1/2^{1/2})$, to make the maximum values equal to 1.0. The x coordinates for other meridians are proportional, and y coordinates are constant for a given latitude.

TABLE 42.—*Mollweide projection: Rectangular coordinates for the 90th meridian*

Latitude	x	y
90°	0.00000	1.00000
85	.20684	0.97837
80	.32593	.94539
75	.42316	.90606
70	.50706	.86191
65	.58111	.81382
60	.64712	.76239
55	.70617	.70804
50	.75894	.65116
45	.80591	.59204
40	.84739	.53097
35	.88362	.46820
30	.91477	.40397
25	.94096	.33850
20	.96229	.27201
15	.97882	.20472
10	.99060	.13681
5	.99765	.06851
0	1.00000	.00000

Radius of sphere: $\frac{1}{2}^{1/2} = 0.707$ unit. For other meridians, use same y, but change x proportionately. Central meridian is zero. For meridians west of central meridian, change sign of x. For southern latitudes, change sign of y.

32. ECKERT IV AND VI PROJECTIONS

SUMMARY

- Pseudocylindrical.
- Equal-area.
- Central meridian is a straight line; 180th meridians of Eckert IV are semi-circles; all other meridians are equally spaced elliptical arcs on Eckert IV and sinusoidal curves on Eckert VI.
- Parallels are unequally spaced straight lines, parallel to each other. Poles are straight lines half as long as the Equator.
- Scale is true along latitudes 40°30' N. and S. on Eckert IV and 49°16' on Eckert VI.
- Used for world maps.
- Presented by Eckert in 1906.

HISTORY AND USAGE

In 1906 Max Eckert (1868–1938) of Kiel, Germany, presented a set of six new projections in which all the poles are lines half as long as the Equator, and in which all parallels of latitude are straight lines parallel to each other. The central meridian on each is also half the length of the Equator (Eckert, 1906). Numbers 4 and 6 are of most significance and are discussed here in detail.

Of the six projections, nos. 2, 4, and 6 are equal-area, and nos. 1, 3, and 5 are not equal-area but have equally spaced parallels. For nos. 1 and 2, the meridians are straight lines broken at the Equator, and those projections are therefore little more than novelties with graticules composed entirely of straight lines, but with unnecessary distortion especially at the Equator. The meridians on nos. 3 and 4 are elliptical arcs, while on 5 and 6 they are sinusoidal curves, with the exception of the straight central meridians, and (on 3 and 4) semicircular outer meridians.

No. 3, with equidistant parallels and elliptical arcs has occasionally been identified as the same as the Ortelius projection, named for the famous cartographer Abraham Ortelius who used a somewhat similar projection in 1570 for his world map. The border, the central meridian, and the parallels of the two projections are shown almost identically, and the meridians on each are equally spaced along the Equator. The shapes of most meridians, however, are different. On the Ortelius, they are circular arcs, semicircles for meridians at or more than 90° from the central meridian, and circular arcs intersecting the central meridian at the poles within 90° of the central meridian.

The most commonly used of Eckert's six projections have been his nos. 4 and 6, which are more often designated with Roman numerals IV and VI, respectively. In the United States, Eckert IV (fig. 56) has been used in several atlases to show climate and other themes. It has also been used as an inset on other maps, such as wall maps of the world by the National Geographic Society. It ranked third as an equal-area world map projection used in U.S. textbooks between 1940 and 1960, after the Goode Homolosine and Sinusoidal (Wong, 1965, p. 101). The Eckert VI (fig. 57) is much less used in the United States, although it has occasionally appeared in textbooks and atlases. It has been more popular in the Soviet Union, having been used for several world distribution maps in the 1937 *Atlas Mira* (World Atlas). An almost identical equal-area projection was presented by Karlheinz Wagner in 1932 and independently by V. V. Kavrayskiy in 1936; theirs does not require the iteration in computations which is required by Eckert VI (Maling, 1960, p. 297; Snyder, 1977, p. 62).

There have been numerous other pseudocylindrical projections with lines for poles, and Eckert's were not the first, but they are the most popular. Some have

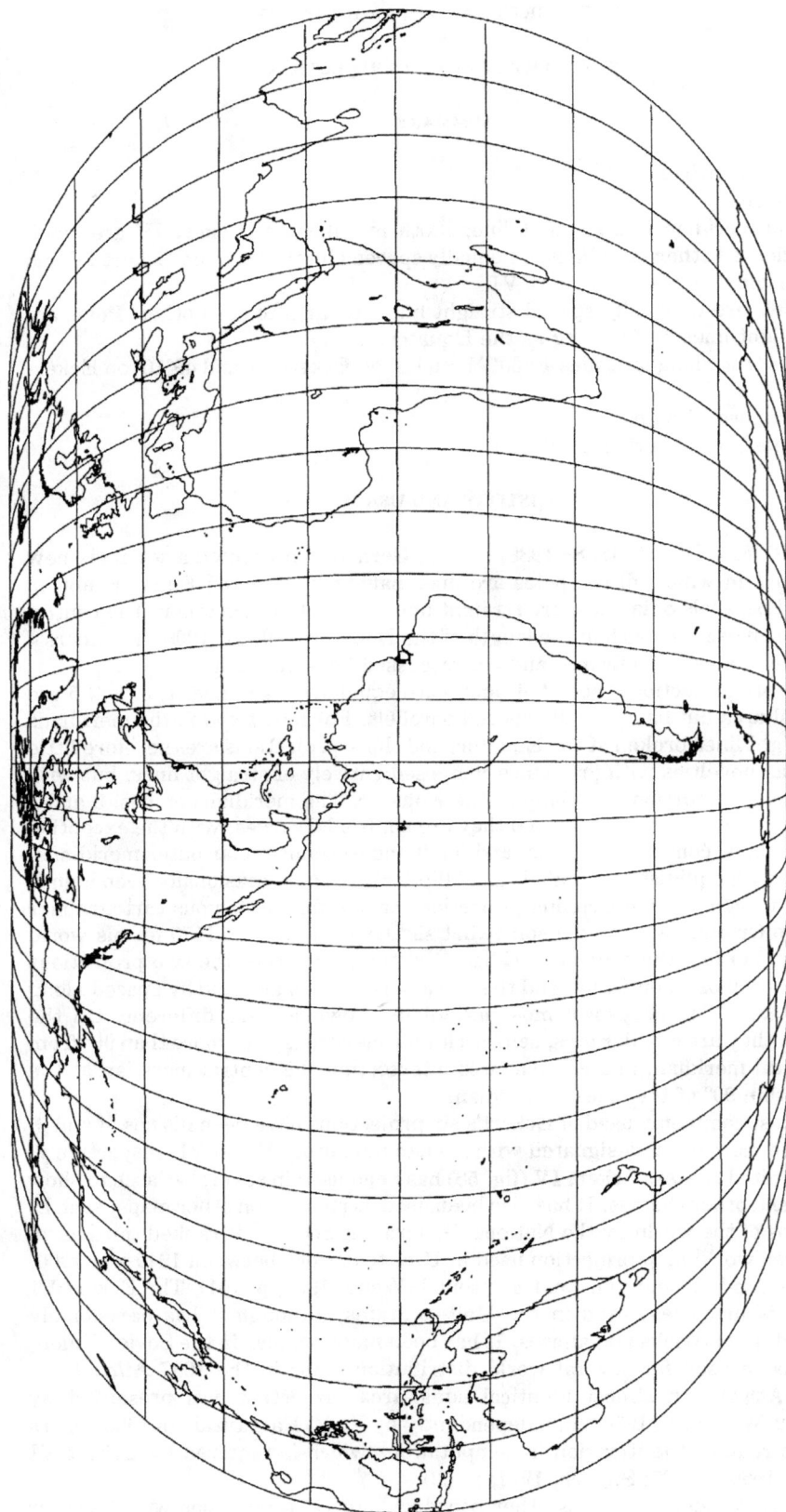

FIGURE 56.—Eckert IV projection. An equal-area pseudocylindrical with poles half the length of the Equator. Outer meridians are semicircles; others are elliptical arcs.

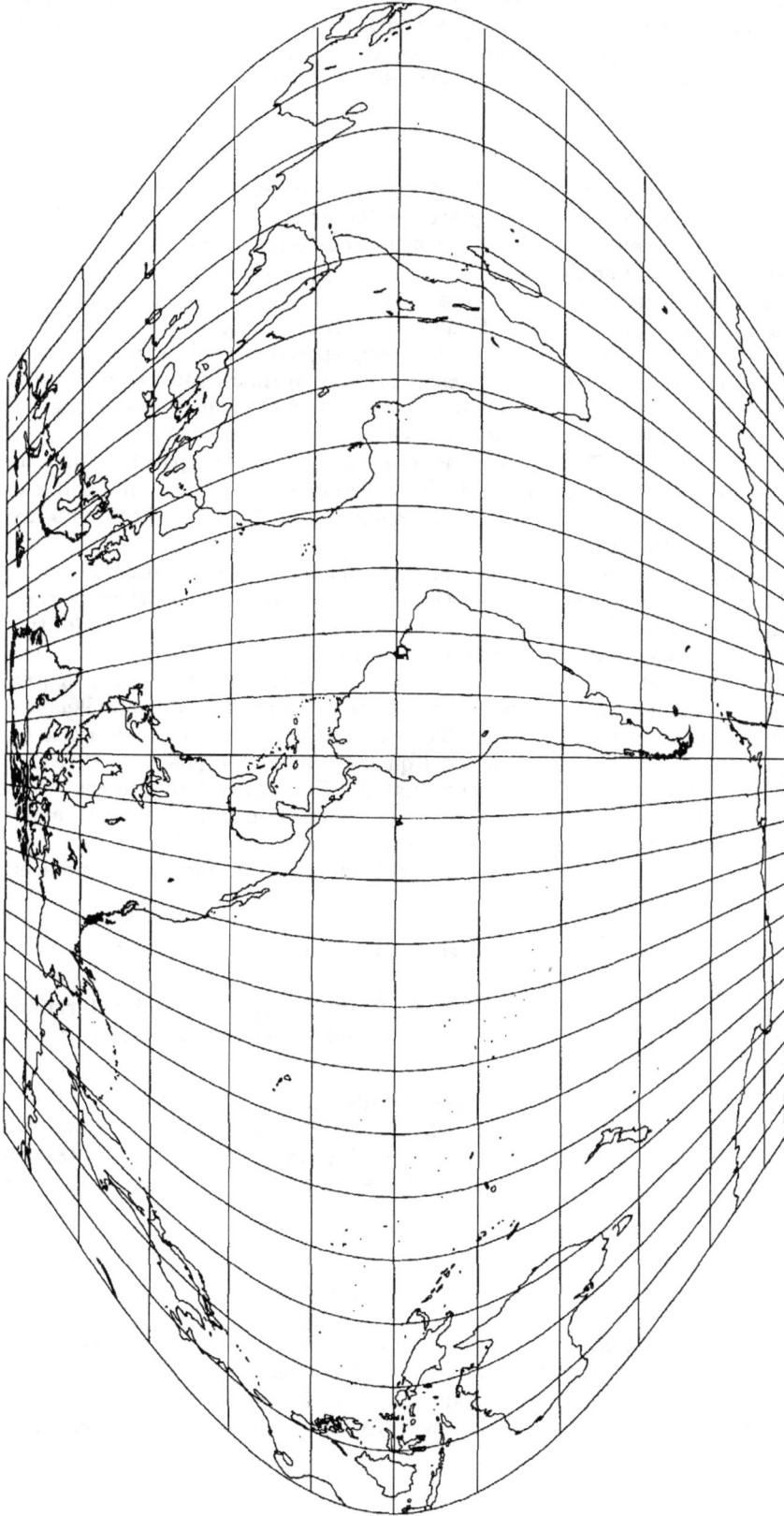

FIGURE 57.—Eckert VI projection. Like figure 56, this is an equal-area pseudocylindrical projection with poles half the length of the Equator. The meridians, however, are sinusoidal curves.

been obtained by averaging a cylindrical projection with the Sinusoidal or Mollweide projection, and others are derived by stipulating that the poles be lines of some other length in proportion to the length of the Equator. Instead of the full sinusoid or full semiellipse, a portion of these curves or of some other mathematical curve has been used for the meridians (Snyder, 1977).

FEATURES

The Eckert IV projection is bounded by two semicircles representing the 180th meridians and two straight lines connecting the ends of the semicircles. These straight lines represent the two poles, which are half the length of the Equator. Meridians are equally spaced semiellipses ranging in eccentricity from zero for the outer circular meridians to 1 for the straight central meridians. The parallels are straight lines parallel to the Equator and spaced to provide correct area within the border. They are therefore unequally spaced and closer together near the poles. There is a north-south stretching of shape at the Equator amounting to 40 percent relative to east-west dimensions. This stretching decreases along the central meridian to zero at latitudes 40°30′ N. and S. and becomes flattening beyond these parallels. The scale is correct only along these two parallels, and the only points free of distortion are at the intersection of these two points with the central meridian. Nearer the poles, the geographical features of the map are flattened in a north-south direction.

The Eckert VI projection of the world is bounded by two sinusoidal curves which have the same shape as the 90th meridians of the Sinusoidal projection. Like the border of the Eckert IV, these curved meridians are connected with two straight lines connecting the ends of the curves. These straight lines, half the length of the Equator, are the poles. The other meridians are equally spaced sinusoids, except for the straight central meridian, and the other parallels are straight and parallel to each other. To preserve area, the parallels must be unequally spaced, farther apart at the Equator than at the poles. As a result, there is a 29 percent north-south stretch at the Equator, relative to east-west dimensions. Other general comments concerning distortion of the Eckert IV apply to Eckert VI, but the latitudes of true scale are 49°16′ N. and S.

These projections are for world maps, not regional maps, and there is no need for the ellipsoidal forms.

FORMULAS FOR THE SPHERE

The forward formulas for both Eckert IV and Eckert VI require iteration. Given R, λ_0, ϕ, and λ, to find x and y (see p. 368 for numerical examples):
Eckert IV:

$$x = \{2/[\pi(4+\pi)]^{1/2}\}R\,(\lambda-\lambda_0)(1+\cos\theta) \qquad (32-1)$$
$$= 0.4222382\,R\,(\lambda-\lambda_0)\,(1+\cos\theta) \qquad (32-1a)$$
$$y = 2[\pi/(4+\pi)]^{1/2}\,R\,\sin\theta \qquad (32-2)$$
$$= 1.3265004\,R\,\sin\theta \qquad (32-2a)$$

where

$$\theta + \sin\theta\cos\theta + 2\sin\theta = (2+\pi/2)\sin\phi \qquad (32-3)$$

The X axis coincides with the Equator, x increasing easterly, and the Y axis coincides with the central meridian, y increasing northerly. Angle θ is a parametric angle, not a polar coordinate. Equation (32−3) may be solved with rapid convergence (but slow at the poles) using a Newton-Raphson iteration consisting of the following instead of (32−3):

$$\Delta\theta = -[\theta + \sin\theta\cos\theta + 2\sin\theta - (2+\pi/2)\sin\phi]/$$
$$[2\cos\theta\,(1+\cos\theta)] \tag{32-4}$$

With $(\phi/2)$ as the first trial θ, $\Delta\theta$ is calculated from (32−4), this value is added to the preceding θ to obtain the next trial θ, and the calculation is repeated with (32−4) until $\Delta\theta$ is less than a predetermined convergence value. Note that all these formulas are in terms of radians.

Eckert VI:

$$x = R\,(\lambda-\lambda_0)\,(1+\cos\theta)/(2+\pi)^{1/2} \tag{32-5}$$
$$y = 2R\theta/(2+\pi)^{1/2} \tag{32-6}$$

where

$$\theta + \sin\theta = (1+\pi/2)\sin\phi \tag{32-7}$$

Axes are as described above for Eckert IV; θ is parametric, but not the same as θ for Eckert IV. Equation (32−7) may be replaced with the following Newton-Raphson iteration, treated in the same manner as equation (32−4) for Eckert IV, but with ϕ as the first trial θ:

$$\Delta\theta = -[\theta + \sin\theta - (1+\pi/2)\sin\phi]/(1+\cos\theta) \tag{32-8}$$

For the *inverse* formulas, given R, λ_0, x, and y, to find ϕ and λ, no iteration is required (see p. 368 for numerical examples):

Eckert IV:

$$\theta = \arcsin\,[y\,(4+\pi)^{1/2}/(2\pi^{1/2}R)] \tag{32-9}$$
$$= \arcsin\,[y/(1.3265004R)] \tag{32-9a}$$
$$\phi = \arcsin\,[(\theta + \sin\theta\cos\theta + 2\sin\theta)/(2+\pi/2)] \tag{32-10}$$
$$\lambda = \lambda_0 + [\pi(4+\pi)]^{1/2}x/[2R(1+\cos\theta)] \tag{32-11}$$
$$= \lambda_0 + x/[0.4222382R\,(1+\cos\theta)] \tag{32-11a}$$

Eckert VI:

$$\theta = (2+\pi)^{1/2}y/(2R) \tag{32-12}$$
$$\phi = \arcsin\,[(\theta + \sin\theta)/(1+\pi/2)] \tag{32-13}$$
$$\lambda = \lambda_0 + (2+\pi)^{1/2}x/[R(1+\cos\theta)] \tag{32-14}$$

Table 43 lists the rectangular coordinates of the 90th meridian for a sphere of radius $[(4+\pi)^{1/2}/(2\pi^{1/2})]$ for Eckert IV and radius $[(2+\pi)^{1/2}/\pi^{1/2}]$ for Eckert VI, to make maximum values equal to 1.0. The x coordinates for other meridians are proportional, and y coordinates are constant for a given latitude.

TABLE 43. — *Eckert IV and VI projections: Rectangular coordinates for 90th meridian*

Latitude	Eckert IV		Eckert VI	
	x	y	x	y
90°	0.50000	1.00000	0.50000	1.00000
85	.55613	0.99368	.50487	0.99380
80	.60820	.97630	.51916	.97560
75	.65656	.94971	.54198	.94648
70	.70141	.91528	.57205	.90794
65	.74291	.87406	.60782	.86164
60	.78117	.82691	.64767	.80913
55	.81625	.77455	.69004	.75180
50	.84822	.71762	.73344	.69075
45	.87709	.65666	.77655	.62689
40	.90291	.59217	.81817	.56090
35	.92567	.52462	.85724	.49332
30	.94539	.45443	.89288	.42454
25	.96208	.38202	.92430	.35488
20	.97573	.30779	.95087	.28457
15	.98635	.23210	.97207	.21379
10	.99393	.15533	.98749	.14269
5	.99848	.07784	.99686	.07140
0	1.00000	.00000	1.00000	.00000

Radius of sphere: $(4+\pi)^{1/2}/(2\pi^{1/2}) = 0.75386$ unit for Eckert IV.

$\quad\quad\quad\quad\quad (2+\pi)^{1/2}/\pi = 0.72177$ unit for Eckert VI.

For other meridians, use same y, but change x proportionately. Central meridian is zero. For meridians west of central meridian, change sign of x. For southern latitudes, change sign of y.

REFERENCES

nowitz, Milton, and Stegun, I.A., 1964, Handbook of mathema-
cal functions: Washington, National Bureau of Standards. (Re-
rinted by Dover Publications, Inc., New York, 1972).

s, O.S., 1918, Lambert projection tables for the United States:
.S. Coast and Geodetic Survey Spec. Pub. 52.

———, 1919, General theory of Polyconic projections: U.S. Coast
nd Geodetic Survey Spec. Pub. 57.

———, 1921, Latitude developments connected with geodesy and
artography with tables, including a table for Lambert Equal-Area
Meridional projection: U.S. Coast and Geodetic Survey Spec.
ub. No. 67.

———, 1927, Tables for Albers projection: U.S. Coast and
eodetic Survey Spec. Pub. 130.

s, H.C., 1805, Beschreibung einer neuen Kegelprojektion: Zach's
Monatliche Correspondenz zur Beförderung der Erd- und Himmels-
unde, Nov., p. 450–459.

ws, H.J., 1935, Note on the use of Oblique Cylindrical Orthomor-
nic projection: Geographical Journal, v. 86, p. 446.

———, 1938, An Oblique Mercator projection for Europe and
sia: Geographical Journal, v. 92, p. 538.

Department of the, 1962, Extension of zone-to-zone transforma-
on tables: U.S. Army Tech. Manual TM 5–241–10.

———, 1973, Universal Transverse Mercator Grid: U.S. Army
ech. Manual TM 5–241–8.

l, G.C., 1984, The derivation of the mapping equations and dis-
ortion formulae for the Satellite Tracking map projections: Blacks-
urg, Va., M.S. thesis, Va. Polytechnic Inst. and State Univ.

a, R.M., 1973, Cartographic products from the Mariner 9 mission:
our. Geophys. Research, v. 78, no. 20, p. 4424–4435.

———, 1976, Cartography of Mars: 1975: Am. Cartographer, v. 3,
. 1, p. 57–63.

a, R.M., Bridges, P.M., Inge, J.L., Isbell, Christopher, Mazur-
ky, Harold, Strobell, M.E., and Tyner, R.L., 1980, Mapping the
alilean satellites of Jupiter with Voyager data: Photogrammetric
ngineering and Remote Sensing, v. 46, no. 10, p. 1303–1312.

n, W.M., 1928, Topographic mapping: U.S. Geol. Survey Bull.
8-E, 218 p.

ann, W., 1910, Die beste bekannte flächentreue Projektion der
nzen Erde: Petermanns Geographische Mitteilungen, v. 56, II,
141–144.

r, L.U., ca. 1970, Analytical photogrammetry: A collinear
eory: Columbus, O., Ohio State Univ.

R.M. and Bormanis, Valdis, 1970, Plane coordinate survey
stem for the Great Lakes based on the Hotine Orthomorphic pro-
ction. U.S. Lake Survey Misc. Paper 70–4, 16 p. Summarized in
oughton, H.W., and Berry, R.M., 1985, Simple algorithms for
lculation of scale factors for plane coordinate systems (1927 NAD
d 1983 NAD): Surveying and Mapping, v. 45, no. 3, p. 247–259.

ye, C.H., compiler, 1929, Formulas and tables for the construc-
n of polyconic projections: U.S. Geol. Survey Bull. 809, 126 p.

S.W., 1929, A new equal-area projection for world maps: Geo-
aphical Journal, v. 73, p. 241–245.

r, J., 1967, Die Projektionen der schweizerischen Plan- und
artenwerke: Winterthur, Switz., Druckerei Winterthur AG.

rd, G., 1971, Geodesy: Oxford, Eng., Clarendon Press.

er, Wilhelm, and Anliker, Ernst, 1930, Heinrich Christian
bers, der Urheber der flächentreuen Kegelrumpfprojektion:
termanns Geographische Mitteilungen, v. 76, p. 238–240.

, F.V., 1951, A new use for the Plate Carrée projection:
ographical Review, v. 41, p. 640–644.

y, R.H., 1965, Mollweide modified: Professional Geographer,
17, no. 3, p. 24.

Brown, L.A., 1949, The story of maps: New York, Bonanza Books,
reprint undated.

Chebyshev, P.L., 1856, Sur la construction des cartes géographiques:
Bulletin de la classe physiq.-math de l'Acad. des Sciences, v. 14,
p. 257–261. St. Petersburg.

Claire, C.N., 1968, State plane coordinates by automatic data pro-
cessing: U.S. Coast and Geodetic Survey Pub. 62–4.

Clark, David, and Clendinning, James, 1944, Plane and geodetic
surveying for engineering: v. 2, Higher surveying, 3rd ed.: Con-
stable & Co., London, Eng., p. 304–306.

Clarke, A.R., and Helmert, F.R., 1911, Figure of the Earth: Encyclopaedia Britannica, 11th ed., v. 8, p. 801–813.

Close, Sir Charles, 1921, Note on a Doubly-Equidistant projection:
Geographical Journal, v. 57, p. 446–448.

———, 1934, A Doubly Equidistant projection of the sphere:
Geographical Journal, v. 83, p. 144–145.

Close, C. F., and Clarke, A.R., 1911, Map projections: Encylopaedia
Britannica, 11th ed., v. 17, p. 653–663.

Cole, J.H., 1943, The use of the conformal sphere for the construction
of map projections: Survey of Egypt paper 46, Giza (Orman).

Colvocoresses, A.P., 1969, A unified plane co-ordinate reference system: World Cartography, v. 9, p. 9–65.

———, 1974, Space Oblique Mercator: Photogrammetric Engineering, v. 40, no. 8, p. 921–926.

Craig, Thomas, 1882, A treatise on projections: U.S. Coast and Geodetic Survey.

Craster, J.E.E., 1938, Oblique Conical Orthomorphic projection for
New Zealand: Geographical Journal, v. 92, p. 537–538.

Dahlberg, R.E., 1962, Evolution of interrupted map projections: Internat. Yearbook of Cartography, v. 2, p. 36–54.

Davies, M.E., and Batson, R.M., 1975, Surface coordinates and cartography of Mercury: Jour. Geophys. Research, v. 80, no. 17,
p. 2417–2430.

Davies, M.E., Abalakin, V.K., Lieske, J.H., Seidelmann, P.K., Sinclair,
A.T., Sinzi, A.M., Smith, B.A., and Tjuflin, Y.S., 1983, Report of
the IAU Working Group on cartographic coordinates and rotational
elements of the planets and satellites: 1982: Celestial Mechanics,
v. 29, p. 309–321.

Debenham, Frank, 1958, The global altas—a new view of the world
from space: New York, Simon and Schuster.

Deetz, C.H., 1918a, Lambert projection tables with conversion tables:
U.S. Coast and Geodetic Survey Spec. Pub. 49.

———, 1918b, The Lambert Conformal Conic projection with two
standard parallels, including a comparison of the Lambert projection with the Bonne and Polyconic projections: U.S. Coast and
Geodetic Survey Spec. Pub. 47.

Deetz, C.H., and Adams, O.S., 1934, Elements of map projection with
applications to map and chart construction (4th ed.): U.S. Coast
and Geodetic Survey Spec. Pub. 68.

Dozier, Jeff, 1980, Improved algorithm for calculation of UTM and geodetic coordinates: NOAA Tech. Rept. NESS 81.

Driencourt, L., and Laborde, J., 1932, Traité des projections des cartes
géographiques: Paris, Hermann et cie.

Eckert, Max, 1906, Neue Entwürfe für Erdkarten: Petermanns Geographische Mitteilungen, v. 52, p. 97–109.

Fite, E.D., and Freeman, Archibald, 1926, A book of old maps delineating American history from the earliest days down to the close
of the Revolutionary War: Cambridge, Harvard Univ. Press,
reprint 1969, Dover Publications, Inc.

Gall, Rev. James, 1885, Use of cylindrical projections for geographical,
astronomical, and scientific purposes: Scottish Geographical Magazine, v. 1, p. 119–123.

Gannett, S.S., compiler, 1904, Geographic tables and formulas (2nd ed.): U.S. Geol. Survey Bull. 234, 311 p.

Germain, A., 1865?, Traité des projections des cartes géographiques, representation plane de la sphère et du sphèroide: Paris.

Goode, J.P., 1925, The Homolosine projection: a new device for portraying the Earth's surface entire: Assoc. Am. Geographers, Annals, v. 15, p. 119–125.

Goussinsky, B., 1951, On the classification of map projections: Empire Survey Review, v. 11, p. 75–79.

Greenhood, David, 1964, Mapping: Univ. Chicago Press.

Haines, G.V., 1981, The Modified Polyconic projection: Cartographica, v. 18, no. 1, p. 49–58.

Harrison, R.E., 1943, The nomograph as an instrument in map making: Geographical Review, v. 33, p. 655–657.

Hassler, F.R., 1825, On the mechanical organisation of a large survey, and the particular application to the Survey of the Coast: Am. Philosophical Soc. Trans., v. 2, new series p. 385–408.

Hayford, J.F., 1909, The figure of the Earth and isostacy from measurements in the United States: U.S. Coast and Geodetic Survey.

Hinks, A.R., 1912, Map projections: Cambridge, Eng., Cambridge Univ. Press.

————, 1940, Maps of the world on an Oblique Mercator projection: Geographical Journal, v. 95, p. 381–383.

————, 1941, More world maps on Oblique Mercator projections: Geographical Journal, v. 97, p. 353–356.

Hotine, Brig. M., 1946–47, The orthomorphic projection of the spheroid: Empire Survey Review, v. 8, p. 300–311; v. 9, p. 25–35, 52–70, 112–123, 157–166.

Keuning, Johannes, 1955, The history of geographical map projections until 1600: Imago Mundi, v. 12, p. 1–25.

Knuth, D.E., 1969, The art of computer programming: v. 2, Seminumerical algorithms: Addison-Wesley Publishing Co., Reading, Mass., p. 423–424.

Laborde, Chef d'escadron, 1928, La nouvelle projection du service géographique de Madagascar: Cahiers due Service géographique de Madagascar, Tananarive, No. 1.

Lallemand, Ch., 1911. Sur les déformations résultant du mode de construction de la Carte internationale du monde au millionième: Comptes Rendus, v. 153, v. 559–567.

Lambert, J.H., 1772, Beiträge zum Gebrauche der Mathematik und deren Anwendung: Part III, section 6: Anmerkungen und Zusätze. zur Entwerfung der Land- und Himmelscharten: Berlin. Translated and introduced by W.R. Tobler, Univ. Michigan, 1972.

Lauf, G.B., 1983, Geodesy and map projections, 2d. ed.: Collingwood, Victoria, Australia, TAFE Publications.

Lee, L.P., 1944, The nomenclature and classification of map projections: Empire Survey Review, v. 7, p. 190–200.

————, 1974, A conformal projection for the map of the Pacific: New Zealand Geographer, v. 30, p. 75–77.

————, 1976, Conformal projections based on elliptic functions: Cartographica, Monograph 16.

Lewis, Brig. Sir C., and Campbell, Col. J.D., ed., 1951, The American Oxford atlas: New York, Oxford Univ. Press.

Maling, D.H., 1960, A review of some Russian map projections: Empire Survey Review, v. 15, no. 115, p. 203–215; no. 116, p. 255–266: no. 117, p. 294–303.

————, 1966, Some notes about the Trystan Edwards projection: Cartographic Journal, v. 3, p. 94–96.

————, 1973, Coordinate systems and map projections: London, George Philip & Son, Ltd.

————, 1974, Personal projections: Geographical Magazine, v. 46, p. 599–600.

Maurer, Hans, 1919: Annalen der Hydrographie, p. 77.

————, 1935, Ebene Kugelbilder: Gotha, Germany, Petermanns Geographische Mitteilungen, Ergänzungsheft 221. Translated with editing and foreword by William Warntz, Harvard papers in theoretical geography, 1968.

Miller, O.M., 1941, A conformal map projection for the Am Geographical Review, v. 31, p. 100–104.

————, 1942, Notes on cylindrical world map projection graphical Review, v. 32, p. 424–430.

————, 1953, A new conformal projection for Europe and A should read Africa]: Geographical Review, v. 43, p. 405–4

————, 1955, Specifications for a projection system for m continuously Africa, Europe, Asia, and Australasia on a 1:5,000,000: New York, American Geographical Society. C report to Army Map Service, May 1955. Also corrections i by Miller, June 2, 1955, and Supplement, January 31, 195

Mitchell, H.C., and Simmons, L.G., 1945, The State coordinate s (a manual for surveyors): U.S. Coast and Geodetic Surve Pub. 235.

Mollweide, C., 1805, Über die vom Prof. Schmidt in Giessen zweyten Abtheilung seines Handbuchs der Naturlehre angegebene Projection der Halbkugelfläche: Zach's Mor Correspondenz ***, v. 13, p. 152–163.

Mugnier, C.J., 1983, Review of U.S. Geol. Survey Bull. 1532: Sur and Mapping, v. 43, no. 4, p. 417–420.

National Academy of Sciences, 1971, North American datum: N Ocean Survey contract rept. E–53–69(N), 80 p., 7 figs.

Newton, G.D., 1985, Computer programs for common map proj U.S. Geol. Survey Bull. 1642, 33 p.

Nordenskiöld, A.E., 1889, Facsimile-atlas: New York, Dover cations, Inc., (reprint 1973).

Nowicki, A.L., 1962, Topographic lunar mapping: The Canadi veyor, v. 16, no. 3, p. 141–148.

O'Keefe, J.A., and Greenberg, Allen, 1977, A note on the Van de ten projection of the whole Earth onto a circular disk: Am. grapher, v. 4, no. 2, p. 127–132.

Pettengill, G.H., Campbell, D.B., and Masursky, Harold, 198 surface of Venus: Scientific American, v. 243, no. 2, p. 54

Philbrick, A.K., 1953, An oblique equal area map for world distri Assoc. Am. Geographers, Annals, v. 43, p. 201–215.

Poole, H., 1934, An oblique Cylindrical Equal-Area map: Geogr Journal, v. 83, p. 142–143.

Quill, Humphrey, 1966, John Harrison—the man who found lor London, John Baker.

Raisz, Erwin, 1962, Principles of cartography: New York, McGra

Reignier, Francois, 1957, Les systèmes de projection et leurs tions ***: Paris, Institut Géographique National.

Reilly, W.I., 1973, A conformal mapping projection with m scale error: Survey Review, v. 22, no. 168, p. 57–71.

Robinson, A.H., Sale, R.D., Morrison, J.L., and Muehrcke, P.C Elements of cartography (5th ed.): New York, John Wi Sons.

Rosenmund, M., 1903, Die Änderung des Projektionssyste schweizerischen Landesvermessung: Bern, Switz.

Royal Society, 1966, Glossary of technical terms in cartograp pendix 1: Named map projections: London, The Royal So

Rubincam, D.P., 1981, Latitude and longitude from Van der grid coordinates: Am. Cartographer, v. 8, no. 2, p. 177–1

Shalowitz, A.L., 1964, Shore and sea boundaries, vol. 2: U.S and Geodetic Survey Pub. 10–1.

Sinnott, R.W., 1984, Virtues of the haversine: Sky and Telescop no. 2, p. 159.

Snyder, J.P., 1977, A comparison of pseudocylindrical map proj Am. Cartographer, v. 4, no. 1, p. 59–81.

————, 1978a, Equidistant Conic map projections: Assc Geographers, Annals, v. 68, no. 3, p. 373–383.

————, 1978b, The Space Oblique Mercator projection: Phot metric Engineering and Remote Sensing, v. 44, no. 5, p. 58

————, 1979a, Calculating map projections for the ellipso Cartographer, v. 6, no. 1, p. 67–76.

————, 1979b, Projection notes: Am. Cartographer, v. 6 p. 81.

——, 1981a, Map projections for satellite tracking: Photogram-
tric Engineering and Remote Sensing, v. 47, no. 2, p. 205–213.

——, 1981b, Space Oblique Mercator projection—mathematical
'elopment: U.S. Geol. Survey Bull. 1518, 108 p.

——, 1981c, The perspective map projection of the Earth: Am.
'tographer, v. 8, no. 2, p. 149–160. Corrections, v. 9, no. 1,
2, p. 84.

——, 1982a, Map projections used by the U.S. Geological Sur-
': U.S. Geol. Survey Bull. 1532, 313 p.

——, 1982b, The modified Polyconic projection for the IMW:
'tographica, v. 19, nos. 3, 4, p. 31–43.

——, 1984a, A low-error conformal map projection for the 50
tes: Am. Cartographer, v. 11, no. 1, p. 27–39.

——, 1984b, Minimum-error map projections bounded by poly-
s: Cartographic Journal, v. 21, no. 2, p. 112–120.

——, 1985a, Computer-assisted map projection research: U.S.
l. Survey Bull. 1629, 157 p.

——, 1985b, The transverse and oblique Cylindrical Equal-Area
jection of the ellipsoid: Assoc. Am. Geographers, Annals, v. 75,
3, p. 431–442.

n, F.A., 1981, Obobshchennaya klassifikatsiya kartogra-
eskikh proektsiy po vidu izobrazheniya meridianov i paralleley:
desiya i Aerofotos'emka, no. 6, p. 111–116.

J.A., 1970, An introduction to the study of map projections
h ed.): London, Univ. London Press.

, P.D., 1952, Conformal projections in geodesy and cartography:
. Coast and Geodetic Survey Spec. Pub. 251.

——, 1970, Spheroidal geodesics, reference systems, and local
metry: U.S. Naval Oceanographic Office.

on, M.M., 1979, Maps for America: U.S. Geol. Survey, 265 p.

waite, C.W., 1927, The Cylindrical Equal-Area projection for a
' map of Eurasia and Africa: Univ. of Calif. Publications in
graphy, v. 2, no. 6, p. 211–230.

A., 1881, Mémoire sur la représentation des surfaces et les
ections des cartes géographiques: Paris, Gauthier Villars.

W.R., 1962, A classification of map projections: Assoc. Am.
graphers, Annals, v. 52, p. 167–175. See also refs. in Maling
'3, p. 98–104) and Maurer (1935, as translated 1968, p. v–vii).

U.S. Coast and Geodetic Survey, 1882, Report of the Superintendent
of the U.S. Coast and Geodetic Survey *** June 1880: Appendix
15: A comparison of the relative value of the Polyconic projection
used on the Coast and Geodetic Survey, with some other projections, by C.A. Schott:

————, 1900, Tables for a Polyconic projection of maps: U.S. Coast
and Geodetic Survey Spec. Pub. 5.

————, 1946, Tables for a Polyconic projection of maps: U.S. Coast
and Geodetic Survey Spec. Pub. 5, 6th ed.

U.S. Geological Survey, 1964, Topographic instructions of the United
States Geological Survey, Book 5, Part 5B, Cartographic tables:
U.S. Geol. Survey.

————, 1970, National Atlas of the United States: U.S. Geol.
Survey.

United Nations, 1963, Specifications of the International Map of the
World of the Millionth Scale, v. 2: New York, United Nations.

Van der Grinten, A.J., 1904, Darstellung der ganzen Erdoberfläche
auf einer Kreisförmigen Projektionsebene: Petermanns Geographische Mitteilungen, v. 50, p. 155–159.

————, 1905, Zur Verebnung der ganzen Erdoberfläche. Nachtrag
zu der Darstellung in Pet. Mitt. 1904 ***: Petermanns Geographische Mitteilungen, v. 51, p. 237.

Van Zandt, F.K., 1976, Boundaries of the United States and the several
States: U.S. Geol. Survey Prof. Paper 909, 191 p.

Wong, F.K.C., 1965, World map projections in the United States from
1940 to 1960: Syracuse, N.Y., M.A. thesis, Syracuse University.

World Geodetic System Committee, 1974, The Department of Defense
World Geodetic System 1972: Washington, Defense Mapping
Agency. Presented by T.O. Seppelin at the Internat. Symposium
on Problems Related to the Redefinition of North American Geodetic Networks, Fredericton, N.B., Canada, May 20–25, 1974.

Wraight, A.J., and Roberts, E.B., 1957, The Coast and Geodetic Survey, 1807–1957: 150 years of history: U.S. Coast and Geodetic
Survey.

Wray, Thomas, 1974, The seven aspects of a general map projection:
Cartographica, Monograph 11.

Young, A.E., 1930, Conformal map projections: Geographical Journal,
v. 76, no. 4, p. 348–351.

APPENDIXES

APPENDIX A

NUMERICAL EXAMPLES

The numerical examples which follow should aid in the use of the many formulas in this study of map projections. Single examples are given for equations for forward and inverse functions of the projections, both spherical and ellipsoidal, when both are given. They are given in the order the projections are given. The order of equations used is based on the order of calculation, even though the equations may be originally listed in a somewhat different order. In some cases, the last digit may vary from check calculations, due to rounding off, or the lack of it.

AUXILIARY LATITUDES (SEE P. 15–18)

For all examples under this heading, the Clarke 1866 ellipsoid is used: a is not needed here, $e^2 = 0.00676866$, or $e = 0.0822719$. Auxiliary latitudes will be calculated for geodetic latitude $\phi = 40°$:

Conformal latitude, using closed equation (3–1):

$$\chi = 2 \arctan \{\tan (45° + 40°/2) [(1 - 0.0822719 \sin 40°)/(1 + 0.0822719 \sin 40°)]^{0.0822719/2}\} - 90°$$
$$= 2 \arctan \{2.1445069 [0.8995456]^{0.0411360}\} - 90°$$
$$= 2 \arctan (2.1351882) - 90°$$
$$= 2 \times 64.9042961° - 90°$$
$$= 39.8085922° = 39°48'30.9''$$

Using series equation (3–2), obtaining χ first in radians, and omitting terms with e^8 for simplicity:

$$\chi = 40° \times \pi/180° - (0.00676866/2 + 5 \times 0.00676866^2/24 + 3 \times 0.00676866^3/32) \times \sin (2 \times 40°) + (5 \times 0.00676866^2/48 + 7 \times 0.00676866^3/80) \times \sin (4 \times 40°) - (13 \times 0.00676866^3/480) \sin (6 \times 40°)$$
$$= 0.6981317 - (0.0033939) \times 0.9848078 + (0.0000048) \times 0.3420201 - (.0000000) \times (-0.8660254)$$
$$= 0.6947910 \text{ radian}$$
$$= 0.6947910 \times 180°/\pi = 39.8085923°$$

For inverse calculations, using closed equation (3–4) with iteration and given $\chi = 39.8085922°$, find ϕ:

First trial:

$$\phi = 2 \arctan \{\tan (45° + 39.8085922°/2) [(1 + 0.0822719 \sin 39.8085922°)/(1 - 0.0822719 \sin 39.8085922°)]^{0.0822719/2}\} - 90°$$
$$= 2 \arctan \{2.1351882 [1.1112023]^{0.0411360}\} - 90°$$
$$= 129.9992366° - 90°$$
$$= 39.9992366°$$

Second trial:

$$\phi = 2 \arctan \{2.1351882 \ [(1 + 0.0822719 \sin 39.9992366°)/(1 - 0.0822719$$
$$\sin 39.9992366°)]^{0.0411360}\} - 90°$$
$$= 2 \arctan (2.1445068) - 90°$$
$$= 39.9999970°$$

The third trial gives $\phi = 40.0000000°$, also given by the fourth trial.
Using series equation (3−5):

$$\phi = 39.8085922° \times \pi/180° + (0.00676866/2 + 5 \times 0.00676866^2/24$$
$$+ 0.00676866^3/12) \sin (2 \times 39.8085922°) + (7 \times 0.00676866^2/48 + 29$$
$$\times 0.00676866^3/240) \sin (4 \times 39.8085922°) + (7 \times 0.00676866^3/120)$$
$$\sin (6 \times 39.8085922°)$$
$$= 0.6947910 + (0.0033939) \times 0.9836256 + (0.0000067) \times 0.3545461$$
$$+ (0.0000000) \times (-0.8558300)$$
$$= 0.6981317 \text{ radian}$$
$$= 0.6981317 \times 180°/\pi = 40.0000000°$$

Isometric latitude, using equation (3−7):

$$\psi = \ln \{\tan (45° + 40°/2) \ [(1 - 0.0822719 \sin 40°)/(1 + 0.0822719$$
$$\sin 40°)]^{0.0822719/2}\}$$
$$= \ln (2.1351882)$$
$$= 0.7585548$$

Using equation (3−8) with the value of χ resulting from the above examples:

$$\psi = \ln \tan (45° + 39.8085923°/2)$$
$$= \ln \tan 64.9042962°$$
$$= 0.7585548$$

For inverse calculations, using equation (3−9) with $\psi = 0.7585548$:

$$\chi = 2 \arctan e^{0.7585548} - 90°$$
$$= 2 \arctan (2.1351882) - 90°$$
$$= 39.8085922°$$

From this value of χ, ϕ may be found from (3−4) or (3−5) as shown above.
Using iterative equation (3−10), with $\psi = 0.7585548$, to find ϕ:

First trial:

$$\phi = 2 \arctan e^{0.7585548} - 90°$$
$$= 39.8085922°, \text{ as just above.}$$

Second trial:

$$\phi = 2 \arctan \{e^{0.7585548} \ [(1 + 0.0822719 \sin 39.8085922°)/(1 - 0.0822719$$
$$\sin 39.8085922°)]^{0.0822719/2}\} - 90°$$
$$= 2 \arctan (2.1351882 \times 1.0043469) - 90°$$
$$= 39.9992365°$$

Third trial:

$$\phi = 2 \arctan \{e^{0.7585548} \ [(1 + 0.0822719 \sin 39.9992365°)/(1 - 0.0822719$$
$$\sin 39.9992365°)]^{0.0822719/2}\} - 90°$$
$$= 39.9999970°$$

Fourth trial, substituting 39.9999970° in place of 39.9992365°:

φ = 40.0000000°, also given by fifth trial.

Authalic latitude, using equations (3−11) and (3−12):

q = (1−0.00676866) {sin 40°/(1−0.00676866 sin² 40°)−
      [1/(2×0.0822719)] ln [(1−0.0822719 sin 40°)/(1+0.0822719 sin
      40°)]}
   = 0.9932313 (0.6445903−6.0774117 ln 0.8995456)
   = 1.2792602

q_p = (1−0.00676866) {sin 90°/(1−0.00676866 sin² 90°)−[1/
      (2×0.0822719)] ln [(1−0.0822719 sin 90°)/(1+0.0822719 sin 90°)]}
   = 1.9954814

β = arcsin (1.2792602/1.9954814)
   = arcsin 0.6410785
   = 39.8722878° = 39°52′20.2″

Determining β from series equation (3−14) involves the same pattern as the example for equation (3−5) given above.

For inverse calculations, using equation (3−17) with iterative equation (3−16), given β = 39.8722878°, and q_p = 1.9954814 as determined above:

q = 1.9954814 sin 39.8722878°
   = 1.2792602

First trial:

φ = arcsin (1.2792602/2)
   = 39.762435°

Second trial:

φ = 39.7642435° + (180°/π) {[(1−0.00676866 sin² 39.7642435°)²/(2 cos
      39.7642435°)] [1.2792602/(1−0.00676866)−sin 39.7642435°/
      (1−0.00676866 sin² 39.7642435°)
      +[1/(2×0.0822719)] ln [(1−0.0822719 sin 39.7642435°)/
      (1+0.0822719 sin 39.7642435°)]]}
   = 39.9996014°

Third trial, substituting 39.9996014° in place of 39.7642435°,

φ = 39.9999992°

Fourth trial gives the same result.

Finding φ from β by series equation (3−18) involves the same pattern as the example for equation (3−5) given above.

Rectifying latitude, using equations (3−20) and (3−21):

M = a[(1−0.00676866/4−3×0.00676866²/64−5×0.00676866³/256)×40°
      ×π/180°−(3×0.00676866/8+3×0.00676866²/32+45×0.00676866³/
      1024) sin (2×40°)+(15×0.00676866²/256+45×0.00676866³/1024)
      sin (4×40°)−(35×0.00676866³/3072) sin (6×40°)]
   = a[0.9983057×0.6981317−0.0025426 sin 80°+0.0000027 sin 160°
      −0.0000000 sin 240°]
   = 0.6944458a

M_p = 1.5681349a, using 90° in place of 40° in the above example.

μ = 90°×0.6944458a/1.5681349a
   = 39.8563451° = 39°51′22.8″

Calculation of μ from series (3−23), and the inverse ϕ from (3−26), is similar to the example for equation (3−2) except that e_1 is used rather than e. From equation (3−24),

e_1 = [1−(1−0.00676866)$^{1/2}$]/[1+(1−0.00676866)$^{1/2}$]
= 0.001697916

Geocentric latitude, using equation (3−28),

ϕ_g = arctan [(1−0.00676866) tan 40°]
= 39.8085032° = 39°48′30.6″

Reduced latitude, using equation (3−31),

η = arctan [(1−0.00676866)$^{1/2}$ tan 40°]
= 39.9042229° = 39°54′15.2″

Series examples for ϕ_g and η follow the pattern of (3−2) and (3−23).

DISTORTION FOR PROJECTIONS OF THE ELLIPSOID (SEE P. 24-25)

Radius of curvature and length of degrees, using the Clarke 1866 ellipsoid at lat. 40° N.:
From equation (4−18),

R' = 6378206.4 (1−0.00676866)/(1−0.00676866 sin^2 40°)$^{3/2}$
= 6,361,703.0 m

From equation (4−19), using the figure just calculated,

L_ϕ = 6361703.0 × π/180° = 111,032.7 m, the length of 1° of latitude at lat. 40° N.

From equation (4−20),

N = 6378206.4/(1−0.00676866 sin^2 40°)$^{1/2}$
= 6,387,143.9 m

From equation (4−21),

L_λ = [6378206.4 cos 40°/(1−0.00676866 sin^2 40°)$^{1/2}$] π/180°
= 85,396.1 m, the length of 1° of longitude at lat. 40° N.

MERCATOR PROJECTION (SPHERE)−FORWARD EQUATIONS (SEE P. 41, 44)

Given: Radius of sphere: $R = 1.0$ unit
           Central meridian: $\lambda_0 = 180°$ W. long.
                     Point: $\phi = 35°$ N. lat.
                             $\lambda = 75°$ W. long.

Find: x, y, k

Using equations (7−1) , (7−2), and (7−3),

x = π×1.0×[(−75°)−(−180°)]/180° = 1.8325957 units
y = 1.0×ln tan (45° + 35°/2) = 1.0×ln tan (62.5°)
   = ln 1.9209821 = 0.6528366 unit
h = k = sec 35° = 1/cos 35° = 1/0.8191520 = 1.2207746

MERCATOR PROJECTION (SPHERE)–INVERSE EQUATIONS (SEE P. 44)

Inversing forward example:
Given: R, λ_0 for forward example
$$x = 1.8325957 \text{ units}$$
$$y = 0.6528366 \text{ unit}$$

Find: ϕ, λ

Using equations (7–4) and (7–5),

$$\phi = 90° - 2 \arctan (e^{-0.6528366/1.0})$$
$$= 90° - 2 \arctan (0.5205670) = 90° - 2 \times 27.5° = 35°$$
$$= 35° \text{ N. lat., since the sign is "+"}$$
$$\lambda = (1.8325957/1.0) \times 180°/\pi + (-180°)$$
$$= 105° - 180° = -75° = 75° \text{ W. long., since the sign is "−"}$$

The scale factor may then be determined as in equation (7–3) using the newly calculated ϕ.

MERCATOR PROJECTION (ELLIPSOID)–FORWARD EQUATIONS (SEE P. 44)

Given: Clarke 1866 ellipsoid: $a = 6378206.4$ m
$$e^2 = 0.00676866$$
or $e = 0.0822719$
Central meridian: $\lambda_0 = 180°$ W. long.
Point: $\phi = 35°$ N. lat.
$\lambda = 75°$ W. long.

Find: x, y, k

Using equations (7–6), (7–7), and (7–8),

$$x = 6378206.4 \times [(-75°) - (-180°)] \times \pi/180° = 11688673.7 \text{ m}$$
$$y = 6378206.4 \ln \left[\tan (45° + 35°/2) \left(\frac{1 - 0.0822719 \sin 35°}{1 + 0.0822719 \sin 35°} \right)^{0.0822719/2} \right]$$
$$= 6378206.4 \ln [1.9209821 \times 0.9961223]$$
$$= 6378206.4 \ln 1.9135331 = 4{,}139{,}145.6 \text{ m}$$
$$k = (1 - 0.00676866 \sin^2 35°)^{1/2}/\cos 35°$$
$$= 1.2194146$$

MERCATOR PROJECTION (ELLIPSOID)–INVERSE EQUATIONS (SEE P. 44–45)

Inversing forward example:

Given: a, e, λ_0 for forward example
$$x = 11688673.7 \text{ m}$$
$$y = 4139145.6 \text{ m}$$

Find: ϕ, λ

Using equation (7–10),

$$t = e^{-4139145.6/6378206.4} = 0.5225935$$

From equation (7−11), the first trial ϕ = 90° − 2 arctan 0.5225935 = 34.8174484°. Using this value on the right side of equation (7−9),

$$\phi = 90° -2 \text{ arctan } \{0.5225935[(1-0.0822719 \sin 34.8174484°)/$$
$$(1+0.0822719 \sin 34.8174484°)]^{0.0822719/2}\}$$
$$= 34.9991687°$$

Replacing 34.8174484° with 34.9991687° for the second trial, recalculation using (7−9) gives ϕ = 34.9999969°. The third trial gives ϕ = 35.0000006°, which does not change (to seven places) with recalculation. If it were not for rounding-off errors in the values of x and y, ϕ would be 35° N. lat.

For λ, using equation (7−12),

$$\lambda = (11688673.7/6378206.4) \times 180°/\pi + (-180°)$$
$$= -75.0000001° = 75.0000001° \text{ W. long.}$$

Using equations (7−13) and (3−5) instead, to find ϕ,

$$\chi = 90° -2 \text{ arctan } 0.5225935$$
$$= 90° -55.1825516°$$
$$= 34.8174484°$$

using t as calculated above from (7−10). Using (3−5), χ is inserted as in the example given above for inverse auxiliary latitude χ:

$$\phi = 35.0000006°$$

TRANSVERSE MERCATOR (SPHERE)−FORWARD EQUATIONS (SEE P. 58)

Given: Radius of sphere: R = 1.0 unit
         Origin: ϕ_0 = 0
                   λ_0 = 75° W. long.
Central scale factor: k_0 = 1.0
         Point: ϕ = 40°30′ N. lat.
                  λ = 73°30′ W. long.

Find: x, y, k

Using equations (8−5), (8−1), (8−3), and (8−4) in order

$$B = \cos 40.5° \sin [(-73.5°)-(-75°)]$$
$$= \cos 40.5° \sin 1.5° = 0.0199051$$
$$x = ½ \times 1.0 \times 1.0 \ln [(1+0.0199051)/(1-0.0199051)]$$
$$= 0.0199077 \text{ unit}$$
$$y = 1.0 \times 1.0 \{\text{arctan } [\tan 40.5°/\cos 1.5°]-0\}$$
$$= 40.5096980° \pi/180° = 0.7070276 \text{ unit}$$
$$k = 1.0/(1-0.0199051^2)^{1/2} = 1.0001982$$

TRANSVERSE MERCATOR (SPHERE)−INVERSE EQUATIONS (SEE P. 60)

Inversing forward example:

Given: R, ϕ_0, λ_0, k_0 for forward example

$$x = 0.0199077 \text{ unit}$$
$$y = 0.7070276 \text{ unit}$$

Find: ϕ, λ

Using equation (8−8),

$$D = 0.7070276/(1.0 \times 1.0) + 0 = 0.7070276 \text{ radian}$$

For the hyperbolic functions of (x/Rk_0), the relationships

$$\sinh\ x = (e^x - e^{-x})/2$$

and

$$\cosh\ x = (e^x + e^{-x})/2$$

are recalled if the function is not directly available on a given computer or calculator. In this case,

$$\begin{aligned}
\sinh (x/Rk_0) &= \sinh\ [0.0199077/(1.0 \times 1.0)] \\
&= (e^{0.0199077} - e^{-0.0199077})/2 \\
&= 0.0199090 \\
\cosh (x/Rk_0) &= (e^{0.0199077} + e^{-0.0199077})/2 \\
&= 1.0001982
\end{aligned}$$

From equation (8−6), with D in radians, not degrees,

$$\begin{aligned}
\phi &= \arcsin (\sin 0.7070276/1.0001982) = \arcsin (0.6495767/1.0001982) \\
&= 40.4999995° \text{ N. lat.}
\end{aligned}$$

From equation (8−7),

$$\begin{aligned}
\lambda &= -75° + \arctan [0.0199090/ \cos 0.7070276] \\
&= -75° + \arctan 0.0261859 = -75° + 1.4999961 = -73.5000039° \\
&= 73.5000039° \text{ W. long.}
\end{aligned}$$

If more decimals were supplied with the x and y calculated from the forward equations, the ϕ and λ here would agree more exactly with the original values.

TRANSVERSE MERCATOR (ELLIPSOID)−FORWARD EQUATIONS (SEE P. 60-61, 63)

Given: Clarke 1866 ellipsoid: $a = 6378206.4$ m
$e^2 = 0.00676866$
Origin (UTM Zone 18): $\phi_0 = 0$
$\lambda_0 = 75°$ W. long.
Central scale factor: $k_0 = 0.9996$
Point $\phi = 40° 30'$ N. lat.
$\lambda = 73° 30'$ W. long.

Find: x, y, k

Using equations (8−12) through (8−15) in order,

$$e'^2 = 0.00676866/(1-0.00676866) = 0.0068148$$
$$N = 6378206.4/(1-0.00676866 \sin^2 40.5°)^{1/2} = 6387330.5 \text{ m}$$
$$T = \tan^2 40.5° = 0.7294538$$
$$C = 0.0068148 \cos^2 40.5° = 0.0039404$$
$$A = (\cos 40.5°) \times [(-73.5°) - (-75°)] \pi/180° = 0.0199074$$

Instead of equation (3−21), we may use (3−22) for the Clarke 1866:

$$M = 111132.0894 \times (40.5°) - 16216.94 \sin (2\times40.5°) + 17.21 \sin (4\times40.5°)$$
$$- 0.02 \sin (6\times40.5°)$$
$$= 4,484,837.67 \text{ m}$$
$$M_0 = 111132.0894 \times 0° - 16216.94 \sin (2\times0°) + 17.21 \sin (4\times0°) - 0.02 \sin (6\times0°)$$
$$= 0.00 \text{ m}$$

Equations (8−9) and (8−10) may now be used:

$$x = 0.9996 \times 6387330.5 \times [0.0199074 + (1-0.7294538+0.0039404)$$
$$\times 0.0199074^3/6 + (5-18\times0.7294538+0.7294538^2+72\times0.0039404$$
$$- 58 \times 0.0068148) \times 0.0199074^5/120]$$
$$= 127,106.5 \text{ m}$$
$$y = 0.9996 \times \{4484837.7-0+6387330.5\times0.8540807\times[0.0199074^2/2$$
$$+ (5-0.7294538+9\times0.0039404+4\times0.0039404^2) \times 0.0199074^4/24$$
$$+ (61-58\times0.7294538+0.7294538^2+600\times0.0039404-330$$
$$\times 0.0068148) \times 0.0199074^6/720]\}$$
$$= 4,484,124.4 \text{ m}$$

These values agree exactly with the UTM tabular values, except that 500,000.0 m must be added to x for "false eastings." To calculate k, using equation (8−11),

$$k = 0.9996 \times [1+(1+0.0039404)\times0.0199074^2/2+(5-4\times0.7294538+42$$
$$\times 0.0039404 + 13 \times 0.0039404^2 - 28 \times 0.0068148) \times 0.0199074^4/24$$
$$+ (61-148\times0.7294538+16\times0.7294538^2) \times 0.0199074^6/720]$$
$$= 0.9997989$$

Using equation (8−16) instead,

$$k = 0.9996 \times [1 + (1 + 0.0068148 \cos^2 40.5°) \times 127106.5^2/$$
$$(2\times0.9996^2 \times 6387330.5^2)]$$
$$= 0.9997989$$

TRANSVERSE MERCATOR (ELLIPSOID)−INVERSE EQUATIONS (SEE P. 63-64)

Inversing forward example:

Given: Clarke 1866 ellipsoid: $a = 6378206.4$ m
                               $e^2 = 0.00676866$

       Origin (UTM Zone 18): $\phi_0 = 0$
                               $\lambda_0 = 75°$ W. long.

       Central scale factor: $k_0 = 0.9996$
                       Point: $x = 127106.5$ m
                               $y = 4484124.4$ m

Find: ϕ, λ

Calculating M_0 from equation (3−22),

M_0 = 111132.089×0°−16216.9 sin (2×0°) + 17.2 sin (4×0°) − 0.02 sin
      (6×0°)
   = 0

From equations (8−12), (8−20), (3−24), and (7−19) in order,

e'^2 = 0.00676866/(1−0.00676866) = 0.0068148
M = 0+4484124.4/0.9996 = 4485918.8 m
e_1 = [1−(1−0.00676866)$^{1/2}$]/[1 + (1−0.00676866)$^{1/2}$]
   = 0.001697916
μ = 4485918.8/[6378206.4×(1−0.00676866/4−3×0.00676866^2/64
      −5×0.00676866^3/256)]
   = 0.7045135 radian

From equation (3−26), using μ in radians, omitting the last term,

ϕ_1 = 0.7045135+(3×0.001697916/2−27×0.001697916^3/32) sin
      (2×0.7045135)+(21×0.001697916^2/16−55×0.001697916^4/32)
      sin (4×0.7045135)+(151×0.001697916^3/96) sin (6×0.7045135)
   = 0.7070283 radian
   = 0.7070283×180°/π
   = 40.5097362°

Now equations (8−21) through (8−25) may be used:

C_1 = 0.0068148 cos^2 40.5097362° = 0.0039393
T_1 = tan^2 40.5097362° = 0.7299560
N_1 = 6378206.4/(1−0.00676866 sin^2 40.5097362°)$^{1/2}$
   = 6387334.2 m
R_1 = 6378206.4×(1−0.00676866)/(1−0.00676866 sin^2 40.5097362°)$^{3/2}$
   = 6,362,271.4 m
D = 127106.5/(6387334.2×0.9996) = 0.0199077

Returning to equation (8−17),

ϕ = 40.5097362°−(6387334.2×0.8543746/6362271.4)×[0.0199077^2/2
      −(5+3×0.7299560+10×0.0039393−4×0.0039393^2−9
      ×0.0068148)×0.0199077^4/24+(61+90×0.7299560+298
      ×0.0039393+45×0.7299560^2−252×0.0068148−3
      ×0.0039393^2)×0.0199077^6/720]×180°/π
   = 40.5000000° = 40°30′ N. lat.

From equation (8−18),

λ = −75° + [0.0199077−(1+2×0.7299560+0.0039393)×0.0199077^3/6
      +(5−2×0.0039393+28×0.7299560−3×0.0039393^2+8
      ×0.0068148+24×0.7299560^2)×0.0199077^5/120]/cos
      40.5097362°} × 180°/π
   = −75° + 1.5000000° = −73.5° = 73°30′ W. long.

OBLIQUE MERCATOR (SPHERE)—FORWARD EQUATIONS (SEE P. 69–70)

Given: Radius of sphere: R = 1.0 unit
Central scale factor: k_0 = 1.0
Central line through: ϕ_1 = 45° N. lat.
ϕ_2 = 0° lat.
λ_1 = 0° long.
λ_2 = 90° W. long.
Point: ϕ = 30° S. lat.
λ = 120° E. long.

Find: x, y, k

Using equation (9–1),

λ_p = arctan {[cos 45° sin 0° cos 0° − sin 45° cos 0° cos (−90°)]/
        [sin 45° cos 0° sin (−90°) − cos 45° sin 0° sin 0°]}
    = arctan {[0−0]/[−0.7071068−0]} = 0°

Since the denominator is negative, add or subtract 180° (the numerator has neither sign, but it doesn't matter). Thus,

$$\lambda_p = 0° + 180° = 180°$$

From equation (9–2),

$$\phi_p = \text{arctan} \left[-\cos (180°-0°)/\tan 45° \right]$$
$$= \text{arctan} \left[+1/0.7071068 \right] = 45°$$

The other pole is then at ϕ = −45°, λ = 0°. From equation (9–6a),

$$\lambda_0 = 180° + 90° = 270°, \text{ equivalent to } 270° -360° \text{ or } -90°.$$

From equation (9–6),

$$A = \sin 45° \sin (-30°) - \cos 45° \cos (-30°) \sin [120° - (-90°)]$$
$$= 0.7071068 \times (-0.5) - 0.7071068 \times 0.8660254 \times (-0.5)$$
$$= -0.0473672$$

From equation (9–3),

$x = -1.0 \times 1.0$ arctan [tan (−30°) cos 45°/cos (120° + 90°) + sin 45° tan (120° + 90°)]
  = 0.7214592

Since cos (120° + 90°) is negative, subtract π, or x = −2.4201335 units

From equation (9–4),

$$y = (1/2) \times 1.0 \times 1.0 \ln [(1-0.0473672)/(1+0.0473672)]$$
$$= -0.0474026 \text{ unit}$$

From equation (9–5),

$$k = 1.0/[1-(-0.0473672)^2]^{1/2} = 1.0011237$$

If the parameters are given in terms of a central point (for equations (9–7) and (9–8)), we shall assume certain artificial parameters (calculated with different formulas) which give the same pole as above:

Given: Radius of sphere: R = 1.0 unit
           Central scale factor: k_0 = 1.0
         Azimuth of central line: β = 48.8062990° east of north
                    Center: ϕ_c = 20° N. lat.
                         λ_c = 68.6557771°W. long.

Using equations (9−7) and (9−8),

ϕ_p = arcsin (cos 20° sin 48.8062990°)
   = 45.0°N. lat.
λ_p = arctan [−cos 48.8062990°/(−sin 20° sin 48.8062990°)]
        −68.6557771°
   = 0°

Since the denominator of the argument of arctan is negative, add −180° to λ_p, using "−" since the numerator is "−":

$$\lambda_p = 180°\text{W. long.}$$

OBLIQUE MERCATOR (SPHERE)−INVERSE EQUATIONS (SEE P. 70)

Inversing forward example:

Given: Radius of sphere: R = 1.0 unit
           Central scale factor: k_0 = 1.0
         Central line through: ϕ_1 = 45° N. lat.
                          ϕ_2 = 0° lat.
                          λ_1 = 0° long.
                          λ_2 = 90° W. long.
              Point: x = −2.4201335 units
                     y = −0.0474026 unit

Find: ϕ, λ

First, ϕ_p and λ_p are determined, exactly as for the forward example, so that λ_0 again is −90°, and ϕ_p = 45°. Determining hyperbolic functions, if not readily available,

$$y/Rk_0 = -0.0747026/(1.0 \times 1.0) = -0.0474026$$
$$e^{-0.0474026} = 0.9537034$$
$$\sinh(y/Rk_0) = (0.9537034 - 1/0.9537034)/2$$
$$= -0.0474203$$
$$\cosh(y/Rk_0) = (0.9537034 + 1/0.9537034)/2$$
$$= 1.0011237$$
$$\tanh(y/Rk_0) = (0.9537034 - 1/0.9537034)/(0.9537034 + 1/0.9537034)$$
$$= -0.0473671$$

From equation (9−9),

ϕ = arcsin {sin 45° × (−0.0473671) + cos 45° sin
     [(−2.4201335/(1.0 × 1.0))×180°/π]/1.0011237
  = arcsin (−0.5000000)
  = −30° = 30°S. lat.

From equation (9−10),

$$\lambda = -90° + \arctan\{[\sin 45° \sin [-2.4201335 \times 180°/(\pi \times 1.0 \times 1.0)]$$
$$- \cos 45° \times (-0.0474203)]/\cos[-2.4201335 \times 180°/$$
$$(\pi \times 1.0 \times 1.0)]\}$$
$$= -90° + 30.0000041°$$
$$= -59.9999959°$$

but the main denominator is −0.7508428, which is negative, while the numerator is also negative. Therefore, add (−180°) to λ, so $\lambda = -59.9999959° - 180° = -239.9999959° = 240°$ W. long. = 120° E. long.

OBLIQUE MERCATOR (HOTINE ELLIPSOID)−FORWARD EQUATIONS
(SEE P. 71−74)

For alternate A:

Given: Clarke 1866 ellipsoid: a = 6378206.4 m
                                e^2 = 0.00676866
          or e = 0.0822719
Central scale factor: k_0 = 0.9996
          Center: ϕ_0 = 40° N. lat.
Central line through: ϕ_1 = 47°30′ N. lat.
                       λ_1 = 122° 18′ W. long. (Seattle, Wash.)
                       ϕ_2 = 25°42′ N. lat.
                       λ_2 = 80°12′ W. long. (Miami, Fla.)
False coordinates: x_0 = 4,000,000.0 m
                    y_0 = 500,000.0 m
          Point: ϕ = 40°48′ N. lat.
                  λ = 74°00′ W. long. (New York City)

Find: x, y, k

Following equations (9−11) through (9−24) in order:

$$B = [1 + 0.00676866 \cos^4 40°/(1-0.00676866)]^{1/2}$$
$$= 1.0011727$$
$$A = 6378206.4 \times 1.0011727 \times 0.9996 \times (1-0.00676866)^{1/2}/$$
$$(1-0.00676866 \sin^2 40°)$$
$$= 6,379,333.2 \text{ m}$$
$$t_0 = \tan (45°-40°/2)/[(1-0.0822719 \sin 40°)/$$
$$(1+0.0822719 \sin 40°)]^{0.0822719/2}$$
$$= 0.4683428$$
$$t_1 = \tan (45°-47.5°/2)/[(1-0.0822719 \sin 47.5°)/$$
$$(1+0.0822719 \sin 47.5°)]^{0.0822719/2}$$
$$= 0.3908266$$
$$t_2 = \tan (45°-25.7°/2)/[(1-0.0822719 \sin 25.7°)/$$
$$(1+0.0822719 \sin 25.7°)]^{0.0822719/2}$$
$$= 0.6303639$$
$$D = 1.0011727 \times (1-0.00676866)^{1/2}/[\cos 40° \times$$
$$(1-0.00676866 \sin^2 40°)^{1/2}]$$
$$= 1.3043327$$
$$E = [1.3043327 + (1.3043327^2-1)^{1/2}] \times 0.4683428^{1.0011727}$$
$$= 1.0021857$$

using the "+" sign, since ϕ_0 is north or positive.

H = $0.3908266^{1.0011727}$ = 0.3903963

L = $0.6303639^{1.0011727}$ = 0.6300229

F = $1.0021857/0.3903963$ = 2.5670986

G = $(2.5670986 - 1/2.5670986)/2$ = 1.0887769

J = $(1.0021857^2 - 0.6300229 \times 0.3903963)/(1.0021857^2 + 0.6300229$
$\times 0.3903963)$ = 0.6065716

P = $(0.6300229 - 0.3903963)/(0.6300229 + 0.3903963)$
= 0.2348315

λ_0 = $\frac{1}{2}[(-122.3°) + (-80.2°)] - \arctan \{0.6065716 \tan [1.0011727$
$\times (-122.3° + 80.2°)/2]/0.2348315\}/1.0011727$
= $-101.25° - \arctan (-0.9953887)/1.0011727$
= $-56.4349628°$

γ_0 = $\arctan \{\sin [1.0011727 \times (-122.3° + 56.4349628°)]/1.0887769\}$
= $-39.9858829°$

α_c = $\arcsin [1.3043327 \sin (-39.9858829°)]$
= $-56.9466070°$

These are constants for the map. For the given ϕ and λ, following equations (9-25) through (9-34) in order:

t = $\tan (45° - 40.8°/2)/[(1 - 0.0822719 \sin 40.8°)/(1 + 0.0822719 \sin$
$40.8°)]^{0.0822719/2}$
= 0.4598671

Q = $1.0021857/0.4598671^{1.0011727}$ = 2.1812805

S = $(2.1812805 - 1/2.1812805)/2$ = 0.8614171

T = $(2.1812805 + 1/2.1812805)/2$ = 1.3198634

V = $\sin [1.0011727 \times (-74° + 56.4349628°)]$
= -0.3021309

U = $[0.3021309 \cos (-39.9858829°) + 0.8614171 \sin (-39.9858829°)]/$
1.3198634
= -0.2440041

v = $6379333.2 \ln [(1 + 0.2440041)/(1 - 0.2440041)]/(2 \times 1.0011727)$
= $1,586,767.3$ m

u = $[[6379333.2 \arctan \{[0.8614171 \cos (-39.9858829°)$
$+ (-0.3021309) \sin (-39.9858829°)]/\cos [1.0011727 \times (-74°$
$+ 56.4349628°)]\}/1.0011727]] \times \pi/180°$
= $4,655,443.7$ m

Note: Since $\cos [1.0011727 \times (-74° + 56.4349628°)] = 0.9532664$, which is positive, no correction is needed to the arctan in the equation for u. The $(\pi/180°)$ is inserted, if arctan is calculated in degrees.

k = $6379333.2 \cos [1.0011727 \times 4655443.7 \times 180°/(\pi \times 6379333.2)]$
$\times (1 - 0.00676866 \sin^2 40.8°)^{1/2}/\{6378206.4 \cos 40.8° \cos$
$[1.0011727 \times (-74° + 56.4349628°)]\}$
= 1.0307554

x = $1586767.3 \cos (-56.9466070°) + 4655443.7 \sin (-56.9466070°)$
$+ 4000000$
= $963,436.1$ m

y = $4655443.7 \cos (-56.9466070°) - 1586767.3 \sin (-56.9466070°)$
$+ 500000$
= $4,369,142.8$ m

For alternate B (forward):

Given: Clarke 1866 ellipsoid: a = 6378206.4 m
                                   e^2 = 0.00676866
                or e = 0.0822719
         Central scale factor: k_0 = 1.0
                       Center: ϕ_0 = 36° N. lat.
                                 λ_c = 77.7610558° W. long.
      Azimuth of central line: α_c = 14.3394883° east of north
                        Point: ϕ = 38°48′33.166″ N. lat.
                                      = 38.8092128°
                                 λ = 76°52′14.863″ W. long.
                                      = −76.8707953°

Find: u, v (example uses center of Zone 2, Path 16, Landsat mapping, with Hotine Oblique Mercator).

Using equations (9−11) through (9−39) in order,

B = $[1+0.00676866 \cos^4 36°/(1-0.00676866)]^{1/2}$
   = 1.0014586
A = 63780206.4 × 1.0014586 × 1.0 × $(1-0.00676866)^{1/2}/(1-0.00676866$
       $\sin^2 36°)$ = 6,380,777.0 m
t_0 = tan $(45°-36°/2)/[(1-0.0822719 \sin 36°)/(1+0.0822719 \sin$
       $36°)]^{0.0822719/2}$
   = 0.5115582
D = 1.0014586 × $(1-0.00676866)^{1/2}/[\cos 36°$
       × $(1-0.00676866 \sin^2 36°)^{1/2}]$
   = 1.2351194
F = 1.2351194 + $(1.2351194^2-1)^{1/2}$ = 1.9600471

using the "+" sign since ϕ_0 is north or positive.

E = $1.9600471×0.5115582^{1.0014586}$ = 1.0016984
G = (1.9600471−1/1.9600471)/2 = 0.7249276
γ_0 = arcsin [(sin 14.3394883°)/1.2351194]
   = 11.5673996°
λ_0 = −77.7610558° − [arcsin (0.7249276 tan 11.5673996°)]/1.0014586
   = −86.2814800°
$u_{(36°, -77.76...°)}$ = + (6380777.0/1.0014586) arctan $[(1.2351194^2-1)^{1/2}/$
       cos 14.3394883°]× $\pi/180°$
   = 4,092,868.9 m

Note: The $\pi/180°$ is inserted, if arctan is calculated in degrees. These are constants for the map. The calculations of u, v, x, and y for (ϕ, λ) follow the same steps as the numerical example for equations (9−25) through (9−34) for alternate A. For ϕ = 38.8092128° and λ = −76.8707953°, it is found that

$$u = 4,414,439.0 \text{ m}$$
$$v = -2,356.3 \text{ m}$$

OBLIQUE MERCATOR (HOTINE ELLIPSOID)−INVERSE EQUATIONS
(SEE P. 74−75)

The above example for alternate A will be inverted, first using equations (9−11) through (9−24), then using equations (9−40) through (9−48). Since no new equations are involved for inverse alternate B, an example of the latter will be omitted. As stated with the inverse equations, the constants for the map are chosen as in the forward examples.

Inversing forward example for alternate A:

Given: Clarke 1866 ellipsoid: a = 6,378,206.4 m
e^2 = 0.00676866
or e = 0.0822719
Central scale factor: k_0 = 0.9996
Center: ϕ_0 = 40° N. lat.
Center line through: ϕ_1 = 47° 30′ N. lat.
λ_1 = 122° 18′ W. long.
ϕ_2 = 25° 42′ N. lat.
λ_2 = 80° 12′ W. long.
False coordinates: x_0 = 4,000,000.0 m
y_0 = 500,000.0 m
Point: x = 963,436.1 m
y = 4,369,142.8 m

Find: ϕ, λ

Using equations (9−11) through (9−24) in order, again gives the following constants:

$$B = 1.0011727$$
$$A = 6,379,333.2 \text{ m}$$
$$E = 1.0021857$$
$$\lambda_0 = -56.4349628°$$
$$\gamma_0 = -39.9858829°$$
$$\alpha_c = -56.9466070°$$

Following equations (9−40) through (9−48) in order:

v = (963436.1−4000000.0) cos (−56.9466070°) − (4369142.8
−500000.0) sin (−56.9466070°)
= 1,586,767.3 m

u = (4369142.8−500000.0) cos (−56.9466070°) + (963436.1
−4000000.0) sin (−56.9466070°)
= 4,655,443.7 m

Q' = $e^{-(1.0011727 \times 1586767.3/6379333.2)}$
= $e^{-0.2490273}$
= 0.7795587

S' = (0.7795587−1/0.7795587)/2 = −0.2516092

T' = (0.7795587+1/0.7795587)/2 = 1.0311679

V' = sin [(1.0011727×4655443.7/6379333.2)×180°/π]
= sin 41.8617535° = 0.6673356

U' = [0.6673356 cos (−39.9858829°) − 0.2516092 sin (−39.9858829°)]/
1.0311679
= 0.6526562

t = $\{1.0021857/[(1+0.6526562)/(1-0.6526562)]^{1/2}\}^{1/1.0011727}$
= 0.4598671

The first trial ϕ for equation (7−9) is

$$\phi = 90° -2 \text{ arctan } (0.4598671) = 40.6077096°$$

Calculating a new trial ϕ:

$$\phi = 90° - 2 \arctan \{0.4598671 \times [(1-0.0822719 \sin 40.6077096°)/$$
$$(1+0.0822719 \sin 40.6077096°)]^{0.0822719/2}\}$$
$$= 40.7992509°$$

Substituting 40.7992509° in place of 40.6077096° and recalculating, $\phi = 40.7999971°$. Using this ϕ for the third trial, $\phi = 40.8000000°$. The next trial gives the same value of ϕ. Thus,

$$\phi = 40.8° = 40°48' \text{ N. lat.}$$
$$\lambda = -56.4349628° - \arctan \{[-0.2516092 \cos (-39.9858829°)$$
$$- 0.6673356 \sin (-39.9858829°)]/\cos [(1.0011727$$
$$\times 4655443.7/6379333.2) \times 180°/\pi]\}/1.0011727$$
$$= -74.0000000° = 74°00' \text{ W. long.}$$

Using series equation (3−5) with (7−13), to avoid iteration of (7−9), and beginning with equation (7−13),

$$\chi = 90° - 2 \arctan 0.4598671$$
$$= 40.6077096°$$

Since equation (3−5) is used in an example under *Auxiliary latitudes*, the calculation will not be shown here.

CYLINDRICAL EQUAL-AREA (SPHERE)–FORWARD EQUATIONS
(SEE P. 77, 80)

Normal aspect:
Given: Radius of sphere: $R = 1.0$ unit
        Central meridian: $\lambda_0 = 75°$ W. long.
        Standard parallel: $\phi_s = 30°$ N. & S. lat.
        Point: $\phi = 35°$ N. lat.
                $\lambda = 80°$ E. long.

Find: x, y

Using equations (10−1) and (10−2),

$$x = \pi \times 1.0 \times [80°-(-75°)] \times (\cos 30°)/180° = 2.3428242 \text{ units}$$
$$y = 1.0 \times \sin 35°/\cos 30° = 0.6623090 \text{ unit}$$

Transverse aspect:
Given: Radius of sphere: $R = 1.0$ unit
               Origin: $\phi_0 = 20°$ S. lat.
                        $\lambda_0 = 75°$ W. long.
          Central scale factor: $h_0 = 0.98$
                Point: $\phi = 25°$ N. Lat.
                        $\lambda = 90°$ W. long.

Find: x, y

Using equations (10−3) and (8−3),

$$x = (1.0/0.98) \times \cos 25° \sin [(-90°)-(-75°)]$$
$$= (1.0/0.98) \times \cos 25° \sin (-15°)$$
$$= -0.2393569 \text{ unit}$$

$$y = 1.0 \times 0.98 \times \{\arctan [\tan 25°/\cos (-15°)] - (-20°)\} \times \pi/180°$$
$$= 0.98 \times 45.7692621° \times \pi/180° = 0.7828478 \text{ unit}$$

Oblique aspect:

Given: Radius of sphere: $R = 1.0$ unit
        Central scale factor: $h_0 = 0.98$
        Central line through: $\phi_1 = 30°$ N. lat.
                                $\phi_2 = 60°$ N. lat.
                                $\lambda_1 = 75°$ W. long.
                                $\lambda_2 = 50°$ W. long.
        Point: $\phi = 30°$ S. lat.
                                $\lambda = 100°$ W. long.

Find: x, y

Using equation (9−1),

$$\lambda_p = \arctan \{[\cos 30° \sin 60° \cos (-75°) - \sin 30° \cos 60° \cos (-50°)]/$$
$$[\sin 30° \cos 60° \sin (-50°) - \cos 30° \sin 60° \sin (-75°)]\}$$
$$= \arctan \{[0.1941143 - 0.1606969]/[-0.1915111 - (-0.7244444)]\}$$
$$= \arctan (0.0334174/0.5329333)$$
$$= 3.5880129° = 3.5880129° \text{ E. long.}$$

Since the denominator is positive, 180° is not added to the result.
From equation (9−6a),

$$\lambda_0 = 3.5880129° + 90° = 93.5880129°$$

From equation (9−2),

$$\phi_p = \arctan \{-\cos [3.5880129° - (-75°)]/\tan 30°\}$$
$$= -18.9169858° = 18.9169858° \text{ S. lat.}$$

The other pole is then at 18.9169858° N. lat. and 176.4119871° W. long. From equations (10−4) and (10−5), calculating the arctan in radians:

x $= 1.0 \times 0.98 \arctan \{[\tan (-30°) \cos (-18.9169858°)$
      $+ \sin (-18.9169858°) \sin (-100° - 93.5880129°)]/$
      $\cos (-100° - 93.5880129°)\}$
   $= 0.98 \times \arctan [-0.6223338/(-0.9720102)]$
   $= 0.98 \times (0.5694937 + \pi)$, adding π since denominator is negative.
   $= 3.6368646$ units
y $= (1.0/0.98) [\sin (-18.9169858°) \sin (-30°) -$
      $\cos (-18.9169858°) \cos (-30°) \sin (-100° - 93.5880129°)]$
   $= -0.0309947$ unit

To locate a pole given a central point using equations (9−7) and (9−8), refer to the numerical example given under the forward spherical equations for the Oblique Mercator projection (p. 000).

CYLINDRICAL EQUAL-AREA (SPHERE)−INVERSE EQUATIONS
(SEE P. 80)

Inversing forward examples:
Normal aspect:
Given: R, λ_0, ϕ_s for forward example

$$x = 2.3428242 \text{ units}$$
$$y = 0.6623090 \text{ unit}$$

Find: ϕ, λ

Using equations (10−6) and (10−7),

$\phi = \arcsin\ [(0.6623090/1.0)\times\cos 30°]$
$\quad = 34.9999988° = 35°$ N. lat, if there were no round-off errors.
$\lambda = [2.3428242/(1.0\times\cos 30°)]\times 180°/\pi + (-75°)$
$\quad = 80°$ E. long., ignoring round-off errors.

Transverse aspect:
Given: R, ϕ_0, λ_0, h_0, for forward example

$$x = -0.2393569 \text{ unit}$$
$$y = 0.7828478 \text{ unit}$$

Find ϕ, λ

Using equation (10−10), (10−8), and (10−9) in order,

$D\ = 0.7828478/(1.0\times 0.98) + (-20°)\times \pi/180°$
$\quad = 0.4497584$
$\phi\ = \arcsin\ \{[1-(0.98\times(-0.2393569)/1.0)^2]^{1/2}$
$\qquad \times \sin\ (0.4497584 \text{ radians})\}$
$\quad = 25°$ N. lat., ignoring round-off errors.
$\lambda\ = -75° + \arctan\ \{[0.98\times(-0.2393569)/1.0]/$
$\qquad [[1-(0.98\times(-0.2393569)/1.0)^2]^{1/2} \cos\ (0.4497584 \text{ radians})]\}$
$\quad = -90° = 90°$ W. long.

Oblique aspect:
Given: R, h_0, and central line through same points as forward example,

$$x\ = 3.6368646 \text{ units}$$
$$y\ = -0.0309947 \text{ unit}$$

Find ϕ, λ

First, ϕ_p and λ_p are determined exactly as for the forward example, so that λ_0 again is 93.5880129°, and ϕ_p is −18.9169858°. Using equations (10−11) and (10−12),

$yh_0/R\ = -0.0309947 \times 0.98/1.0$
$\qquad = -0.0303748$
$x/(Rh_0)\ = 3.6368646/(1.0\times 0.98)$
$\qquad = 3.7110863$
$\phi\ = \arcsin\ \{-0.0303748\times\sin\ (-18.9169858°)$
$\qquad + [1-(-0.0303748)^2]^{1/2}$
$\qquad \times \cos\ (-18.9169858°)$
$\qquad \times \sin\ (3.7110863 \text{ radians})\}$
$\quad = \arcsin\ (-0.5) = -30° = 30°$ S. lat.
$\lambda\ = 93.5880129° + \arctan\ \{[[1-(-0.0303748)^2]^{1/2}$
$\qquad \times \sin\ (-18.9169858°)$
$\qquad \times \sin\ (3.7110863 \text{ radians})$
$\qquad - (-0.0303748)\times\cos\ (-18.9169858°)]/$
$\qquad [[1-(-0.0303748)^2]^{1/2} \times \cos\ (3.7110863 \text{ radians})]\}$
$\quad = 260°$ or $-100° = 100°$ W. long.

CYLINDRICAL EQUAL-AREA (ELLIPSOID)–FORWARD EQUATIONS
(SEE P. 81–82)

Normal aspect:
Given: Clarke 1866 ellipsoid: $a = 6{,}378{,}206.4$ m
                                      $e^2 = 0.00676866$
                         or $e = 0.0822719$
           Standard parallel: $\phi_s = 5°$ N. & S. lat.
           Central meridian: $\lambda_0 = 75°$ W. long.
                        Point: $\phi = 10°$
                                  $\lambda = 78°$ W. long.

Using equations (10–13), (3–12), (10–14), and (10–15) in order,

k_0 $= \cos 5°/[1-0.00676866\times\sin^2 5°]^{1/2}$
      $= 0.9962203$
q $= (1-0.00676866) \times \{\sin 5°/(1-0.00676866\times\sin^2 5°)$
        $-[1/(2\times0.0822719)] \times \ln [(1-0.0822719\times\sin 5°)/$
        $(1+0.0822719\times\sin 5°)]\}$
      $= 0.1731376$
x $= 6{,}378{,}206.4 \times 0.9962203 \times [-78°-(-75°)] \times \pi/180°$
      $= -332{,}699.8$ m
y $= 6{,}378{,}206.4 \times 0.1731376/(2\times0.9962203)$
      $= 554{,}248.5$ m

Transverse aspect:
Given: Clarke 1866 ellipsoid: $a = 6{,}378{,}206.4$ m
                                      $e^2 = 0.00676866$
                         or $e = 0.0822719$
           Central meridian: $\lambda_0 = 75°$ W. long.
           Latitude of origin: $\phi_0 = 30°$ N. lat.
           Scale factor at λ_0: $h_0 = 0.99$
                        Point: $\phi = 40°$ N. lat.
                                  $\lambda = 83°$ W. long.

Find: x, y

Using equations (3–12) and (3–11),

q $= (1-0.00676866) \times \{\sin 40°/(1-0.00676866\times\sin^2 40°)$
        $-[1/(2\times0.0822719)] \times \ln [(1-0.0822719\times\sin 5°)/$
        $(1+0.0822719\times\sin 5°)]\}$
      $= 1.2792602$

Inserting 90° in place of 40° in the same equation,

$q_p = 1.9954814$
$\beta = \arcsin (1.2792602/1.9954814)$
    $= 39.8722878°$

Using equations (10–16) and (10–17),

$\beta_c = \arctan [\tan 39.8722878°/\cos [-83°-(-75°)]$
      $= 40.1482125°$
$q_c = 1.9954814 \times \sin 40.1482125°$
    $= 1.2866207$

For the first trial ϕ_c in equation (3–16),

$$\phi_c = \arcsin\ (1.2866207/2)$$
$$= 40.0391089°$$

Substituting into equation (3–16),

$$\phi_c = 40.0391089° + [(1-0.00676866\ \sin^2\ 40.0391089°)^2/$$
$$(2\ \cos\ 40.0391089°)] \times \{1.2866207/(1-0.00676866)$$
$$-\sin\ 40.0391089°/(1-0.00676866\ \sin^2\ 40.0391089°)$$
$$+[1/(2\times 0.0822719)]\ \ln\ [(1-0.0822719$$
$$\sin\ 40.0391089°)/(1+0.0822719\ \sin\ 40.0391089°)]\} \times 180°/\pi$$
$$= 40.2757321°$$

Substituting 40.2757321° in place of 40.0391089° in the same equation, the new trial ϕ_c is found to be 40.2761382°. The next iteration results in no change to seven decimal places. Thus,

$$\phi_c = 40.2761382°$$

Using equation (10–18),

$$x\ = 6,378,206.4 \times \cos\ 39.8722878° \times \cos\ 40.2761382°$$
$$\times \sin\ [-83°-(-75°)]/[0.99\times \cos\ 40.1482125°$$
$$\times (1-0.00676866\times \sin^2\ 40.2761382°)^{1/2}]$$
$$= -687,825.8\ \text{m}$$

Using equation (3–21),

$$M_c\ = 6,378,206.4 \times [(1-0.00676866/4-3\times 0.00676866^2/64$$
$$-5 \times 0.00676866^3/256) \times 40.2761382° \times \pi/180°$$
$$- (3\times 0.00676866/8+3\times 0.00676866^2/32$$
$$+ 45 \times 0.00676866^3/1024) \times \sin\ (2\times 40.2761382°)$$
$$+ (15\times 0.00676866^2/256+45\times 0.00676866^3/1024)$$
$$\times \sin\ (4\times 40.2761382°) - (35\times 0.00676866^3/$$
$$3072) \times \sin\ (6\times 40.2761382°)]$$
$$= 4,459,980.0\ \text{m}$$

Substituting $\phi_0 = 30°$ in the same equation in place of 40.2761382°,

$$M_0 = 3,319,933.3\ \text{m}$$

Using equation (10–19),

$$y = 0.99 \times (4,459,980.0-3,319,933.3)$$
$$= 1,128,646.2\ \text{m}$$

Oblique aspect:
Given: Clarke 1866 ellipsoid: $a = 6,378,206.4$ m
                                     $e^2 = 0.00676866$
                        or $e = 0.0822719$
    Central scale factor: · $h_0 = 1.0$
    Central line through: $\phi_1 = 30°$ N. lat.
                                     $\phi_2 = 40°$ N. lat.
                                     $\lambda_1 = 75°$ W. long.
                                     $\lambda_2 = 80°$ W. long.

Point: $\phi = 42°$ N. lat.
          $\lambda = 77°$ W. long.

Find: x, y

To find the position of the pole, equations (3−12) and (3−11) are used as in the examples for the normal and transverse aspects just above, to determine β_1 from ϕ_1 and β_2 from ϕ_2. The results are

$$\beta_1 = 29.8877623°$$
$$\beta_2 = 39.8722878°$$

Inserting these values in place of ϕ_1 and ϕ_2 in equations (9−1) and (9−2), listed under spherical formulas for the projection,

λ_p = arctan [(cos 29.8877623° sin 39.8722878°−
        sin 29.8877623° cos 39.8722878° cos (−80°))/
        (sin 29.8877623° cos 39.8722878° sin (−80°)
        − cos 29.8877623° sin 39.8722878° sin (−75°))]
     = arctan (0.4894080/0.1602532)
     = 71.8693268°, not adding 180° since denominator is positive.

β_p = arctan [−cos (71.8693268°−(−75°))/tan 29.8877623°]
     = 55.5374608°

Using equations (10−17) and (3−16), with subscript p instead of c, ϕ_p is found by iteration as in the example for ϕ_c under the transverse aspect. Finally,

$$\phi_p = 55.6583959°$$

Using equations (10−20) and (10−21), and table 13 for the Clarke 1866 ellipsoid, the specific Fourier coefficients are calculated:

B = 0.9991507126 + (−0.0008471537) cos (2 × 55.6583959°)
      + (0.0000021283) cos (4 × 55.6583959°)
      + (−0.0000000054) cos (6 × 55.6583959°)
   = 0.9994571
A_2 = −0.0001412090 + (−0.0001411258) cos (2 × 55.6583959°)
      + (0.0000000839) cos (4 × 55.6583959°)
      + (0.0000000006) cos (6 × 55.6583959°)
   = −0.0000900
A_4 = −0.0000000435 + (−0.0000000579) cos (2 × 55.6583959°)
      + (−0.0000000144) cos (4 × 55.6583959°)
      + (0) cos (6×55.6583959°)
   = −0.0000000

Equations (3−12) and (3−11) are again used to determine β from ϕ, giving

$$\beta = 41.8710109°$$

From equation (10−22),

$$
\begin{aligned}
\lambda' &= \arctan \{[\cos 55.5374608° \sin 41.8710109° \\
&\quad - \sin 55.5374608° \cos 41.8710109° \cos (-77° \\
&\quad -71.8693268°)]/\cos 41.8710109° \sin (-77°-71.8693268°)]\} \\
&= \arctan [0.9032359/(-0.3849775)] \\
&= -66.9153117° + 180° = 113.0846883°
\end{aligned}
$$

adding 180° because the denominator is negative.

Using equations (10−23) through (10−25), using q_p as computed above for the transverse aspect,

$$
\begin{aligned}
x &= 6,378,206.4 \times 1.0 \times [0.9994571 \times 113.0846883° \times \pi/180° \\
&\quad + (-0.0000900) \times \sin (2 \times 113.0846883°) \\
&\quad + (-0.0000000) \times \sin (4 \times 113.0846883°) \\
&= 12,582,246.4 \text{ m} \\
F &= 0.9994571 + 2 \times (-0.0000900) \times \cos (2 \times 113.0846883°) \\
&\quad + 4 \times (-0.0000000) \times \cos (4 \times 113.0846883°) \\
&= 0.9995817 \\
y &= (6,378,206.4 \times 1.9954814/2) \times [\sin 55.5374608° \\
&\quad \times \sin 41.8710109° + \cos 55.5374608° \times \cos 41.8710109° \\
&\quad \times \cos (-77°-71.8693268°)]/(1.0 \times 0.9995817) \\
&= 1,207,233.0 \text{ m}
\end{aligned}
$$

CYLINDRICAL EQUAL-AREA (ELLIPSOID)−INVERSE EQUATIONS
(SEE P. 82-84)

Inversing forward examples:
Normal aspect:
Given: a, e^2, ϕ_s, and λ_0 as in forward ellipsoid examples

$$
\begin{aligned}
x &= -332,699.8 \text{ m} \\
y &= 554,248.5 \text{ m}
\end{aligned}
$$

Find: ϕ, λ

After k_0 and q_p are determined from (10−13) and (3−12) as in the forward normal and transverse examples,

$$
\begin{aligned}
k_0 &= 0.9962203 \\
q_p &= 1.9954814
\end{aligned}
$$

then, from (10−26),

$$
\begin{aligned}
\beta &= \arcsin [2 \times 554,248.5 \times 0.9962203/(6,378,206.4 \times 1.9954814)] \\
&= 4.9775164°
\end{aligned}
$$

Using equations (10−17) and (3−16), with subscript c omitted, ϕ is found from β by iteration as in the example for ϕ_c under the forward transverse ellipsoid example. Finally,

$$
\phi = 5° \text{ N. lat.}
$$

From (10−27),

$$\lambda = -75° + [-332,699.8/(6,378,206.4 \times 0.9962203)] \times 180°/\pi$$
$$= -78° = 78° \text{ W. long.}$$

Transverse aspect:
Given: a, e^2, λ_0, ϕ_0, h_0 as in forward ellipsoid example:

$$x = -687,825.8 \text{ m}$$
$$y = 1,128,646.2 \text{ m}$$

Find: ϕ, λ

After M_0 is calculated from (3−21), using $\phi_0 = 30°$ in place of ϕ_c, as in the forward ellipsoid example,

$$M_0 = 3,319,933.3 \text{ m}$$

From (10−28),

$$M_c = 3,319,933.3 + 1,128,646.2/0.99$$
$$= 4,459,980.0 \text{ m}$$

From (7−19), (3−24) and (3−26),

$$\mu_c = 4,459,980.0/[6,378,206.4 \times (1-0.00676866/4$$
$$-3 \times 0.00676866^2/64 - 5 \times 0.00676866^3/256)]$$
$$= 0.7004398 \text{ radians} = 40.1322426°$$
$$e_1 = [1-(1-0.00676866)^{1/2}]/[1+(1-0.00676866)^{1/2}]$$
$$= 0.0016979$$
$$\phi_c = [0.7004398 + (3 \times 0.0016979/2 - 27 \times 0.0016979^3/32)$$
$$\sin(2 \times 40.1322426°) + (21 \times 0.0016979^2/16$$
$$-55 \times 0.0016979/32) \sin(4 \times 40.1322426°)$$
$$+ (151 \times 0.0016979^3/96) \sin(6 \times 40.1322426°)$$
$$+ (1097 \times 0.0016979^4/512) \sin(8 \times 40.1322426°)] \times 180°/\pi$$
$$= 40.2761378°$$

Using (3−12) and (3−11), with q_p calculated as in the forward example,

$$q_c = (1-0.00676866) \times \{\sin 40.2761378°/(1$$
$$-0.00676866 \times \sin^2 40.2761378°) - [1/(2$$
$$\times 0.0822719)] \ln [(1-0.0822719 \times \sin 40.2761378°)/$$
$$(1+0.0822719 \times \sin 40.2761378°)]\}$$
$$= 1.2866207$$
$$\beta_c = \arcsin(1.2866207/1.9954814)$$
$$= 40.1482122°$$

From equations (10−29) through (10−31),

$$\beta' = -\arcsin [0.99 \times (-687,825.8) \times \cos(40.1482122°)$$
$$\times (1-0.00676866 \times \sin^2 40.2761378°)^{1/2}/$$
$$(6,378,206.4 \times \cos 40.2761378°)]$$
$$= 6.1315692°$$

β = arcsin (cos 6.1315692° sin 40.1482122°)
     = 39.8722875°
λ = −75°−arctan (tan 6.1315692°/cos 40.1482122°)
     = −75°−8° = −83° = 83° W. long.

Using (10−17) and (3−16), with subscript c omitted, ϕ is found from β by iteration as in the example for ϕ_c under the forward transverse ellipsoid example. Finally,

$$\phi = 40° \text{ N. lat.}$$

Oblique aspect:
Given: a, e^2, h_0, calculated pole location (ϕ_p, λ_p), calculated Fourier coefficients B, A_2, and A_4 as in the forward oblique ellipsoid example, and R_q as calculated for the forward normal ellipsoid example,

$$x = 12,582,246.4 \text{ m}$$
$$y = 1,207,233.0 \text{ m}$$

Find ϕ, λ

First q_p = 1.9954814, as found from (3−12) in the forward transverse example.

To solve for λ' from (10−32), the first trial λ' is found as described:

λ' = [12,582,246.4/(6,378,206.4×1.0×0.9994571)]×180°/π
     = 113.0884082°

Using equation (10−32),

λ' = [12,582,246.4/(6,378,206.4×1.0)×180°/π
         −(−0.0000900) × sin (2×113.0884082°)
         −(−0.0000000) × sin (4×113.0884082°)]/
         0.9994571
     = 113.0846878°

Substituting 113.0846878° in place of 113.0884082° in this equation, λ' is calculated to be 113.0846883°. The next iteration yields no change to seven decimal places, so that

$$\lambda' = 113.0846883°$$

Equation (10−24) is used to calculate F just as it was in the forward oblique example, so F is again

$$F = 0.9995817$$

From equations (10−33) through (10−35),

β' = arcsin [2×0.9995817×1.0×1,207,233.0/
       (6,378,206.4×1.9954814)]
     = 10.93083763°

β = arcsin (sin 55.5374608° sin 10.93083763°
       + cos 55.5374608° cos 10.93083763°
       sin 113.0846883°)
     = 41.8710109°

λ = 71.8693268° + arctan [cos 10.93083763°
    cos 113.0846883°/(cos 55.5374608°
    sin 10.93083763° − sin 55.5374608°
    cos 10.93083763° sin 113.0846883°)]
  = 71.8693268° + arctan [−0.3849775/(−0.6374127)]
  = 71.8693268° + 31.1306732° + 180°, adding 180°
    because of the negative denominator. Thus,

λ = 283°, or −77°, or 77° W. long.

Using (10−17) and (3−16), ϕ is found from β as previously, dropping subscript c and with iteration, to produce

$$\phi = 42° \text{ N. lat.}$$

The computation of Fourier coefficients is not shown here, since it is lengthy and is not needed unless a different ellipsoid is desired. An example of computation of Fourier coefficients is given under the Space Oblique Mercator projection.

MILLER CYLINDRICAL (SPHERE)−FORWARD EQUATIONS (SEE P. 88)

Given: Radius of sphere: R = 1.0 unit
     Central meridian: λ_0 = 0° long.
            Point: ϕ = 50° N. lat.
                 λ = 75° W. long.

Find x, y, h, k

Using equations (11−1) through (11−5) in order,

$$x = 1.0 \times [-75° - 0°] \times \pi/180°$$
$$= -1.3089969 \text{ units}$$
$$y = 1.0 \times [\ln \tan (45° + 0.4 \times 50°)]/0.8$$
$$= (\ln \tan 65°)/0.8$$
$$= 0.9536371 \text{ unit}$$

or

$$y = 1.0 \times \{\text{arcsinh } [\tan (0.8 \times 50°)]\}/0.8$$
$$= \text{arcsinh } 0.8390996/0.8$$
$$= 0.9536371 \text{ unit}$$
$$h = \sec (0.8 \times 50°) = 1/\cos 40° = 1.3054073$$
$$k = \sec 50° = 1/\cos 50° = 1.5557238$$
$$\sin \tfrac{1}{2}\omega = (\cos 40° - \cos 50°)/(\cos 40° + \cos 50°)$$
$$= 0.0874887$$
$$\omega = 10.0382962°$$

MILLER CYLINDRICAL (SPHERE)−INVERSE EQUATIONS (SEE P. 88)

Inversing forward example:

Given: R, λ_0 for forward example
$$x = -1.3089969 \text{ units}$$
$$y = 0.9536371 \text{ unit}$$

Find: ϕ, λ

Using equations (11−6) and (11−7),

$$\phi = 2.5 \arctan e^{(0.8\times0.9536371/1.0)} - (5\pi/8) \times 180°/\pi$$
$$= 2.5 \arctan e^{0.7629096} - 1.9634954 \times 180°/\pi$$
$$= 2.5 \arctan (2.1445069) - 1.9634954 \times 180°/\pi$$
$$= 2.5 \times 65.0000006° - 112.5000000°$$
$$= 50.0000015° = 50° \text{ N. lat.}$$

or

$$\phi = \arctan [\sinh (0.8\times0.9536371/1.0)]/0.8$$
$$= (\arctan 0.8390997)/0.8$$
$$= 50.0000015° = 50° \text{ N. lat.}$$
$$\lambda = 0° - (1.3089969/1.0) \times 180°/\pi$$
$$= 0° - 74.9999978° = 75° \text{ W. long.}$$

CASSINI (SPHERE)–FORWARD EQUATIONS (SEE P. 94)

Given: Radius of sphere: $R = 1.0$ unit

Origin: $\phi_0 = 20°$ S. lat.
$\lambda_0 = 75°$ W. long.
Point: $\phi = 25°$ N. lat.
$\lambda = 90°$ W. long.

Find: x, y, h'

Using equations (8−5), and (13−1) through (13−3) in order,

$$B = \cos 25° \sin [-90°-(-75°)]$$
$$= -0.2345697$$
$$x = 1.0 \times \arcsin (-0.2345697) \times \pi/180°$$
$$= -0.2367759 \text{ unit}$$
$$y = 1.0 \times \{\arctan [\tan 25°/\cos [-90°-(-75°)]]-(-20°)\} \times \pi/180°$$
$$= 1.0 \times 45.7692621° \times \pi/180° = 0.7988243 \text{ unit}$$
$$h' = 1/[1-(-0.2345697)^2]^{1/2}$$
$$= 1.0287015$$

CASSINI (SPHERE)–INVERSE EQUATIONS (SEE P. 94−95)

Inversing forward example:

Given: R, ϕ_0, λ_0 for forward example

$$x = -0.2367759 \text{ unit}$$
$$y = 0.7988243 \text{ unit}$$

Find: ϕ, λ

Using equations (13−6), (13−4), and (13−5) in order,

$$D = (0.7988243/1.0)\times180°/\pi + (-20°)$$
$$= 25.7692610°$$

ϕ = arcsin {sin 25.7692610° cos [(−0.2367759/1.0) × 180°/π]}
   = arcsin 0.4226182
   = 25° N. lat.

λ = −75° + arctan{tan[(−0.2367759/1.0)×180°/π]/cos 25.7692610°}
   = −75° + arctan (−0.2679492)
   = −75°+(−15°) = −90° = 90° W. long.

CASSINI (ELLIPSOID)–FORWARD EQUATIONS (SEE P. 95)

Given: Clarke 1866 ellipsoid: a = 6,378,206.4 m
                               e^2 = 0.00676866
          Origin: ϕ_0 = 40° N. lat.
                   λ_0 = 75° W. long.
          Point: ϕ = 43° N. lat.
                   λ = 73° W. long.

Find: x, y, s at Az = 30° east of north

Using equations (4−20), (8−13), (8−15), (8−14), and (3−21) in order,

N = 6,378,206.4/(1−0.00676866² × sin² 43°)^{1/2}
   = 6,388,270.3 m
T = tan² 43° = 0.8695844
A = [−73° − (−75°)]×(π/180°)× cos 43°
   = 0.02552906
C = 0.00676866 × cos² 43°/(1−0.00676866)
   = 0.003645081
M = 6,378,206.4 × [(1−0.00676866/4−3×0.00676866²/64
       −5×0.00676866³/256)×43°× π/180°−(3×0.00676866/
       8+3×0.00676866²/32+45×0.00676866³/1024)
       sin (2×43°)+(15×0.00676866²/256+45
       ×0.00676866³/1024) sin (4×43°)−(35×0.00676866³/
       3072) sin (6×43°)]
   = 4,762,504.8 m

Substituting 40° for 43° in equation (3−21),

M_0 = 4,429,318.9 m

Using equations (13−7) through (13−9) in order,

x = 6,388,270.3 × [0.02552906−0.8695844×0.02552906³/
       6−(8−0.8695844+8×0.003645081)×0.8695844
       × 0.02552906⁵/120]
   = 163,071.1 m

y = 4,762,504.8 − 4,429,318.9 + 6,388,270.3 × tan 43°
       × [0.02552906²/2 + (5−0.8695844+6×0.003645081)
       × 0.02552906⁴/24]
   = 335,127.6 m

s = 1 + 163,071.1² cos² 30° × (1−0.00676866 × sin² 43°)²/
       [2 × 6,378,206.4² × (1−0.00676866)]
   = 1.0002452

CASSINI (ELLIPSOID)—INVERSE EQUATIONS (SEE P. 95)

Inversing forward example:

Given: a, e^2, ϕ_0, λ_0 as in forward example

$$x = 163{,}071.1 \text{ m}$$
$$y = 335{,}127.6 \text{ m}$$

Find: ϕ, λ

Calculating M_0 from equation (3−21) as in the forward example for $\phi_0 = 40°$,

$$M_0 = 4{,}429{,}318.9 \text{ m}$$

Using equations (13−12), (7−19), and (3−24) in order,

$$\begin{aligned}
M_1 &= 4{,}429{,}318.9 + 335{,}127.6 \\
&= 4{,}764{,}446.5 \text{ m} \\
\mu_1 &= 4{,}764{,}446.5/[6{,}378{,}206.4 \times (1-0.00676866/4 \\
&\quad - 3 \times 0.00676866^2/64 - 5 \times 0.00676866^3/256)] \\
&= 0.7482562 \text{ radians} = 42.8719240° \\
e_1 &= [1-(1-0.00676866)^{1/2}]/[1+(1-0.00676866)^{1/2}] \\
&= 0.001697916
\end{aligned}$$

Using equations (3−26), (8−22), (8−23), (8−24), and (13−13) in order,

$$\begin{aligned}
\phi_1 &= 42.8719240° + [(3\times0.001697916/2-27\times0.001697916^3/ \\
&\quad 32) \sin (2\times42.8719240°) + (21\times0.001697916^2/16 \\
&\quad - 55 \times 0.001697916^4/32) \sin (4\times42.8719240°) \\
&\quad + (151\times0.001697916^3/96) \sin (6\times42.8719240°) \\
&\quad + (1097\times0.001697916^4/512) \sin (8\times42.8719240°)] \times 180°/\pi \\
&= 43.0174782° \\
T_1 &= \tan^2 43.0174782° \\
&= 0.8706487 \\
N_1 &= 6{,}378{,}206.4/(1-0.00676866 \sin^2 43.0174782°)^{1/2} \\
&= 6{,}388{,}276.9 \text{ m} \\
R_1 &= 6{,}378{,}206.4 \times (1-0.00676866)/(1-0.00676866 \\
&\quad \times \sin^2 43.0174782°)^{3/2} \\
&= 6{,}365{,}088.8 \text{ m} \\
D &= 163{,}071.1/6{,}388{,}276.9 \\
&= 0.0255266
\end{aligned}$$

Using equations (13−10) and (13−11) in order,

$$\begin{aligned}
\phi &= 43.0174782° - (6{,}388{,}276.9\times\tan 43.0174782°/ \\
&\quad 6{,}365{,}088.8) \times [0.0255266^2/2 - (1+3\times0.8706487) \\
&\quad \times 0.0255266^4/24] \times 180°/\pi \\
&= 43° \text{ N. lat.} \\
\lambda &= -75° + [[0.0255266-0.8706487\times0.0255266^3/3 \\
&\quad + (1+3\times0.8706487) \times 0.8706487 \times 0.0255266^5/15]/ \\
&\quad \cos 43.0174782°] \times 180°/\pi \\
&= -73° = 73° \text{ W. long.}
\end{aligned}$$

ALBERS CONICAL EQUAL-AREA (SPHERE)—FORWARD EQUATIONS (SEE P. 100)

Given: Radius of sphere: R = 1.0 unit

Standard parallels: ϕ_1 = 29° 30′ N. lat.

ϕ_2 = 45° 30′ N. lat.

Origin: ϕ_0 = 23° N. lat.

λ_0 = 96° W. long.

Point: ϕ = 35° N. lat.

λ = 75° W. long.

Find: ρ, θ, x, y, k, h, ω

From equations (14−6), (14−5), (14−3), (14−3a), and (14−4) in order,

n = (sin 29.5° + sin 45.5°)/2

= 0.6028370

C = cos² 29.5° + 2 × 0.6028370 sin 29.5°

= 1.3512213

ρ = 1.0 × (1.3512213−2×0.6028370 sin 35°)$^{1/2}$/0.6028370

= 1.3473026 units

ρ_0 = 1.0 × (1.3512213−2×0.6028370 sin 23°)$^{1/2}$/0.6028370

= 1.5562263 units

θ = 0.6028370×[(−75°)−(−96°)]

= 12.6595771°

From equations (14−1), (14−2), and (14−7) in order,

x = 1.3473026 sin 12.6595771°

= 0.2952720 unit

y = 1.5562263 − 1.3473026 cos 12.6595771°

= 0.2416774 unit

h = cos 35°/(1.3512213−2×0.6028370 sin 35°)$^{1/2}$

= 1.0085547

and

k = 1/1.0085547 = 0.9915178

From equation (4−9),

sin ½ω = |1.0085547−0.9915178|/(1.0085547+0.9915178)

ω = 0.9761189°

ALBERS CONICAL EQUAL-AREA (SPHERE)—INVERSE EQUATIONS (SEE P. 101)

Inversing forward example:

Given: R, ϕ_1, ϕ_2, ϕ_0, λ_0 for forward example

x = 0.2952720 unit

y = 0.2416774 unit

Find: ρ, θ, ϕ, λ

As in the forward example, from equations (14−6), (14−5), and (14−3a) in order,

$$n = (\sin 29.5° + \sin 45.5°)/2$$
$$= 0.6028370$$
$$C = \cos^2 29.5° + 2 \times 0.6028370 \sin 29.5°$$
$$= 1.3512213$$
$$\rho_0 = 1.0 \times (1.3512213 - 2 \times 0.6028370$$
$$\sin 23°)^{1/2}/0.6028370$$
$$= 1.5562263 \text{ units}$$

From equations (14−10), (14−11), (14−8), and (14−9) in order,

$$\rho = [0.2952720^2 + (1.5562263 - 0.2416774)^2]^{1/2}$$
$$= 1.3473026 \text{ units}$$
$$\theta = \arctan [0.2952720/(1.5562263 - 0.2416774)]$$
$$= 12.6595766°. \text{ Since the denominator is positive,}$$
there is no adjustment to θ.
$$\phi = \arcsin \{[1.3512213 - (1.3473026 \times 0.6028370/1.0)^2]/$$
$$(2 \times 0.6028370)\}$$
$$= \arcsin 0.5735764$$
$$= 35.0000007° \approx 35° \text{ N. lat.}$$
$$\lambda = 12.6595766°/0.6028370 + (-96°)$$
$$= 20.9999992 - 96°$$
$$= -75.0000008° = 75° \text{ W. long.}$$

ALBERS CONICAL EQUAL-AREA (ELLIPSOID)—FORWARD EQUATIONS (SEE P. 101)

Given: Clarke 1866 ellipsoid: a = 6378206.4 m
e^2 = 0.00676866
or e = 0.0822719
Standard parallels: ϕ_1 = 29° 30′ N. lat.
ϕ_2 = 45° 30′ N. lat.
Origin: ϕ_0 = 23° N. lat.
λ_0 = 96° W. long.
Point: ϕ = 35° N. lat.
λ = 75° W. long.

Find: ρ, θ, x, y, k, h, ω

From equation (14−15),

$$m_1 = \cos 29.5°/(1 - 0.00676866 \sin^2 29.5°)^{1/2}$$
$$= 0.8710708$$
$$m_2 = \cos 45.5°/(1 - 0.00676866 \sin^2 45.5°)^{1/2}$$
$$= 0.7021191$$

From equation (3−12),

$$q_1 = (1 - 0.00676866) \{\sin 29.5°/(1 - 0.00676866 \sin^2 29.5°)$$
$$- [1/(2 \times 0.0822719)] \ln [(1 - 0.0822719 \sin 29.5°)/$$
$$(1 + 0.0822719 \sin 29.5°)]\}$$
$$= 0.9792529$$

Using the same formula for q_2 (with ϕ_2 instead of ϕ_1),

$$q_2 = 1.4201080$$

Using the same formula for q_0 (with ϕ_0 instead of ϕ_1),

$$q_0 = 0.7767080$$

From equations (14−14), (14−13), and (14−12a) in order,

$$
\begin{aligned}
n &= (0.8710708^2 - 0.7021191^2)/(1.4201080 - 0.9792529) \\
&= 0.6029035 \\
C &= 0.8710708^2 + 0.6029035 \times 0.9792529 \\
&= 1.3491594 \\
\rho_0 &= 6378206.4 \times (1.3491594 - 0.6029035 \times 0.7767080)^{1/2}/0.6029035 \\
&= 9{,}929{,}079.6 \text{ m}
\end{aligned}
$$

These are the constants for the map. For $\phi = 35°$ N. lat. and $\lambda = 75°$ W. long.:
Using equation (3−12), but with ϕ in place of ϕ_1,

$$q = 1.1410831$$

From equations (14−12), (14−4), (14−1), and (14−2) in order,

$$
\begin{aligned}
\rho{'} &= 6378206.4 \times (1.3491594 - 0.6029035 \times 1.1410831)^{1/2}/0.6029035 \\
&= 8{,}602{,}328.2 \text{ m} \\
\theta &= 0.6029035 \times [-75° - (-96°)] = 12.6609735° \\
x &= 8602328.2 \sin 12.6609735° = 1{,}885{,}472.7 \text{ m} \\
y &= 9929079.6 - 8602328.2 \cos 12.6609735° \\
&= 1{,}535{,}925.0 \text{ m}
\end{aligned}
$$

From equations (14−15), (14−16), (14−18), and (4−9) in order,

$$
\begin{aligned}
m &= \cos 35°/(1 - 0.00676866 \sin^2 35°)^{1/2} \\
&= 0.8200656 \\
k &= 8602328.2 \times 0.6029035/(6378206.4 \times 0.8200656) \\
&= 0.9915546 \\
h &= 1/0.9915546 = 1.0085173 \\
\sin \tfrac{1}{2}\omega &= |1.0085173 - 0.9915546|/(1.0085173 + 0.9915546) \\
\omega &= 0.9718678°
\end{aligned}
$$

ALBERS CONICAL EQUAL-AREA (ELLIPSOID)−INVERSE EQUATIONS
(SEE P. 102)

Inversing forward example:

Given: Clarke 1866 ellipsoid: $a = 6378206.4$ m
                                  $e^2 = 0.00676866$
                           or $e = 0.0822719$
       Standard parallels: $\phi_1 = 29° 30'$ N. lat.
                            $\phi_2 = 45° 30'$ N. lat.
              Origin: $\phi_0 = 23°$ N. lat.
                            $\lambda_0 = 96°$ W. long.
              Point: $x = 1{,}885{,}472.7$ m
                            $y = 1{,}535{,}925.0$ m

Find: ρ, θ, ϕ, λ

The same constants n, C, ρ_0 are calculated with the same equations as those used for the forward example. For the particular point:

From equation (14−10),

$$\rho = [1885472.7^2 + (9929079.6 - 1535925.0)^2]^{1/2}$$
$$= 8,602,328.3 \text{ m}$$

From equation (14−11),

$$\theta = \arctan [1885472.7/(9929079.6 - 1535925.0)]$$
$$= \arctan 0.2246441$$
$$= 12.6609733°. \text{ The denominator is positive; therefore } \theta \text{ is not}$$
$$\text{adjusted. From equation (14−19),}$$
$$q = [1.3491594 - (8602328.3 \times 0.6029035/6378206.4)^2]/0.6029035$$
$$= 1.1410831$$

Using for the first trial ϕ the arcsin of (1.1410831/2), or 34.7879983°, calculate a new ϕ from equation (3−16),

$$\phi = 34.7879983° + [(1-0.00676866 \sin^2 34.7879983°)^2/(2 \cos$$
$$34.7879983°)] \times \{1.1410831/(1-0.00676866) - \sin 34.7879983°/$$
$$(1-0.00676866 \sin^2 34.7879983°) + [1/(2 \times 0.0822719)] \ln$$
$$[(1-0.0822719 \sin 34.7879983°)/(1 + 0.0822719 \sin$$
$$34.7879983°)]\} \times 180°/\pi$$
$$= 34.9997335°$$

Note that $180°/\pi$ is included to convert to degrees. Replacing 34.7879983° by 34.9997335° for the second trial, the calculation using equation (14−19) now provides a third ϕ of 35.0000015°. A recalculation with this value results in no change to seven decimal places. (This does not give exactly 35° due to rounding-off errors in x and y.) Thus,

$$\phi = 35.0000015° \text{ N. lat.}$$

For the longitude use equation (14−9),

$$\lambda = (-96°) + 12.6609733°/0.6029035$$
$$= -75.0000003° \text{ or } 75.0000003° \text{ W. long.}$$

For scale factors, we revert to the forward example, since ϕ and λ are now known.
   Series equation (3−18) may be used to avoid the iteration above. Beginning with equation (14−21),

$$\beta = \arcsin [1.1410831/\{1-[(1-0.00676866)/(2 \times 0.0822719)] \ln$$
$$[(1-0.0822719)/(1 + 0.0822719)]\}]$$
$$= 34.8781793°$$

An example is not shown for equation (3−18), since it is similar to the example for (3−5).

LAMBERT CONFORMAL CONIC (SPHERE)–FORWARD EQUATIONS
(SEE P. 106–107)

Given: Radius of sphere: $R = 1.0$ unit
        Standard parallels: $\phi_1 = 33°$ N. lat.
                             $\phi_2 = 45°$ N. lat.
               Origin: $\phi_0 = 23°$ N. lat.
                             $\lambda_0 = 96°$ W. long.
                Point: $\phi = 35°$ N. lat.
                             $\lambda = 75°$ W. long.

Find: ρ, θ, x, y, k

From equations (15–3), (15–2), and (15–1a) in order,

$$n = \ln(\cos 33°/\cos 45°)/\ln[\tan(45° + 45°/2)/\tan(45° + 33°/2)]$$
$$= 0.6304777$$
$$F = [\cos 33° \tan^{0.6304777}(45° + 33°/2)]/0.6304777$$
$$= 1.9550002 \text{ units}$$
$$\rho_0 = 1.0 \times 1.9550002/\tan^{0.6304777}(45° + 23°/2)$$
$$= 1.5071429 \text{ units}$$

The above constants apply to the map generally. For the specific ϕ and λ, using equations (15–1), (14–4), (14–1), and (14–2) in order,

$$\rho = 1.0 \times 1.9550002/\tan^{0.6304777}(45° + 35°/2)$$
$$= 1.2953636 \text{ units}$$
$$\theta = 0.6304777 \times [(-75°)-(-96°)]$$
$$= 13.2400316°$$
$$x = 1.2953636 \sin 13.2400316°$$
$$= 0.2966785 \text{ unit}$$
$$y = 1.5071429 - 1.2953636 \cos 13.2400316°$$
$$= 0.2462112 \text{ unit}$$

From equation (15–4),

$$k = \cos 33° \tan^{0.6304777}(45° + 33°/2)/[\cos 35° \tan^{0.6304777}$$
$$(45° + 35°/2)]$$
$$= 0.9970040$$

or from equation (4–5),

$$k = 0.6304777 \times 1.2953636/(1.0 \cos 35°)$$
$$= 0.9970040$$

LAMBERT CONFORMAL CONIC (SPHERE)–INVERSE EQUATIONS
(SEE P. 107)

Inversing forward example:

Given: R, ϕ_1, ϕ_2, ϕ_0, λ_0 for forward example
$$x = 0.2966785 \text{ unit}$$
$$y = 0.2462112 \text{ unit}$$

Find: ρ, θ, ϕ, λ

After calculating n, F, and ρ_0 as in the forward example, obtaining the same values, equation (14−10) is used:

$$\rho = [0.2966785^2 + (1.5071429 - 0.2462112)^2]^{1/2}$$
$$= 1.2953636 \text{ units}$$

From equation (14−11),

$$\theta = \arctan [0.2966785/(1.5071429 - 0.2462112)]$$
$$= 13.2400329°. \text{ Since the denominator is positive, } \theta \text{ is not adjusted.}$$

From equation (14−9),

$$\lambda = 13.2400329°/0.6304777 + (-96°)$$
$$= -74.9999981° = 74.9999981° \text{ W. long.}$$

From equation (15−5),

$$\phi = 2 \arctan (1.0 \times 1.9550002/1.2953636)^{1/0.6304777} - 90°$$
$$= 34.9999974° \text{ N. lat.}$$

LAMBERT CONFORMAL CONIC (ELLIPSOID)−FORWARD EQUATIONS
(SEE P. 107−108)

Given: Clarke 1866 ellipsoid: $a = 6,378,206.4$ m
                              $e^2 = 0.00676866$
        or $e = 0.0822719$
Standard parallels: $\phi_1 = 33°$ N. lat.
                    $\phi_2 = 45°$ N. lat.
Origin: $\phi_0 = 23°$ N. lat.
          $\lambda_0 = 96°$ W. long.
Point: $\phi = 35°$ N. lat.
        $\lambda = 75°$ W. long.

Find: ρ, θ, x, y, k

From equation (14−15),

$$m_1 = \cos 33°/(1 - 0.00676866 \sin^2 33°)^{1/2}$$
$$= 0.8395138$$
$$m_2 = \cos 45°/(1 - 0.00676866 \sin^2 45°)^{1/2}$$
$$= 0.7083064$$

From equation (15−9),

$$t_1 = \tan (45° - 33°/2)/[(1 - 0.0822719 \sin 33°)/(1 + 0.0822719 \sin 33°)]^{0.0822719/2}$$
$$= 0.5449623$$
$$t_2 = 0.4162031, \text{ using above with } 45° \text{ in place of } 33°.$$
$$t_0 = 0.6636390, \text{ using above with } 23° \text{ in place of } 33°.$$

From equations (15−8), (15−10), and (15−7a) in order,

$$n = \ln (0.8395138/0.7083064)/\ln (0.5449623/0.4162031)$$
$$= 0.6304965$$

F = 0.8395138/(0.6304965 × 0.5449623^{0.6304965})

= 1.9523837

ρ_0 = 6378206.4 × 1.9523837 × 0.6636390^{0.6304965}

= 9,615,955.2 m

The above are constants for the map. For the specific ϕ, λ, using equation (15−9),

t = 0.5225935, using above calculation with 35° in place of 33°.

From equations (15−7), (14−4), (14−1), and (14−2) in order,

ρ = 6378206.4 × 1.9523837 × 0.5225935^{0.6304965}

= 8,271,173.9 m

θ = 0.6304965 × [−75°−(−96°)] = 13.2404256°

x = 8271173.9 sin 13.2404256°

= 1,894,410.9 m

y = 9615955.2−8271173.9 cos 13.2404256°

= 1,564,649.5 m

From equations (14−15) and (14−16),

m = cos 35°/(1−0.00676866 sin² 35°)^{1/2}

= 0.8200656

k = 8271173.9 × 0.6304965/(6378206.4 × 0.8200656)

= 0.9970171

LAMBERT CONFORMAL CONIC (ELLIPSOID)−INVERSE EQUATIONS
(SEE P. 109)

Inversing forward example:

Given: Clarke 1866 ellipsoid: a = 6,378,206.4 m

e^2 = 0.00676866

or e = 0.0822719

Standard parallels: ϕ_1 = 33° N. lat.

ϕ_2 = 45° N. lat.

Origin: ϕ_0 = 23° N. lat.

λ_0 = 96° W. long.

Point: x = 1,894,410.9 m

y = 1,564,649.5 m

The map constants n, F, and ρ_0 are calculated as in the forward example, obtaining the same values. Then, from equation (14−10),

ρ = [1894410.9² + (9615955.2 − 1564649.5)²]^{1/2}

= 8,271,173.8 m

From equation (14−11),

θ = arctan [1894410.9/(9615955.2 − 1564649.5)]

= 13.2404257°. The denominator is positive; therefore θ is not adjusted.

From equation (15−11),

t = [8271173.8/(6378206.4 × 1.9523837)]^{1/0.6304965}

= 0.5225935

To use equation (7–9), an initial trial ϕ is found as follows:

$$\phi = 90° - 2 \arctan 0.5225935$$
$$= 34.8174484°$$

Inserting this into the right side of equation (7–9),

$$\phi = 90° - 2 \arctan \{0.5225935 \times [(1-0.0822719 \sin 34.8174484°)/$$
$$(1 + 0.0822719 \sin 34.8174484°)]^{0.0822719/2}$$
$$= 34.9991687°$$

Replacing 34.8174484° with 34.9991687° for the second trial, a ϕ of 34.9999969° is obtained. Recalculation with the new ϕ results in $\phi = 35.0000006°$, which does not change to seven decimals with a fourth trial. (This is not exactly 35°, due to rounding-off errors.) Therefore,

$$\phi = 35.0000006° \text{ N. lat.}$$

From equation (14–9),

$$\lambda = 13.2404257°/0.6304965 + (-96°)$$
$$= -75.0000013° = 75.0000013° \text{ W. long.}$$

Examples using equations (3–5) and (7–13) are omitted here, since comparable examples for these equations have been given above.

EQUIDISTANT CONIC (SPHERE)–FORWARD EQUATIONS (SEE P. 113)

Given: Radius of sphere: $R = 1.0$ unit
      Standard parallels: $\phi_1 = 29° 30'$ N. lat.
                            $\phi_2 = 45° 30'$ N. lat.
             Origin: $\phi_0 = 23°$ N. lat.
                         $\lambda_0 = 96°$ W. long.
            Point: $\phi = 35°$ N. lat.
                         $\lambda = 75°$ W. long.

Find: ρ, θ, x, y, k

From equations (16–4), (16–3), (16–2), (16–1), and (14–4) in order,

$$n = (\cos 29.5° - \cos 45.5°)/[(45.5° - 29.5°) \times \pi/180°]$$
$$= 0.6067853$$
$$G = (\cos 29.5°)/0.6067853 + 29.5° \times \pi/180°$$
$$= 1.9492438$$
$$\rho_0 = 1.0 \times (1.9492438 - 23° \times \pi/180°)$$
$$= 1.5478181 \text{ units}$$
$$\rho = 1.0 \times (1.9492438 - 35° \times \pi/180°)$$
$$= 1.3383786 \text{ units}$$
$$\theta = 0.6067853 \times [-75° - (-96°)]$$
$$= 12.7424921°$$

Using equations (14–1), (14–2), and (16–5) in order,

$$x = 1.3383786 \sin 12.7424921°$$
$$= 0.2952057 \text{ unit}$$
$$y = 1.5478181 - 1.3383786 \cos 12.7424921°$$
$$= 0.2424021 \text{ unit}$$

$$k = (1.9492438 - 35° \times \pi/180°) \times 0.6067853/\cos 35°$$
$$= 0.99140$$

EQUIDISTANT CONIC (SPHERE)—INVERSE EQUATIONS (SEE P. 113)

Inversing forward example:

Given: R, ϕ_1, ϕ_2, ϕ_0, λ_0 for forward example

$$x = 0.2952057 \text{ unit}$$
$$y = 0.2424021 \text{ unit}$$

Find: ρ, θ, ϕ, λ

Calculating n, G, and ρ_0 as in the forward example,

$$n = 0.6067853$$
$$G = 1.9492438$$
$$\rho_0 = 1.5478181 \text{ units}$$

Using equations (14−10) and (14−11) in order,

$$\rho = + [0.2952057^2 + (1.5478181 - 0.2424021)^2]^{1/2}$$
$$= 1.3383786 \text{ units, positive because } n \text{ is positive}$$
$$\theta = \arctan [0.2952057/(1.5478181 - 0.2424021)]$$
$$= 12.7424933°, \text{ not adding } 180° \text{ since denominator is positive}$$

Using equations (16−6) and (14−9) in order,

$$\phi = [1.9492438 - 1.3383786/1.0] \times 180°/\pi$$
$$= 35° \text{ N. lat.}$$
$$\lambda = -96° + 12.7424933°/0.6067853$$
$$= -75° = 75° \text{ W. long.}$$

EQUIDISTANT CONIC (ELLIPSOID)—FORWARD EQUATIONS (SEE P. 114)

Given: Clarke 1866 ellipsoid: $a = 6,378,206.4 \text{ m}$
                                       $e^2 = 0.00676866$
           Standard parallels: $\phi_1 = 29° 30' \text{ N. lat.}$
                                    $\phi_2 = 45° 30' \text{ N. lat.}$
                     Origin: $\phi_0 = 23° \text{ N. lat.}$
                                 $\lambda_0 = 96° \text{ W. long.}$
                      Point: $\phi = 35° \text{ N. lat.}$
                                 $\lambda = 75° \text{ W. long.}$

Find: ρ, θ, x, y, k

From equations (14−15) and (3−21),

$$m = \cos 35°/(1 - 0.00676866 \sin^2 35°)^{1/2}$$
$$= 0.8200656$$
$$M = 6,378,206.4 \times [(1 - 0.00676866/4 - 3 \times 0.00676866^2/64$$
$$- 5 \times 0.00676866^3/256) \times 35° \times \pi/180° - (3 \times 0.00676866/$$
$$8 + 3 \times 0.00676866^2/32 + 45 \times 0.00676866^3/1024)$$
$$\sin (2 \times 35°) + (15 \times 0.00676866^2/256 + 45 \times 0.00676866^3/$$
$$1024) \sin (4 \times 35°) - (35 \times 0.00676866^3/3072)$$
$$\sin (6 \times 35°)]$$
$$= 3,874,395.2 \text{ m}$$

Using the same equations, but with $\phi_1 = 29.5°$ in place of $35°$,

$$m_1 = 0.8710708$$
$$M_1 = 3,264,511.2 \text{ m}$$

Similarly, with $\phi_2 = 45.5°$ in place of $35°$,

$$m_2 = 0.7021191$$
$$M_2 = 5,040,295.0 \text{ m}$$

and with $\phi_0 = 23°$ in place of $35°$,

$$M_0 = 2,544,389.8 \text{ m}$$

Using equations (16–10), (16–11), (16–9), (16–8), and (14–4) in order,

$n = 6,378,206.4 \times (0.8710708 - 0.7021191)/(5,040,295.0 - 3,264,511.2)$
$\quad = 0.6068355$
$G = 0.8710708/0.6068355 + 3,264,511.2/6,378,206.4$
$\quad = 1.9472543$
$\rho_0 = 6,378,206.4 \times 1.9472543 - 2,544,389.8$
$\quad = 9,875,599.9 \text{ m}$
$\rho = 6,378,206.4 \times 1.9472543 - 3,874,395.2$
$\quad = 8,545,594.4 \text{ m}$
$\theta = 0.6068355 \times [-75° - (-96°)]$
$\quad = 12.7435458°$

Constants n, G, and ρ_0 apply to the entire map.
Using equations (14–1), (14–2), and (16–7) in order,

$x = 8,545,594.4 \times \sin 12.7435458°$
$\quad = 1,885,051.9 \text{ m}$
$y = 9,875,599.9 - 8,545,594.4 \times \cos 12.7435458°$
$\quad = 1,540,507.6 \text{ m}$
$k = (1.9472543 - 3,874,395.2/6,378,206.4)$
$\quad\quad \times 0.6068355/0.8200656$
$\quad = 0.99144$

EQUIDISTANT CONIC (ELLIPSOID)–INVERSE EQUATIONS (SEE P. 114)

Inversing forward example:
Given: a, e^2, ϕ_1, ϕ_2, ϕ_0, λ_0 for forward example

$$x = 1,885,051.9 \text{ m}$$
$$y = 1,540,507.6 \text{ m}$$

Calculating n, G, and ρ_0 as in the forward example,

$$n = 0.6068355$$
$$G = 1.9472543$$
$$\rho_0 = 9,875,599.9 \text{ m}$$

Using equations $(14-10)$, $(14-11)$, $(16-12)$, $(7-19)$, $(3-24)$, and $(3-26)$ in order,

$\rho = + [1,885,051.9^2 + (9,875,599.9 - 1,540,507.6)^2]^{1/2}$
$\quad = 8,545,594.4$ m
$\theta = \arctan [1,885,051.9/(9,875,599.9 - 1,540,507.6)]$
$\quad = 12.7435461°$, not adding $180°$ since denominator is positive.
$M = 6,378,206.4 \times 1.9472543 - 8,545,594.4$
$\quad = 3,874,395.4$ m
$\mu = 3,874,395.4/[6,378,206.4 \times (1 - 0.00676866/4$
$\quad\quad - 3 \times 0.00676866^2/64 - 5 \times 0.00676866^3/256)]$
$\quad = 0.6084737$ radians $= 34.8629767°$
$e_1 = [1 - (1 - 0.00676866)^{1/2}]/[1 + (1 - 0.00676866)^{1/2}]$
$\quad = 0.001697916$
$\phi = 34.8629767° + [(3 \times 0.001697916/2 - 27 \times 0.001697916^3/$
$\quad\quad 32) \sin (2 \times 34.8629767°) + (21 \times 0.001697916^2/16$
$\quad\quad - 55 \times 0.001697916^4/32) \sin (4 \times 34.8629767°)$
$\quad\quad + (151 \times 0.001697916^3/96) \sin (6 \times 34.8629767°)$
$\quad\quad + (1097 \times 0.001697916^4/512) \sin (8 \times 34.8629767°)]$
$\quad\quad \times 180°/\pi$
$\quad = 35°$ N. lat.

Using equation $(14-9)$,

$\lambda = -96° + 12.7435461°/0.6068355$
$\quad = -75° = 75°$ W. long.

BIPOLAR OBLIQUE CONIC CONFORMAL (SPHERE)–FORWARD EQUATIONS
(SEE P. 118–120)

This example will illustrate equations $(17-14)$ through $(17-23)$, assuming prior calculation of the constants from equations $(17-1)$ through $(17-13)$.

Given: Radius of sphere: $R = 6,370,997$ m
$\quad\quad\quad\quad$ Point: $\phi = 40°$ N. lat.
$\quad\quad\quad\quad\quad\quad\quad$ $\lambda = 90°$ W. long.

Find: x, y, k

From equations $(17-14)$ and $(17-15)$,

$z_B = \arccos \{\sin 45° \sin 40° + \cos 45° \cos 40° \cos [(-19°59'36'')$
$\quad\quad - (-90°)]\}$
$\quad = 50.22875°$
$Az_B = \arctan \{\sin (-19°59'36'' + 90°)/[\cos 45° \tan 40° - \sin 45° \cos$
$\quad\quad (-19°59'36'' + 90°)]\}$
$\quad = 69.48856°$

Since $69.48856°$ is less than $104.42834°$, proceed to equation $(17-16)$.
From equations $(17-16)$ through $(17-22)$,

$\quad\quad\quad \rho_B = 1.89725 \times 6370997 \tan^{0.63056} (\frac{1}{2} \times 50.22875°)$
$\quad\quad\quad\quad = 7,496,100$ m
$\quad\quad\quad k = 7,496,100 \times 0.63056/(6370997 \sin 50.22875°)$
$\quad\quad\quad\quad = 0.96527$

$$\alpha = \text{arccos}\{[\tan^{0.63056}(\tfrac{1}{2}\times 50.22875°) + \tan^{0.63056} \tfrac{1}{2}(104°$$
$$-50.22875°)]/1.27247\}$$
$$= 1.88279°$$
$$n(Az_{BA}-Az_B) = 0.63056 \times (104.42834° - 69.48856°) = 22.03163°$$

This is greater than α, so $\rho_B' = \rho_B$.

$$x' = 7,496,100 \sin [0.63056 (104.42834°-69.48855°)]$$
$$= 2,811,900 \text{ m}$$
$$y' = 7,496,100 \cos [0.63056 (104.42834°-69.48855°)]$$
$$-1.20709 \times 6,370,997$$
$$= -741,670 \text{ m}$$

From equations (17−32) and (17−33),

$$x = -2,811,900 \cos 45.81997° + 741670 \sin 45.81997°$$
$$= -1,427,800 \text{ m}$$
$$y = 741,670 \cos 45.81997° + 2811900 \sin 45.81997°$$
$$= 2,533,500 \text{ m}$$

BIPOLAR OBLIQUE CONIC CONFORMAL (SPHERE)−INVERSE EQUATIONS
(SEE P. 120−121)

Inversing the forward example:

Given: Radius of sphere: $R = 6,370,997$ m
                    Point: $x = -1,427,800$ m
                                  $y = 2,533,500$ m

Find: ϕ, λ

From equations (17−34) and (17−35),

$$x' = -(-1,427,800) \cos 45.81997° + 2,533,500 \sin 45.81997°$$
$$= 2,811,900 \text{ m}$$
$$y' = -(-1,427,800) \sin 45.81997° - 2,533,500 \cos 45.81997°$$
$$= -741,670 \text{ m}$$

Since x' is positive, go to equations (17−36) through (17−44) in order:

$$\rho'_B = [2,811,900^2 + (1.20709\times 6,370,997 - 741,670)^2]^{1/2}$$
$$= 7,496,100 \text{ m}$$
$$Az'_B = \arctan [2,811,900/(1.20709 \times 6,370,997 - 741,670)]$$
$$= 22.03150° \text{ (The denominator is positive, so there is no}$$
$$\text{quadrant correction.)}$$
$$\rho_B = 7,496,100 \text{ m}$$
$$z_B = 2 \arctan [7,496,100/(1.89725\times 6,370,997)]^{1/0.63056}$$
$$= 50.22873°$$
$$\alpha = \text{arccos}\{[\tan^{0.63056} (\tfrac{1}{2}\times 50.22873°)$$
$$+ \tan^{0.63056} \tfrac{1}{2}(104°-50.22873°)]/ 1.27247\}$$
$$= 1.88279°$$

Since Az'_B is greater than α, go to equation (17−42).

$$Az_B = 104.42834° - 22.03150°/0.63056$$
$$= 69.48876°$$

ϕ = arcsin (sin 45° cos 50.22873° + cos 45° sin 50.22873° cos 69.48876°)

   = 39.99987° or 40° N. lat., if rounding off had not accumulated errors.

λ = (−19°59′36″)−arctan {sin 69.48876°/[cos 45°/tan 50.22873° − sin 45° cos 69.48876°]}

   = −89.99987° or 90° W. long., if rounding off had not accumulated errors.

POLYCONIC (SPHERE)−FORWARD EQUATIONS (SEE P. 128-129)

Given: Radius of sphere: R = 1.0 unit
            Origin: ϕ_0 = 30° N. lat.
                   λ_0 = 96° W. long.
            Point: ϕ = 40° N. lat.
                   λ = 75° W. long.

Find: x, y, h

From equations (18−2) through (18−4),

E = (−75° + 96°) sin 40°
   = 13.4985398°
x = 1.0 cot 40° sin 13.4985398°
   = 0.2781798 unit
y = 1.0 × [40° × π/180° − 30° × π/180° + cot 40° (1−cos 13.4985398°)]
   = 0.2074541 unit

From equations (18−6) and (18−5),

D = arctan {(13.4985398° × π/180° − sin 13.4985398°)/(sec² 40° − cos 13.4985398°)]
   = 0.17018327°
h = (1−cos² 40° cos 13.4985398°)/sin² 40° cos 0.17018327°
   = 1.0392385

POLYCONIC (SPHERE)−INVERSE EQUATIONS (SEE P. 129)

Inversing the forward example:

Given: Radius of sphere: R = 1.0 unit
            Origin: ϕ_0 = 30° N. lat.
                   λ_0 = 96° W. long.
            Point: x = 0.2781798 unit
                   y = 0.2074541 unit

Find: ϕ, λ

Since $y \neq$ −1.0 × 30° × π/180°, use equations (18−7) and (18−8):

A = 30° × π/180° + 0.2074541/1.0
   = 0.7310529
B = 0.2781798²/1.0² + 0.7310529²
   = 0.6118223

Assuming an initial ϕ_n = A = 0.7310529 radians, it is simplest to work with equation (18−9) in *radians:*

$$\phi_{n+1} = 0.7310529 - [\mathbf{0.7310529} \times (0.7310529 \tan 0.7310529 + 1)$$
$$-0.7310529 - \tfrac{1}{2}(0.7310529^2 + 0.6118223) \tan 0.7310529]/$$
$$[(0.7310529 - \mathbf{0.7310529})/\tan 0.7310529 - 1]$$
$$= 0.6963533 \text{ radian}$$

Using 0.6963533 in place of 0.7310529 (except that the boldface retains the value of A) a new ϕ_{n+1} of 0.6981266 radian is obtained. Again substituting this value, 0.6981317 radian is obtained. The fourth iteration results in the same answer to seven decimal places. Therefore,

$$\phi = 0.6981317 \times 180°/\pi = 40.0000004° \text{ or } 40° \text{ N. lat.}$$

From equation (18−10),

$$\lambda = [\arcsin (0.2781798 \tan 40°/1.0)]/\sin 40° + (-96°)$$
$$= -75.0000014° = 75° \text{ W. long.}$$

POLYCONIC (ELLIPSOID)−FORWARD EQUATIONS (SEE P. 129−130)

Given: Clarke 1866 ellipsoid: $a = 6,378,206.4$ m
                                              $e^2 = 0.00676866$
                 Origin: $\phi_0 = 30°$ N. lat.
                            $\lambda_0 = 96°$ W. long.
                 Point: $\phi = 40°$ N. lat.
                            $\lambda = 75°$ W. long.

Find: x, y, h

From equation (3−21),

$$M = 6,378,206.4 \times [(1-0.00676866/4 - 3 \times 0.00676866^2/64$$
$$-5 \times 0.00676866^3/256) \times 40° \times \pi/180° - (3 \times 0.00676866/8$$
$$+3 \times 0.00676866^2/32 + 45 \times 0.00676866^3/1024)$$
$$\sin (2 \times 40°) + (15 \times 0.00676866^2/256 + 45 \times 0.00676866^3/1024)$$
$$\sin (4 \times 40°) - (35 \times 0.00676866^3/3072) \sin (6 \times 40°)]$$
$$= 4,429,318.9 \text{ m}$$

Using 30° in place of 40°,

$$M_0 = 3,319,933.3 \text{ m}$$

From equation (4−20),

$$N = 6,378,206.4/(1-0.00676866 \sin^2 40°)^{1/2}$$
$$= 6,378,143.9 \text{ m}$$

From equations (18−2), (18−12), and (18−13),

$$E = (-75° + 96°) \sin 40°$$
$$= 13.4985398°$$
$$x = 6,387,143.9 \cot 40° \sin 13.4985398°$$
$$= 1,776,774.5 \text{ m}$$
$$y = 4,429,318.9 - 3,319,933.3 + 6,387,143.9 \cot 40°$$
$$(1 - \cos 13.4985398°)$$
$$= 1,319,657.8 \text{ m}$$

To calculate scale factor h, from equations (18-16) and (18-15),

D = arctan $\{(13.4985398° \times \pi/180° - \sin 13.4985398°)/[\sec^2 40°$
$- \cos 13.4985398° - 0.00676866 \sin^2 40°/(1-0.00676866$
$\sin^2 40°)]\}$
$= 0.1708380522°$
h = $[1-0.00676866 + 2(1-0.00676866 \sin^2 40°) \sin^2$
$\frac{1}{2}(13.4985398°)/\tan^2 40°]/(1-0.00676866) \cos 0.1708380522°$
$= 1.0393954$

POLYCONIC (ELLIPSOID)—INVERSE EQUATIONS (SEE P. 130-131)

Inversing the forward example:

Given: Clarke 1866 ellipsoid: $a = 6,378,206.4$ m
$e^2 = 0.00676866$
Origin: $\phi_0 = 30°$ N. lat.
$\lambda_0 = 96°$ W. long.
Point: $x = 1,776,774.5$ m
$y = 1,319,657.8$ m

Find: ϕ, λ

First calculating M_0 from equation (3-21), as in the forward example,

$$M_0 = 3,319,933.3 \text{ m}$$

Since $y \neq M_0$, from equations (18-18) and (18-19),

A = $(3,319,933.3 + 1,319,657.8)/6,378,206.4$
$= 0.7274131$
B = $1,776,774.5^2/6,378,206.4^2 + 0.7274131^2$
$= 0.6067309$

Assuming an initial value of $\phi_n = 0.7274131$ radian, the following calculations are made in radians from equations (18-20), (3-21), (18-17), and (18-21):

C = $(1-0.00676866 \sin^2 0.7274131)^{1/2} \tan 0.7274131$
$= 0.8889365$
M_n = $4,615,626.1$ m
M'_n = $1 - 0.00676866/4 - 3\times0.00676866^2/64 - 5 \times 0.00676866^3/256$
$- 2 \times (3\times0.00676866/8 + 3 \times 0.00676866^2/32 + 45$
$\times 0.00676866^3/1024) \cos (2\times0.7274131) + 4 \times (15$
$\times 0.00676866^2/256 + 45 \times 0.00676866^3/1024) \cos (4$
$\times 0.7274131) - 6 \times (35\times0.00676866^3/3072) \cos (6$
$\times 0.7274131)$
$= 0.9977068$
M_a = $4,615,626.1/6,378,206.4 = 0.7236558$
ϕ_{n+1} = $0.7274131 - [0.7274131 \times (0.8889365 \times 0.7236558 + 1)$
$- 0.7236558 - \frac{1}{2}(0.7236558^2 + 0.6067309) \times 0.8889365]/$
$[0.00676866 \sin (2\times0.7274131) \times (0.7236558^2 + 0.6067309$
$- 2 \times 0.7274131 \times 0.7236558)/(4\times0.8889365)$
$+ (0.7274131 - 0.7236558) \times (0.8889365 \times 0.9977068$
$- 2/\sin (2\times0.7274131)) - 0.9977068]$
$= 0.6967280$ radian

Substitution of 0.6967280 in place of 0.7274131 in equations (18−20), (3−21), (18−17), and (18−21), except for boldface values, which are A, not ϕ_n, a new ϕ_{n+1} of 0.6981286 is obtained. Using this in place of the previous value results in a third ϕ_{n+1} of 0.6981317, which is unchanged by recalculation to seven decimals. Thus,

$$\phi = 0.6981317 \times 180°/\pi = 40.0000005° = 40° \text{ N. lat.}$$

From equation (18−22), using the finally calculated C of 0.8379255,

$$\lambda = [\arcsin (1,776,774.5 \times 0.8379255/6,378,206.4)]/\sin 40° + (−96°)$$
$$= −75° = 75° \text{ W. long.}$$

MODIFIED POLYCONIC (IMW)−FORWARD EQUATIONS (SEE P. 131, 134-135)

Given: International ellipsoid: $a = 6,378,388.0$ m
$e^2 = 0.00672267$
Northernmost lat. of quad: $\phi_2 = 40°$ N. lat.
Southernmost lat. of quad: $\phi_1 = 36°$ N. lat.
Central meridian: $\lambda_0 = 75°$ W. long.
Meridian true to scale: $\lambda_1 = 73°$ W. long.
Point: $\phi = 39°$ N. lat.
$\lambda = 76°$ W. long.

For constants applying to entire map, using equations (18−26) and (18−27) for $n = 1$,

$$R_1 = 6,378,388.0 \times \cot 36°/(1−0.00672267 \times \sin^2 36°)^{1/2}$$
$$= 8,789,311.0 \text{ m}$$
$$F_1 = [−73°−(−75°)] \sin 36°$$
$$= 1.1755705°$$

Using $\phi_2 = 40°$ for $n = 2$ in the same equations,

$$R_2 = 7,612,045.9 \text{ m}$$
$$F_2 = 1.2855752°$$

Using equations (18−23) through (18−25) for $n = 1$ and 2,

$$x_1 = 8,789,311.0 \times \sin 1.1755705°$$
$$= 180,322.7 \text{ m}$$
$$x_2 = 7,612,045.9 \times \sin 1.2855752°$$
$$= 170,781.1 \text{ m}$$
$$y_1 = 8,789,311.0 \times (1−\cos 1.1755705°)$$
$$= 1,849.957 \text{ m}$$
$$T_2 = 7,612,045.9 \times (1−\cos 1.2855752°)$$
$$= 1,916.033 \text{ m}$$

Using equation (3−21) for $n = 1$,

$$M_1 = 6,378,388 \times [(1−0.00672267/4 − 3 \times 0.00672267^2/4$$
$$− 5 \times 0.00672267^3/256) \times 36° \times \pi/180° − (3 \times 0.00672267/$$
$$8 + 3 \times 0.00672267^2/32 + 45 \times 0.00672267^3/1024)$$
$$\sin (2 \times 36°) + (15 \times 0.00672267^2/256 + 45$$
$$\times 0.00672267^3/1024) \sin (4 \times 36°) − (35 \times 0.00672267^3/$$
$$3072) \sin (6 \times 36°)]$$
$$= 3,985,606.6 \text{ m}$$

Repeating the calculation for $n = 2$ and $\phi_2 = 40°$,

$$M_2 = 4,429,605.0 \text{ m}$$

Using equations (18−28) through (18−33) in order,

$$
\begin{aligned}
y_2 &= [(4,429,605.0-3,985,606.6)^2 - (170,781.1 - 180,322.7)^2]^{1/2} \\
&\quad + 1,849.957 \\
&= 445,745.8 \text{ m}
\end{aligned}
$$

$$
\begin{aligned}
C_2 &= 445,745.8 - 1,916.033 \\
&= 443,829.8 \text{ m}
\end{aligned}
$$

$$
\begin{aligned}
P &= (4,429,605.0 \times 1,849.957 - 3,985,606.6 \times 445,745.8)/ \\
&\quad (4,429,605.0 - 3,985,606.6) \\
&= -3,982,836.2 \text{ m}
\end{aligned}
$$

$$
\begin{aligned}
Q &= (445,745.8 - 1,849.957)/(4,429,605.0 - 3,985,606.6) \\
&= 0.9997691
\end{aligned}
$$

$$
\begin{aligned}
P' &= (4,429,605.0 \times 180,322.7 - 3,985,606.6 \times 170,781.1)/ \\
&\quad (4,429,605.0 - 3,985,606.6) \\
&= 265,974.0 \text{ m}
\end{aligned}
$$

$$
\begin{aligned}
Q' &= (170,781.1 - 180,322.7)/(4,429,605.0 - 3,985,606.6) \\
&= -0.02149016
\end{aligned}
$$

The above constants apply to the entire quadrangle. The following values are for the specific point. Using equations (3−21) and (18−26) without subscripts, for $\phi = 39°$,

$$
\begin{aligned}
M &= 4,318,576.8 \text{ m} \\
R &= 7,887,159.9 \text{ m}
\end{aligned}
$$

Using equations (18−34) through (18−40) in order,

$$
\begin{aligned}
x_a &= 265,974.0 + (-0.02149016) \times 4,318,576.8 \\
&= 173,167.1 \text{ m}
\end{aligned}
$$

$$
\begin{aligned}
y_a &= -3,982,836.2 + 0.9997691 \times 4,318,576.8 \\
&= 334,743.2 \text{ m}
\end{aligned}
$$

$$
\begin{aligned}
C &= 334,743.2 - 7,887,159.9 + (7,887,159.9^2-173,167.1^2)^{1/2} \\
&= 332,842.0 \text{ m}
\end{aligned}
$$

$$
\begin{aligned}
x_b &= 7,612,045.9 \sin [(-76°-(-75°)) \sin 40°] \\
&= -85,395.9 \text{ m}
\end{aligned}
$$

$$
\begin{aligned}
y_b &= 443,829.8 + 7,612,045.9 \times \{1-\cos [(-76°-(-75°)) \sin 40°]\} \\
&= 444,308.8 \text{ m}
\end{aligned}
$$

$$
\begin{aligned}
x_c &= 8,789,311.0 \sin [(-76°-(-75°)) \sin 36°] \\
&= -90,166.1 \text{ m}
\end{aligned}
$$

$$
\begin{aligned}
y_c &= 8,789,311.0 \times \{1- \cos [(-76°-(-75°)) \sin 36°]\} \\
&= 462.5 \text{ m}
\end{aligned}
$$

Using equations (18−41) through (18−44),

$$
\begin{aligned}
D &= [-85,395.9-(-90,166.1)]/[444,308.8-462.5] \\
&= 0.01074735
\end{aligned}
$$

$$
\begin{aligned}
B &= -90,166.1 + 0.01074735 \times (332,842.0+7,887,159.9-462.5) \\
&= -1,827.9 \text{ m}
\end{aligned}
$$

$$
\begin{aligned}
x &= \{-1,827.9-0.01074735 \times [7,887,159.9^2 \times (1+0.01074735^2) - \\
&\quad (-1,827.9)^2]^{1/2}\}/(1+0.01074735^2) \\
&= -86,588.8 \text{ m}
\end{aligned}
$$

$$
\begin{aligned}
y &= 332,842.0 + 7,887,159.9 - [7,887,159.9^2 - (-86,588.8)^2]^{1/2} \\
&= 333,317.3 \text{ m}
\end{aligned}
$$

MODIFIED POLYCONIC (IMW)—INVERSE EQUATIONS (SEE P. 135)

Inversing forward example:

Given: a, e^2, ϕ_2, ϕ_1, λ_0, λ_1 for forward example
$$x = -86{,}588.8 \text{ m}$$
$$y = 333{,}317.3 \text{ m}$$

These constants are calculated exactly as in the forward case, and have the same values for this example: x_1, x_2, y_1, M_1, M_2, y_2, C_2, P, Q, P', Q'. The first trial ϕ and λ, or ϕ_{t1} and λ_{t1}, are found from equations (18–47) and (18–48):

$$\phi_{t1} = 40°$$
$$\lambda_{t1} = [-86{,}588.8/(6{,}378{,}388.0 \times \cos 40°)] \times 180°/\pi + (-75°)$$
$$= -76.0153586°$$

Calculating x, y for these trial values of ϕ, λ, exactly as in the forward case, results in the following test values:

$$y_c = 476.8 \text{ m}$$
$$x_{t1} = -86{,}707.4 \text{ m}$$
$$y_{t1} = 444{,}323.6 \text{ m}$$

The new trial ϕ and λ are found from equations (18–49) and (18–50):

$$\phi_{t2} = [(40°-36°) \times (333{,}317.3-476.8)/(444{,}323.6-476.8)] + 36°$$
$$= 38.999598°$$
$$\lambda_{t2} = [(-76.0153586°-(-75°)) \times (-86{,}588.8)/(-86{,}707.4)] + (-75°)$$
$$= -76.0139694°$$

Calculating x, y from these trial values, and then recalculating ϕ, λ:

$$y_c = 475.5 \text{ m}$$
$$x_{t2} = -87{,}798.8 \text{ m}$$
$$y_{t2} = 333{,}286.1 \text{ m}$$
$$\phi_{t3} = 38.9998792°$$
$$\lambda_{t3} = -75.9999952°$$

The next iteration produces the following:

$$y_c = 462.5 \text{ m}$$
$$x_{t3} = -86{,}588.5 \text{ m}$$
$$y_{t3} = 333{,}303.9 \text{ m}$$
$$\phi_{t4} = 38.9999997°$$
$$\lambda_{t4} = -75.9999984°$$

Then

$$y_c = 462.5 \text{ m}$$
$$x_{t4} = -86{,}588.7 \text{ m}$$
$$y_{t4} = 333{,}317.3 \text{ m}$$
$$\phi_{t5} = 38.9999996°$$
$$\lambda_{t5} = -76.0000001°$$

And finally, since there is no significant change,

$$y_c = 462.5 \text{ m}$$
$$x_{t5} = -86,588.8 \text{ m}$$
$$y_{t5} = 333,317.3 \text{ m}$$
$$\phi_{t6} = 38.9999996°$$
$$\lambda_{t6} = -76.0000001°$$

Thus, $\phi = 39°$ N. lat. and $\lambda = 76°$ W. long.

BONNE (SPHERE)–FORWARD EQUATIONS (SEE P. 139–140)

Given: Radius of sphere: $R = 1.0$ unit
Standard parallel: $\phi_1 = 40°$ N. lat.
Central meridian: $\lambda_0 = 75°$ W. long.
Point: $\phi = 30°$ N. lat.
$\lambda = 85°$ W. long.

Find: x, y

Using equations (19–1) through (19–4) in order,

$\rho = 1.0 \times [\cot 40° + (40°-30°) \times \pi/180°]$
$= 1.3662865$ units
$E = 1.0 \times [-85° - (-75°)] \cos 30°/1.3662865$
$= -6.3385344°$
$x = 1.3662865 \sin (-6.3385344°)$
$= -0.1508418$ unit
$y = 1.0 \cot 40° - 1.3662865 \cos (-6.3385344°)$
$= -0.1661807$ unit

BONNE (SPHERE)–INVERSE EQUATIONS (SEE P. 140)

Inversing forward example:

Given: R, ϕ_1, λ_0 for forward example

$$x = -0.1508418 \text{ unit}$$
$$y = -0.1661807 \text{ unit}$$

Find: ϕ, λ

Using equations (19–5) through (19–7) in order,

$\rho = [(-0.1508418)^2 + (1.0 \cot 40°-(-0.1661807))^2]^{1/2}$
$= 1.3662865$ units
$\phi = (\cot 40°) \times 180°/\pi + 40°-(1.3662865/1.0) \times 180°/\pi$
$= 30°$ N. lat.
$\lambda = -75° + 1.3662865 \times \{\arctan [-0.1508418/(1.0 \cot 40°$
$- (-0.1661807))]\}/(1.0 \cos 30°)$
$= -75° + 1.3662865 \times \{\arctan [-0.1508418/1.3579343]\}/$
$\cos 30°$
$= -85° = 85°$ W. long., not adding 180° to the arctan because the denominator is positive.

BONNE (ELLIPSOID)—FORWARD EQUATIONS (SEE P.140)

Given: Clarke 1866 ellipsoid: a = 6,378,206.4 m
                              e^2 = 0.00676866
             Standard parallel: ϕ_1 = 40° N. lat.
             Central meridian: λ_0 = 75° W. long.
                        Point: ϕ = 30° N. lat.
                               λ = 85° W. long.

Find: x, y

Using equations (14−15) and (3−21),

m = cos 30°/(1−0.00676866 sin^2 30°)$^{1/2}$
      = 0.8667591
M = 6,378,206.4 × [(1−0.00676866/4−3×0.00676866^2/64
         −5×0.00676866^3/256) × 30° × π/180° − (3×0.00676866/8
         + 3 × 0.00676866^2/32 + 45 × 0.00676866^3/1024)
         sin (2×30°) + (15×0.00676866^2/256 + 45 × 0.00676866^3/
         1024) sin (4×30°) − (35×0.00676866^3/3072) sin (6×30°)]
      = 3,319,933.3 m

Using the same equations, but with ϕ_1 = 40° in place of 30°,

$$m_1 = 0.7671179$$
$$M_1 = 4,429,318.9 \text{ m}$$

Using equations (19−8) through (19−11) in order,

ρ = 6,378,206.4 × 0.7671179/sin 40° + 4,429,318.9 − 3319933.3
      = 8,721,287.6 m
E = 6,378,206.4 × 0.8667591 × [−85° − (−75°)]/8,721,287.6
      = −6.3389360°
x = 8,721,287.6 sin (−6.3389360°)
      = −962,915.1 m
y = 6,378,206.4 × 0.7671179/sin 40°−8,721,287.6 cos (−6.3389360°)
      = −1,056,065.0 m

BONNE (ELLIPSOID)—INVERSE EQUATIONS (SEE P. 140)

Inversing forward example:

Given: a, e^2, ϕ_1, λ_0 for forward example

$$x = -962,915.1 \text{ m}$$
$$y = -1,056,065.0 \text{ m}$$

Find: ϕ, λ

Using equations (14−15) and (3−21), m_1 and M_1 are calculated as in the forward example:

$$m_1 = 0.7671179$$
$$M_1 = 4,429,318.9 \text{ m}$$

Using equations (19−12), (19−13), (7−19), (3−24), and (3−26) in order,

$$\rho \quad = [(-962,915.1)^2 + (6,378,206.4 \times 0.7671179/\sin 40° - (-1,056,065.0))^2]^{1/2}$$
$$= 8,721,287.6 \text{ m}$$
$$M \quad = 6,378,206.4 \times 0.7671179/\sin 40° + 4,429,318.9 - 8,721,287.6$$
$$= 3,319,933.3 \text{ m}$$
$$\mu \quad = \{3,319,933.3/[6,378,206.4 \times (1 - 0.00676866/4$$
$$- 3 \times 0.00676866^2/64 - 5 \times 0.00676866^3/256)]\} \times 180°/\pi$$
$$= 29.8737595°$$
$$e_1 \quad = [1 - (1 - 0.00676866)^{1/2}]/[1 + (1 - 0.00676866)^{1/2}]$$
$$= 0.001697916$$
$$\phi \quad = 29.8737595° + [(3 \times 0.001697916/2 - 27 \times 0.001697916^3/32)$$
$$\sin (2 \times 29.8737595°) + (21 \times 0.001697916^2/16$$
$$- 55 \times 0.001697916^4/32) \sin (4 \times 29.8737595°)$$
$$+ (151 \times 0.001697916^3/96) \sin (6 \times 29.8737595°)$$
$$+ (1097 \times 0.001697916^4/512) \sin (8 \times 29.8737595°)] \times 180°/\pi$$
$$= 30° \text{ N. lat.}$$

Using equation (14-15),

$$m = \cos 30°/(1 - 0.00676866 \times \sin^2 30°)^{1/2}$$
$$= 0.8667591$$

Using equation (19-14),

$$\lambda = -75° + 8721287.6 \times \{\arctan [-962,915.1/$$
$$(6,378,206.4 \times 0.7671179/\sin 40° - (-1,056,065.0))]\}/$$
$$(6,378,206.4 \times 0.8667591)$$
$$= -85° = 85° \text{ W. long.}$$

ORTHOGRAPHIC (SPHERE)-FORWARD EQUATIONS (SEE P. 148-149)

Given: Radius of sphere: $R = 1.0$ unit
Center: $\phi_1 = 40°$ N. lat.
$\lambda_0 = 100°$ W. long
Point: $\phi = 30°$ N. lat.
$\lambda = 110°$ W. long.

Find: x, y

In general calculations, to determine whether this point is beyond viewing, using equation (5-3),

$$\cos c = \sin 40° \sin 30° + \cos 40° \cos 30° \cos (-110° + 100°)$$
$$= 0.9747290$$

Since this is positive, the point is within view.
Using equations (20-3) and (20-4),

$$x = 1.0 \cos 30° \sin (-110° + 100°)$$
$$= -0.1503837$$
$$y = 1.0 [\cos 40° \sin 30° - \sin 40° \cos 30° \cos (-110° + 100°)]$$
$$= -0.1651911$$

Examples of other forward equations are omitted, since the formulas for the oblique aspect apply generally.

ORTHOGRAPHIC (SPHERE)—INVERSE EQUATIONS (SEE P. 150)

Inversing forward example:

Given: Radius of sphere: $R = 1.0$ unit
                Center: $\phi_1 = 40°$ N. lat.
                          $\lambda_0 = 100°$ W. long.
                 Point: $x = -0.1503837$ unit
                          $y = -0.1651911$ unit

Find: ϕ, λ

Using equations (20−18) and (20−19),

$$\rho = [(-0.1503837)^2 + (-0.1651911)^2]^{1/2}$$
$$= 0.2233906$$
$$c = \arcsin (0.2233906/1.0)$$
$$= 12.9082572°$$

Using equations (20−14) and (20−15),

$$\phi = \arcsin [\cos 12.9082572° \sin 40° + (-0.1651911 \sin$$
$$12.9082572° \cos 40°/0.2233906)]$$
$$= 30.0000007°, \text{ or } 30° \text{ N. lat. if rounding off did not occur.}$$
$$\lambda = -100° + \arctan [-0.1503837 \sin 12.9082572°/(0.2233906$$
$$\cos 40° \cos 12.9082572° + 0.1651911 \sin 40° \sin$$
$$12.9082572°)]$$
$$= -100° + \arctan [-0.0335943/0.1905228]$$
$$= -100° + (-9.9999964°)$$
$$= -109.9999964°, \text{ or } 110° \text{ W. long. if rounding off did not}$$
$$\text{occur}$$

Since the denominator of the argument of arctan is positive, no adjustment for quadrant is necessary.

STEREOGRAPHIC (SPHERE)—FORWARD EQUATIONS (SEE P. 157–158)

Given: Radius of sphere: $R = 1.0$ unit
                     Center: $\phi_1 = 40°$ N. lat.
                               $\lambda_0 = 100$ W. long.
      Central scale factor: $k_0 = 1.0$
                      Point: $\phi = 30°$ N. lat.
                               $\lambda = 75°$ W. long.

Find: x, y, k

Using equations (21−4), (21−2), and (21−3) in order,

$$k = 2 \times 1.0/[1 + \sin 40° \sin 30° + \cos 40° \cos 30° \cos (-75° + 100°)]$$
$$= 1.0402304$$
$$x = 1.0 \times 1.0402304 \cos 30° \sin (-75° + 100°)$$
$$= 0.3807224 \text{ unit}$$
$$y = 1.0 \times 1.0402304 [\cos 40° \sin 30° - \sin 40° \cos 30° \cos (-75° + 100°)]$$
$$= -0.1263802 \text{ unit}$$

Examples of other forward equations are omitted, since the above equations are general.

STEREOGRAPHIC (SPHERE)−INVERSE EQUATIONS (SEE P. 158-159)

Inversing forward example:

Given: Radius of sphere: $R = 1.0$ unit
                        Center: $\phi_1 = 40°$ N. lat.
                                      $\lambda_0 = 100$ W. long.
          Central scale factor: $k_0 = 1.0$
                         Point: $x = 0.3807224$ unit
                                      $y = -0.1263802$ unit

Find: ϕ, λ

Using equations (20−18) and (21−15),

$$\rho = [0.3807224^2 + (-0.1263802)^2]^{1/2} = 0.4011502 \text{ units}$$
$$c = 2 \arctan [0.4011502/(2 \times 1.0 \times 1.0)]$$
$$= 22.6832261°$$

Using equations (20−14) and (20−15),

ϕ $= \arcsin [\cos 22.6832261° \sin 40° + (-0.1263802)$
              $\sin 22.6832261° \cos 40°/0.4011502]$
       $= \arcsin 0.5000000 = 30° = 30°$ N. lat.
λ $= -100° + \arctan [0.3807224 \sin 22.6832261°/(0.4011502$
              $\cos 40° \cos 22.6832261° + 0.1263802 \sin 40° \sin 22.6832261°)]$
       $= -100° + \arctan (0.1468202/0.3148570)$
       $= -100° + 25.0000013°$
       $= -74.9999987° = 75°$ W. long.

except for effect of rounding-off input data. Since the denominator of the argument of arctan is positive, no quadrant adjustment is necessary. If it were negative, 180° should be added.

STEREOGRAPHIC (ELLIPSOID)−FORWARD EQUATIONS (SEE P. 160-161)

Oblique aspect:

Given: Clarke 1866 ellipsoid: $a = 6,378,206.4$ m
                                          $e^2 = 0.00676866$
                             or $e = 0.0822719$
                        Center: $\phi_1 = 40°$ N. lat.
                                          $\lambda_0 = 100°$ W. long.
          Central scale factor: $k_0 = 0.9999$
                         Point: $\phi = 30°$ N. lat.
                                          $\lambda = 90°$ W. long.

Find: x, y, k

From equation (3−1),

χ_1 $= 2 \arctan \{\tan (45° + 40°/2) [(1 - 0.0822719 \sin 40°)/$
              $(1 + 0.0822719 \sin 40°)]^{0.0822719/2}\} - 90°$
       $= 2 \arctan 2.1351882 - 90°$
       $\doteq 39.8085922°$
χ $= 2 \arctan \{\tan (45° + 30°/2) [(1 - 0.0822719 \sin 30°)/$
              $(1 + 0.0822719 \sin 30°)]^{0.0822719/2}\} - 90°$

$$= 2 \arctan 1.7261956 - 90°$$
$$= 29.8318339°$$

From equation (14−15),

$$m_1 = \cos 40°/(1 - 0.00676866 \sin^2 40°)^{1/2}$$
$$= 0.7671179$$
$$m = \cos 30°/(1 - 0.00676866 \sin^2 30°)^{1/2}$$
$$= 0.8667591$$

From equation (21−27),

$$A = 2 \times 6{,}378{,}206.4 \times 0.9999 \times 0.7671179/\{\cos 39.8085922°$$
$$[1 \times \sin 39.8085922° \sin 29.8318339° + \cos 39.8085922°$$
$$\cos 29.8318339° \cos (-90° + 100°)]\}$$
$$= 6{,}450{,}107.7 \text{ m}$$

From equations (21−24), (21−25), and (21−26),

$$x = 6{,}450{,}107.7 \cos 29.8318339° \sin (-90° + 100°)$$
$$= 971{,}630.8 \text{ m}$$
$$y = 6{,}450{,}107.7 [\cos 39.8085922° \sin 29.8318339°$$
$$- \sin 39.8085922° \cos 29.8318339° \cos (-90° + 100°)]$$
$$= -1{,}063{,}049.3 \text{ m}$$
$$k = 6{,}450{,}107.7 \cos 29.8318339°/[6{,}378{,}206.4 \times 0.8667591]$$
$$= 1.0121248$$

Polar aspect with known k_0:

Given: International ellipsoid: $a = 6{,}378{,}388.0$ m
$\qquad\qquad\qquad\qquad\qquad\qquad e^2 = 0.00672267$
$\qquad\qquad\qquad$ or $\qquad e = 0.0819919$
$\qquad\qquad$ Center: South Pole $\phi_1 = 90°$ S. lat.
$\qquad\qquad\qquad\qquad\qquad\qquad \lambda_0 = 100°$ W. long. (meridian
$\qquad\qquad\qquad\qquad\qquad\qquad\qquad\qquad$ along pos. Y axis)
$\qquad$ Central scale factor: $k_0 = 0.994$
$\qquad\qquad\qquad\qquad$ Point: $\phi = 75°$ S. lat.
$\qquad\qquad\qquad\qquad\qquad\qquad \lambda = 150°$ E. long.

Find: x, y, k

Since this is the south polar aspect, for calculations change signs of x, y, ϕ, λ, and λ_0 (ϕ_c is not used): $\lambda_0 = 100°$ E. long., $\phi = 75°$ N. lat., $\lambda = 150°$ W. long. Using equations (15−9) and (21−33),

$$t = \tan(45° - 75°/2)/[(1 - 0.0819919 \sin 75°)/(1 + 0.0819919 \sin 75°)]^{0.0819919/2}$$
$$= 0.1325120$$
$$\rho = 2 \times 6{,}378{,}388.0 \times 0.994 \times 0.1325120/[(1 + 0.0819919)^{[1 + 0.0819919]}$$
$$\times (1 - 0.0819919)^{[1 - 0.0819919]}]^{1/2}$$
$$= 1{,}674{,}638.5 \text{ m}$$

Using equations (21−30) and (21−31), changing signs of x and y for the south polar aspect,

$$x = -1{,}674{,}638.5 \sin (-150° - 100°)$$
$$= -1{,}573{,}645.4 \text{ m}$$

$$y = + \ 1{,}674{,}638.5 \cos (- \ 150° - 100°)$$
$$= -572{,}760.1 \text{ m}$$

From equation (14−15),

$$m = \cos 75°/(1 - 0.00672267 \sin^2 75°)^{1/2}$$
$$= 0.2596346$$

From equation (21−32),

$$k = 1{,}674{,}638.5/(6{,}378{,}388 \times 0.2596346)$$
$$= 1.0112245$$

Polar aspect with known ϕ_c not at the pole:

Given: International ellipsoid: $a = 6{,}378{,}388.0$ m
                                  $e^2 = 0.00672267$
                          or $e = 0.0819919$
        Standard parallel: $\phi_c = 71°$ S. lat.
                            $\lambda_0 = 100$ W. long. (meridian
                                 along pos. Y axis)
        Point: $\phi = 75°$ S. lat.
                $\lambda = 150°$ E. long.

Find: x, y, k

Since ϕ_c is southern, for calculations change signs of x, y, ϕ_c, ϕ, λ, and λ_0: $\phi_c = 71°$ N. lat., $\phi = 75°$ N. lat., $\lambda = 150°$ W. long., $\lambda_0 = 100°$ E. long. Using equation (15−9), t for 75° has been calculated in the preceding example, or

$$t = 0.1325120$$

For t_c, substitute 71° in place of 75° in (15−9), and

$$t_c = 0.1684118$$

From equations (14−15) and (21−34),

$$m_c = \cos 71°/(1 - 0.00672267 \sin^2 71°)^{1/2}$$
$$= 0.3265509$$
$$\rho = 6{,}378{,}388.0 \times 0.3265509 \times 0.1325120/0.1684118$$
$$= 1{,}638{,}869.6 \text{ m}$$

Equations (21−30), (21−31), and (21−32) are used as in the preceding south polar example, changing signs of x and y.

$$x = -1{,}638{,}869.6 \sin (- \ 150° - 100°)$$
$$= -1{,}540{,}033.6 \text{ m}$$
$$y = + \ 1{,}638{,}869.6 \cos (- \ 150° - 100°)$$
$$= -560{,}526.4 \text{ m}$$
$$k = 1{,}638{,}869.6/(6{,}378{,}388.0 \times 0.2596346)$$
$$= 0.9896255$$

where m is calculated in the preceding example.

STEREOGRAPHIC (ELLIPSOID)–INVERSE EQUATIONS (SEE P. 161–162)

Oblique aspect (inversing forward example):

Given: Clarke 1866 ellipsoid: $a = 6{,}378{,}206.4$ m
                                             $e^2 = 0.00676866$
                          or $e = 0.0822719$
               Central: $\phi_1 = 40°$ N. lat.
                              $\lambda_0 = 100°$ W. long.
      Central scale factor: $k_0 = 0.9999$
                   Point: $x = 971{,}630.8$ m
                                $y = -1{,}063{,}049.3$ m

Find: ϕ, λ

From equation (14–15),

$$m_1 = \cos 40°/(1-0.00676866 \sin^2 40°)^{1/2}$$
$$= 0.7671179$$

From equation (3–1), as in the forward oblique example,

$$\chi_1 = 39.8085922°$$

From equations (20–18) and (21–38),

$\rho = [971{,}630.8^2 + (-1{,}063{,}049.3)^2]^{1/2}$
$\quad = 1{,}440{,}187.6$ m
$c_e = 2 \arctan [1{,}440{,}187.6 \cos 39.8085922°/(2\times6{,}378{,}206.4$
$\qquad \times\ 0.9999 \times 0.7671179)]$
$\quad = 12.9018251°$

From equation (21–37),

$\chi = \arcsin [\cos 12.9018251° \sin 39.8085922°$
$\qquad + (-1{,}063{,}049.3 \sin 12.9018251° \cos 39.8085922°/1{,}440{,}187.6)]$
$\quad = 29.8318337°$

Using χ as the first trial ϕ in equation (3–4),

$\phi = 2 \arctan \{\tan (45° + 29.8318337°/2) \times [(1 + 0.0822719$
$\qquad \sin 29.8318337°)/(1-0.0822719 \sin 29.8318337°)]^{0.0822719/2}\}$
$\qquad -90°$
$\quad = 29.9991438°$

Using this new trial value in the same equation for ϕ, not for χ,

$\phi = 2 \arctan \{\tan (45° + 29.8318337°/2) \times [(1 + 0.0822719$
$\qquad \sin 29.9991438°)/(1-0.0822719 \sin 29.9991438°)]^{0.0822719/2}\}$
$\qquad -90°$
$\quad = 29.9999953°$

Repeating with 29.9999953° in place of 29.9991438°, the next trial ϕ is

$$\phi = 29.9999997°$$

The next trial calculation produces the same ϕ to seven decimals. Therefore, this is ϕ.

Using equation (21–36),

$$
\begin{aligned}
\lambda &= -100° + \arctan\,[971{,}630.8 \sin 12.9018251°/ \\
&\quad (1{,}440{,}187.6 \cos 39.8085922° \cos 12.9018251° \\
&\quad + 1{,}063{,}049.3 \sin 39.8085922° \sin 12.9018251°)] \\
&= -100° + \arctan\,(216{,}946.9/1{,}230{,}366.8) \\
&= -100° + 10.0000000° \\
&= -90.0000000° = 90° \text{ W. long.}
\end{aligned}
$$

Since the denominator of the arctan argument is positive, no quadrant adjustment is necessary. If it were negative, it would be necessary to add or subtract 180°, whichever would place the final λ between $+180°$ and $-180°$.

Instead of the iterative equation (3–4), series equation (3–5) may be used (omitting terms with e^8 here for simplicity):

$$
\begin{aligned}
\phi &= 29.8318337° \times \pi/180° + (0.00676866/2 + 5 \times 0.00676866^2/24 \\
&\quad + 0.00676866^3/12) \sin (2 \times 29.8318337°) + (7 \times 0.00676866^2/48 \\
&\quad + 29 \times 0.00676866^3/240) \sin (4 \times 29.8318337°) + (7 \\
&\quad \times 0.00676866^3/120) \sin (6 \times 29.8318337°) \\
&= 0.5235988 \text{ radian} \\
&= 29.9999997°
\end{aligned}
$$

Polar aspect with known k_0 (inversing forward example):

Given: International ellipsoid:
$$a = 6{,}378{,}388.0 \text{ m}$$
$$e^2 = 0.00672267$$
or $e = 0.0819919$

Center: South Pole
$$\phi_1 = 90° \text{ S. lat.}$$
$$\lambda_0 = 100° \text{ W. long. (meridian along pos. } Y \text{ axis)}$$

Central scale factor: $k_0 = 0.994$

Point:
$$x = -1{,}573{,}645.4 \text{ m}$$
$$y = -572{,}760.1 \text{ m}$$

Find: ϕ, λ

Since this is the south polar aspect, change signs as stated in text: For calculation, use $\phi_c = 90°$, $\lambda_0 = 100°$ E. long., $x = 1{,}573{,}645.4$ m, $y = 572{,}760.1$ m. From equations (20–18) and (21–39),

$$
\begin{aligned}
\rho &= (1{,}573{,}645.4^2 + 572{,}760.1^2)^{1/2} \\
&= 1{,}674{,}638.5 \text{ m} \\
t &= 1{,}674{,}638.5 \times [(1 + 0.0819919)^{[1+0.0819919]} \\
&\quad (1 - 0.0819919)^{[1-0.0819919]}]^{1/2}/(2 \times 6{,}378{,}388.0 \times 0.994) \\
&= 0.1325120
\end{aligned}
$$

To iterate with equation (7–9), use as the first trial ϕ,

$$
\begin{aligned}
\phi &= 90° - 2 \arctan 0.1325120 \\
&= 74.9031975°
\end{aligned}
$$

Substituting in (7–9),

$$
\begin{aligned}
\phi &= 90° - 2 \arctan \{0.1325120 \times [(1 - 0.0819919 \sin 74.9031975°)/ \\
&\quad (1 + 0.0819919 \sin 74.9031975°)]^{0.0819919/2}\} \\
&= 74.9999546°
\end{aligned}
$$

Using this second trial ϕ in the same equation instead of 74.9031975°,

$$\phi = 74.9999986°.$$

The third trial gives the same value to seven places, so, since the sign of ϕ must be reversed for the south polar aspect,

$$\phi = -74.9999986°, = 75° \text{ S. lat., disregarding effects of rounding off.}$$

If the series equation (3−5) is used instead of (7−9), χ is first found from (7−13):

$$\chi = 90° - 2 \arctan 0.1325120$$
$$= 74.9031975°$$

Substituting this into (3−5), after converting χ to radians for the first term, ϕ is found in radians and is converted to degrees, then given a reversal of sign for the south polar aspect, giving the same result as the iteration.

From equation (20−16),

$$\lambda = +100° + \arctan[1{,}573{,}645.4/(-572{,}760.1)]$$
$$= 100° + (-69.9999995°)$$
$$= 30.0000005°$$

However, since the denominator of the argument of arctan is negative, 180° must be added to λ (added, not subtracted, since the numerator is positive), *then* the sign of λ must be changed for the south polar aspect:

$$\lambda = -(30.0000005° + 180°)$$
$$= -210.0000005°$$

To place this between +180° and −180°, add 360°, so

$$\lambda = +149.9999995° \text{ or } 150° \text{ E. long., disregarding effects of rounding off.}$$

Polar aspect with known ϕ_c not at the pole (inversing forward example):

Given: International ellipsoid: $a = 6{,}378{,}388.0$ m
$\qquad\qquad\qquad\qquad\qquad e^2 = 0.00672267$
$\qquad\qquad\qquad$ or $e = 0.0819919$
$\qquad$ Standard parallel: $\phi_c = 71°$ S. lat.
$\qquad\qquad\qquad\qquad\qquad \lambda_0 = 100°$ W. long. (meridian along pos. Y axis)
$\qquad\qquad\qquad$ Point: $x = -1{,}540{,}033.6$ m
$\qquad\qquad\qquad\qquad\qquad y = -560{,}526.4$ m

Find: ϕ, λ

Since this is south polar, change signs as stated in text: For calculation, $\phi_c = 71°$ N. lat., $\lambda_0 = 100°$ E. long., $x = 1{,}540{,}033.6$, $y = 560{,}526.4$. From equations (15−9) and (14−15), as calculated in the corresponding forward example,

$$t_c = \tan(45° - 71°/2)/[(1 - 0.0819919 \sin 71°)/$$
$$(1 + 0.0819919 \sin 71°)]^{0.0819919/2}$$
$$= 0.1684118$$

$$m_c = \cos 71°/(1-0.00672267 \sin^2 71°)^{1/2}$$
$$= 0.3265509$$

From equations (20−18) and (21−40),

$$\rho = (1{,}540{,}033.6^2 + 560{,}526.4^2)^{1/2}$$
$$= 1{,}638{,}869.5 \text{ m}$$
$$t = 1{,}638{,}869.5 \times 0.1684118/(6{,}378{,}388.0 \times 0.3265509)$$
$$= 0.1325120$$

For the first trial ϕ in equation (7−9),

$$\phi = 90° - 2 \arctan 0.1325120$$
$$= 74.903197°$$

Substituting in (7−9),

$$\phi = 90° - 2 \arctan \big\{0.1325120\,[(1 - 0.0819919 \sin 74.903197°)/$$
$$(1 + 0.0819919 \sin 74.903197°)]^{0.0819919/2}\big\}$$
$$= 74.9999586°$$

Replacing 74.903197° with 74.9999586°, the next trial ϕ is

$$\phi = 75.0000026°$$

The next iteration results in the same ϕ to seven places, so changing signs,

$$\phi = -75.0000026° = 75° \text{ S. lat., disregarding effects of rounding off.}$$

The use of series equation (3−5) with (7−13) to avoid iteration follows the same procedure as the preceding example. For λ, equation (20−16) is used, calculating with reversed signs:

$$\lambda = +100° + \arctan[1{,}540{,}033.6/(-560{,}526.4)]$$
$$= 100° + (-69.9999997°)$$
$$= 30.0000003°$$

Since the denominator in the argument for arctan is negative, add 180°:

$$\lambda = 210.0000003°$$

Now subtract 360° to place λ between +180° and −180°:

$$\lambda = -149.9999997°$$

Finally, reverse the sign to account for the south polar aspect:

$$\lambda = +149.9999997° = 150° \text{ E. long., disregarding rounding off in the input.}$$

GNOMONIC (SPHERE)−FORWARD EQUATIONS (SEE P. 165, 167)

Given: Radius of sphere: $R = 1.0$ unit
Center: $\phi_1 = 40°$ N. lat.
$\lambda_0 = 100°$ W. long.
Point: $\phi = 30°$ N. lat.
$\lambda = 110°$ W. long.

Find: x, y

Using equation (5–3),

$$\cos c = \sin 40° \sin 30° + \cos 40° \cos 30° \cos [-110°-(-100°)]$$
$$= 0.9747290$$

Since $\cos c$ is positive (not zero or negative), the point is in view and may be plotted. Using equations (22–3) through (22–5) in order,

$$k' = 1/0.9747290$$
$$= 1.0259262$$
$$x = 1.0 \times 1.0259262 \cos 30° \sin [-110° - (-100°)]$$
$$= -0.1542826 \text{ unit}$$
$$y = 1.0 \times 1.0259262 \times \{\cos 40° \sin 30° - \sin 40°$$
$$\cos 30° \cos [-110° - (-100°)]\}$$
$$= -0.1694739 \text{ unit}$$

Examples of other forward equations are omitted, since the above equations are general.

GNOMONIC (SPHERE)–INVERSE EQUATIONS (SEE P. 167)

Inversing forward example:

Given: R, ϕ_1, λ_0 for forward example

$$x = -0.1542826 \text{ unit}$$
$$y = -0.1694739 \text{ unit}$$

Find: ϕ, λ

Using equations (20–18) and (22–16),

$$\rho = [(-0.1542826)^2 + (-0.1694739)^2]^{1/2}$$
$$= 0.2291823 \text{ unit}$$
$$c = \arctan (0.2291823/1.0)$$
$$= 12.9082593°$$

Using equations (20–14) and (20–15),

$$\phi = \arcsin [\cos 12.9082593° \sin 40° + (-0.1694739$$
$$\sin 12.9082593° \cos 40°/0.2291823)]$$
$$= 30° \text{ N. lat.}$$
$$\lambda = -100° + \arctan [-0.1542826 \sin 12.9082593°/$$
$$(0.2291823 \cos 40° \cos 12.9082593° - (-0.1694739)$$
$$\sin 40° \sin 12.9082593°)]$$
$$= -100° + \arctan (-0.03446529/0.1954624)$$
$$= -110° = 110° \text{ W. long., not adding } 180° \text{ to the arctan, because the denominator is positive.}$$

VERTICAL PERSPECTIVE (SPHERE)–FORWARD EQUATIONS (SEE P. 173)

Given: Radius of sphere: $R = 6,371$ km
            Height of perspective point: $H = 500$ km
            Center of projection: $\phi_1 = 39°$ N. lat.
                                   $\lambda_0 = 77°$ W. long.

Point: ϕ = 41° N. lat.
         λ = 74° W. long.

Find: x, y

First H is converted to P as described after equation (23–3):

$$P = 500/6{,}371 + 1$$
$$= 1.0784806$$

Using equation (5–3),

$$\cos c = \sin 39° \sin 41° + \cos 39° \cos 41° \cos [-74°-(-77°)]$$
$$= 0.99858702$$

Since cos c is greater than $1/P$, the point is within range and may be plotted. Using equations (23–3), (22–4), and (22–5) in order,

k' $= (1.0784806-1)/(1.0784806-0.99858702)$
$= 0.98231426$
x $= 6{,}371 \times 0.98231426 \cos 41° \sin [-74°-(-77°)]$
$= 247.19409$ km
y $= 6{,}371 \times 0.98231426 \times \{\cos 39° \sin 41° - \sin 39° \cos 41° \cos[-74°-(-77°)]\}$
$= 222.48596$ km

VERTICAL PERSPECTIVE (SPHERE)–INVERSE EQUATIONS (SEE P. 175)

Inversing forward example:

Given: R, H, ϕ_1, λ_0 for forward example
       x = 247.19409 km
       y = 222.48596 km

Find: ϕ, λ

The conversion of H to P is made as in the forward example, so that

$$P = 1.0784806$$

Using equations (20–18), (23–4), (20–14), and (20–15) in order,

ρ $= [247.19409^2 + 222.48596^2]^{1/2}$
$= 332.57318$ km
c $= \arcsin \{[1.0784806 - (1-332.57318^2 \times (1.0784806 + 1)/$
$(6371^2 \times (1.0784806-1)))^{1/2}]/[6{,}371 \times (1.0784806-1)/$
$332.57318 + 332.57318/(6{,}371 \times (1.0784806-1))]\}$
$= 3.0461860°$
ϕ $= \arcsin [\cos 3.0461860° \sin 39° + (222.48596 \sin 3.0461860°$
$\cos 39°/332.57318)]$
$= 41°$ N. lat.
λ $= -77° + \arctan [247.19409 \sin 3.0461860°/(332.57318$
$\cos 39° \cos 3.0461860° - 222.48596 \sin 39° \sin 3.0461860°)]$
$= -77° + \arctan (13.1361245/250.652184)$
$= -74° = 74°$ W. long., not adding 180° to the arctan because the denominator is positive.

TILTED PERSPECTIVE (SPHERE)—FORWARD EQUATIONS (SEE P. 175–176)

Using forward example for Vertical Perspective (sphere), but applying tilt:

Given:	Radius of sphere:	R	$= 6{,}371$ km
	Height of perspective point:	H	$= 500$ km
	Center of projection:	ϕ_1	$= 39°$ N. lat.
		λ_0	$= 77°$ W. long.
	Tilt of plane:	ω	$= 30°$
	Azimuth of upward tilt:	γ	$= 50°$ east of north
	Point:	ϕ	$= 41°$ N. lat.
		λ	$= 74°$ W. long.

Find: x_t, y_t

First, x,y is calculated exactly as in the forward Vertical Perspective (sphere) example, so that

$$x = 247.19409 \text{ km}$$
$$y = 222.48596 \text{ km}$$

Using equations (23−7), (23−5), and (23−6) in order,

$$A = \{(222.48596 \cos 50° + 247.19409 \sin 50°) \sin 30°/500\} + \cos 30°$$
$$= 1.1983983$$
$$x_t = (247.19409 \cos 50° − 222.48596 \sin 50°) \cos 30°/1.1983983$$
$$= − 8.3400123 \text{ km}$$
$$y_t = (222.48596 \cos 50° + 247.19409 \sin 50°)/1.1983983$$
$$= 277.34759 \text{ km}$$

TILTED PERSPECTIVE (SPHERE)—INVERSE EQUATIONS (SEE P. 176)

Inversing forward example:

Given: R, H, ϕ_1, λ_0, ω, γ for forward example
$$x_t = −8.3400123 \text{ km}$$
$$y_t = 277.34759 \text{ km}$$

Find: ϕ, λ

Using equations (23−11) through (23−14) in order,

$$M = 500 \times (−8.3400123)/(500−277.34759 \sin 30°)$$
$$= −11.5408351$$
$$Q = 500 \times 277.34759 \cos 30°/(500−277.34759 \sin 30°)$$
$$= 332.372874$$
$$x = −11.5408351 \cos 50° + 332.372874 \sin 50°$$
$$= 247.19409 \text{ km}$$
$$y = 332.372874 \cos 50° − (−11.5408351) \sin 50°$$
$$= 222.48596 \text{ km}$$

These values of x and y are used to calculate ϕ and λ exactly as for the Vertical Perspective (sphere) inverse equations, so

$$\phi = 41° \text{ N. lat.}$$
$$\lambda = 74° \text{ W. long.}$$

VERTICAL PERSPECTIVE (ELLIPSOID)–FORWARD EQUATIONS (SEE P. 176-177)

Given: Clarke 1866 ellipsoid: a = 6,378,206.4 m
                                                      e^2 = 0.00676866
           Height of perspective point: H = 500,000 m
           Center of projection: ϕ_1 = 39° N. lat.
                                                      λ_0 = 77° W. long.
           Height of center above ellipsoid: h_0 = 200 m
                                      Point: ϕ = 41° N. lat.
                                                      λ = 74° W. long.
                                                      h = 100 m

Find: x, y

Since H is given, P is computed from equations (8–23), (23–21) and (23–17), using ϕ_1 as the first trial ϕ_g:

$$N_1 = 6,378,206.4/(1-0.00676866 \sin^2 39°)^{1/2}$$
$$= 6,386,772.6 \text{ m}$$
$$P = (\cos 39°/\cos 39°)(500,000 + 6,386,772.6 + 200)/6,378,206.4$$
$$= 1.0797664$$
$$\phi_g = 39° - \arcsin[6,386,772.6 \times 0.00676866 \sin 39° \cos 39°/$$
$$(1.0797664 \times 6,378,206.4)]$$
$$= 38.8241050°$$

Substituting 38.8241050° in place of the second 39° only in the equation for P, the second iterations produce

$$P = 1.0770938$$
$$\phi_g = 38.8236686°$$

The next iterations produce

$$P = 1.0770872$$
$$\phi_g = 38.8236675°$$

There is no change in the next iteration; therefore, these values are final. Using equations (4–20), (23–15), (23–16), (23–19), (23–19a), and (23–20) in order,

$$N = 6,378,206.4/(1-0.00676866 \sin^2 41°)^{1/2}$$
$$= 6,387,517.6 \text{ m}$$
$$C = [(6,387,517.6+100)/6,378,206.4] \cos 41°$$
$$= 0.7558232$$
$$S = [[6,387,517.6 \times (1-0.00676866) + 100]/6,378,206.4] \sin 41°$$
$$= 0.6525799$$
$$K = 500,000/[1.0770872 \cos (39°-38.8236675°)$$
$$-0.6525799 \sin 39° - 0.7558232 \cos 39° \cos (-74°-(-77°))]$$
$$= 6,264,070.9 \text{ m}$$
$$x = 6,264,070.9 \times 0.7558232 \sin [-74°-(-77°)]$$
$$= 247,786.2 \text{ m}$$
$$y = 6,264,070.9 \times [1.0770872 \sin (39°-38.8236675°)$$
$$+ 0.6525799 \cos 39° - 0.7558232 \sin 39° \cos (-74°-(-77°))]$$
$$= 222,134.1 \text{ m}$$

VERTICAL PERSPECTIVE (ELLIPSOID)–INVERSE EQUATIONS (SEE P. 177–178)

Inversing forward example:

Given: a, e^2, H, ϕ_1, λ_0, h_0 for forward example

$$x = 247{,}786.2 \text{ m}$$
$$y = 222{,}134.1 \text{ m}$$

Find: ϕ, λ

Equations (23–21) and (23–17) are used to compute P and ϕ_g, just as in the forward equations, so that

$$P = 1.0770872$$
$$\phi_g = 38.8236675°$$

Then, using equations (23–22) through (23–28) in order,

$$
\begin{aligned}
B &= 1.0770872 \cos (39° - 38.8236675°) \\
&= 1.0770821 \\
D &= 1.0770872 \sin (39° - 38.8236675°) \\
&= 0.00331482 \\
L &= 1 - 0.00676866 \cos^2 39° \\
&= 0.9959120 \\
G &= 1 - 0.00676866 \sin^2 39° \\
&= 0.9973193 \\
J &= 2 \times 0.00676866 \sin 39° \cos 39° \\
&= 0.00662075 \\
u &= -2 \times 1.0770821 \times 0.9959120 \times 500{,}000 - 2 \times 0.00331482 \\
&\quad \times 0.9973193 \times 222{,}134.1 + 1.0770821 \times 0.00662075 \\
&\quad \times 222{,}134.1 + 0.00331482 \times 500{,}000 \times 0.00662075 \\
&= -1{,}072{,}553.2 \text{ m} \\
v &= 0.9959120 \times 500{,}000^2 + 0.9973193 \times 222{,}134.1^2 \\
&\quad - 500{,}000 \times 0.00662075 \times 222{,}134.1 + (1-0.00676866) \\
&\quad \times 247{,}786.2^2 \\
&= 3.584366 \times 10^{11} \text{ m}^2
\end{aligned}
$$

For the initial trial, since h may not be zero, $E = 1$. Using equations (23–29) through (23–34) in order,

$$
\begin{aligned}
t &= 1.0770872^2 \times (1-0.00676866 \cos^2 38.8236675°) \\
&\quad - 1.0 \times (1-0.00676866) \\
&= 0.1621193 \\
K' &= \{-(-1{,}072{,}553.2) + [(-1{,}072{,}553.2)^2 - 4 \times 0.1621193 \\
&\quad \times 3.584366 \times 10^{11}]^{1/2}\}/(2\times 0.1621193) \\
&= 6{,}262{,}797.2 \text{ m} \\
X &= 6{,}378{,}206.4 \times [(1.0770821-500{,}000/6{,}262{,}797.2) \cos 39° \\
&\quad -(222{,}134.1/6{,}262{,}797.2-0.00331482) \sin 39°] \\
&= 4{,}814{,}079.9 \text{ m} \\
Y &= 6{,}378{,}206.4 \times 247{,}786.2/6{,}262{,}797.2 \\
&= 252{,}352.3 \text{ m} \\
S &= (222{,}134.1/6{,}262{,}795.7-0.00331482) \cos 39° \\
&\quad + (1.0770821-500{,}000/6{,}262{,}795.7) \sin 39° \\
&= 0.6525753
\end{aligned}
$$

$\lambda = -77° + \arctan(252,352.3/4,814,079.9)$
$= -73.9993222°$

For a first trial ϕ, use arcsin S, or $40.7360514°$. For this trial ϕ and λ, select h. It will be taken as 100 m, for the sake of this example, in order to repeat the forward example. Using equations (23−35) and (23−36),

$\phi = \arcsin \{0.6525753/[(1-0.00676866)/(1-0.00676866$
$\sin^2 40.7360514°)^{1/2} + 100/6,378,206.4]\}$
$= 41.0004168°$
$E = [1/(1-0.00676866 \sin^2 41.0004168°)^{1/2} + 100/6,378,206.4]^2$
$- 0.00676866 \sin^2 41.0004168° \times [1/(1-0.00676866 \sin^2 41.0004168°)$
$- 100^2/(6,378,206.4^2-6,378,206.4^2 \times 0.00676866)]$
$= 1.0000314$

Using this value of E in equation (23−29), and the above value of ϕ (41.0004168°) in the right side of equation (23−35), each equation (23−29) through (23−35) is recomputed, with the following results:

$$t = 0.1620882$$
$$K' = 6,264,074.3 \text{ m}$$
$$X = 4,814,189.6 \text{ m}$$
$$Y = 252,300.9 \text{ m}$$
$$S = 0.6525799$$
$$\lambda = -74.0000011°$$
$$\phi = 40.9999978°$$
$$E = 1.0000314$$

The next iteration produces

$$\lambda = -74.0000011°$$
$$\phi = 40.9999991°$$

The next produces no change in λ or ϕ to seven decimal places. Thus,

$$\lambda = 74° \text{ W. long.}$$
$$\phi = 41° \text{ N. lat.}$$

TILTED PERSPECTIVE (ELLIPSOID)−"CAMERA" PARAMETERS FROM PROJECTIVE CONSTANTS (SEE P. 178)

Using forward example for Vertical Perspective (ellipsoid), but applying tilt:

Given: a, e^2, H, ϕ_1, λ_0, h_0 for forward Vertical Perspective (ellipsoid) example

Tilt of plane:	ω	$= 30°$
Azimuth of upward tilt:	γ	$= 50°$ east of north
Point:	ϕ	$= 41°$ N. lat.
	λ	$= 74°$ W. long.
	h	$= 100$ m

Find: x_t, y_t

First, x and y are calculated exactly as for the forward Vertical Perspective (ellipsoid) example, giving

$$x = 247,786.2 \text{ m}$$
$$y = 222,134.1 \text{ m}$$

Using equations (23−7), (23−5), and (23−6) in order,

$A = \{(222,134.1 \cos 50° + 247,786.2 \sin 50°) \sin 30°/500,000\} + \cos 30°$
  $= 1.1986257$
$x_t = (247,786.2 \cos 50°−222,134.1 \sin 50°) \cos 30°/1.1986257$
  $= -7,868.693 \text{ m}$
$y_t = (222,134.1 \cos 50°+247,786.2 \sin 50°)/1.1986257$
  $= 277,484.7 \text{ m}$

TILTED PERSPECTIVE (ELLIPSOID WITH "CAMERA" PARAMETERS)−INVERSE EQUATIONS (SEE P. 178)

Inversing forward example:

Given: a, e^2, H, ϕ_1, λ_0, h_0, ω, γ for forward example

$$x_t = -7,868.693 \text{ m}$$
$$y_t = 277,484.7 \text{ m}$$

Using equations (23−11) through (23−14) in order,

$M = 500,000 \times (-7,868.693)/(500,000−277,484.7 \sin 30°)$
  $= -10,890.694 \text{ m}$
$Q = 500,000 \times 277,484.7 \cos 30°/(500,000−277,484.7 \sin 30°)$
  $= 332,600.29 \text{ m}$
$x = -10,890.694 \cos 50° + 332,600.29 \sin 50°$
  $= 247,786.2 \text{ m}$
$y = 332,600.29 \cos 50° − (-10,890.694) \sin 50°$
  $= 222,134.1 \text{ m}$

Then ϕ and λ are calculated from x and y exactly as for the inverse Vertical Perspective (ellipsoid) example, giving

$$\lambda = 74° \text{ W. long.}$$
$$\phi = 41° \text{ N. lat.}$$

TILTED PERSPECTIVE (ELLIPSOID WITH PROJECTIVE EQUATIONS)−FORWARD (SEE P. 178−180)

An example is not given to solve equations (23−43) and (23−44), solving 11 simultaneous equations, since it is tedious but also fairly standard in approach. The examples below determine constants $K_1−K_{11}$ for the example used above, and then apply them to find rectangular coordinates.

Given: parameters for forward Tilted Perspective (ellipsoid) example, repeated here:

Clarke 1866 ellipsoid: a = 6,378,206.4 m

e^2 = 0.00676866

Height of perspective point: H = 500,000 m

Center of projection: ϕ_1 = 39° N. lat.

λ_0 = 77° W. long.

Height of center above ellipsoid: h_0 = 200 m

Tilt of plane: ω = 30°

Azimuth of upward tilt: γ = 50° east of north

To produce the same rectangular coordinates, the (X_t, Y_t) axes are assumed to coincide with the (X_t', Y_t') axes; thus,

$$\theta = 0°$$
$$x_0 = 0$$
$$y_0 = 0$$

First P and ϕ_g are calculated by iteration from H, etc., exactly as they are in the forward Vertical Perspective (ellipsoid) example above, resulting in

$$P = 1.0770872$$
$$\phi_g = 38.8236675°$$

Using equations (23−45) through (23−62) in order,

U = 1.0770872 × [sin (39°−38.8236675°) cos 50° sin 30°

+ cos (39°−38.8236675°) cos 30°]

= 0.9338458

F = [sin 39° sin (−77°) cos 50° − cos (−77°) sin 50°]/0.9338458

= −0.6066034

V = [sin 39° sin (−77°) sin 50° + cos (−77°) cos 50°] cos 30°/0.9338458

= −0.3015228

M = [sin 39° cos (−77°) sin 50°−sin (−77°) cos 50°] cos 30°/0.9338458

= 0.6813973

N = [sin 39° cos (−77°) cos 50° + sin (−77°) sin 50°]/0.9338458

= −0.7018436

W = [−sin 50° cos 30° cos 0° − cos 50° sin 0°]/0.9338458

= −0.7104106

T = [−sin 50° cos 30° sin 0° + cos 50° cos 0°]/0.9338458

= 0.6883231

K_5 = −(−0.7018436) sin 30° − cos 39° cos (−77°) cos 30°/0.9338458

= 0.1887983

K_6 = −(−0.6066034) sin 30° − cos 39° sin (−77°) cos 30°/0.9338458

= 1.0055359

K_7 = (cos 39° cos 50° sin 30° − sin 39° cos 30°)/0.9338458

= −0.3161523

K_1 = 500,000 × [0.6813973 cos 0° + (−0.7018436) sin 0°]

+ 0.1887983 × 0

= 340,698.6 m

K_2 = 500,000 × [−0.3015228 cos 0° + (−0.6066034) sin 0°]

+ 1.0055359 × 0

= −150,761.5 m

K_3 = 500,000 × (−0.7104106) cos 39° + (−0.3161523) × 0

= −276,046.4 m

K_4 = 500,000 × (−0.7104106) × 1.0770872 sin (39° − 38.8236675°) + 0
    = −1,177.4 m

K_8 = 500,000 × [0.6813973 sin 0° − (−0.7018436) cos 0°] + 0.1887983 × 0
    = 350,921.8 m

K_9 = 500,000 × [−0.3015228 sin 0° − (−0.6066034) cos 0°] + 1.0055359 × 0
    = 303,301.7 m

K_{10} = 500,000 × 0.6883231 cos 39° + (−0.3161523) × 0
    = 267,463.7 m

K_{11} = 500,000 × 0.6883231 × 1.0770872 sin (39°−38.8236675°) + 0
    = 1,140.8 m

To test these constants K_1-K_{11}, equations (23−15), (23−16), and (23−38) through (23−42) may be used, remembering that $x'_t = x_t$ and $y'_t = y_t$ in this example.

Using the same point previously used,

$$\phi = 41° \text{ N. lat.}$$
$$\lambda = 74° \text{ W. long.}$$
$$h = 100 \text{ m}$$

Find: x_t, y_t

Calculating C and S exactly as in the forward Vertical Perspective (ellipsoidal) example,

$$C = 0.7558232$$
$$S = 0.6525799$$

Using (23−38) through (23−40),

$$X = 0.7558232 \cos (−74°)$$
$$= 0.2083331$$
$$Y = 0.7558232 \sin (−74°)$$
$$= −0.7265439$$
$$Z = 0.6525799$$

Using equations (23−41) and (23−42), first calculating the denominator,

den. = 0.1887983 × 0.2083331 + 1.0055359 × (−0.7265439)
    + (−0.3161523) × 0.6525799 + 1
    = 0.1024523

x_t = [340,698.6 × 0.2083331 + (−150,761.5) × (−0.7265439)
    + (−276,046.4) × 0.6525799 + (−1,177.4)]/0.1024523
    = −7,868.7 m

y_t = [350,921.8 × 0.2083331 + (303,301.7) × (−0.7265439)
    + 267,463.7 × 0.6525799 + 1,140.8]/0.1024523
    = 277,484.8 m

These values agree with the results in the forward Tilted Perspective (ellipsoid) example.

TILTED PERSPECTIVE (ELLIPSOID WITH
PROJECTIVE EQUATIONS)−INVERSE (SEE P. 180)

Inversing forward example:
Given: K_1-K_{11} as determined just above

$$x_t = -7,868.7 \text{ m}$$
$$y_t = 277,484.8 \text{ m}$$

Find: ϕ, λ

Using equations (23–63) through (23–77) in order, since $x_t = x'_t$ and $y_t = y'_t$ by choice in the example for calculating K_n,

A_1 $= -7,868.7 \times 0.1887983 - 340,698.6$
       $= -342,184.2 \text{ m}$

A_2 $= -7,868.7 \times 1.0055359 - (-150,761.5)$
       $= 142,849.2 \text{ m}$

A_3 $= -7,868.7 \times (-0.3161523) - (-276,046.4)$
       $= 278,534.1 \text{ m}$

A_4 $= -1,177.4 - (-7,868.7)$
       $= 6,691.3 \text{ m}$

A_5 $= 277,484.8 \times 0.1887983 - 350,921.8$
       $= -298,533.1 \text{ m}$

A_6 $= 277,484.8 \times 1.0055359 - 303,301.7$
       $= -24,280.8 \text{ m}$

A_7 $= 277,484.8 \times (-0.3161523) - 267,463.7$
       $= -355,191.2 \text{ m}$

A_8 $= 1,140.8 - 277,484.8$
       $= -276,344.0 \text{ m}$

A_9 $= -342,184.2 \times (-276,344.0) - 6,691.3 \times (-298,533.1)$
       $= 9.655812 \times 10^{10} \text{ m}^2$

A_{10} $= -342,184.2 \times (-355,191.2) - 278,534.1 \times (-298,533.1)$
       $= 2.046925 \times 10^{11} \text{ m}^2$

A_{11} $= 142,849.2 \times (-298,533.1) - (-342,184.2) \times (-24,280.8)$
       $= -5.095372 \times 10^{10} \text{ m}^2$

A_{12} $= 142,849.2 \times (-355,191.2) - 278,534.1 \times (-24,280.8)$
       $= -4.397575 \times 10^{10} \text{ m}^2$

A_{13} $= 142,849.2 \times (-276,344.0) - 6,691.3 \times (-24,280.8)$
       $= -3.931305 \times 10^{10} \text{ m}^2$

A_{14} $= (2.046925 \times 10^{11})^2 + (-5.095372 \times 10^{10})^2/(1 - 0.00676866)$
         $+ (-4.397575 \times 10^{10})^2$
       $= 4.644686 \times 10^{22} \text{ m}^4$

A_{15} $= 9.655812 \times 10^{10} \times 2.046925 \times 10^{11} + (-4.397575) \times 10^{10}$
         $\times (-3.931305) \times 10^{10}$
       $= 2.149354 \times 10^{22} \text{ m}^4$

Assuming $E = 1$ for the first trial, using equations (23–78), (23–79), (23–80), and (23–35), with a trial "+" sign for the "±" in equation (23–79),

A_{16} $= (9.655812 \times 10^{10})^2 - 1.0 \times (-5.095372 \times 10^{10})^2 + (-3.931305 \times 10^{10})^2$
       $= 8.272705 \times 10^{21} \text{ m}^4$

S $= [2.149354 \times 10^{22}/(4.644686 \times 10^{22})]$
       $+ \{[2.149354 \times 10^{22}/(4.644686 \times 10^{22})]^2$
       $- 8.272705 \times 10^{21}/(4.644686 \times 10^{22})\}^{1/2}$
       $= 0.6525751$

λ $= \arctan \{(9.655812 \times 10^{10} - 2.046925 \times 10^{11} \times 0.6525751)/$
       $[-4.397575 \times 10^{10} \times 0.6525751 - (-3.931305 \times 10^{10})]\}$
       $= \arctan [-3.7019109 \times 10^{10}/(1.0615571 \times 10^{10})]$
       $= -73.9992678°$

The first trial ϕ is arcsin $S = 40.7360359°$. It is assumed that $h = 100$ for this example based on ϕ and λ.

$$\begin{aligned}
\phi &= \arcsin \left[0.6525751/[(1-0.00676866)/(1-0.00676866 \right. \\
&\quad \left. \times \sin^2 40.7360359°)^{1/2} + 100/6378206.4\right] \\
&= 41.0004013°
\end{aligned}$$

Since ϕ and λ place the approximate point at a reasonable location, the trial "+" sign is satisfactory.

A second trial E is now calculated from equation (23−36):

$$\begin{aligned}
E &= [1/(1-0.00676866 \times \sin^2 41.0004013°)^{1/2} + 100/6,378,206.4]^2 \\
&\quad -0.00676866 \times \sin^2 41.0004013° \, [1/(1-0.00676866 \times \sin^2 41.0004013°) \\
&\quad -100^2/(6,378,206.4^2-6,378,206.4^2 \times 0.00676866)] \\
&= 1.0000314
\end{aligned}$$

This is substituted in place of 1.0 for E in equation (23−78) and A_{16}, S, λ, and ϕ are recalculated until ϕ changes by a negligible amount. Finally, disregarding round-off errors in the above example,

$$\phi = 41° \text{ N. lat.}$$
$$\lambda = 74° \text{ W. long.}$$

TILTED PERSPECTIVE (ELLIPSOID)−"CAMERA" PARAMETERS FROM PROJECTIVE CONSTANTS (SEE P. 180−181)

Using constants calculated in forward example for Tilted Perspective (ellipsoid with projective equations):

Given: Clarke 1866 ellipsoid: $a = 6,378,206.4$ m
$e^2 = 0.00676866$

Height of center above ellipsoid: $h_0 = 200$ m
Constants K_1-K_{11} previously calculated

Find: H, ϕ_1, λ_0, ω, γ, θ, x_0, y_0

The three simultaneous equations (23−81) are set up as follows:

$$340,698.6 \, X_0 + (-150,761.5) \, Y_0 + (-276,046.4) \, Z_0 = -(-1,177.4)$$
$$350,921.8 \, X_0 + 303,301.7 \, Y_0 + 267,463.7 \, Z_0 = -1,140.8$$
$$0.1887983 \, X_0 + 1.0055359 \, Y_0 + (-0.3161523) \, Z_0 = -1$$

Solving these three equations for the three unknowns,

$$X_0 = 0.1887645$$
$$Y_0 = -0.8176291$$
$$Z_0 = 0.6752538$$

Using equations (23−82) through (23−86),

$$\begin{aligned}
x_p &= [340,698.6 \times 0.1887983 + (-150,761.5) \times 1.0055359 \\
&\quad + (-276,046.4) \times (-0.3161523)]/[0.1887983^2 \\
&\quad + 1.0055359^2 + (-0.3161523)^2] \\
&= -0.06961613 \text{ m} \\
y_p &= [0.1887983 \times 350,921.8 + 1.0055359 \times 303,301.7 \\
&\quad + (-0.3161523) \times 267,463.7]/[0.1887983^2 \\
&\quad + 1.0055359^2 + (-0.3161523)^2] \\
&= 250,000.04 \text{ m}
\end{aligned}$$

$\lambda_0 = \arctan\ (-0.8176291/0.1887645)$
   $= -77° = 77°$ W. long., not adding 180° since the denominator is positive.
$P = [0.1887645^2 + (-0.8176291)^2 + 0.6752538^2]^{1/2}$
   $= 1.0770873$
$\phi_g = \arcsin\ (0.6752538/1.0770873)$
   $= 38.8236777°$

Using equation (23−87), with ϕ_g as the first approximation for ϕ_1,

$\phi_1 = 38.8236777° + \arcsin\ [0.00676866 \sin 38.8236777°$
   $\cos 38.8236777°/[1.0770873 \times (1-0.00676866 \sin^2 38.8236777°)^{1/2}]]$
   $= 38.9997744°$

Substituting this value for ϕ_1 in the same equation, and leaving the first use of ϕ_g intact, since it is part of the equation, the second iteration gives,

$$\phi_1 = 39.0000099°$$

The next iteration gives

$$\phi_1 = 39.0000102°$$

and the next gives no change to seven decimals. Therefore, disregarding round-off errors,

$$\phi_1 = 39° \text{ N. lat.}$$

Using equation (23−88),

$H = 6,378,206.4 \times [1.0770873 \cos 38.8236777°/\cos 39°$
   $-1/(1-0.00676866 \sin^2 39°)^{1/2} - 200/6,378,206.4]$
   $= 500,000.0$ m

Using equations (23−15), (23−16), (23−38) through (23−40), (23−41), and (23−42), coordinates x_0 and y_0 are found for ϕ_1 and λ_0:

$C = [1/(1-0.00676866 \sin^2 39°)^{1/2} + 200/6,378,206.4] \cos 39°$
   $= 0.7782141$
$S = [(1-0.00676866)/(1-0.00676866 \sin^2 39°)^{1/2}$
   $+ 200/6,378,206.4] \sin 39°$
   $= 0.6259200$
$X = 0.7782141 \cos\ (-77^0)$
   $= 0.1750601$
$Y = 0.7782141 \sin\ (-77^0)$
   $= -0.7582685$
$Z = 0.6259200$
$x_0 = [340,698.6 \times 0.1750601 + (-150,761.5) \times (-0.7582685)$
   $+ (-276,046.4) \times (0.6259200) + (-1,177.4)]/[0.1887983$
   $\times 0.1750601 + 1.0055359 \times (-0.7582685) + (-0.3161523)$
   $\times (0.6259200) + 1]$
   $= 0.04$ m, actually zero if round-off had not occurred.
$y_0 = -0.03$ m similarly from (23−42) as y_t', actually zero

Using equations (23−89) and (23−90),

$$\omega = \arcsin \{[(0.04-(-0.06961613))^2 + (250,000.04-(-0.03))^2]^{1/2}/$$
$$500,000\}$$
$$= 29.9999447°, \text{ actually } 30° \text{ without round-off.}$$
$$\theta = \arctan [(0.04-(-0.06961613))/(250,000.04-(-0.03))]$$
$$= 0.0000256°, \text{ actually } 0° \text{ without round-off.}$$

Calculating (x_t', y_t') for $(\phi_1 + 0.02°, \lambda_0)$ just as coordinates (x_0, y_0) were calculated above,

$$x_t' = -1,698.034 \text{ m}$$
$$y_t' = 1,645.247 \text{ m}$$

Using equations (23−91) through (23−93),

$$x_t = [-1,698.034-0.04] \cos 0° + [1,645.247-(-0.03)] \sin 0°$$
$$= -1,698.07 \text{ m}$$
$$y_t = [1,645.247-(-0.03)] \cos 0° - [-1,698.034-0.04] \sin 0°$$
$$= 1,645.28 \text{ m}$$
$$\gamma = - \arctan [-1,698.07/(1,645.28 \cos 30°)]$$
$$= 49.99997°, \text{ actually } 50° \text{ without round-off.}$$

LAMBERT AZIMUTHAL EQUAL-AREA (SPHERE)−FORWARD EQUATIONS
(SEE P. 186)

Given: Radius of sphere: $R = 3.0$ units
            Center: $\phi_1 = 40°$ N. lat.
                     $\lambda_0 = 100°$ W. long.
            Point: $\phi = 20°$ S. lat.
                    $\lambda = 100°$ E. long.

Find: x, y

Using equation (24−2),

$$k' = \{2/[1 + \sin 40° \sin (-20°) + \cos 40° \cos (-20°) \cos (100° + 100°)]\}^{1/2}$$
$$= 4.3912175$$

Using equations (22−4) and (22−5),

$$x = 3.0 \times 4.3912175 \cos (-20°) \sin (100° + 100°)$$
$$= -4.2339303 \text{ units}$$
$$y = 3.0 \times 4.3912175 [\cos 40° \sin (-20°) - \sin 40° \cos (-20°) \cos (100° + 100°)]$$
$$= 4.0257775 \text{ units}$$

Examples for the polar and equatorial reductions, equations (24−3) through (24−14), are omitted, since the above general equations give the same results.

LAMBERT AZIMUTHAL EQUAL-AREA (SPHERE)−INVERSE EQUATIONS
(SEE P. 186−187)

Inversing forward example:

Given: Radius of sphere: $R = 3.0$ units
            Center: $\phi_1 = 40°$ N. lat.
                     $\lambda_0 = 100°$ W. long.

Point: $x = -4.2339303$ units
$y = 4.0257775$ units

Find: ϕ, λ

Using equations (20–18) and (24–16),

$$\rho = [(-4.2339303)^2 + 4.0257775^2]^{1/2}$$
$$= 5.8423497 \text{ units}$$
$$c = 2 \arcsin [5.8423497/(2 \times 3.0)]$$
$$= 153.6733917°$$

From equation (20–14),

$\phi = \arcsin [\cos 153.6733917° \sin 40° + 4.0257775$
$\sin 153.6733917° \cos 40°/5.8423497]$
$= -19.9999993° = 20°$ S. lat., disregarding rounding-off effects.

From equation (20–15),

$\lambda = -100° + \arctan [-4.2339303 \sin 153.6733917°/$
$(5.8423497 \cos 40° \cos 153.6733917°$
$-4.0257775 \sin 40° \sin 153.6733917°)]$
$= -100° + \arctan [-1.8776951/(-5.1589246)]$
$= -100° + 20.0000005°$
$= -79.9999995°$

Since the denominator of the argument of arctan is negative, add 180°:

$\lambda = 100.0000005° = 100°$ E. long., disregarding rounding-off effects.

In polar spherical cases, the calculation of λ from equations (20–16) or (20–17) is simpler than the above, but the quadrant adjustment follows the same rules.

LAMBERT AZIMUTHAL EQUAL-AREA (ELLIPSOID)–FORWARD EQUATIONS
(SEE P. 187-188)

Oblique aspect:

Given: Clarke 1866 ellipsoid: $a = 6{,}378{,}206.4$ m
                                      $e^2 = 0.00676866$
                  or $e = 0.0822719$
        Center: $\phi_1 = 40°$ N. lat.
                      $\lambda_0 = 100°$ W. long.
          Point: $\phi = 30°$ N. lat.
                     $\lambda = 110°$ W. long.

Find: x, y

Using equation (3–12),

$q = (1-0.00676866) \{\sin 30°/(1-0.00676866 \sin^2 30°) - [1/$
$(2 \times 0.0822719)] \ln [(1-0.0822719 \sin 30°)/$
$(1+0.0822719 \sin 30°)]\}$
$= 0.9943535$

Inserting $\phi_1 = 40°$ in place of 30° in the same equation,

$$q_1 = 1.2792602$$

Inserting 90° in place of 30°,

$$q_p = 1.9954814$$

Using equation (3−11),

$$\beta = \arcsin (0.9943535/1.9954814)$$
$$= 29.8877622°$$
$$\beta_1 = \arcsin (1.2792602/1.9954814)$$
$$= 39.8722878°$$

Using equation (3−13),

$$R_q = 6{,}378{,}206.4 \times (1.9954814/2)^{1/2}$$
$$= 6{,}370{,}997.2 \text{ m}$$

Using equation (14−15),

$$m_1 = \cos 40°/(1-0.00676866 \sin^2 40°)^{1/2}$$
$$= 0.7671179$$

Using equations (24−19) and (24−20),

$$B = 6{,}370{,}997.2 \times \{2/[1 + \sin 39.8722878° \sin 29.8877622°$$
$$+ \cos 39.8722878° \cos 29.8877622° \cos (-110° + 100°)]\}^{1/2}$$
$$= 6{,}411{,}606.1 \text{ m}$$
$$D = 6{,}378{,}206.4 \times 0.7671179/(6{,}370{,}997.2 \cos 39.8722878°)$$
$$= 1.0006653$$

Using equations (24−17) and (24−18),

$$x = 6{,}411{,}606.1 \times 1.0006653 \cos 29.8877622° \sin (-110° + 100°)$$
$$= -965{,}932.1 \text{ m}$$
$$y = (6{,}411{,}606.1/1.0006653)[\cos 39.8722878° \sin 29.8877622°$$
$$- \sin 39.8722878° \cos 29.8877622° \cos (-110° + 100°)]$$
$$= -1{,}056{,}814.9 \text{ m}$$

Polar aspect:
Given: International ellipsoid: $a = 6{,}378{,}388.0$ m
$$e^2 = 0.00672267$$
or $e = 0.0819919$
Center: North Pole $\phi_1 = 90°$ N. lat.
$$\lambda_0 = 100° \text{ W. long. (meridian along}$$
$$\text{neg. } Y \text{ axis)}$$
Point: $\phi = 80°$ N. lat.
$$\lambda = 5° \text{ E. long.}$$

Find: ϕ, λ, h, k

From equation (3−12),

$$q = (1-0.00672267) \{\sin 80°/(1-0.00672267 \sin^2 80°)$$
$$-[1/(2 \times 0.0819919)] \ln [(1-0.0819919 \sin 80°)/$$
$$(1 + 0.0819919 \sin 80°)]\}$$
$$= 1.9649283$$

Using the same equation with 90° in place of 80°,

$$q_p = 1.9955122$$

From equation (14–15),

$$m = \cos 80°/(1 - 0.00672267 \sin^2 80°)^{1/2}$$
$$= 0.1742171$$

Using equations (24–23), (21–30), (21–31), and (21–32),

$$\rho = 6{,}378{,}388.0 \times (1.9955122 - 1.9649283)^{1/2}$$
$$= 1{,}115{,}468.3 \text{ m}$$
$$x = 1{,}115{,}468.3 \sin (5° + 100°)$$
$$= 1{,}077{,}459.7 \text{ m}$$
$$y = -1{,}115{,}468.3 \cos (5° + 100°)$$
$$= 288{,}704.5 \text{ m}$$
$$k = 1{,}115{,}468.3/(6{,}378{,}388.0 \times 0.1742171)$$
$$= 1.0038193$$
$$h = 1/1.0038193 = 0.9961952$$

LAMBERT AZIMUTHAL EQUAL-AREA (ELLIPSOID)–INVERSE EQUATIONS
(SEE P. 188–190)

Oblique aspect (inversing forward example):

Given: Clarke 1866 ellipsoid: $a = 6{,}378{,}206.4$ m
$$e^2 = 0.00676866$$
or $e = 0.0822719$
Center: $\phi_1 = 40°$ N. lat.
$$\lambda_0 = 100° \text{ W. long.}$$
Point: $x = -965{,}932.1$ m
$$y = -1{,}056{,}814.9 \text{ m}$$

Find: ϕ, λ

Since these are the same map parameters as those used in the forward example, calculations of map constants not affected by ϕ and λ are not repeated here.

$$q_p = 1.9954814$$
$$\beta_1 = 39.8722878°$$
$$R_q = 6{,}370{,}997.2 \text{ m}$$
$$D = 1.0006653$$

Using equations (24–28), (24–29), and (24–27),

$$\rho = \{[-965{,}932.1/1.0006653]^2 + [1.0006653 \times (-1{,}056{,}814.9)]^2\}^{1/2}$$
$$= 1{,}431{,}827.1 \text{ m}$$
$$c_e = 2 \arcsin [1{,}431{,}827.1/(2 \times 6{,}370{,}997.2)]$$
$$= 12.9039908°$$
$$q = 1.9954814 [\cos 12.9039908° \sin 39.8722878°$$
$$+ 1.0006653 \times (-1{,}056{,}814.9) \sin 12.9039908°$$
$$\cos 39.8722878°/1{,}431{,}827.1]$$
$$= 0.9943535$$

For the first trial ϕ in equation (3−16),

$$\phi = \arcsin\ (0.9943535/2)$$
$$= 29.8133914°$$

Substituting into equation (3−16),

$$\phi = 29.8133914° + [(1-0.00676866\ \sin^2\ 29.8133914°)^2/$$
$$(2\ \cos\ 29.8133914°)] \times \{0.9943535/(1-0.00676866)$$
$$-\ \sin\ 29.8133914°/(1-0.00676866\ \sin^2\ 29.8133914)$$
$$+\ [1/(2 \times 0.0822719)]\ \ln\ [(1-0.0822719$$
$$\sin\ 29.8133914°)/(1\ +\ 0.0822719\ \sin\ 29.8133914°)]\} \times 180°/\pi$$
$$= 29.9998293°$$

Substituting 29.9998293° in place of 29.8133914° in the same equation, the new trial ϕ is found to be

$$\phi = 30.0000002°$$

The next iteration results in no change to seven decimal places; therefore,

$$\phi = 30°\ \text{N. lat.}$$

Using equation (24−26),

$$\lambda = -100° + \arctan\ \{-965,932.1\ \sin\ 12.9039908°/[1.0006653$$
$$\times\ 1,431,827.1\ \cos\ 39.8722878°\ \cos\ 12.9039908°$$
$$-1.0006653^2\ (-1,056,814.9)\ \sin\ 39.8722878°$$
$$\sin\ 12.9039908°]\}$$
$$= -100° + \arctan\ (-215,710.0/1,223,352.4)$$
$$= -100° - 9.9999999°$$
$$= -109.9999999° = 110°\ \text{W. long.}$$

Since the denominator of the argument for arctan is positive, no quadrant adjustment is necessary.

Polar aspect (inversing forward example):

Given: International ellipsoid: $a = 6,378,388.0$ m
                                        $e^2 = 0.00672267$
                                 or $e = 0.0819919$
            Center: North Pole $\phi_1 = 90°$ N. lat.
                                 $\lambda_0 = 100°$ W. long. (meridian
                                        along neg. Y axis)
            Point: $x = 1,077,459.7$ m
                        $y = 288,704.5$ m

Find: ϕ, λ

First q_p is found to be 1.9955122 from equation (3−12), as in the corresponding forward example for the polar aspect. From equations (20−18) and (24−31),

$$\rho = (1,077,459.7^2\ +\ 288,704.5^2)^{1/2}$$
$$= 1,115,468.4\ \text{m}$$
$$q = +\ [1.9955122\ -\ (1,115,468.4/6,378,388.0)^2]$$
$$= 1.9649283$$

Iterative equation (3−16) may be used to find ϕ. The first trial ϕ is

$$\phi = \text{arcsin} \ (1.9649283/2)$$
$$= 79.2542275°$$

When this is used in equation (3−16) as in the oblique inverse example, the next trial ϕ is found to be

$$\phi = 79.9744304°$$

Using this value instead, the next trial is

$$\phi = 79.9999713°$$

and the next,

$$\phi = 80.0000036°$$

The next value is the same, so

$$\phi = 80° \text{ N. lat.}$$

From equation (20−16),

$$\lambda = -100° + \text{arctan} \ [1,077,459.7/(-288,704.5)]$$
$$= -174.9999978°$$

Since the denominator of the argument for arctan is negative, add 180°, or

$$\lambda = 5.0000022° = 5° \text{ E. long.}$$

AZIMUTHAL EQUIDISTANT (SPHERE)−FORWARD EQUATIONS
(SEE P. 195-196)

Given: Radius of sphere: $R = 3.0$ units
        Center: $\phi_1 = 40°$ N. lat.
                 $\lambda_0 = 100°$ W. long.
         Point: $\phi = 20°$ S. lat.
                 $\lambda = 100°$ E. long.

Find: x, y

Using equations (5−3) and (25−2),

$$\cos c = \sin 40° \sin (-20°) + \cos 40° \cos (-20°) \cos (100° + 100°)$$
$$= -0.8962806$$
$$c = 153.6733925°$$
$$k' = (153.6733925° \times \pi/180°)/\sin 153.6733925°$$
$$= 6.0477621$$

Using equations (22−4) and (22−5),

x $= 3.0 \times 6.0477621 \cos (-20°) \sin (100° + 100°)$
    $= -5.8311398$ units
y $= 3.0 \times 6.0477621 \ [\cos 40° \sin (-20°) - \sin 40° \cos (-20°)$
        $\cos (100° + 100°)]$
    $= 5.5444634$ units

Since the above equations are general, examples of other forward formulas are not given.

AZIMUTHAL EQUIDISTANT (SPHERE)–INVERSE EQUATIONS (SEE P. 196-197)

Inversing forward example:

Given: Radius of sphere: $R = 3.0$ units
                  Center: $\phi_1 = 40°$ N. lat.
                           $\lambda_0 = 100°$ W. long.
                   Point: $x = -5.8311398$ units
                           $y = 5.5444634$ units

Find: ϕ, λ

Using equations (20−18) and (25−15),

$$\rho = [(-5.8311398)^2 + 5.5444634^2]^{1/2}$$
$$= 8.0463200 \text{ units}$$
$$c = 8.0463200/3.0$$
$$= 2.6821067 \text{ radians}$$
$$= 2.6821067 \times 180°/\pi = 153.6733925°$$

Using equation (20−14),

ϕ = arcsin (cos 153.6733925° sin 40° + 5.5444634 sin
     153.6733925° cos 40°/8.0463200)
     = −19.9999999°
     = 20° S. lat., disregarding effects of rounding off.

Using equation (20−15),

λ = −100° + arctan [(−5.8311398) sin 153.6733925°/(8.0463200
      cos 40° cos 153.6733925° − 5.5444634 sin 40°
      sin 153.6733925°)]
    = −100° + arctan [(−2.5860374)/(−7.1050794)]
    = −100° − arctan 0.3639702
    = −80.0000001°

but since the denominator of the argument of arctan is negative, add or subtract 180°, whichever places the final result between +180° and −180°:

$$\lambda = -80.0000001° + 180°$$
$$= 99.9999999°$$
$$= 100° \text{ E. long., disregarding effects of rounding off.}$$

AZIMUTHAL EQUIDISTANT (ELLIPSOID)–FORWARD EQUATIONS
(SEE P. 197-201)

Polar aspect:

Given: International ellipsoid: $a = 6,378,388.0$ m
                                        $e^2 = 0.00672267$
              Center: North Pole $\phi_1 = 90°$ N. lat.
                                  $\lambda_0 = 100°$ W. long. (meridian
                                           along neg. Y axis)
                     Point: $\phi = 80°$ N. lat.
                             $\lambda = 5°$ E. long.

Find: x, y, k

Using equation (3–21),

M = 6,378,388.0 × [(1−0.00672267/4 − 3 × 0.00672267²/64 − 5
$\qquad$ × 0.00672267³/256) × 80° × π/180° − (3 × 0.00672267/8
$\qquad$ + 3 × 0.00672267²/32 + 45 × 0.00672267³/1024) sin (2 × 80°)
$\qquad$ + (15 × 0.00672267²/256 + 45 × 0.00672267³/1024) sin (4 × 80°)
$\qquad$ − (35 × 0.00672267³/3072) sin (6 × 80°)]
$\quad$ = 8,885,403.1 m

Using the same equation (3–21), but with 90° in place of 80°,

$$M_p = 10,002,288.3 \text{ m}$$

Using equation (14–15),

$$m = \cos 80°/(1−0.00672267 \sin^2 80°)^{1/2}$$
$$= 0.1742171$$

Using equations (25–16), (21–30), (21–31), and (21–32),

$$\rho = 10,002,288.3 − 8,885,403.1$$
$$= 1,116,885.2 \text{ m}$$
$$x = 1,116,885.2 \sin (5° + 100°)$$
$$= 1,078,828.3 \text{ m}$$
$$y = −1,116,885.2 \cos (5° + 100°)$$
$$= 289,071.2 \text{ m}$$
$$k = 1,116,885.2/(6,378,388.0 × 0.1742171)$$
$$= 1.0050946$$

Oblique aspect (Guam projection):

Given:$\quad$ Clarke 1866 ellipsoid:$\quad$ a = 6,378,206.4 m
$\qquad\qquad\qquad\qquad\qquad\qquad$ e^2 = 0.00676866
$\qquad\qquad$ Center:$\quad$ ϕ_1 = 13°28′20.87887″ N. lat.
$\qquad\qquad\qquad\qquad\quad$ λ_0 = 144°44′55.50254″ E. long.
$\qquad\quad$ False origin:$\quad$ x_0 = 50,000 m
$\qquad\qquad\qquad\qquad\quad$ y_0 = 50,000 m
$\qquad\qquad\quad$ Point:$\quad$ ϕ = 13°20′20.53846″ N. lat.
$\qquad\qquad\qquad\qquad\quad$ λ = 144°38′07.19265″ E. long.

Find: x, y

Using equation (25–18), after converting angles to degrees and decimals: (ϕ_1 = 13.472466353°, λ_0 = 144.748750706°, ϕ = 13.339038461°, λ = 144.635331292°),

x $\quad$= [6,378,206.4 × (144.635331292° − 144.748750706°)
$\qquad$ cos 13.339038461°/(1−0.00676866 sin² 13.339038461°)^{1/2}]
$\qquad$ × π/180°
$\quad$= −12,287.52 m

Since 50,000 m is added to the origin for the Guam projection,

$$x = −12,287.52 + 50,000.0$$
$$= 37,712.48 \text{ m}$$

From equation (3−21),

$$M = 6,378,206.4 \times [(1-0.00676866/4 - 3 \times 0.00676866^2/64 - 5 \\ \times 0.00676866^3/256) \times 13.339038461° \times \pi/180° - (3 \\ \times 0.00676866/8 + 3 \times 0.00676866^2/32 + 45 \times 0.00676866^3/ \\ 1024) \sin (2 \times 13.339038461°) + (15 \times 0.00676866^2/256 \\ + 45 \times 0.00676866^3/1024) \sin (4 \times 13.339038461°) \\ - (35 \times 0.00676866^3/3072) \sin (6 \times 13.339038461°)] \\ = 1,475,127.96 \text{ m}$$

Substituting $\phi_1 = 13.472466353°$ in place of $13.339038461°$ in the same equation,

$$M_1 = 1,489,888.76 \text{ m}$$

Using equation (25−19), and using the x without false origin,

$$y = 1,475,127.96 - 1,489,888.76 + (-12,287.52)^2 \tan 13.339038461° \\ \times (1-0.00676866 \sin^2 13.339038461°)^{1/2}/(2 \times 6,378,206.4) \\ = -14,758.00 \text{ m}$$

Adding 50,000 meters for the false origin,

$$y = 35,242.00 \text{ m}$$

Oblique aspect (Micronesia form):

Given: Clarke 1866 ellipsoid: $a = 6,378,206.4$ m
$e^2 = 0.00676866$
Center: Saipan Island: $\phi_1 = 15°11'05.6830''$ N. lat.
$\lambda_0 = 145°44'29.9720''$ E. long.
False origin: $x_0 = 28,657.52$ m
$y_0 = 67,199.99$ m
Point: Station Petosukara $\phi = 15°14'47.4930''$ N. lat.
$\lambda = 145°47'34.9080''$ E. long.

Find: x, y

First convert angles to degrees and decimals:

$$\phi_1 = 15.18491194°$$
$$\lambda_0 = 145.7416589°$$
$$\phi = 15.24652583°$$
$$\lambda = 145.7930300°$$

From equations (4−20a), (4−20), (25−20), and (25−21) in order,

$$N_1 = 6,378,206.4/(1-0.00676866 \times \sin^2 15.18491194°)^{1/2} \\ = 6,379,687.9 \text{ m}$$
$$N = 6,378,206.4/(1-0.00676866 \times \sin^2 15.24652583°)^{1/2} \\ = 6,379,699.7 \text{ m}$$
$$\psi = \arctan [(1-0.00676866) \tan 15.24652583° \\ + 0.00676866 \times 6379687.9 \sin 15.18491194°/ \\ (6,379,699.7 \times \cos 15.24652583°)] \\ = 15.2461374°$$
$$Az = \arctan \{\sin (145.79303° - 145.7416589°)/ \\ [\cos 15.18491194° \times \tan 15.2461374° \\ - \sin 15.18491194° \times \cos (145.79303° - 145.7416589°)]\} \\ = 38.9881345°$$

Since $\sin Az \neq 0$, from equation (25–22a),

s = arcsin [sin (145.79303° − 145.7416589°) × cos 15.2461374°/
sin 38.9881345°]
= 0.001374913 radians, since s is used only in radians.

From equations (25–23) through (25–27) in order,

G = 0.00676866$^{1/2}$ sin 15.18491194°/(1−0.00676866)$^{1/2}$
= 0.02162319
H = 0.00676866$^{1/2}$ cos 15.18491194° cos 38.9881345°/
(1−0.00676866)$^{1/2}$
= 0.06192519
c = 6,379,687.9 × 0.001374913 × {1−0.001374913^2 ×0.06192519^2
× (1−0.06192519^2)/6 + (0.001374913^3/8) × 0.02162319
× 0.06192519 × (1−2×0.06192519^2) + (0.001374913^4/120)
× [0.06192519^2 × (4−7×0.06192519^2) − 3 × 0.02162319^2
× (1−7×0.06192519^2)] − (0.001374913^5/48) × 0.02162319
× 0.06192519}
= 8,771.52 m
x = 8,771.52 × sin 38.9881345° + 28,657.52
= 34,176.20 m
y = 8,771.52 × cos 38.9881345° + 67,199.99
= 74,017.88 m

AZIMUTHAL EQUIDISTANT (ELLIPSOID)–INVERSE EQUATIONS
(SEE P. 201-202)

Polar aspect (inversing forward example):

Given: International ellipsoid: a = 6,378,388.0 m
e^2 = 0.00672267
Center: North Pole: ϕ_1 = 90° N. lat.
λ_0 = 100° W. long. (meridian along
neg. Y axis)
Point: x = 1,078,828.3 m
y = 289,071.2 m

Find: ϕ, λ

Using equation (3–21), as in the corresponding forward example,
$$M_p = 10,002,288.3 \text{ m}$$

Using equations (20–18), (25–28), and (7–19),

ρ = (1,078,828.3^2 +289,071.2^2)$^{1/2}$
= 1,116,885.2 m
M = 10,002,288.3 − 1,116,885.2
= 8,885,403.1 m
μ = 8,885,403.1/[6,378,388.0 × (1−0.00672267/4−3×0.00672267^2/64
− 5 ×0.00672267^3/256)]
= 1.3953965 radians
= 1.3953965 × 180°/π = 79.9503324°

Using equations (3–24) and (3–26),

e_1 = [1−(1−0.00672267)$^{1/2}$]/[1 +(1−0.00672267)$^{1/2}$]
= 0.0016863

$$\phi = 1.3953965 \text{ radians} + (3 \times 0.0016863/2 - 27 \times 0.0016863^3/32)$$
$$\sin (2 \times 79.9503324°) + (21 \times 0.0016863^2/16 - 55$$
$$\times 0.0016863^4/32) \sin (4 \times 79.9503324°) + (151$$
$$\times 0.0016863^3/96) \sin (6 \times 79.9503324°)$$
$$= 1.3962634 \text{ radians}$$
$$= 1.3962634 \times 180°/\pi = 79.9999999°$$
$$= 80° \text{ N. lat., rounding off.}$$

Using equation (20−16),

$$\lambda = -100° + \arctan [1,078,828.3/(-289,071.2)]$$
$$= -100° - 74.9999986° + 180°$$
$$= 5.0000014°$$
$$= 5° \text{ E. long., rounding off.}$$

The 180° is added because the denominator in the argument for arctan is negative.

Oblique aspect (Guam projection, inversing forward example):

Given: Clarke 1866 ellipsoid: $a = 6,378,206.4$ m
$e^2 = 0.00676866$
Center: $\phi_1 = 13.472466353°$ N. lat.
$\lambda_0 = 144.748750706°$ E. long.
False origin: $x_0 = 50,000$ m
$y_0 = 50,000$ m
Point: $x = 37,712.48$ m
$y = 35,242.00$ m

Find: ϕ, λ

First subtract 50,000 m from x and y to relate them to actual projection origin: $x = -12,287.52$ m, $y = -14,758.00$ m. Calculation of M_1 from equation (3−21) is exactly the same as in the forward example, or

$$M_1 = 1,489,888.76 \text{ m}$$

From equation (25−30), the first trial M is found from an assumed $\phi = \phi_1$:

$$M = 1,489,888.76 + (-14,758.00) - (-12,287.52)^2 \tan 13.472466353°$$
$$\times (1 - 0.00676866 \sin^2 13.472466353°)^{1/2}/(2 \times 6,378,206.4)$$
$$= 1,475,127.92 \text{ m}$$

Using equation (7−19) and the above trial M,

$$\mu = 1,475,127.92/[6,378,206.4 \times (1 - 0.00676866/4 - 3 \times 0.00676866^2/$$
$$64 - 5 \times 0.00676866^3/256)]$$
$$= 0.2316688 \text{ radian}$$

Using equation (3−24),

$$e_1 = [1 - (1 - 0.00676866)^{1/2}]/[1 + (1 - 0.00676866)^{1/2}]$$
$$= 0.0016979$$

Using equation (3−26) in *radians*, although it could be converted to degrees,

ϕ = 0.2316688 + (3 × 0.0016979/2 − 27 × 0.0016979^3/32)
        sin (2×0.2316688) + (21×0.0016979^2/16−55
        × 0.0016979^4/32) sin (4×0.2316688) + (151
        × 0.0016979^3/96) sin (6×0.2316688)
   = 0.2328101 radian
   = 0.2328101 × 180°/π = 13.3390381°

If this new trial value of ϕ is used in place of ϕ_1 in equation (25−30), a new value of M is found:

$$M = 1,475,127.95 \text{ m}$$

This in turn, used in (7−19), gives

$$\mu = 0.2316688 \text{ radian}$$

and from (3−26),

$$\phi = 13.3390384°$$

The third trial, through the above equations and starting with this value of ϕ, produces no change to seven decimal places. Thus, this is the final value of ϕ. Converting to degrees, minutes, and seconds,

$$\phi = 13°20'20.538'' \text{ N. lat.}$$

Using equation (25−31) for longitude,

λ = 144.748750706° + [(−12,287.52) × (1−0.00676866
        $\sin^2$ 13.3390384°)$^{1/2}$/(6,378,206.4 cos 13.3390384°)] × 180°/π
   = 144.6353313°
   = 144°38'07.193'' E. long.

Oblique aspect (Micronesia form, inversing forward example):

Given: Clarke 1866 ellipsoid: a = 6,378,206.4 m
                                  e^2 = 0.00676866
          Center: Saipan Island ϕ_1 = 15.18491194° N. lat.
                                  λ_0 = 145.7416589° E. long.
          False origin: x_0 = 28,657.52 m
                         y_0 = 67,199.99 m
                 Point: x = 34,176.20 m
                         y = 74,017.88 m

Find: ϕ, λ

From equations (25−32) through (25−41) in order,

c = [(34,176.20 − 28,657.52)^2 + (74,017.88−67,199.99)^2]$^{1/2}$
     = 8,771.51 m
Az = arctan [(34,176.20−28,657.52)/(74,017.88−67,199.99)]
     = 38.9881292°
N_1 = 6,378,206.4/(1−0.00676866 $\sin^2$ 15.18491194°)$^{1/2}$
     = 6,379,687.9 m

A = $-0.00676866 \cos^2 15.18491194° \cos^2 38.9881292°/$
$(1-0.00676866)$

= -0.003834730

B = $3 \times 0.00676866 \times (1+0.003834730) \sin 15.18491194° \cos$
$15.18491194° \times \cos 38.9881292°/(1-0.00676866)$

= 0.004032465

D = $8,771.51/6,379,687.9$

= 0.001374913

E = $0.001374913 + 0.003834730 \times (1-0.003834730) \times 0.001374913^3/6$
$- 0.004032465 \times (1-3 \times 0.003834730) \times 0.001374913^4/24$

= 0.001374913. This is in radians for use in equation (25-38).

For use as degrees in equations (25-39) and (25-40),

E = $0.001374913 \times 180°/\pi = 0.07877669°$

F = $1 + 0.003834730 \times 0.001374913^2/2 - 0.004032465$
$\times 0.001374913^3/6$

= 1.000000004

ψ = $\arcsin (\sin 15.18491194° \cos 0.07877669° + \cos 15.18491194°$
$\times \sin 0.07877669° \cos 38.9881292°)$

= $15.2461374°$

λ = $145.7416589° + \arcsin (\sin 38.9881292° \sin 0.07877669°/$
$\cos 15.2461374°)$

= $145.7416589° + 0.0513711°$

= $145.7930300°$

= $145° 47'34.908''$ E. long.

ϕ = $\arctan [(1-0.00676866 \times 1.000000004 \sin 15.18491194°/\sin$
$15.2461374°) \times \tan 15.2461374°/(1-0.00676866)]$

= $15.2465258°$

= $15°14'47.493''$ N. lat.

MODIFIED-STEREOGRAPHIC CONFORMAL (SPHERE)—FORWARD EQUATIONS
(SEE P. 207-208)

Using Modified-Stereographic Conformal projection of Alaska (spherical form) as example:

Given: Radius of sphere: $R = 1.0$ unit
Order of equation: $m = 6$
Center: $\phi_1 = 64°$ N. lat.
$\lambda_0 = 152°$ W. long.
Constants A_1-A_6: See Table 33, using constants for sphere.
B_1-B_6: See Table 33, using constants for sphere.
Point: $\phi = 60°$ N. lat.
$\lambda = 150°$ W. long.

Find: x, y, k

Using equations (26-1) through (26-3) in order,

k' = $2/\{1 + \sin 64° \sin 60° + \cos 64° \cos 60° \cos [-150°-(-152°)]\}$

= 1.0012864

x' = $1.0012864 \cos 60° \sin [-150°-(-152°)]$

= 0.01747220

y' = $1.0012864 \times \{\cos 64° \sin 60° - \sin 64° \cos 60° \cos [-150° - (-152°)]\}$

= -0.06957209

Using equations in (26−6), with $j=2$, in order,

$$r = 2 \times 0.01747220$$
$$= 0.03494439$$
$$s' = 0.01747220^2 + (-0.06957209)^2$$
$$= 0.00514555$$
$$g_0 = 0$$
$$a_1 = A_6 + iB_6$$
$$= 0.3660976 + (-0.2937382)i$$
$$b_1 = A_5 + iB_5$$
$$= 0.0636871 + (-0.1408027)i$$
$$c_1 = 6 \times (A_6 + iB_6)$$
$$= 2.1965856 + (-1.7624292)i$$
$$d_1 = 5 \times (A_5 + iB_5)$$
$$= 0.3184355 + (-0.7040135)i$$
$$a_2 = b_1 + ra_1$$
$$= 0.0636871 + (-0.1408027)i + 0.03494439 \times [0.3660976 + (-0.2937382)i]$$
$$= 0.07648016 + (-0.15106720)i$$
$$b_2 = A_4 + iB_4 - s'a_1$$
$$= -0.0153783 + (-0.1968253)i - 0.00514555 \times [0.3660976 + (-0.2937382)i]$$
$$= -0.01726207 + (-0.19531385)i$$
$$c_2 = d_1 + rc_1$$
$$= 0.3184335 + (-0.7040135)i + 0.03494439 \times [2.1965856 + (-1.7624292)i]$$
$$= 0.39519385 + (-0.76560052)i$$
$$d_2 = 4 \times (A_4 + iB_4) - s'c_1$$
$$= 4 \times [-0.0153783 + (-0.1968253)i] - 0.00514555 \times [2.1965856 + (-1.7624292)i]$$
$$= -0.07281585 + (-0.77823253)i$$

Incrementing j to 3, 4, and 5 for the four variables a_j, b_j, c_j, and d_j in the same set of equations,

$$a_3 = b_2 + ra_2 = -0.01458952 + (-0.20059281)i$$
$$b_3 = A_3 + iB_3 - s'a_2 = 0.00706707 + 0.00558982\, i$$
$$c_3 = d_2 + rc_2 = -0.05900604 + (-0.80498597)i$$
$$d_3 = 3 \times (A_3 + iB_3) - s'c_2 = 0.02034831 + 0.01837694i$$
$$a_4 = b_3 + ra_3 = 0.00655725 + (-0.00141977)i$$
$$b_4 = A_2 + iB_2 - s'a_3 = 0.00532637 + (-0.00308534)i$$
$$c_4 = d_3 + rc_3 = 0.01828638 + (-0.00975281)i$$
$$d_4 = 2 \times (A_2 + iB_2) - s'c_3 = 0.01080622 + (-0.00409290)i$$
$$a_5 = b_4 + ra_4 = 0.00555551 + (-0.00313495)i$$
$$b_5 = A_1 + iB_1 - s'a_4 = 0.99721856 + 0.00000731i$$
$$c_5 = d_4 + rc_4 = 0.01144523 + (-0.00443371)i$$
$$d_5 = 1 \times (A_1 + iB_1) - s'c_4 = 0.99715821 + 0.00005018i$$

Incrementing j to 6 for a_j and b_j only,

$$a_6 = b_5 + ra_5 = 0.99741269 + (-0.00010224)i$$
$$b_6 = g_0 - s'a_5 = -0.00002859 + 0.00001613i$$

Using equations (26−7) through (26−9) in order, and with the relationship $i^2 = -1$,

$$x + iy = 1 \times \{[0.01747220 + (-0.06957209)i][0.99741269 + (-0.00010224)i] + (-0.000002859) + 0.00001613i\}$$

$$= 0.01742699 + 0.00000711i^2 - 0.06939387i - 0.00002859 + 0.00001613i$$
$$= 0.01739129 - 0.06937775i$$
$$x = 0.01739129 \text{ unit}$$
$$y = -0.06937775 \text{ unit}$$
$$F_2 + iF_1 = [0.01747220 + (-0.06957209)i][0.01144523$$
$$+ (-0.00443371)i] + 0.99715821 + 0.00005018i$$
$$= 0.99704972 + (-0.00082355)i$$
$$k = [0.99704972^2 + (-0.00082355)^2]^{1/2} \times 1.0012864$$
$$= 0.9983327$$

MODIFIED-STEREOGRAPHIC CONFORMAL (SPHERE)–
INVERSE EQUATIONS (SEE P. 208)

Inversing forward example:

Given: R, m, ϕ_1, λ_0, $A_1 - A_6$, and $B_1 - B_6$ for forward example

$$x = 0.01739129 \text{ unit}$$
$$y = -0.06937775 \text{ unit}$$

Find: ϕ, λ

Using the Knuth algorithm equations (26−6) with (26−10), (26−13), and (26−8), but not in that order, the first trial $x' = 0.01739129/1$, and trial $y' = -0.06937775/1$. Except for the values of x' and y', equations (26−6) are used in the same manner as they were in the forward example, resulting in

$$a_6 = 0.99741192 + (-0.00010209) i$$
$$b_6 = -0.00002841 + 0.00001606 i$$
$$c_5 = 0.01144135 + (-0.00445277) i$$
$$d_5 = 0.99715864 + 0.00004934 i$$

Using equations (26−13), (26−8), and (26−10) in order,

$$f(x' + iy') = [0.01739129 + (-0.06937775) i] [0.99741192$$
$$+ (-0.00010209) i] + (-0.00002841) + 0.00001606 i$$
$$- [0.01739129 + (-0.06937775) i]/1$$
$$= -0.00008051 + 0.00019384 i$$
$$F_2 + iF_1 = [0.01739129 + (-0.06937775) i] [0.01144135$$
$$+ (-0.00445277) i] + 0.99715864 + 0.00004934 i$$
$$= 0.99704869 + (-0.00082188) i$$
$$\Delta (x' + iy') = - [-0.00008051 + 0.00019384 i]/[0.99704869 + (-0.00082188) i]$$
$$= -0.00008091 + 0.00019435 i$$

The division in equation (26−10) uses the relationship that

$$(a + bi)/(c + di) = (ac + bd)/(c^2 + d^2) + [(bc - ad)/(c^2 + d^2)] i$$

Adding $\Delta (x' + iy')$ to $(x' + iy')$,

$$x' = 0.01739129 - 0.00008091$$
$$= 0.01747220$$
$$y' = -0.06937775 + 0.00019435$$
$$= -0.06957210$$

Repeating the above steps with the new values of (x', y'), the new

$$\Delta (x' + iy') = 0.00000000 + 0.00000000\ i$$

Thus there is no change to eight decimals, so equations (26−14) through (26−17) may be used in order,

ρ = $[0.01747220^2 + (-0.06957210)^2]^{1/2}$
   = 0.07173252
c = 2 arctan $(0.07173252/2)$
   = $4.1082095°$
ϕ = arcsin $[\cos 4.1082095° \sin 64° + (-0.06957210$
     $\sin 4.1082095° \cos 64°/0.07173252)]$
   = $60°$ N. lat.
λ = $-152°$ + arctan $[0.01747220 \sin 4.1082095°/$
     $(0.07173252 \cos 64° \cos 4.1082095°$
     $- (-0.06957210) \sin 64° \sin 4.1082095°)]$
   = $-150°$ = $150°$ W. long., not adding $180°$ to the arctan because the denominator
     is positive.

SPACE OBLIQUE MERCATOR (SPHERE)−FORWARD EQUATIONS
(SEE P. 218–219)

Given: Radius of sphere: $R = 6,370,997.0$ m
         Landsat 1, 2, 3 orbit: $i = 99.092°$
                              $P_2/P_1 = 18/251$
                              Path = 15
         Point: $\phi = 40°$ N. lat.
                              $\lambda = 73°$ W. long.

Find: x, y for point taken during daylight northern (first) quadrant of orbit.

Assuming that this is only one of several points to be located, the Fourier constants should first be calculated. Simpson's rule may be written as follows, using λ' as the main variable:

If

$$F = \int_a^b f(\lambda')d\lambda'$$

a close approximation of the integral is

$$F = (\Delta\lambda'/3)[f(\lambda'_a) + 4f(\lambda'_a + \Delta\lambda') + 2f(\lambda'_a + 2\Delta\lambda') + 4f(\lambda'_a + 3\Delta\lambda')$$
$$+ 2f(\lambda'_a + 4\Delta\lambda') + \ldots + 4f(\lambda'_b - \Delta\lambda') + f(\lambda'_b)]$$

where $f(\lambda')$ is calculated for λ' equal to a, and for λ' at each equal interval $\Delta\lambda'$ until $\lambda' = b$. The values $f(\lambda')$ are alternately multiplied by 4 and 2 as the formula indicates, except for the two end values, and all the resulting values are added and multiplied by one-third of the interval. The interval $\Delta\lambda'$ must be chosen so there is an even number of intervals.

Applying this rule to equation (27−1) with the suggested 9° interval in λ', the function $f(\lambda') = (H - S^2)/(1 + S^2)^{1/2}$ is calculated for a λ' of 0°, 9°, 18°, 27°, 36°, . . ., 81°, and 90°, with ten 9° intervals. The calculation for $\lambda' = 9°$ is as follows, using equations (27−4) and (27−5):

$$H = 1 - (18/251) \cos 99.092°$$
$$= 1.0113321$$
$$S = (18/251) \sin 99.092° \cos 9°$$
$$= 0.0699403$$
$$f(\lambda') = (1.0113321 - 0.0699403^2)/(1 + 0.0699403^2)^{1/2}$$
$$= 1.0039879$$

To calculate B, the following table may be figuratively prepared, although a computer or calculator program would normally be used instead (H is a constant):

λ'		S	$f(\lambda')$	Multiplier	Summation
0°	-----------------	0.0708121	1.0038042	×1 =	1.0038042
9	-----------------	.0699403	1.0039879	×4 =	4.0159516
18	-----------------	.0673463	1.0045212	×2 =	2.0090423
27	-----------------	.0630941	1.0053522	×4 =	4.0214087
36	-----------------	.0572882	1.0064001	×2 =	2.0128001
45	-----------------	.0500717	1.0075627	×4 =	4.0302507
54	-----------------	.0416223	1.0087263	×2 =	2.0174526
63	-----------------	.0321480	1.0097770	×4 =	4.0391079
72	-----------------	.0218822	1.0106114	×2 =	2.0212227
81	-----------------	.0110775	1.0111474	×4 =	4.0445895
90	-----------------	.0000000	1.0113321	×1 =	1.0113321
				Total =	30.2269624

To convert to B, again referring to equation (27−1) and remaining in degrees for the final multipliers, since they cancel,

$$B = (2/180°) \times (9°/3) \times 30.2269624$$
$$= 1.0075654$$

This is the Fourier coefficient B for equation (27−6) with λ' in radians. To use λ' in degrees, multiply B by $\pi/180°$:

$$B = 1.0075654 \times \pi/180$$
$$= 0.017585334$$

Calculations of A_n and C_n are similar, except that the calculations of the function involve an additional trigonometric term at each step. For example, to calculate C_3 for $\lambda' = 9°$, using equation (27−3) and the S found above from equation (27−5),

$$f(\lambda') = [S/(1 + S^2)^{1/2}] \cos 3\lambda'$$
$$= [0.0699403/(1 + 0.0699403^2)^{1/2}] \cos (3 \times 9°)$$
$$= 0.06216542$$

The sums for A_n corresponding to 30.2269624 for B are as follows:

$$\text{for } A_2: \quad -0.0564594$$
$$\text{for } A_4: \quad 0.000041208$$

To convert to the desired constants,

$$A_2 = [4/(180° \times 2)] \times (9°/3) \times (-0.0564594)$$
$$= -0.00188198$$
$$A_4 = [4/(180° \times 4)] \times (9°/3) \times (0.000041208)$$
$$= 0.0000006868$$

The sums for C_n:

$$\text{for } C_1: \quad 1.0601909$$
$$\text{for } C_3: \quad -0.0006626541$$

To convert,

$$C_1 = [4 \times (1.0113321 + 1)/(180° \times 1)] \times (9°/3) \times (1.0601909)$$
$$= 0.1421597$$
$$C_3 = [4 \times (1.0113321 + 1)/(180° \times 3)] \times (9°/3) \times (-0.0006626541)$$
$$= -0.0000296182$$

These constants, rounded to seven decimal places except for B, will be used below:

Using equation $(27-11)$,

$$\lambda_0 = 128.87° - (360°/251) \times 15$$
$$= 107.36°$$

To solve equations $(27-8)$ and $(27-9)$ by iteration, determine λ'_p from equation $(27-12)$ and the discussion following the equation, with $N = 0$:

$$\lambda'_p = 90° \times (4 \times 0 + 2 - 1)$$
$$= 90°$$

Then

$$\lambda_{t_p} = -73° - 107.36° + (18/251) \times 90°$$
$$= -173.9058167°$$
$$\cos \lambda_{t_p} = -0.9943487$$

Using λ'_p as the first trial value of λ' in equation $(27-9)$, using extra decimal places for illustration:

$$\lambda_t = -73° - 107.36° + (18/251) \times 90°$$
$$= -173.9058167°, \text{ as before.}$$

Using equation $(27-8)$,

$$\lambda' = \arctan [\cos 99.092° \tan (-173.9058167°) + \sin 99.092°$$
$$\tan 40°/\cos (-173.9058167°)]$$
$$= -40.36910525°$$

For quadrant correction, from the discussion following equation $(27-12)$, using the sign of $\cos \lambda_{t_p}$ as calculated above,

$$\lambda' = -40.36910525° + 90° - 90° \sin 90° \times (-1)$$
$$= -40.36910525° + 180°$$
$$= 139.6308947°$$

This is the next trial λ'. Using equation $(27-9)$,

$$\lambda_t = -73° - 107.36° + (18/251) \times 139.6308947°$$
$$= -170.3466291°$$

Substituting this value of λ_t in place of $-173.9058167°$ in equation (27–8),

$$\lambda' = -40.9362858°$$

The same quadrant adjustment applies:

$$\begin{aligned}\lambda' &= -40.9362858° + 180°\\ &= 139.0637142°\end{aligned}$$

Substituting this in equation (27–9),

$$\lambda_t = -170.3873034°$$

and from equation (27–8),

$$\lambda' = 139.0707998°$$

From the 4th iteration,

$$\begin{aligned}\lambda_t &= -170.3867952°\\ \lambda' &= 139.0707113°\end{aligned}$$

From the 5th iteration,

$$\begin{aligned}\lambda_t &= -170.3868016°\\ \lambda' &= 139.0707124°\end{aligned}$$

From the 6th iteration,

$$\begin{aligned}\lambda_t &= -170.3868015°\\ \lambda' &= 139.0707124°\end{aligned}$$

Since λ' has not changed to seven decimal places, the last iteration is taken as the final value. Using equation (27–10), with the final value of λ_t,

$$\begin{aligned}\phi' &= \arcsin\,[\cos 99.092° \sin 40° - \sin 99.092° \cos 40° \sin\\ &\quad (-170.3868015°)]\\ &= 1.4179606°\end{aligned}$$

From equation (27–5),

$$\begin{aligned}S &= (18/251)\sin 99.092° \cos 139.0707124°\\ &= -0.0534999\end{aligned}$$

From equations (27–6) and (27–7),

$$\begin{aligned}x &= 6{,}370{,}997 \times \{0.017585334 \times 139.0707124° + (-0.0018820)\\ &\quad \sin(2 \times 139.0707124°) + 0.0000007 \sin(4 \times 139.0707124°)\\ &\quad -[-0.0534999/(1 + (-0.0534999)^2)^{1/2}]\ \ln \tan\\ &\quad (45° + 1.4179606°/2)\}\\ &= 15{,}601{,}233.74\ \text{m}\\ y &= 6{,}370{,}997 \times \{0.1421597 \sin 139.0707124° + (-0.0000296)\\ &\quad \sin(3 \times 139.0707124°) + [1/(1 + (-0.0534999)^2)^{1/2}]\\ &\quad \ln \tan(45° + 1.4179606°/2)\}\\ &= 750{,}650.37\ \text{m}\end{aligned}$$

SPACE OBLIQUE MERCATOR (SPHERE)–INVERSE EQUATIONS
(SEE P. 219–221)

Inversing forward example:

Given: Radius of sphere: R = 6,370,997.0 m
    Landsat 1, 2, 3 orbit: i = 99.092°
                      P_2/P_1 = 18/251
                    Path = 15
        Point: x = 15,601,233.74 m
                    y = 750,650.37 m

Find: ϕ, λ

Constants A_2, A_4, B, C_1, C_3, and λ_0 are calculated exactly and have the same values as in the forward example above. To solve equation (27–15) by iteration, the first trial λ' is x/BR, using the value of B for λ' in degrees in this example:

$$\lambda' = 15,601,233.74/(0.017585334 \times 6370997.0)$$
$$= 139.2518341°$$

Using equation (27–5) to find S for this trial λ',

$$S = (18/251) \sin 99.092° \cos 139.2518341°$$
$$= -0.0536463$$

Inserting these values in the right side of equation (27–15),

$$\lambda' = \{15,601,233.74/6,370,997.0 + (-0.0536463)$$
$$\times 750,650.37/6,370,997.0 - (-0.0018820) \sin (2 \times 139.2518341°)$$
$$- 0.0000007 \sin (4 \times 139.2518341°) - (-0.0536463)$$
$$\times [0.1421597 \sin 139.2518341° + (-0.0000296)$$
$$\sin (3 \times 139.2518341°)]\}/0.017585334$$
$$= 139.0695675°$$

Substituting this new trial value of λ' in (27–5) for a new S, then both in (27–15) for a new λ', the next trial value is

$$\lambda' = 139.0707197°$$

The fourth value is

$$\lambda' = 139.0707124°$$

and the fifth does not change to seven decimal places. Therefore, this λ' is the final value. The corresponding S last calculated from (27–5) is

$$S = (18/251) \sin 99.092° \cos 139.0707124°$$
$$= -0.0534999$$

Using equation (27–16),

$$\ln \tan (45° + \phi'/2) = [1 + (-0.0534999)^2]^{1/2} \times [750650.37/$$
$$6370997.0 - 0.1421597 \sin 139.0707124°$$
$$- (-0.0000296) \sin (3 \times 139.0707124°)]$$
$$= 0.02475061$$

$$\tan (45° + \phi'/2) = e^{0.02475061}$$
$$= 1.0250594$$
$$45° + \phi'/2 = \arctan 1.0250594$$
$$= 45.7089803°$$
$$\phi' = 2 \times (45.7089803° - 45°)$$
$$= 1.4179606°$$

Using equation (27−13),

$$\lambda = \arctan [(\cos 99.092° \sin 139.0707124° - \sin 99.092°$$
$$\tan 1.4179606°)/\cos 139.0707124°] - (18/251)$$
$$139.0707124° + 107.36°$$
$$= \arctan [-0.1279654/(-0.7555187)] + 97.3868015°$$
$$= 9.6131985° + 97.3868015°$$
$$= 107.0000000°$$

Since the denominator of the argument of arctan is negative, and the numerator is negative, 180° must be subtracted from λ, or

$$\lambda = 107.0000000° - 180° = -73.0000000°$$
$$= 73° \text{ W. long.}$$

Using equation (27−14),

$$\phi = \arcsin (\cos 99.092° \sin 1.4179606° + \sin 99.092°$$
$$\cos 1.4179606° \sin 139.0707124°)$$
$$= 40.0000000°$$
$$= 40° \text{ N. lat.}$$

For groundtrack calculations, equations (27−17) through (27−20) are used, given the same Landsat parameters as above for R, i, P_2/P_1, and path 15, with $\lambda_0 = 107.36°$, and $\phi = 40°$ S. lat. on the daylight (descending) part of the orbit. Using equation (27−17),

$$\lambda' = \arcsin [\sin (-40°)/\sin 99.092°]$$
$$= -40.6145062°$$

To adjust for quadrant, subtract from 180°, which is the λ' of the descending node:

$$\lambda' = 180° - (-40.6145062°)$$
$$= 220.6145062°$$

Using equation (27−18),

$$\lambda = \arctan [(\cos 99.092° \sin 220.6145062°)/\cos 220.6145062°]$$
$$- (18/251) \times 220.6145062° + 107.36°$$
$$= \arctan [0.1028658/(-0.7591065)] + 91.5390394°$$
$$= 83.8219462°$$

Since the denominator of the argument for arctan is negative, add 180°, but 360° must be then subtracted to place λ between +180° and −180°:

$$\lambda = 83.8219462 + 180° - 360°$$
$$= -96.1780538°$$
$$= 96°10'40.99'' \text{ W. long.}$$

If λ is given instead, with the above λ used for the example, equations (27−19) and (27−9) are iterated together using the same type of initial trial λ' as that used in the forward example for equations (27−8) and (27−9). In this case, as described following equation (27−12), λ'_p is 270°, but this is only known from the final results. If $\lambda'_p = 90°$ is chosen, the same answer will be obtained, since there is considerable overlap in actual regions for which two adjacent λ'_p's may be used. If $\lambda'_p = 450°$ is chosen, the λ' calculated will be about 487.9°, or the position on the next orbit for this λ. Using $\lambda'_p = 270°$ and the equation for λ_{tp} following equation (27−12),

$$\lambda_{tp} = -96.1780538° - 107.36° + (18/251) \times 270°$$
$$= -184.1755040°$$

for which the cosine is negative. From equation (27−9), the first trial λ_t is the same as λ_{tp}. From equation (27−19),

$$\lambda' \quad = \arctan\,[\tan\,(-184.1755040°)/\cos\,99.092°]$$
$$= 24.7970120°$$

For quadrant adjustment, using the procedure following (27−12),

$$\lambda' \quad = 24.7970120 + 270° - 90° \sin 270° \times (-1)$$
$$= 204.7970120°$$

where the (−1) takes the sign of $\cos \lambda_{tp}$.

Substituting this as the trial λ' in (27−9),

$$\lambda_t \quad = -96.1780538° - 107.36° + (18/251) \times 204.7970120°$$
$$= -188.8514155°$$

Substituting this in place of $-184.1755040°$ in (27−19),

$$\lambda' = 44.5812628°$$

but with the same quadrant adjustment as before,

$$\lambda' = 224.5812628°$$

Repeating the iteration, successive values of λ' are

$$\lambda' \quad = 219.5419815°, \text{ then}$$
$$= 220.8989682°, \text{ then}$$
$$= 220.5386678°, \text{ then}$$
$$= 220.6346973°, \text{ then}$$
$$= 220.6091287°, \text{ then}$$
$$= 220.6159384°, \text{ etc.}$$

After a total of about 16 iterations, a value which does not change to seven decimal places is obtained:

$$\lambda' = 220.6145063°$$

Using equation (27−20),

$$\phi = \arcsin \,(\sin 99.092° \,\sin 220.6145063°)$$
$$= -40.0000000°$$
$$= 40° \text{ S. lat.}$$

SPACE OBLIQUE MERCATOR (ELLIPSOID)−FORWARD EQUATIONS
(SEE P. 222–224)

While equations are also given for orbits of small eccentricity, the calculations are so lengthy that examples will only be given for the circular Landsat 1, 2, or 3 orbit.

Given: Clarke 1866 ellipsoid: $a = 6{,}378{,}206.4 \text{ m}$
$e^2 = 0.00676866$
Landsat 1, 2, 3 orbit: $i = 99.092°$
$P_2/P_1 = 18/251$
$R_0 = 7{,}294{,}690.0 \text{ m}$
Path $= 15$
Point: $\phi = 40° \text{ N. lat.}$
$\lambda = 73° \text{ W. long.}$

Find: x, y for point taken during daylight northern (first) quadrant of orbit.

The calculation of Fourier constants for the map follows the same basic procedure as that given for the forward example for the spherical form, except for greater complications in computing each step for the Simpson's numerical integration. The formula for Simpson's rule (see above) is not repeated here, but an example of calculation of a function $f(\lambda'')$ for constant A_2 at $\lambda'' = 18°$ is given below, as represented in equation (27−22).

$$f(\lambda'') = [(HJ-S^2)/(J^2 + S^2)^{1/2}] \cos 2\lambda''$$

Using equations (27−24) through (27−27) in order,

$$J = (1-0.00676866)^3$$
$$= 0.9798312$$
$$W = [(1-0.00676866 \cos^2 99.092°)^2/(1-0.00676866)^2]-1$$
$$= 0.0133334$$
$$Q = 0.00676866 \sin^2 99.092°/(1-0.00676866)$$
$$= 0.0066446$$
$$T = 0.00676866 \sin^2 99.092° \times (2-0.00676866)/(1-0.00676866)^2$$
$$= 0.0133345$$

Using equations (27−30) and (27−31),

$$S = (18/251) \sin 99.092° \cos 18° \times [(1 + 0.0133345$$
$$\sin^2 18°)/(1 + 0.0133334 \sin^2 18°) (1 + 0.0066446 \sin^2 18°)]^{1/2}$$
$$= 0.0673250$$
$$H = [(1 + 0.0066446 \sin^2 18°)/(1 + 0.0133334 \sin^2 18°)]^{1/2}$$
$$\times [(1 + 0.0133334 \sin^2 18°)/(1 + 0.0066446 \sin^2 18°)^2$$
$$-(18/251) \times 1.0 \cos 99.092°]$$
$$= 1.0110133$$

Calculating the function $f(\lambda'')$ as given above,

$$f(\lambda'') = [(1.0110133 \times 0.9798312 - 0.0673250^2)/(0.9798312^2$$
$$+ 0.0673250^2)^{1/2}] \cos(2 \times 18°)$$
$$= 0.8122693$$

In tabular form, using 9° intervals in λ'', the calculation of A_2 proceeds as follows, integrating only to 90° for the circular orbit:

λ''	H	S	$f(\lambda'')$	Multiplier	Summation
0°	1.0113321	0.0708121	1.0035971	$\times 1 =$	1.0035971
9	1.0112504	0.0699346	0.9545807	$\times 4 =$	3.8183229
18	1.0110133	0.0673250	0.8122693	$\times 2 =$	1.6245386
27	1.0106439	0.0630509	0.5904356	$\times 4 =$	2.3617425
36	1.0101782	0.0572226	0.3106003	$\times 2 =$	0.6212007
45	1.0096617	0.0499888	0.0000000	$\times 4 =$	0.0000000
54	1.0091450	0.0415321	−0.3110197	$\times 2 =$	−0.6220394
63	1.0086787	0.0320636	−0.5919529	$\times 4 =$	−2.3678116
72	1.0083085	0.0218167	−0.8151437	$\times 2 =$	−1.6302874
81	1.0080708	0.0110417	−0.9585531	$\times 4 =$	−3.8342122
90	1.0079888	0.0000000	−1.0079888	$\times 1 =$	−1.0079888
				Total $=$	−0.0329376

To convert to A_2, referring to equation (27−22),

$$A_2 = [4/(180° \times 2)] \times (9°/3) \times (-0.0329376)$$
$$= -0.0010979$$

Similar calculations of A_4, B, C_1, and C_3 lead to the values given in the text following equation (27−54):

$$B = 0.0175544891 \text{ for } \lambda'' \text{ in degrees}$$
$$A_4 = -0.0000013$$
$$C_1 = 0.1434410$$
$$C_3 = 0.0000285$$

Since the calculations of j_n and m_n are not necessary for calculation of x and y from ϕ and λ, or the inverse, and are also lengthy, they will be omitted in these examples. The examples given will, however, assist in the understanding of the text concerning their calculations. The other general constant needed is λ_0, determined from (27−37), as in the forward spherical formulas and example:

$$\lambda_0 = 128.87° - (360°/251) \times 15$$
$$= 107.36°$$

For coordinates of the specific point, equations (27−34) and (27−35) are iterated together. Except for the additional factor of $(1-e^2)$ in (27−34), the procedure is identical to the forward spherical example for solving (27−8) and (27−9). The calculations of λ'_p and the first trial λ_t are identical with that example since ϕ and λ have been made the same. The sign of $\cos \lambda_{tp}$ is also negative.

$$\lambda'_p = 90°$$
$$\lambda_t = -173.9058167°$$

Using equation (27−34),

λ'' = arctan [cos 99.092° tan (−173.9058167°)+(1−0.00676866)
 sin 99.092° tan 40°/cos (−173.9058167°)]
 = −40.1810005°

For quadrant correction,

λ'' = −40.1810005° + 90° − 90° sin 90° × (−1)
    = 139.8189995°

Successive iterations give

(2) λ_t = −170.3331395°
    λ'' = 139.2478915°
(3) λ_t = −170.3740954°
    λ'' = 139.2550483°
(4) λ_t = −170.3735822°
    λ'' = 139.2549587°
(5) λ_t = −170.3735886°
    λ'' = 139.2549598°
(6) λ_t = −170.3735885°
    λ'' = 139.2549598°

These last values do not change within seven decimal places in subsequent iterations.

Using equation (27−36) with the final value of λ_t,

ϕ'' = arcsin {[(1−0.00676866) cos 99.092° sin 40°−sin 99.092°
 cos 40° sin (−170.3735885°)]/(1−0.00676866
 sin^2 40°)$^{1/2}$}
 = 1.4692784°

From equation (27−30), using 139.2549598° in place of 18° in the example for calculation of Fourier constants,

$$S = -0.0535730$$

From equations (27−32) and (27−33),

x = 6,378,206.4×{0.0175544891 × 139.2549598° + (−0.0010979)
 sin (2×139.2549598°) + (−0.0000013) sin (4 × 139.2549598°)
 −[−0.0535730/(0.9798312^2 + (−0.0535730)^2)$^{1/2}$] ln tan (45°
 + 1.4692784°/2)}
 = 15,607,700.94 m
y = 6,378,206.4×{0.1434410 sin 139.2549598° + 0.0000285
 sin (3 × 139.2549598°) + [0.9798312/(0.9798312^2
 + (−0.0535730)^2)$^{1/2}$] ln tan (45° + 1.4692784°/2)}
 = 760,636.33 m

For calculation of positions along the groundtrack for a circular orbit, these examples use the same basic Landsat parameters as those in the preceding example, except that ϕ = 40° S. lat. on the daylight (descending) part of the orbit. To find λ', ϕ_g is first calculated from equation (27−41):

$\phi_g = (-40°) - \arcsin \{6,378,206.4 \times 0.00676866 \sin (-40°) \cos$
$\qquad (-40°)/[7,294,690.0 \times (1-0.00676866 \sin^2 (-40°))^{1/2}]$
$\qquad = -40° - (-0.1672042°)$
$\qquad = -39.8327958°$

From equation (27−42),

$$\lambda' = \arcsin [\sin (-39.8327958°)/\sin 99.092°]$$
$$= -40.4436361°$$

To adjust for quadrant, since the satellite is traveling south, subtract from $\frac{1}{2} \times 360°$:

$$\lambda' = 180° - (-40.4436361°)$$
$$= 220.4436361°$$

Using equation (27−43),

$\lambda = \arctan [(\cos 99.092° \sin 220.4436361°)/\cos 220.4436361°]$
$\qquad -(18/251) \times 220.4436361° + 107.36°$
$\qquad = \arctan [0.1025077/(-0.7610445)] + 91.5512930°$
$\qquad = 83.8800995°$

Since the denominator of the argument for arctan is negative, add 180°, but 360° must also be subtracted to place λ between $+ 180°$ and $- 180°$:

$$\lambda = 83.8800995° + 180° - 360°$$
$$= -96.1199005°$$
$$= 96°07'11.64'' \text{ W. long.}$$

If λ is given instead, with the above λ used in the example, equations (27−19) and (27−35) are iterated together with λ' in place of λ'' in the latter. The technique is the same as that used previously for solving (27−8) and (27−9) in the forward spherical example. See also the discussion for the corresponding spherical groundtrack example, using equations (27−19) and (27−9), near the end of the inverse example. Since the formulas for the circular orbit are the same for ellipsoid or sphere for this particular calculation, the various iterations are not shown here. With $\lambda = -96.1199005°$, λ' is found to be 220.4436361°. To find the corresponding ϕ from equation (27−44), a trial $\phi = \arcsin (\sin 99.092° \sin 220.4436361°) = -39.8327958°$ is inserted:

$\phi = \arcsin (\sin 99.092° \sin 220.4436361°) + \arcsin \{6,378,206.4$
$\qquad \times 0.00676866 \sin (-39.8327958°) \cos (-39.8327958°)/$
$\qquad [7,294,690.0 \times (1-0.00676866$
$\qquad \sin^2 (-39.8327958°))^{1/2}]\}$
$\qquad = -39.9998234°$

Substituting $-39.9998234°$ in place of $-39.8327958°$ in the same equation, a new value of ϕ is obtained:

$$\phi = -39.9999998°$$

With the next iteration,

$$\phi = -40.0000000°$$

which does not change to seven decimal places. Thus,

$$\phi = 40° \text{ S. lat.}$$

SPACE OBLIQUE MERCATOR (ELLIPSOID)–INVERSE EQUATIONS
(SEE P. 224–225)

This example is also limited to the circular Landsat orbit, using the parameters of the forward example.

Inversing forward example:

Given: Clarke 1866 ellipsoid: a = 6,378,206.4 m
                                  e^2 = 0.00676866
Landsat 1, 2, 3 orbit: i = 99.092°
                                  P_2/P_1 = 18/251
                                  R_0 = 7,294,690.0 m
                                  Path = 15 (thus λ_0 = 107.36° as in forward example)
Point: x = 15,607,700.94 m
                                  y = 760,636.33 m

Find: ϕ, λ

All constants J, W, Q, T, A_n, B, and C_n, as calculated in the forward example, must be calculated or otherwise provided for use for inverse calculations.

To find λ'' from equation (27–51) by iteration, the procedure is identical to that given for (27–15) in the inverse spherical example, except for the use of different constants. For the initial $\lambda'' = x/aB$,

$$\begin{aligned}\lambda'' &= 15,607,700.94/(6,378,206.4 \times 0.0175544891)\\ &= 139.3965968°\end{aligned}$$

Using equation (27–30) to find S for this value of λ'',

$$\begin{aligned}S &= (18/251) \sin 99.092° \cos 139.3965968° \times [(1 + 0.0133345\\ &\quad \sin^2 139.3965968°)/(1 + 0.0133334 \sin^2 139.3965968°)(1\\ &\quad + 0.0066446 \sin^2 139.3965968°)]^{1/2}\\ &= -0.0536874\end{aligned}$$

Inserting these values into (27–51),

$$\begin{aligned}\lambda'' &= \{15,607,700.94/6,378,206.4 + (-0.0536874/0.9798312)\\ &\quad \times (760,636.33/6,378,206.4) - (-0.0010979) \sin (2\\ &\quad \times 139.3965968°)-(-0.0000013) \sin (4 \times 139.3965968°)\\ &\quad - (-0.0536874/0.9798312) \times [0.1434410 \sin 139.3965968°\\ &\quad + 0.0000285 \sin (3 \times 139.3965968°)]\}/0.0175544891\\ &= 139.2539963°\end{aligned}$$

Substituting this new trial value of λ'' into (27–30) for a new S, then both into (27–51), the next trial value is

$$\lambda'' = 139.2549663°$$

and the fourth trial value is

$$\lambda'' = 139.2549597°$$

The fifth trial value is

$$\lambda'' = 139.2549598°$$

which does not change with another iteration to seven decimal places. Therefore, this is the final value of λ''. The corresponding S last calculated from (27−30) using this value of λ'' is −0.0535730. Using equation (27−52),

$$
\begin{aligned}
\ln \tan(45° + \phi''/2) &= [1 + (-0.0535730)^2/0.9798312^2]^{1/2} \\
&\quad \times [760{,}636.33/6{,}378{,}206.4 - 0.1434410 \sin \\
&\quad 139.2549598° - 0.0000285 \sin (3 \times 139.2549598°)] \\
&= 0.0256466 \\
\tan (45° + \phi''/2) &= e^{0.0256466} \\
&= 1.0259783 \\
45° + \phi''/2 &= \arctan 1.0259783 \\
&= 45.7346392° \\
\phi'' &= 2 \times (45.7346392° - 45°) \\
&= 1.4692784°
\end{aligned}
$$

Using equations (27−48), (27−47), and (27−46) in order,

$$
\begin{aligned}
U &= 0.00676866 \cos^2 99.092°/(1-0.00676866) \\
&= 0.0001702 \\
V &= \{[1-\sin^2 1.4692784°/(1-0.00676866)] \cos 99.092° \\
&\quad \sin 139.2549598° - \sin 99.092° \sin 1.4692784° \\
&\quad \times [(1+0.0066446 \sin^2 139.2549598°) \times (1-\sin^2 1.4692784°) \\
&\quad -0.0001702 \sin^2 1.4692784°]^{1/2}\}/ \\
&\quad [1-\sin^2 1.4692784° (1+0.0001702)] \\
&= -0.1285013 \\
\lambda_t &= \arctan (-0.1285013/\cos 139.2549598°) \\
&= \arctan [-0.1285013/(-0.7576215)] \\
&= 9.6264115°
\end{aligned}
$$

Since the denominator of the argument for arctan is negative, and the numerator is negative, subtract 180°:

$$
\begin{aligned}
\lambda_t &= 9.6264115° - 180° \\
&= -170.3735885°
\end{aligned}
$$

Using equation (27−45),

$$
\begin{aligned}
\lambda &= -170.3735885° - (18/251) \times 139.2549598° + 107.36° \\
&= -73.0000000° \\
&= 73° \text{ W. long.}
\end{aligned}
$$

Using equation (27−49),

$$
\begin{aligned}
\phi &= \arctan \{[\tan 139.2549598° \cos (-170.3735885°) - \cos 99.092° \\
&\quad \sin (-170.3735885°)]/[(1-0.00676866) \sin 99.092°]\} \\
&= 40.0000000° \\
&= 40° \text{ W. lat.}
\end{aligned}
$$

SATELLITE-TRACKING (SPHERE)—FORWARD EQUATIONS (SEE P. 231-232, 236)

Cylindrical form:

Given: Radius of sphere: R = 1.0 unit
       Landsat 1, 2, 3 orbit: i = 99.092°
                                 P_2/P_1 = 18/251
       Map parameters: λ_0 = 90° W. long.
                                 ϕ_1 = 30° N. and S. lat.
       Point: ϕ = 40° N. lat.
                                 λ = 75° W. long.

Find: x, y, h, k

Using equation (28–1),

$$F_1' = [(18/251) \cos^2 30° - \cos 99.092°]/(\cos^2 30° - \cos^2 99.092°)^{1/2}$$
$$= 0.2487473$$

Repeating this for 40° in place of 30°, using equation (28–1a),

$$F' = 0.2669577$$

Using equations (28–2) through (28–8) in order,

$$\lambda' = -\arcsin (\sin 40°/\sin 99.092°)$$
$$= -40.6145062°$$
$$\lambda_t = \arctan [\tan (-40.6145062°) \cos 99.092°]$$
$$= 7.7170932°$$
$$L = 7.7170932° - (18/251) \times (-40.6145062°)$$
$$= 10.6296873°$$
$$x = 1.0 \times [-75° - (-90°)] \cos 30° \times \pi/180°$$
$$= 0.2267249 \text{ unit}$$
$$y = 1.0 \times 10.6296873° \times (\pi/180°) \cos 30°/0.2487473$$
$$= 0.6459071 \text{ unit}$$
$$k = \cos 30°/\cos 40°$$
$$= 1.1305159$$
$$h = 1.1305159 \times 0.2669577/0.2487473$$
$$= 1.2132788$$

Conic Form (two parallels with conformality):

Given: Radius of sphere: R = 1.0 unit
       Landsat 1, 2, 3 orbit: i = 99.092°
                                 P_2/P_1 = 18/251
       Map parameters: λ_0 = 90° W. long.
                                 ϕ_0 = 30° N. lat.
                                 ϕ_1 = 45° N. lat.
                                 ϕ_2 = 70° N. lat.
       Point: ϕ = 40° N. lat.
                                 λ = 75° W. long.

Find: x, y, ρ_s, k, h
Using equation (28–9) for an n of zero,

$$F_0 = \arctan \{[(18/251) \cos^2 30° - \cos 99.092°]/(\cos^2 30° - \cos^2 99.092°)^{1/2}\}$$
$$= 13.9686735°$$

Repeating this for ϕ_1 (45°) and ϕ_2 (70°), in place of 30°,

$$F_1 = 15.7111447°$$
$$F_2 = 28.7497148°$$

Using equations (28−2a) through (28−4a) for an n of zero,

$$\lambda'_0 = -\arcsin (\sin 30°/\sin 99.092°)$$
$$= -30.4218063°$$
$$\lambda_{t0} = \arctan [\tan (-30.4218063°)\cos 99.092°]$$
$$= 5.3013386°$$
$$L_0 = 5.3013386° - (18/251) \times (-30.4218063°)$$
$$= 7.4829821°$$

Repeating these equations for an n of 1 and then 2,

$$\lambda'_1 = -45.7337490°$$
$$\lambda_{t1} = 9.2086865°$$
$$L_1 = 12.4883976°$$
$$\lambda'_2 = -72.1102281°$$
$$\lambda_{t2} = 26.0835377°$$
$$L_2 = 31.2547891°$$

Using equations (28−10) through (28−12),

$$n = (28.7497148° - 15.7111447°)/(31.2547891° - 12.4883976°)$$
$$= 0.6947830$$
$$s_0 = 15.7111447° - 0.6974830 \times 12.4883976°$$
$$= 7.0344182°$$
$$\rho_0 = 1.0 \cos 45° \sin 15.7111447°/[0.6947830 \sin (0.6947830 \times 7.4829821°$$
$$+ 7.0344182°)]$$
$$= 1.3005967 \text{ units}$$

These constants apply to the entire map. For the point (ϕ, λ), using equations (28−9) and (28−2a) through (28−4a) in order for an omitted n, or a ϕ of 40°

$$F = 14.9469825°$$
$$\lambda' = -40.6145062°$$
$$\lambda_t = 7.7170932°$$
$$L = 10.6296873°$$

Since n is positive and L is greater than $(-s_0/n)$, the point may be plotted. Using equation (28−13), the calculation is the same as that for ρ_0, except that L is used in place of L_0:

$$\rho = 1.1066853 \text{ units}$$

Using equations (14−4), (14−1), and (14−2) in order,

$$\theta = 0.6947830 \times [-75° - (-90°)]$$
$$= 10.4217452°$$
$$x = 1.1066853 \sin 10.4217452°$$
$$= 0.2001910 \text{ unit}$$
$$y = 1.3005967 - 1.1066853 \cos 10.4217452°$$
$$= 0.2121685 \text{ unit}$$

Using equations (28−14) through (28−16) in order,

$$\rho_s = 1.0 \cos 45° \sin 15.7111447°/0.6947830$$
$$= 0.2755908 \text{ unit}$$
$$k = 1.1066853 \times 0.6947830/(1.0 \times \cos 40°)$$
$$= 1.0037357$$
$$h = 1.0037357 \tan 14.9469825°/\tan (0.6947830 \times 10.6296873° + 7.0344182°)$$
$$= 1.0421246$$

SATELLITE-TRACKING (SPHERE)−INVERSE EQUATIONS (SEE P. 236−237)

Inversing forward examples:

Cylindrical form:

Given: R, i, P_2/P_1, λ_0, ϕ_1 as in forward example

$$x = 0.2267249 \text{ unit}$$
$$y = 0.6459071 \text{ unit}$$

Find: ϕ, λ

Calculate F_1' from (28−1), exactly as in the forward example:

$$F_1' = 0.2487473$$

Using equation (28−19),

$$L = [0.6459071 \times 0.2487473/(1.0 \cos 30°)] \times 180°/\pi$$
$$= 10.6296860°$$

Using equations (28−24) and (28−25) rather than (28−20) and (28−21), and a first trial λ' of (−90°),

$$A = \tan [10.6296860° + (18/251) \times (−90°)]/\cos 99.092°$$
$$= −0.4620014$$
$$\Delta\lambda' = −[−90° − \arctan (−0.4620014)]/[1 − ((−0.4620014)^2 + 1/\cos^2 99.092°)$$
$$(18/251) \cos 99.092°/((−0.4620014)^2 + 1)]$$
$$= 47.3862943°$$
$$\lambda' = −90° + 47.3862943°$$
$$= −42.6137057°$$

Replacing (−90°) in (28−24) and (28−25) with (−42.6137057°),

$$\Delta\lambda' = 1.9959795°$$
$$\lambda' = −40.6177262°$$

Repeating the iteration successively gives

$$\Delta\lambda' = 0.0032237°$$
$$\lambda' = −40.6145026°$$
$$\Delta\lambda' = −0.0000000$$

Since there is no change to seven decimals,

$$\lambda' = −40.6145026°$$

Using equations (28−22) and (28−23),

$$\begin{aligned}
\phi &= - \arcsin\left[\sin\left(-40.6145026°\right)\sin 99.092°\right] \\
&= 40° \text{ N. lat., neglecting round-off errors} \\
\lambda &= -90° + \left[0.2267249/(1.0\cos 30°)\right] \times 180°/\pi \\
&= -75° = 75° \text{ W. long.}
\end{aligned}$$

Conic form (two parallels with conformality):

Given: R, i, P_2/P_1, λ_0, ϕ_0, ϕ_1, ϕ_2 as in forward example

$$\begin{aligned}
x &= 0.2001910 \text{ unit} \\
y &= 0.2121685 \text{ unit}
\end{aligned}$$

Find: ϕ, λ

Calculate F_0, F_1, F_2, λ_0', λ_{t0}, L_0, λ_1', λ_{t1}, L_1, λ_2', λ_{t2}, L_2, n, s_0, and ρ_0 exactly as in the forward example. Using equations (14−10), (14−11), and (28−26) in order

$$\begin{aligned}
\rho &= \left[0.2001910^2 + (1.3005967-0.2121685)^2\right]^{1/2} \\
&= 1.1066853 \text{ units} \\
\theta &= \arctan\left[0.2001910/(1.3005967-0.2121685)\right] \\
&= 10.4217462° \\
L &= \left[\arcsin\left(1.0\sin 45°\sin 15.7111447°/\right.\right. \\
&\qquad \left.\left.(1.1066853\times 0.6947830)\right) -7.0344182°\right]/ \\
&\qquad 0.6947830 \\
&= 10.6296877°
\end{aligned}$$

With $(-90°)$ as the first trial λ' in (28−24) and (28−25), calculating as in the inverse cylindrical example,

$$\begin{aligned}
A &= -0.4620016 \\
\Delta\lambda' &= 47.3862896° \\
\lambda' &= -42.6137104°
\end{aligned}$$

Replacing $(-90°)$ as the trial λ' with $(-42.6137104°)$, and successively iterating, the result converges to

$$\lambda' = -40.6145076°$$

Using equations (28−22) and (14−9),

$$\begin{aligned}
\phi &= - \arcsin\left[\sin\left(-40.6145076°\right)\sin 99.092°\right] \\
&= 40° \text{ N. lat., disregarding round-off errors.} \\
\lambda &= -90° + 10.4217462°/0.6947830 \\
&= -75° = 75° \text{ W. long.}
\end{aligned}$$

VAN DER GRINTEN (SPHERE)−FORWARD EQUATIONS (SEE P. 241-242)

Given: Radius of sphere: $R = 1.0$ unit
Central meridian: $\lambda_0 = 85°$ W. long.
Point: $\phi = 50°$ S. lat.
$\lambda = 160°$ W. long.

Find: x, y

From equations (29–6), (29–3), (29–4), (29–5), and (29–6a) in order,

$$\theta = \arcsin |2 \times (-50°)/180°|$$
$$= \arcsin 0.5555556$$
$$= 33.7489886°$$
$$A = \frac{1}{2} | 180°/[(-160°)-(-85°)]-[(-160°)-(-85°)]/180°|$$
$$= \frac{1}{2} | -2.4000000-(-0.4166667) |$$
$$= 0.9916667$$
$$G = \cos 33.7489886°/(\sin 33.7489886° + \cos 33.7489886° - 1)$$
$$= 2.1483315$$
$$P = 2.1483315 \times (2/\sin 33.7489886°-1)$$
$$= 5.5856618$$
$$Q = 0.9916667^2 + 2.1483315 = 3.1317342$$

From equation (29–1),

$$x = -\pi \times 1.0 \times \{0.9916667 \times (2.1483315 - 5.5856618^2)$$
$$+ [0.9916667^2 \times (2.1483315 - 5.5856618^2)^2$$
$$- (5.5856618^2 + 0.9916667^2) \times (2.1483315^2 - 5.5856618^2)]^{1/2}\}/$$
$$(5.5856618^2 + 0.9916667^2)$$
$$= -1.1954154 \text{ units}$$

taking the initial "−" sign because $(\lambda - \lambda_0)$ is negative. Note that π is not converted to 180° here, since there is no angle in degrees to offset it. From equation (29–2),

$$y = -\pi \times 1.0 \times \{5.5856618 \times 3.1317342 - 0.9916667$$
$$\times [(0.9916667^2 + 1) \times (5.5856618^2 + 0.9916667^2)$$
$$- 3.1317342^2]^{1/2}/(5.5856618^2 + 0.9916667^2)$$
$$= -0.9960733 \text{ units, taking the initial "−" sign because } \phi \text{ is negative.}$$

VAN DER GRINTEN (SPHERE)–INVERSE EQUATIONS (SEE P. 242)

Inversing forward example:

Given: Radius of sphere: $R = 1.0$ unit
         Central meridian: $\lambda_0 = 85°$ W. long.
                    Point: $x = -1.1954154$ units
                             $y = -0.9960733$ unit

Find: ϕ, λ

Using equations (29–9) through (29–19) in order,

$$X = -1.1954154/(\pi \times 1.0)$$
$$= -0.3805125$$
$$Y = -0.9960733/(\pi \times 1.0)$$
$$= -0.3170600$$
$$c_1 = -0.3170600 \times [1 + (-0.3805125)^2 + (-0.3170600)^2]$$
$$= -0.3948401$$
$$c_2 = -0.3948401 - 2 \times (-0.3170600)^2 + (-0.3805125)^2$$
$$= -0.4511044$$

$$c_3 = -2 \times (-0.3948401) + 1 + 2 \times (-0.3170600)^2$$
$$+ [(-0.3805125)^2 + (-0.3170600)^2]^2$$
$$= 2.0509147$$

$$d = (-0.3170600)^2/2.0509147 + [2 \times (-0.4511044)^3/2.0509147^3$$
$$- 9 \times (-0.3948401) \times (-0.4511044)/2.0509147^2]/27$$
$$= 0.0341124$$

$$a_1 = [-0.3948401 - (-0.4511044)^2/(3 \times 2.0509147)]/2.0509147$$
$$= -0.2086455$$

$$m_1 = 2 \times (0.2086455/3)^{1/2}$$
$$= 0.5274409$$

$$\theta_1 = (1/3) \operatorname{arccos} [3 \times 0.0341124/(-0.2086455 \times 0.5274409)]$$
$$= (1/3) \operatorname{arccos} (-0.9299322)$$
$$= 52.8080831°$$

$$\phi = -180° \times [-0.5274409 \times \cos (52.8080831° + 60°)$$
$$- (-0.4511044)/(3 \times 2.0509147)]$$
$$= -50° = 50° \text{ S. lat., taking the initial “−” sign because } y \text{ is}$$
negative.

$$\lambda = 180° \times \{(-0.3805125)^2 + (-0.3170600)^2 - 1 +$$
$$[1 + 2 \times ((-0.3805125)^2 - (-0.3170600)^2)$$
$$+ ((-0.3805125)^2 + (-0.3170600)^2)^2]^{1/2}\}/$$
$$[2 \times (-0.3805125)] + (-85°)$$
$$= -160° = 160° \text{ W. long.}$$

SINUSOIDAL (SPHERE)–FORWARD EQUATIONS (SEE P. 247)

Given: Radius of sphere: R = 1.0 unit
Central meridian: λ_0 = 90° W. long.
Point: ϕ = 50° S. lat.
λ = 75° W. long.

Find: x, y, h, k, θ', ω

From equations (30−1) through (30−5) in order,

$$x = 1.0 \times [-75° - (-90°)] \times \cos (-50°) \times \pi/180°$$
$$= 0.1682814 \text{ unit}$$

$$y = 1.0 \times (-50°) \times \pi/180°$$
$$= -0.8726646 \text{ unit}$$

$$h = \{1 + [-75° - (-90°)]^2 \times (\pi/180°)^2 \times \sin^2 (-50°)\}^{1/2}$$
$$= 1.0199119$$

$$k = 1.0$$

$$\theta' = \operatorname{arcsin} (1/1.0199119)$$
$$= 78.6597719°$$

$$\omega = 2 \arctan |(1/2)[-75° - (-90°)] \times (\pi/180°) \times \sin (-50°) |$$
$$= 11.4523842°$$

SINUSOIDAL (SPHERE)–INVERSE EQUATIONS (SEE P. 248)

Inversing forward example:

Given: Radius of sphere: R = 1.0 unit
Central meridian: λ_0 = 90° W. long.
Point: x = 0.1682814 unit
y = −0.8726646 unit

Find: ϕ, λ

From equations (30−6) and (30−7),

$$\phi = (-0.8726646/1.0) \times 180°/\pi$$
$$= -49.9999985°$$
$$= 50° \text{ S. lat. rounding off.}$$
$$\lambda = -90° + [0.1682814/(1.0 \times \cos(-49.9999985°))] \times 180°/\pi$$
$$= -75.0000007°$$
$$= 75° \text{ W. long.}$$

SINUSOIDAL (ELLIPSOID)−FORWARD EQUATIONS (SEE P. 248)

Given: Clarke 1866 ellipsoid: $a = 6,378,206.4$ m
                                      $e^2 = 0.00676866$
          Central meridian: $\lambda_0 = 90°$ W. long.
                   Point: $\phi = 50°$ S. lat.
                          $\lambda = 75°$ W. long.

Find: x, y

Using equations (30−8), (3−21), and (30−9) in order,

$$x = 6,378,206.4 \times [-75° - (-90°)] \times (\pi/180°) \cos(-50°) /$$
$$(1-0.00676866 \sin^2(-50°))^{1/2}$$
$$= 1,075,471.5 \text{ m}$$
$$M = 6,378,206.4 \times [(1-0.00676866/4-3\times0.00676866^2/64$$
$$-5 \times 0.00676866^3/256) \times (-50°) \times \pi/180° - (3 \times 0.00676866/8$$
$$+ 3 \times 0.00676866^2/32 + 45 \times 0.00676866^3/1024) \sin(2 \times (-50°))$$
$$+ (15 \times 0.00676866^2/256 + 45 \times 0.00676866^3/1024) \sin(4 \times (-50°))$$
$$- (35 \times 0.00676866^3/3072) \sin(6 \times (-50°))]$$
$$= -5,540,628.0 \text{ m}$$
$$y = -5,540,628.0 \text{ m}$$

SINUSOIDAL (ELLIPSOID)−INVERSE EQUATIONS (SEE P. 248)

Inversing forward example:

Given: a, e^2, λ_0 for forward example

$$x = 1,075,471.5 \text{ m}$$
$$y = -5,540,628.0 \text{ m}$$

Using equations (30−10), (7−19), (3−24), (3−26), and (30−11) in order,

$$M = -5,540,628.0$$
$$\mu = -5,540,628.0/[6378206.4 \times (1-0.00676866/4$$
$$-3 \times 0.00676866^2/64-5\times0.00676866^3/256)]$$
$$= -0.8701555 \text{ radians} = -49.8562390°$$
$$e_1 = [1-(1-0.00676866)^{1/2}]/[1+(1-0.00676866)^{1/2}]$$
$$= 0.001697916$$
$$\phi = -49.8562390° + [(3\times0.001697916/2-27\times0.001697916^3/32)$$
$$\sin(2\times(-49.8562390°)) + (21\times0.001697916^2/16$$
$$-55\times0.001697916^4/32) \sin(4\times(-49.8562390°))$$
$$+ (151\times0.001697916^3/96) \sin(6\times(-49.8562390°))$$
$$+ (1097\times0.001697916^4/512) \sin(8\times(-49.8562390°))]\times180°/\pi$$
$$= -50° = 50° \text{ S. lat.}$$
$$\lambda = -90° + [1075471.5\times(1-0.00676866 \sin^2(-50°))^{1/2}/$$
$$(6,378,206.4 \times \cos(-50°))] \times 180°/\pi$$
$$= -75° = 75° \text{ W. long.}$$

MOLLWEIDE (SPHERE)–FORWARD EQUATIONS (SEE P. 251)

Given: Radius of sphere: R = 1.0 unit
Central meridian: λ_0 = 90° W. long.
Point: ϕ = 50° S. lat.
λ = 75° W. long.

Find: x, y

From equation (31–4), using ϕ or −50° as the first trial θ',

$\Delta\theta'$ = −[(−50°) × π/180° + sin (−50°) − π sin (−50°)]/
[1 + cos (−50°)] × 180°/π
= −26.7818469°

The next trial θ' = −50°−26.7818469° = −76.7818469°. Using this in place of −50° for θ' (not ϕ) in equation (31–4), subsequent iterations produce the following:

$$\Delta\theta' = -4.3367097°$$
$$\theta' = -81.1185566°$$
$$\Delta\theta' = -0.1391597°$$
$$\theta' = -81.2577163°$$
$$\Delta\theta' = -0.0001450°$$
$$\theta' = -81.2578612°$$
$$\Delta\theta' = -0.0000000°$$

Since there is no change to seven decimal places, using (31–5),

$$\theta = -81.2578612°/2$$
$$= -40.6289306°$$

Using (31–1) and (31–2),

x = (8$^{1/2}$/π) × 1.0 × [−75°−(−90°)] cos (−40.6289306°) × π/180°
= 0.1788845 unit
y = 2$^{1/2}$ × 1.0 sin (−40.6289306°)
= −0.9208758 unit

MOLLWEIDE (SPHERE)–INVERSE EQUATIONS (SEE P. 251-252)

Inversing forward example:

Given: Radius of sphere: R = 1.0 unit
Central meridian: λ_0 = 90° W. long.
Point: x = 0.1788845 unit
y = −0.9208758 unit

Find: ϕ, λ

Using equations (31–6) through (31–8) in order,

θ = arcsin [−0.9208758/(2$^{1/2}$ × 1.0)]
= −40.6289311°
ϕ = arcsin{[2 × (−40.6289311°) × π/180° + sin[2 × (−40.6289311°)]]/π}
= −50° ≈ 50° S. lat., neglecting round-off errors
λ = −90° + {π × 0.1788845/[8$^{1/2}$ × 1.0 cos (−40.6289311°)]} × 180°/π
= −75° ≈ 75° W. long.

ECKERT IV (SPHERE)—FORWARD EQUATIONS (SEE P. 256–257)

Given: Radius of sphere: R = 1.0 unit
         Central meridian: λ_0 = 90° W. long.
              Point: ϕ = 50° S. lat.
                        λ = 75° W. long.

Find: x, y

From equation (32–4), using ($\phi/2$) or −25° as the first trial θ,

$$\Delta\theta = -[(-25°) \times \pi/180° + \sin(-25°)\cos(-25°) + 2\sin(-25°)$$
$$-(2+\pi/2)\sin(-50°)]/[2\cos(-25°) \times (1+\cos(-25°))]$$
$$= -17.7554344°$$

The next trial θ = −25°−17.7554344° = −42.7554344°. Using this in place of −25° for θ in equation (32–4), subsequent iterations produce the following:

$$\Delta\theta = -2.9912099°$$
$$\theta = -45.7466443°$$
$$\Delta\theta = -0.1113894°$$
$$\theta = -45.8580337°$$
$$\Delta\theta = -0.0001573°$$
$$\theta = -45.8581910°$$
$$\Delta\theta = -0.0000000°$$

Since there is no change to seven decimal places, θ = −45.8581910°. Using (32–1a) and (32–2a),

$$x = 0.4222382 \times 1 \times [-75°-(-90°)] \times (\pi/180°) \times [1 + \cos(-45.8581910°)]$$
$$= 0.1875270 \text{ unit}$$
$$y = 1.3265004 \times 1 \times \sin(-45.8581910°)$$
$$= -0.9519210 \text{ unit}$$

ECKERT IV (SPHERE)—INVERSE EQUATIONS (SEE P. 257)

Inversing forward example:

Given: Radius of sphere: R = 1.0 unit
         Central meridian: λ_0 = 90° W. long.
              Point: x = 0.1875270 unit
                        y = −0.9519210 unit

Find: ϕ, λ

Using equations (32–9a), (32–10), and (32–11a) in order,

$$\theta = \arcsin[-0.9519210/(1.3265004\times1)]$$
$$= -45.8581937°$$
$$\phi = \arcsin[(-45.8581937° \times \pi/180° + \sin(-45.8581937°)$$
$$\cos(-45.8581937°) + 2\sin(-45.8581937°))/$$
$$(2+\pi/2)]$$
$$= -50.0000027° = 50° \text{ S. lat., disregarding round-off errors.}$$
$$\lambda = -90° + \{0.1875270/[0.4222382\times1$$
$$\times (1 + \cos(-45.8581937°))]\} \times 180°/\pi$$
$$= -74.9999991° = 75° \text{ W. long.}$$

ECKERT VI (SPHERE)—FORWARD EQUATIONS (SEE P. 257)

Given: Radius of sphere: R = 1.0 unit
         Central meridian: λ_0 = 90° W. long.
                   Point: ϕ = 50° S. lat.
                            λ = 75° W. long.

Find: x, y

From equation (32−8), using ϕ or −50° as the first trial θ,

$$\Delta\theta = -\{[(-50° \times \pi/180° + \sin(-50°) - (1 + \pi/2)\sin(-50°)]/$$
$$[1 + \cos(-50°)]\} \times 180°/\pi$$
$$= -11.5316184°$$

The next trial $\theta = -50° - 11.5316184° = -61.5316184°$. Using this in place of −50° for θ (but not ϕ) in equation (32−8), subsequent iterations produce the following:

$$\Delta\theta = -0.6337921°$$
$$\theta = -62.1654105°$$
$$\Delta\theta = -0.0021049°$$
$$\theta = -62.1675154°$$
$$\Delta\theta = -0.0000000°$$

Since there is no change to seven decimal places, $\theta = -62.1675154°$. Using (32−5) and (32−6),

$$x = 1 \times [-75° - (-90°)] \times (\pi/180°) \times [1 + \cos(-62.1675154°)]/(2 + \pi)^{1/2}$$
$$= 0.1693623 \text{ unit}$$
$$y = 2 \times 1 \times (-62.1675154°) \times (\pi/180°)/(2 + \pi)^{1/2}$$
$$= -0.9570223 \text{ unit}$$

ECKERT VI (SPHERE)—INVERSE EQUATIONS (SEE P. 257)

Inversing forward example:

Given: Radius of sphere: R = 1.0 unit
         Central meridian: λ_0 = 90° W. long.
                   Point: x = 0.1693623 unit
                            y = −0.9570223 unit

Find: ϕ, λ

Using equations (32−12), (32−13), and (32−14) in order,

θ = $(2+\pi)^{1/2} \times (-0.9570223) \times (180°/\pi)/(2 \times 1)$
         = −62.1675178°
ϕ = arcsin $[(-62.1675178° \times \pi/180° + \sin(-62.1675178°))/(1 + \pi/2)]$
         = −50.0000021° = 50° S. lat., disregarding round-off errors.
λ = $-90° + (2+\pi)^{1/2} \times 0.1693623 \times (180°/\pi)/[1 \times (1 + \cos(-62.1675178°))]$
         = −75° = 75° W. long.

APPENDIX B
USE OF MAP PROJECTIONS BY U.S. GEOLOGICAL SURVEY—SUMMARY

Note This list is not exhaustive. For further details, see text.

Class/Projection   *Maps*

Cylindrical

Mercator ------------------------------------- Northeast Equatorial Pacific
                                              Indonesia (Tectonic)
                                              Other planets and satellites

Transverse Mercator ----------------------- 7½' and 15' quadrangles for
                                              22 States
                                              North America

Universal Transverse Mercator ---------- 1° lat. × 2° long. quadrangles
                                              of U.S. metric quadrangles and
                                              County maps.

"Modified Transverse Mercator" ---------- Alaska

Oblique Mercator --------------------------- Grids in southeast
                                              Alaska
                                              Landsat Satellite Imagery

Miller Cylindrical -------------------------- World

Equidistant Cylindrical --------------------- United States and some State Index
                                              Maps

Conic

Albers Equal-Area Conic -------------------- United States and sections

Lambert Conformal Conic ------------------- 7½' and 15' quadrangles for
                                              32 States
                                              Quadrangles for Puerto Rico, Virgin
                                                Islands, and Samoa
                                              State Base Maps
                                              Quadrangles for International
                                                Map of the World
                                              Some other planets and satellites
                                              Some State Index Maps

Bipolar Oblique Conic
  Conformal ------------------------------------- North America (Geologic)

Polyconic -------------------------------------- Quadrangles for all States

Modified Polyconic ------------------------- Quadrangles for International
                                              Map of the World

Azimuthal

Orthographic (oblique) --------------------- Pictorial views of Earth
                                              or portions

Stereographic (oblique) --------------------- Other planets and satellites
              (polar) ------------------------- Antarctica
                                              Arctic regions
                                              Other planets and satellites

Lambert Azimuthal Equal-Area
  (oblique) ------------------------------------- Pacific Ocean
  (polar) --------------------------------------- Arctic regions (Hydrocarbon
                                              Provinces)
                                              North and South Polar regions
                                              (polar expeditions)

Azimuthal Equidistant (oblique) .... World

Quadrangles for Guam and
Micronesia

Space

Space Oblique Mercator .... Satellite image mapping

Miscellaneous

Van der Grinten .... World (Subsea Mineral Resources,
misc.)

Sinusoidal (interrupted) .... World (Hydrocarbon Provinces)

APPENDIX C
STATE PLANE COORDINATE SYSTEMS—CHANGES FOR 1983 DATUM

isting indicates changes for the NAD 1983 datum rojections, parameters, and origins of zones as ed in table 8 for the NAD 1927 datum. It is im- c to understand that State plane coordinates based datum *cannot* be correctly converted to coordinates 1983 datum merely by using inverse formulas to t from 1927 rectangular coordinates to latitude and ide, and then using forward formulas with this e and longitude to convert to 1983 rectangular coor- s. Due to readjustment of the survey control net- and to the change of ellipsoid, the latitude and ide also change slightly from one datum to the

These changes have been approved by the National Geodetic Survey (William M. Kaula, James Stem, pers. comm., 1986). They are given in the same order as the entries in table 8, except that *only the changes are shown*. All parameters not listed remain as before, except for the different ellipsoid and datum. Because all coordinates at the origin have been changed, and because they vary considerably, they are presented in the body of the table rather than as footnotes. Samoa is not being changed to the new datum.

[L *indicates Lambert Conformal Conic*]

Area	Projection	Zones
rnia	L	6
na	L	1
ska	L	1
Rico and Virgin Islands	L	1
Carolina	L	1
ing	Unresolved	

Transverse Mercator projection

Zone	Coordinates of origin (meters) x	y	Other Changes
ma			
ast	200,000	0	
est	600,000	0	
a, 2–9	500,000	0	
ia, all	213,360	0	Origin in Intl. feet[1]
are	200,000	0	
a			
ast, West	200,000	0	
ia			
ast	200,000	0	
est	700,000	0	
i, all	500,000	0	
ast	200,000	0	
entral	500,000	0	
est	800,000	0	
s			
ast	300,000	0	
est	700,000	0	
a			
ast	100,000	250,000	
est	900,000	250,000	
ast	300,000	0	Lat. of origin 43°40′ N.
est	900,000	0	
ssippi			
ast	300,000	0	Scale reduction 1:20,000, Lat. of origin 29°30′ N.
est	700,000	0	Scale reduction 1:20,000, Lat. of origin 29°30′ N.

	Transverse Mercator projection		
	Coordinates of origin (meters)		
Zone	*x*	*y*	*Other changes*
Missouri			
East	250,000	0	
Central	500,000	0	
West	850,000	0	
Nevada			
East	200,000	8,000,000	
Central	500,000	6,000,000	
West	800,000	4,000,000	
New Hampshire	300,000	0	
New Jersey	150,000	0	Central meridian 74°30′
			Scale reduction 1:10,000
New Mexico			
East	165,000	0	
Central	500,000	0	
West	830,000	0	
New York			
East	All parameters identical with above New Jersey zone.		
Central	250,000	0	
West	350,000	0	
Rhode Island	100,000	0	
Vermont	500,000	0	
Wyoming	Unresolved		

Lambert Conformal Conic projection

| one | Coordinates of origin (meters) | | Other changes |
---	x	y	
10	1,000,000	0	
as			
rth	400,000	0	
ith	400,000	400,000	
nia			Zone 7 deleted.
5	2,000,000	500,000	
do, all	914,401.8289	304,800.6096	
ticut	304,800.6096	152,400.3048	
, North	600,000	0	
rth	1,500,000	1,000,000	
ith	500,000	0	
s			
rth	400,000	0	
ith	400,000	400,000	
ky			
rth	500,000	0	
ith	500,000	500,000	
na			
rth	1,000,000	0	Lat. of origin 30°30′ N.
ith	1,000,000	0	Lat. of origin 28°30′ N.
shore	1,000,000	0	Lat. of origin 25°30′ N.
nd	400,000	0	Lat. of origin 37°40′ N.
chusetts			
inland	200,000	750,000	
and	500,000	0	
an			GRS 80 ellipsoid used without alteration.
rth	8,000,000	0	
ntral	6,000,000	0	Long. of origin 84°22′ W.
ith	4,000,000	0	Long. of origin 84°22′ W.
sota, all	800,000	100,000	
na	600,000	0	Standard parallels, 45°00′ and 49°00′ N. Long. of origin 109°30′ W. Lat. of origin 44°15′ N.
ngle zone)			

	Lambert Conformal Conic projection		
	Coordinates of origin (meters)		
Zone	x	y	
Nebraska (single zone)	500,000	0	Standard parallels, 40°0 43°00′ N. Long. of origin 100°00′ Lat. of origin 39°50′ N.
New York			
Long Island	300,000	0	Lat. of origin 40°10′ N.
North Carolina	609,621.22	0	
North Dakota, all	600,000	0	
Ohio, all	600,000	0	
Oklahoma, all	600,000	0	
Oregon			
North	2,500,000	0	
South	1,500,000	0	
Pennsylvania, all	600,000	0	
Puerto Rico and Virgin Islands	200,000	200,000	(Two previous zones ide except for x and y of or
South Carolina (single zone)	609,600	0	Standard parallels, 32°30′ and 34°50′ N. Long. of origin 81°00′ W Lat. of origin 31°50′ N.
South Dakota, all	600,000	0	
Tennessee	600,000	0	Lat. of origin 34°20′ N.
Texas			
North	200,000	1,000,000	
North Central	600,000	2,000,000	Central meridian 98°30′
Central	700,000	3,000,000	
South Central	600,000	4,000,000	
South	300,000	5,000,000	
Utah			
North	500,000	1,000,000	
Central	500,000	2,000,000	
South	500,000	3,000,000	
Virginia			
North	3,500,000	2,000,000	
South	3,500,000	1,000,000	
Washington, all	500,000	0	
West Virginia, all	600,000	0	
Wisconsin, all	600,000	0	

NOTE: All these systems are based on the GRS 80 ellipsoid.
[1]For the International foot, 1 in=2.54 cm, or 1 ft=30.48 cm.

INDEX

[Italic page numbers indicate major references]

☆U.S. G.P.O. 1987- 181-407:60010

Type of map projection	Cylinders		
	Mercator	**Oblique Mercator**	**Transverse Mercator**
	Conformal		
Lines of longitude (meridians)	Meridians are straight and parallel.	Meridians are complex curves concave toward the line of tangency, except each 180th meridian is straight.	Meridians are complex curves concave toward a straight central meridian that is tangent to the globe. The straight central meridian intersects the equator and one meridian at a 90° angle.
Lines of latitude (parallels)	Latitude lines are straight and parallel.	Parallels are complex curves concave toward the nearest pole.	Parallels are complex curves concave toward the nearest pole; the equator is straight.
Graticule spacing	Meridian spacing is equal, and the parallel spacing increases away from the equator. The graticule spacing retains the property of conformality. The graticule is symmetrical. Meridians and parallels intersect at right angles.	Graticule spacing increases away from the line of tangency and retains the property of conformality.	Parallels are spaced at their true distances on the straight central meridian. Graticule spacing increases away from the tangent meridian. The graticule retains the property of conformality.
Linear scale	Linear scale is true along the equator only (line of tangency), or along two parallels equidistant from the equator (the secant form). Scale can be determined by measuring one degree of latitude, which equals 60 nautical miles, 69 statute miles, or 111 kilometers.	Linear scale is true along the line of tangency, or along two lines equidistant from and parallel to the line of tangency.	Linear scale is true along the line of tangency, or along two lines equidistant from and parallel to the line of tangency.
Notes	Projection can be thought of as being mathematically based on a cylinder tangent at the equator. Any straight line is a constant-azimuth (rhumb) line. Areal enlargement is extreme away from the equator; poles cannot be represented. Shape is true only within any small area. Reasonably accurate projection within a 15° band along the line of tangency.	Projection is mathematically based on a cylinder tangent along any great circle other than the equator or a meridian. Shape is true only within any small area. Areal enlargement increases away from the line of tangency. Reasonably accurate projection within a 15° band along the line of tangency.	Projection is mathematically based on a cylinder tangent to a meridian. Shape is true only within any small area. Areal enlargment increases away from the tangent meridian. Reasonably accurate projection within a 15° band along the line of tangency. Cannot be edge-joined in an east-west direction if each sheet has its own central meridian.
Uses	An excellent projection for equatorial regions. Otherwise the Mercator is a special-purpose map best suited for navigation. Secant constructions are used for large-scale coastal charts. The use of the Mercator map projection as the base for nautical charts is universal. Examples are the charts published by the National Ocean Survey, U.S. Dept. of Commerce.	Useful for plotting linear configurations that are situated along a line oblique to the earth's equator. Examples are: NASA Surveyor Satellite tracking charts, ERTS flight indexes, strip charts for navigation, and the National Geographic Society's maps "West Indies", "Countries of the Caribbean", "Hawaii", and "New Zealand".	Used where the north-south dimension is greater than the east-west dimension. Used as the base for the U.S. Geological Survey's 1:250,000-scale series and for some of the 7½-minute and 15-minute quadrangles of the National Topographic Map Series.
Examples		←— Line of tangency 	

Type of map projection	Cylinders	Cones	
	Modified Transverse Mercator	**Equidistant Conic (or Simple Conic)**	**Lambert Conformal Conic**
		Equidistant	Conformal
Lines of longitude (meridians)	On pre-1973 editions of the Alaska Map E, meridians are curved concave toward the center of the projection. On post-1973 editions the meridians are straight.	Meridians are straight lines converging on a polar axis but not at the pole.	Meridians are straight lines converging at a pole.
Lines of latitude (parallels)	Parallels are arcs concave to the pole.	Parallels are arcs of concentric circles concave toward a pole.	Parallels are arcs of concentric circles concave toward a pole a centered at the pole.
Graticule spacing	Meridian spacing is approximately equal and decreases toward the pole. Parallels are approximately equally spaced. The graticule is symmetrical on post-1973 editions of the Alaska Map E.	Meridian spacing is true on the standard parallels and decreases toward the pole. Parallels are spaced at true scale along the meridians. Meridians and parallels intersect each other at right angles. The graticule is symmetrical.	Meridian spacing is true on the standard parallels and decreas toward the pole. Parallel spacing increases away from th standard parallels and decreases between them. Meridians ar parallels intersect each other at right angles. The graticule spacin retains the property of conformality. The graticule is symmetric
Linear scale	Linear scale is more nearly correct along the meridians than along the parallels.	Linear scale is true along all meridians and along the standard parallel or parallels.	Linear scale is true on standard parallels. Maximum scale error is 2 percent on a map of the United States (48 states) with standar parallels at 33° N. and 45 ° N.
Notes	The Alaska Map E was adapted from a set of transverse Mercator projections 8° wide and approximately 18° long, repeated east and west of an arbitrary point of origin until a projection 72° wide was obtained. The post-1973 editions of the Alaska Map E more nearly approximate an equidistant conic map projection.	Projection is mathematically based on a cone that is tangent at one parallel or conceptually secant at two parallels. North or South Pole is represented by an arc.	Projection is mathematically based on a cone that is tangent at on parallel or (more often) that is conceptually secant on tw parallels. Areal distortion is minimal but increases away from th standard parallels. North or South Pole is represented by a point the other pole cannot be shown. Great circle lines ar approximately straight. Retains its properties at various scales sheets can be joined along their edges.
Uses	The U.S. Geological Survey's Alaska Map E at the scale of 1:2,500,000. The figure below represents the 1954 edition. The 1973 edition is similar, but the meridians are straight. The Bathymetric Maps Eastern Continental Margin U.S.A., published by the American Association of Petroleum Geologists, uses these straight meridians on its Modified Transverse Mercator and is more equivalent to the Equidistant Conic map projection.	The Equidistant Conic projection is used in atlases for portraying mid-latitude areas. It is good for representing regions with a few degrees of latitude lying on one side of the Equator. The Kavraisky No. 4 map projection is an Equidistant conic map projection, in which standard parallels are chosen to minimize overall error.	Used for large countries in the mid-latitudes having an east-wes orientation. The United States (50 states) Base Map uses standar parallels at 37° N. and 65° N. Some of the National Topographi Map Series 7½-minute and 15-minute quadrangles and the State Base Map Series are constructed on the Lambert Conformal Coni map projection. The latter series uses standard parallels of 33° N and 45° N. Aeronautical charts for Alaska use standard parallels a 55° N. and 65° N. The National Atlas of Canada uses standar parallels at 49° N. and 77° N. In the figure below, the outline represents the United States (50 states) Base Map.
Examples			

Type of map projection	Cones		
	Albers Conic Equal-Area	**American Polyconic**	**Bipolar Oblique Conic Conformal**
	Equal Area		Conformal
Lines of longitude (meridians)	Meridians are straight lines converging on the polar axis, but not at the pole.	Meridians are complex curves concave toward a straight central meridian.	Meridians are complex curves concave toward the center of the projection.
Lines of latitude (parallels)	Parallels are arcs of concentric circles concave toward a pole.	Parallels are nonconcentric circles except for a straight equator.	Parallels are complex curves concave toward the nearest pole.
Graticule spacing	Meridian spacing is equal on the standard parallels and decreases toward the poles. Parallel spacing decreases away from the standard parallels and increases between them. Meridians and parallels intersect each other at right angles. The graticule spacing preserves the property of equivalence of area. The graticule is symmetrical.	Meridian spacing is equal and decreases toward the poles. Parallels are spaced true to scale on the central meridian, and the spacing increases toward the east and west borders. The graticule spacing results in a compromise of all properties.	Graticule spacing increases away from the lines of true scale and retains the property of conformality.
Linear scale	Linear scale is true on the standard parallels. Maximum scale error is 1¼ percent on a map of the United States (48 states) with standard parallels of 29½° N. and 45½° N.	Linear scale is true along each parallel and along the central meridian. Maximum scale error is 7 percent on a map of the United States (48 states).	Linear scale is true along two lines that do not lie along any meridian or parallel. Scale is compressed between these lines and expanded beyond them. Linear scale is generally good, but there is as much as a 10 percent error at the edge of the projection as used.
Notes	Projection is mathematically based on a cone that is conceptually secant on two parallels. No areal deformation. North or South Pole is represented by an arc. Retains its properties at various scales; individual sheets can be joined along their edges.	Projection is mathematically based on an infinite number of cones tangent to an infinite number of parallels. Distortion increases away from the central meridian. Has both areal and angular deformation.	Projection is mathematically based on two cones whose apexes are 104° apart, and which conceptually are obliquely secant to the sphere along lines following the trend of North and South America.
Uses	Used for thematic maps. Used for large countries with an east-west orientation. Maps based on the Albers equal-area conic for Alaska use standard parallels 55° N. and 65° N.; for Hawaii, the standard parallels are 8° N. and 18° N. The National Atlas of the United States, United States Base Map (48 states), and the Geologic map of the United States (outlined below) are based on the standard parallels of 29½° N. and 45½° N.	Used for areas with a north-south orientation. Only along central meridian does it portray true shape, area, distance, and direction. Formerly used as the base of the 7½- and 15-minute quadrangles of the National Topographic Map Series. Individual sheets of this series can be edge-joined since they are drawn with straight meridians for convenience. They cannot be mosaicked beyond a few sheets.	Used to represent one or both of the American continents. Examples are the Basement map of North America and the Tectonic map of North America.
Examples			

Basement map

Tectonic map

Lines of tangency

Type of map projection	Pseudo-Cylinders		Miscellaneous
	Sinusoidal	**Eckert No. 6**	**Van Der Grinten**
	Equal Area	Equal Area	Compromise
Lines of longitude (meridians)	Meridians are sinusoidal curves, curved concave toward a straight central meridian.	Meridians are sinusoidal curves concave toward a straight central meridian.	Meridians are circular arcs concave toward a straight central meridian.
Lines of latitude (parallels)	All parallels are straight, parallel lines.	All parallels are straight, parallel lines.	Parallels are circular arcs concave toward the poles except for a straight equator.
Graticule spacing	Meridian spacing is equal and decreases toward the poles. Parallel spacing is equal. The graticule spacing retains the property of equivalence of area.	Meridian and parallel spacing decreases toward the poles. The graticule spacing retains the property of equivalence of area.	Meridian spacing is equal at the equator. The parallels are spaced farther apart toward the poles. Central meridian and equator are straight lines. The poles commonly are not represented. The graticule spacing results in a compromise of all properties.
Linear scale	Linear scale is true on the parallels and the central meridian.	Linear scale is true along parallel 49° 16' north and south of the equator.	Linear scale is true along the equator. Scale increases rapidly toward the poles.
Notes	Projection is mathematically based on a cylinder tangent on the equator. The sinusoidal projection may have several central meridians and may be interrupted on any meridian to help reduce distortion at high latitudes. There is no angular deformation along the central meridian and the equator.	Projection is mathematically based on a cylinder tangent at the equator. Poles are represented by straight lines half the length of the equator. Distortion of shape is extreme at high latitudes.	The projection has both areal and angular deformation. It was conceived as a compromise between the Mercator and the Mollweide, which shows the world in an ellipse. The Van der Grinten shows the world in a circle.
Uses	Used as an equal-area projection to portray areas that have a maximum extent in a north-south direction. Used as a world equal-area projection in atlases to show distribution patterns. The figure below represents an interrupted version of the sinusoidal projection with three central meridians. Used by the U.S. Geological Survey as the base for maps showing prospective hydrocarbon provinces of the world and sedimentary basins of the world.	Used as an equal-area map projection of the world in atlases such as the Great Soviet World Atlas, 1937. Kavraisky No. 6 map projection closely resembles Eckert No. 6 and is used in the Ocean Atlas, 1953, Vol. 2.	The Van der Grinten projection is used by the National Geographic Society for world maps. Used by the U.S. Geological Survey to show distribution of mineral resources on the sea floor (McKelvey and Wang, 1970).
Examples			

Type of map projection	Azimuthal Equidistant	
	Equidistant	
Lines of longitude (meridians)	Polar aspect: the meridians are straight lines radiating from the point of tangency. Oblique aspect: the meridians are complex curves concave toward the point of tangency. Equatorial aspect: the meridians are complex curves concave toward a straight central meridian, except the outer meridian of a hemisphere, which is a circle.	
Lines of latitude (parallels)	Polar aspect: the parallels are concentric circles. Oblique aspect: the parallels are complex curves. Equatorial aspect: the parallels are complex curves concave toward the nearest pole; the equator is straight.	
Graticule spacing	Polar aspect: the meridian spacing is equal and increases away from the point of tangency. Parallel spacing is equidistant. Angular and areal deformation increase away from the point of tangency.	
Linear scale	Polar aspect: linear scale is true from the point of tangency along the meridians only. Oblique and equatorial aspects: linear scale is true from the point of tangency. In all aspects the Azimuthal Equidistant shows distances true to scale when measured between the point of tangency and any other point on the map.	
Notes and uses	Projection is mathematically based on a plane tangent to the earth. The entire earth can be represented. Generally the Azimuthal Equidistant map projection portrays less than one hemisphere, though the other hemisphere can be portrayed but is much distorted. Has true direction and true distance scaling from the point of tangency. The Azimuthal Equidistant projection is used for radio and seismic work, as every place in the world will be shown at its true distance and direction from the point of tangency. The U.S. Geological Survey uses the oblique aspect of the Azimuthal Equidistant in the National Atlas and for large-scale mapping of Micronesia. The polar aspect is used as the emblem of the United Nations.	
Examples		

Type of map projection	Lambert Azimuthal Equal-Area
	Equal Area
Lines of longitude (meridians)	Polar aspect: the meridians are straight lines radiating from the point of tangency. Oblique and equatorial aspects: meridians are complex curves concave toward a straight central meridian, except the outer meridian of a hemisphere, which is a circle.
Lines of latitude (parallels)	Polar aspect: parallels are concentric circles. Oblique and equatorial aspects: the parallels are complex curves. The equator on the equatorial aspect is a straight line.
Graticule spacing	Polar aspect: the meridian spacing is equal and increases, and the parallel spacing is unequal and decreases toward the periphery of the projection. The graticule spacing in all aspects retains the property of equivalence of area.
Linear scale	Linear scale is better than most azimuthals but not as good as the equidistant. Angular deformation increases toward the periphery of the projection. Scale decreases radially toward the periphery of the map projection. Scale increases perpendicular to the radii toward the periphery.
Notes and uses	The Lambert Azimuthal Equal-Area projection is mathematically based on a plane tangent to the earth. It is the only projection that can accurately represent both areas and true direction from the center of the projection. This projection generally represents only one hemisphere. The polar aspect is used by the U.S. Geological Survey in the National Atlas. The polar, oblique, and equatorial aspects are used by the U.S. Geological Survey for the Circum-Pacific Map.
Examples	

Type of map projection	Orthographic
Lines of longitude (meridians)	Polar aspect: the meridians are straight lines radiating from the point of tangency. Oblique aspect: the meridians are ellipses, concave toward the center of the projection. Equatorial aspect: the meridians are ellipses concave toward the straight central meridian.
Lines of latitude (parallels)	Polar aspect: the parallels are concentric circles. Oblique aspect: the parallels are ellipses concave toward the poles. Equatorial aspect: the parallels are straight and parallel.
Graticule spacing	Polar aspect: meridian spacing is equal and increases, and the parallel spacing decreases from the point of tangency. Oblique and equatorial aspects: the graticule spacing decreases away from the center of the projection.
Linear scale	Scale is true on the parallels in the polar aspect and on all circles centered at the pole of the projection in all aspects. Scale decreases along lines radiating from the center of the projection.
Notes and uses	The Orthographic projection is geometrically based on a plane tangent to the earth, and the point of projection is at infinity. The earth appears as it would from outer space. This projection is a truly graphic representation of the earth and is a projection in which distortion becomes a visual aid. It is the most familiar of the azimuthal map projections. Directions from the center of the Orthographic map projection are true. The U.S. Geological Survey uses the Orthographic map projection in the National Atlas.
Examples	

Type of map projection	Stereographic
	Conformal
Lines of longitude (meridians)	Polar aspect: the meridians are straight lines radiating from the point of tangency. Oblique and equatorial aspects: the meridians are arcs of circles concave toward a straight central meridian. In the equatorial aspect the outer meridian of the hemisphere is a circle centered at the projection center.
Lines of latitude (parallels)	Polar aspect: the parallels are concentric circles. Oblique aspect: the parallels are nonconcentric arcs of circles concave toward one of the poles with one parallel being a straight line. Equatorial aspect: parallels are nonconcentric arcs of circles concave toward the poles; the equator is straight.
Graticule spacing	The graticule spacing increases away from the center of the projection in all aspects, and it retains the property of conformality.
Linear scale	Scale increases toward the periphery of the projection.
Notes and uses	The Stereographic projection is geometrically projected onto a plane, and the point of the projection is on the surface of the sphere opposite the point of tangency. Circles on the earth appear as straight lines, parts of circles, or circles on the projection. Directions from the center of the stereographic map projection are true. Generally only one hemisphere is portrayed. The Stereographic projection is the most widely used azimuthal projection, mainly used for portraying large, continent-size areas of similar extent in all directions. It is used in geophysics for solving problems in spherical geometry. The polar aspect is used for topographic maps and navigational charts. The American Geographical Society uses the stereographic map projection as the basis for its "Map of the Arctic". The U.S. Geological Survey uses the stereographic map projection as the basis for maps of Antarctica.
Examples	

Type of map projection	Gnomonic
Lines of longitude (meridians)	Polar aspect: the meridians are straight lines radiating from the point of tangency. Oblique and equatorial aspects: the meridians are straight lines.
Lines of latitude (parallels)	Polar aspect: the parallels are concentric circles. Oblique and equatorial aspects: parallels are ellipses, parabolas, or hyperbolas concave toward the poles (except for the equator, which is straight).
Graticule spacing	Polar aspect: the meridian spacing is equal and increases away from the pole. The parallel spacing increases very rapidly from the pole. Oblique and equatorial aspects: the graticule spacing increases very rapidly away from the center of the projection.
Linear scale	Linear scale and angular and areal deformation are extreme, rapidly increasing away from the center of the projection.
Notes and uses	The Gnomonic projection is geometrically projected onto a plane, and the point of projection is at the center of the earth. It is impossible to show a full hemisphere with one Gnomonic map. It is the only projection in which any straight line is a great circle, and it is the only projection that shows the shortest distance between any two points as a straight line. Consequently, it is used in seismic work because seismic waves travel in approximately great circles. The Gnomonic projection is used with the Mercator projection for navigation.
Examples	

INTRODUCTION

Most map users give little thought to the map projection used for a large-scale map of a small area. As the map scale becomes smaller and the area shown increases, however, the properties of the map projection become increasingly important. The brief descriptions of the properties and uses of map projections in this report are intended to help the user compare these projections and choose the one best suited to a particular purpose.

This report is a revision of U.S. Geological Survey Map I-1096, "A survey of the properties and uses of selected map projections" (Alpha and Gerin, 1978). Principal differences between this and the earlier version are that (1) new terms are included, (2) a new example of the Albers equal-area projection is provided, and (3) the Kavraisky No. 4 projection has been deleted (mainly because it is rarely used).

NATURAL PROPERTIES OF THE EARTH'S GRATICULE[1]

1. Parallels are parallel.
2. Parallels are spaced equally on meridians.
3. Meridians and other great circle arcs are straight lines (if looked at perpendicularly to the earth's surface).
4. Meridians converge toward the poles and diverge toward the equator.
5. Meridians are equally spaced on the parallels, but their distance apart decreases from the equator to the pole.
6. Meridians at the equator are spaced the same as parallels.
7. Meridians at 60° are half as far apart as parallels.
8. Parallels and meridians cross one another at right angles.
9. The area of the surface bounded by any two parallels and two meridians (a given distance apart) is the same anywhere between the same two parallels.
10. The scale factor at each point is the same in any direction.

[1]From Robinson (1969, p. 212)

DEFINITION OF TERMS

ASPECT—Individual azimuthal map projections are divided into three aspects: the polar aspect which is tangent at the pole, the equatorial aspect which is tangent at the equator, and the oblique aspect which is tangent anywhere else. (The word "aspect" has replaced the word "case" in the modern cartographic literature).

CONFORMALITY—A map projection is conformal when (1) meridians and parallels intersect at right angles, and (2) at any point the scale is the same in every direction. The shapes of very small areas and angles with very short sides are preserved.

DEVELOPABLE SURFACE—A developable surface is a simple geometric form capable of being flattened without stretching. Many map projections can then be grouped by a particular developmental surface: cylinder, cone, or plane.

EQUAL AREA—A map projection is equal area when every part, as well as the whole, has the same area as the corresponding part on the earth, at the same reduced scale.

GRATICULE—The graticule is the spherical coordinate system based on lines of latitude and longitude.

LINEAR SCALE—Linear scale is the relation between a distance on a map projection and the corresponding distance on the earth.

MAP PROJECTION—A map projection is a systematic representation of a round body such as the earth on a flat (plane) surface. Each map projection has specific properties that make it useful for specific objectives.

SOURCES

Alpha, T. R., and Gerin, Marybeth, 1978, A survey of the properties and uses of selected map projections: U.S. Geological Survey Miscellaneous Investigations Series Map I-1096.

Central Intelligence Agency, 1973, Projection handbook: Washington, D.C., Central Intelligence Agency, 14 p.

Deetz, C. H., and Adams, O. S., 1944, Elements of map projection: U.S. Coast and Geodetic Survey Special Publication 68, 5th ed., 266 p.

Greenwood, David, 1951, Mapping: Chicago, University of Chicago Press, 289 p.

King, P. B., 1969, Tectonic map of North America: U.S. Geological Survey, scale 1:5,000,000.

King, P. B., and Beikman, H. M., 1974, Geologic map of the United States (exclusive of Alaska and Hawaii): U.S. Geological Survey, scale 1:2,500,000.

Maling, D. H., 1960, A review of some Russian map projections: Empire Survey Rev., v. 15, no. 115, p. 203–215; no. 116, p. 255–266; no. 117, p. 294–303.

McKelvey, V. E., and Wang, F. H., 1970, World subsea mineral resources: U.S. Geological Survey Miscellaneous Geologic Investigations Map I-632.

Miller, O. M., 1941, A conformal map projection for the Americas: Geographical Review, v. 31, no. 1, p. 100–104.

Ministerstvo Oborony S.S.S.R., 1953, Morskio Atlas, Vol. 2: Fisiko-Geograficheskii Izdanie Glavnogo Shtaba Voenno-Morskikh Sil, 76 plates (Ministry of Defense, U.S.S.R., 1953, Ocean Atlas, Vol. 2, Physiogeographic publication of the major headquarters of naval strength).

Raisz, Erwin, 1962, Principles of cartography: New York, McGraw Hill, 315 p.

Richards, Peter, and Adler, R. K., 1972, Map projections for geodesists, cartographers, and geographers: Amsterdam, North Holland, 174 p.

Robinson, A. H., 1949, An analytical approach to map projections: Annals of the Association of American Geographers, vol. 39, p. 283–290.

Robinson, A. H., Sale, R. D., and Morrison, J. L., 1978, Elements of cartography: New York, John Wiley & Sons, 448 p.

Snyder, John P., 1982, Map projections used by the U.S. Geological Survey: U.S. Geological Survey Bulletin 1532.

Steers, J. A., 1970, An introduction to the study of map projections, 15th ed.: London, University of London Press, 294 p.

Stepler, P. F., 1974, CAM Cartographic Automatic Mapping program documentation, version 4: Washington, D.C., Central Intelligence Agency, 111 p.

U.S. Geological Survey, 1954, Alaska Map E, scale 1:2,500,000 (base map).

_____ 1962, Tectonic map of the United States, scale 1:2,500,000.

_____ 1967, Basement map of North America, scale 1:5,000,000.

_____ 1970, The national atlas of the United States of America: Washington, D.C., 417 p.

_____ 1972, United States, scale 1:2,500,000 (base map, 2 sheets).

_____ 1973, Alaska Map E, scale 1:2,500,000 (base map).

_____ 1975, The United States, scale 1:6,000,000 (base map).